A LABORATORY GUIDE TO

TENTH EDITION

# Human Physiology

## Concepts and Clinical Applications

A LABORATORY GUIDE TO

TENTH EDITION

# Human Physiology

## Concepts and Clinical Applications

*Stuart Ira Fox*

PIERCE COLLEGE

Boston Burr Ridge, IL Dubuque, IA Madison, WI New York San Francisco St. Louis
Bangkok Bogotá Caracas Kuala Lumpur Lisbon London Madrid Mexico City
Milan Montreal New Delhi Santiago Seoul Singapore Sydney Taipei Toronto

A LABORATORY GUIDE TO HUMAN PHYSIOLOGY: CONCEPTS AND CLINICAL APPLICATIONS, TENTH EDITION

Published by McGraw-Hill, a business unit of The McGraw-Hill Companies, Inc., 1221 Avenue of the Americas, New York, NY 10020. 

This book is printed on recycled, acid-free paper containing 10% postconsumer waste.

1 2 3 4 5 6 7 8 9 0 QPD/QPD 0 9 8 7 6 5 4 3

ISBN 0–07–244087–2

Publisher: *Martin J. Lange*
Sponsoring edditor: *Michelle Watnick*
Developmental editor: *Kristine A. Queck*
Director of development: *Kristine Tibbetts*
Marketing manager: *James F. Connely*
Project manager: *Sheila M. Frank*
Production supervisor: *Sherry L. Kane*
Senior media project manager: *Stacy A. Patch*
Senior media technology producer: *Barbara R. Block*
Coordinator of freelance design: *Michelle D. Whitaker*
Cover designer: *Diane Beasley*
Cover illustration: *William Westwood*
Senior photo research coordinator: *John C. Leland*
Photo research: *Chris Hammond/PhotoFind, LLC*
Compositor: *Shepherd-Imagineering Media Services, Inc.*
Typeface: *10.2/12 Goudy*
Printer: *Quebecor World Dubuque, IA*

# Contents

# 4 The Endocrine System 145



# 5 Skeletal Muscles 171



# 6 Blood Gas Transport, Immunity, and Clotting Functions 201



# 7 The Cardiovascular System 233

## 8

## Respiration and Metabolism 287



## 9

## Renal Function and Homeostasis 327



## 10

## Digestion and Nutrition 351



## 11

## Reproductive System 387

# Preface

The tenth edition, like the previous editions, is a stand-alone human physiology manual that can be used in conjunction with any human physiology textbook. It includes a wide variety of exercises that support most areas covered in a human physiology course, allowing instructors the flexibility to choose those exercises best suited to meet their particular instructional goals. Background information that is needed to understand the principles and significance of each exercise is presented in a concise manner, so that little or no support is needed from the lecture text.

However, lecture and laboratory segments of a human physiology course are most effectively wedded when they cover topics in a similar manner and sequence. Thus, this laboratory guide is best used in conjunction with the textbook *Human Physiology*, eighth edition, by Stuart Ira Fox (McGraw-Hill, © 2004).

The laboratory experiences provided by this guide allow students to become familiar—in an intimate way that cannot be achieved by lecture and text alone—with many fundamental concepts of physiology. In addition to providing hands-on experience in applying physiological concepts, the laboratory sessions allow students to interact with the subject matter, with other students, and with the instructor in a personal, less formal way. Active participation is required to carry out the exercise procedures, collect data, and to complete the laboratory report.

The questions in the laboratory reports, like those at the end of each chapter in the textbook *Human Physiology*, by Stuart Ira Fox, are organized into three levels. These are (1) *Test Your Knowledge of Terms and Facts*, (2) *Test Your Understanding of Concepts*, and (3) *Test Your Ability to Analyze and Apply Your Knowledge*. This organization promotes higher-order learning and understanding in the laboratory, and helps to better integrate the laboratory with information learned in the lecture portion of the physiology course.

Clinically oriented laboratory exercises that heighten student interest and demonstrate the health applications of physiology have been a hallmark of previous editions and continue to be featured in this latest edition. Change is required, however, because vendors change and available laboratory equipment and supplies change. This tenth edition accommodates such changes and makes new advances in improving the ability of students to benefit from the physiology laboratory experience.

## New to the Tenth Edition

### Integration of the Laboratory Guide with the Textbook

This laboratory guide contains all of the information students need to understand and perform the laboratory exercises. It is thus a self-contained, stand-alone laboratory guide. This benefits students because (1) they don't have to bring the larger and heavier textbook to the laboratory section, and (2) don't have to sift through the textbook to find the information that is particularly relevant to the laboratory exercise.

However, students benefit when the laboratory is well integrated with the lecture portion of the physiology course. In order to facilitate the interaction between lecture and laboratory, this guide uses two devices to allow students to cross-reference the material in the laboratory to the information in the lecture textbook, *Human Physiology*, eighth edition, by Stuart Ira Fox. These two devices are:

1. **Textbook Correlations** boxes, introduced in the last edition, are found at the beginning of each laboratory exercise. These provide specific page numbers in the textbook that correspond to the laboratory exercise. Students don't need this information to answer the questions in their laboratory report, but will benefit from greater depth and wider perspective when their textbook is used in conjunction with the laboratory guide.
2. **Figure Cross-References** between the laboratory guide and the textbook are new to the tenth edition. Whenever a figure in this laboratory guide has a full-color counterpart in the textbook, the specific number of the full-color text figure is provided in the caption of the laboratory manual figure. This allows students to better integrate the laboratory exercise with the concepts discussed in the textbook and lecture portion of their course.

### Changes in the Laboratory Exercises

The tenth edition retains the new procedures utilizing the Biopac and Intelitool systems that were introduced with the previous edition, and now adds the iWorx system, where appropriate. These are systems for performing computerized data acquisition and analysis that can be adapted for use with this laboratory guide. Also retained from the previous editions are the multimedia correlations for the exercises, so instructors can use the relevant programs for A.D.A.M. *Interactive Physiology* and others in the laboratory for additional instruction. In the present edition, *MediaPhys* has been added to these multimedia correlations. Here are some additional changes that are specific for the exercises:

- **Exercise 1.1** Change in Materials section; table 1.3 updated
- **Exercise 1.3** Alternative equipment and procedures for Exercise 1.3A; discussion of diabetes mellitus added to Exercise 1.3B; Laboratory Report for Exercise 1.3A revised, with new instructions for calculations
- **Exercise 2.1C** Data table changed
- **Exercise 2.3** Materials sections modified
- **Exercise 2.4C** New procedure for measurement of lactate dehydrogenase
- **Exercise 2.6** Changes and additions to Materials section, Procedure section, and data table
- **Exercise 3.2** Biopac added to the multimedia correlations
- **Exercise 3.3** Intelitool's Flexicomp added to the multimedia correlations
- **Exercise 3.5** A new section (Section H) added on color vision and color blindness
- **Exercise 4.2** Data table modified
- **Exercise 4.3** Information about type 2 diabetes mellitus added; also, a new table comparing type 1 and type 2 diabetes added
- **Exercise 5.1** Biopac and iWorx equipment added to Materials section, and instructions added to Procedure section; also, information regarding troponin and tropomyosin added
- **Exercise 5.3** Biopac added to multimedia correlations
- **Exercise 7.1** Biopac and iWorx added to Materials section, and instructions in their use added to Procedure section
- **Exercise 8.1** iWorx system added to Materials section
- **Exercise 8.2** iWorx system added to Materials section
- **Exercise 8.3** Alternative sources of blood added to Materials section
- **Exercise 9.3** New source for Ictotest in Materials section; new procedure for bilirubin test in Procedure section
- **Exercise 11.1** Introduction section expanded and alternative procedures added
- **Exercise 11.2** Over-the-counter home pregnancy test kits added to Materials section; instructions in their use added to Procedures section; new and expanded information regarding pregnancy testing added to introduction
- ***Appendix* 2** *Sources of Equipment and Solutions* extensively updated to reflect changes in exercises listed here, changes in company ownership, and changes of phone numbers and websites
- ***Appendix* 3** *Multimedia Correlations to the Laboratory Exercises* reorganized and updated to include *MediaPhys* correlations to exercises

### New and Revised Figures in the Tenth Edition

The tenth edition contains more than 25 figures that are new to this edition. Some of these depict newer laboratory equipment or techniques. Most, however, are figures that help students to better understand the physiological concepts related to the exercises. These improve the self-contained aspect of this laboratory guide while also increasing the degree of cross-referencing between lecture and laboratory.

In addition to figures that are unique to the tenth edition, this laboratory guide contains many figures that have been revised from previous editions. This was done to improve the usefulness of these figures to students studying the laboratory exercises, and to increase the correspondence between figures in the laboratory guide and lecture textbook.

## Safety

Special effort has been made to address concerns about the safe use and disposal of body fluids. For example, normal and abnormal artificial serum can be used as a substitute for blood in Section 2 (plasma chemistry); artificial saliva is suggested in exercise 10.2 (digestion); and in Section 9 (renal function) both normal and abnormal artificial urine is now available. In the interest of safety, a substitute for the use of benzene (previously required in two exercises) is now provided.

⚠ The international symbol for caution is used throughout the laboratory guide to alert the reader when special attention is necessary while preparing for or performing a laboratory exercise. For reference, laboratory safety guidelines appear on the inside front cover.

## Technology

Computer-assisted and computer-guided instruction in human physiology laboratories has greatly increased in recent years. Computer programs provide a number of benefits: some experiments that require animal sacrifice can be simulated; data can be analyzed against a data bank and displayed in an appealing and informative manner; class data records can be analyzed; and costs can be reduced by eliminating the use of some of the most expensive equipment.

This edition continues to reference programs offered by Biopac, Intelitool, and A.D.A.M. Benjamin/Cummings *InterActive Physiology* Modules (800-755-2326; www.adam.com). In addition, correlations to McGraw-Hill's *MediaPhys* and to the iWorx system have been incorporated into the tenth edition.

## Organization of the Laboratory Guide

The exercises in this guide are organized in this manner:

1. Each exercise begins with a list of **materials** needed to perform the exercise, so that it is easier to set up the laboratory. This section is identified by a materials icon.
2. Following the materials section is an overview paragraph describing the **concept** behind the laboratory exercise.
3. Following the concept paragraph is a list of **learning objectives,** to help students guide their learning while performing the exercise.
4. A box providing **textbook correlations** is placed near the beginning of each exercise. This section can be used to help integrate the lecture textbook (*Human Physiology*, eighth edition, by Stuart Ira Fox with the laboratory material.
5. A brief **introduction** to the exercise presents the essential information for understanding the physiological significance of the exercise. This concisely written section eliminates the need to consult the lecture text.
6. Boxed information, set off as screened insets, provide the **clinical significance** of different aspects of the laboratory exercise. This approach was pioneered by this laboratory manual and the current edition continues that tradition.
7. The **procedure** is stated in the form of easy-to-follow steps. These directions are set off from the textual material through the use of a distinctive typeface, making it easier for students to locate them as they perform the exercise.
8. **Normal values** boxes are placed following the procedures, in cases where students obtain measurements that are clinically applicable. These boxes are indicated with a scales icon to emphasize the relationship between clinical measurements for diagnosis and physiological regulation of homeostasis.
9. A **laboratory report** follows each exercise. Students enter data here when appropriate, and answer questions. The questions in the laboratory report begin with the most simple form (objective questions) in most exercises and progress to essay questions. The essay questions are designed to stimulate conceptual learning and to maximize the educational opportunity provided by the laboratory experience.

## Supplemental Materials

In addition to this laboratory manual, a comprehensive selection of supplemental materials is available for use in conjunction with *Human Physiology*, eighth edition, by Stuart Ira Fox. Students can order supplemental study materials by contacting their campus bookstore. Instructors can obtain teaching aids by calling the McGraw-Hill Customer Service Department at (800)338–3987, visiting our A&P website at www.mhhe.com/ap, or contacting a local McGraw-Hill sales representative.

### Online Learning Center

The *Human Physiology*, eighth edition, Online Learning Center (OLC) at www.mhhe.com/fox8 offers an extensive array of learning and teaching tools.

**Essential Study Partner** A collection of interactive study modules that contains hundreds of animations, learning activities, and quizzes designed to help students grasp complex concepts.

**Monitored News Feeds** Online access to course-specific current articles refereed by content experts, course-specific real-time news, weekly course updates, refereed and updated research links, daily news, and the Northernlight.com Special Collection™ of journals and articles.

**Online Tutoring** A 24-hour tutorial service moderated by qualified instructors. Help with difficult concepts is only an e-mail away.

Along with these outstanding online tools, the OLC features specialized content for both students and instructors using the eighth edition of *Human Physiology*. The Student Center of the OLC features quizzes, interactive learning games, and study tools tailored to coincide with each chapter of the textbook. The Instructor Center is an online repository of teaching aids. It houses downloadable and printable versions of traditional ancillaries plus a wealth of online content.

### Instructor's Manual for the Laboratory Guide

Accessed via the Instructor Center of the *Human Physiology* OLC (www.mhhe.com/fox8), this helpful manual created by Laurence G. Thouin, Jr. of Pierce College includes suggestions for coordinating lab exercises with the textbook, introductions to each exercise, materials lists, approximate completion times, and solutions to the laboratory reports for each exercise. A listing of laboratory supply houses and instructions for mixing commonly used solutions are also provided.

Instructors using this lab manual independently of *Human Physiology*, eighth edition, by Stuart Ira Fox, can access the Instructor's Manual on McGraw-Hill's Lab Central at www.mhhe.com/biosci/ap/labcentral.

### Other Offerings

In addition to the materials specifically designed to accompany *Human Physiology*, McGraw-Hill offers these supplemental resources to enrich the study and instruction of human physiology.

**MediaPhys** McGraw-Hill's ***MediaPhys*** is an interactive CD-ROM that offers 13 complete modules (including Muscular, Nervous, Cardiovascular, Respiratory, Urinary, Digestive, Endocrine, and Reproductive systems) that feature detailed explanations, high-quality illustrations, and animations to provide students with a thorough introduction to the world of physiology. *MediaPhys* is filled with interactive activities and quizzes to help reinforce concepts that are often difficult to grasp, making *MediaPhys* a tool that helps students truly understand the concepts of the human body rather than simply memorize them. Contact your campus bookstore to order.

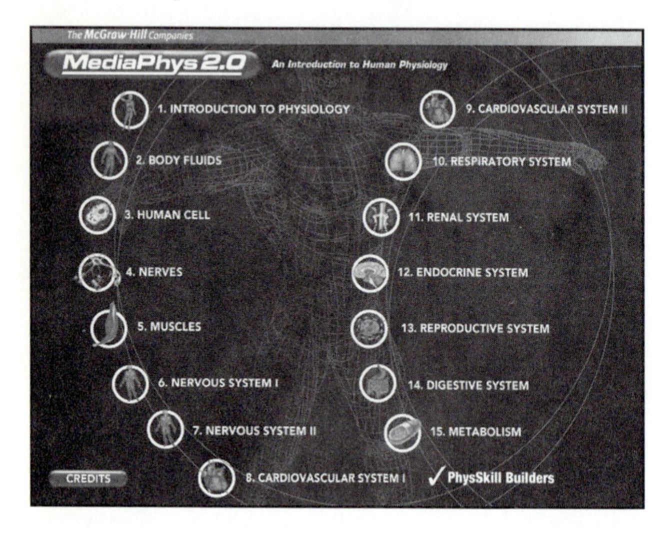


**Dynamic Human Version 2.0** A set of two interactive CD-ROMs that cover each body system and demonstrate clinical concepts, histology, and physiology with animated three-dimensional and other images.

**Case Histories in Human Physiology,** third edition, by Donna Van Wynsberghe and Gregory Cooley (print or Internet-based). Stimulate analytical thinking using case studies and problem solving. Includes an instructor's answer key.

**Life Science Animations Library CD-ROM** More than 400 animations in an easy-to-use program that enables instructors to quickly view the animations and import them into multimedia classroom presentations or Web-based course materials.

**Laboratory Atlas of Anatomy and Physiology,** third edition, by Eder et al. A comprehensive full-color atlas that covers histology, human skeletal anatomy, and human muscular anatomy using dissections and reference tables.

## Acknowledgments

The tenth edition was greatly benefited by input from my colleague Dr. Laurence G. Thouin, Jr. His numerous suggestions helped to make the tenth edition more accurate and student friendly.

The shaping of the tenth edition was also aided by suggestions from other colleagues and students. Ms. Karen Gebhardt was particularly instrumental in checking laboratory sources for materials and reworking some of the procedures that are new to this edition. I greatly appreciate the support of the editors at McGraw-Hill, particularly Kristine Queck. Their contributions help to make this the best edition yet of *A Laboratory Guide to Human Physiology: Concepts and Clinical Applications*.

# Introduction: Structure and Physiological Control Systems

# Section 1

The cell is the basic unit of structure and function in the body. Each cell is surrounded by a *plasma* (or *cell*) *membrane* and contains specialized structures called *organelles* within the cell fluid, or *cytoplasm.* The structure and functions of a cell are largely determined by genetic information contained within the membrane-bound *nucleus.* This genetic information is coded by the specific chemical structure of *deoxyribonucleic acid (DNA)* molecules, the major component of *chromosomes.* Through genetic control of *ribonucleic acid (RNA)* and the synthesis of proteins (such as enzymes described in section 2), DNA within the cell nucleus directs the functions of the cell and, ultimately, those of the entire body.

Cells with similar specializations are grouped together to form **tissues,** and tissues are grouped together to form larger units of structure and function known as **organs.** Organs that are located in different parts of the body but that cooperate in the service of a common function are called **organ systems** (e.g., the cardiovascular system).

The complex activities of cells, tissues, organs, and systems are coordinated by a wide variety of regulatory mechanisms that act to maintain **homeostasis**—a state of dynamic constancy in the internal environment. **Physiology** is largely the study of the control mechanisms that participate in maintaining homeostasis.

# Microscopic Examination of Cells

EXERCISE 1.1

### Materials

1. Compound microscopes
2. Prepared microscope slides, including whitefish blastula (early embryo), clean slides, and cover slips (Note: Slides with dots, lines, or the letter *e* can be prepared with dry transfer patterns used in artwork.)
3. Lens paper
4. Methylene blue stain
5. Cotton-tipped applicator sticks or toothpicks

### Textbook Correlations

Before performing this exercise, you should study the introductory material presented here. Further information relating to this exercise can be found in these pages of *Human Physiology,* eighth edition, by Stuart I. Fox:

- *Cytoplasm and Its Organelles.* Chapter 3, pp. 55–61
- *DNA Synthesis and Cell Division.* Chapter 3, pp. 69–78.

The microscope and the metric system are important tools in the study of cells. Cells contain numerous organelles with specific functions and are capable of reproducing themselves by mitosis. However, there is also a special type of cell division called meiosis that is used in the gonads to produce sperm or ova.

### OBJECTIVES

1. Identify the major parts of a microscope and demonstrate proper technique in the care and handling of this instrument.
2. Define and interconvert units of measure in the metric system; and estimate the size of microscopic objects.
3. Describe the general structure of a cell and the specific functions of the principal organelles.
4. Describe the processes of mitosis and meiosis and explain their significance.

The **microscope** is the most basic and widely used instrument in the life science laboratory. The average binocular microscope for student use, as shown in figure 1.1, includes these parts:

1. eyepieces each with an ocular lens (usually 10× magnification, and may have a pointer)
2. a stage platform with manual or mechanical stage controls
3. a substage condenser lens and iris diaphragm, each with controls
4. coarse focus and fine focus adjustment controls
5. objective lenses on a revolving nosepiece (usually include: a scanning lens, 4×; a low-power lens, 10×; and a high-power lens, 45×)

### Care and Cleaning

The microscope is an expensive, delicate instrument. To maintain it in good condition, always take these precautions:

1. Carry the microscope with two hands.
2. Use the *coarse* focus knob *only* with low power and always move the objective lens *away from the slide*, never toward the slide.
3. Clean the ocular and objective lenses with lens paper moistened with distilled water before and after use. (Use alcohol only if oil has been used with an oil-immersion, 100× lens.)
4. Always leave the lowest power objective lens (usually 4× or 10× facing the stage before putting the microscope away.

## A. The Inverted Image

Obtain a slide with the letter *e* mounted on it. Place the slide on the microscope stage, and rotate the nosepiece until the 10× objective clicks into the down position. Using the coarse adjustment, carefully lower the objective

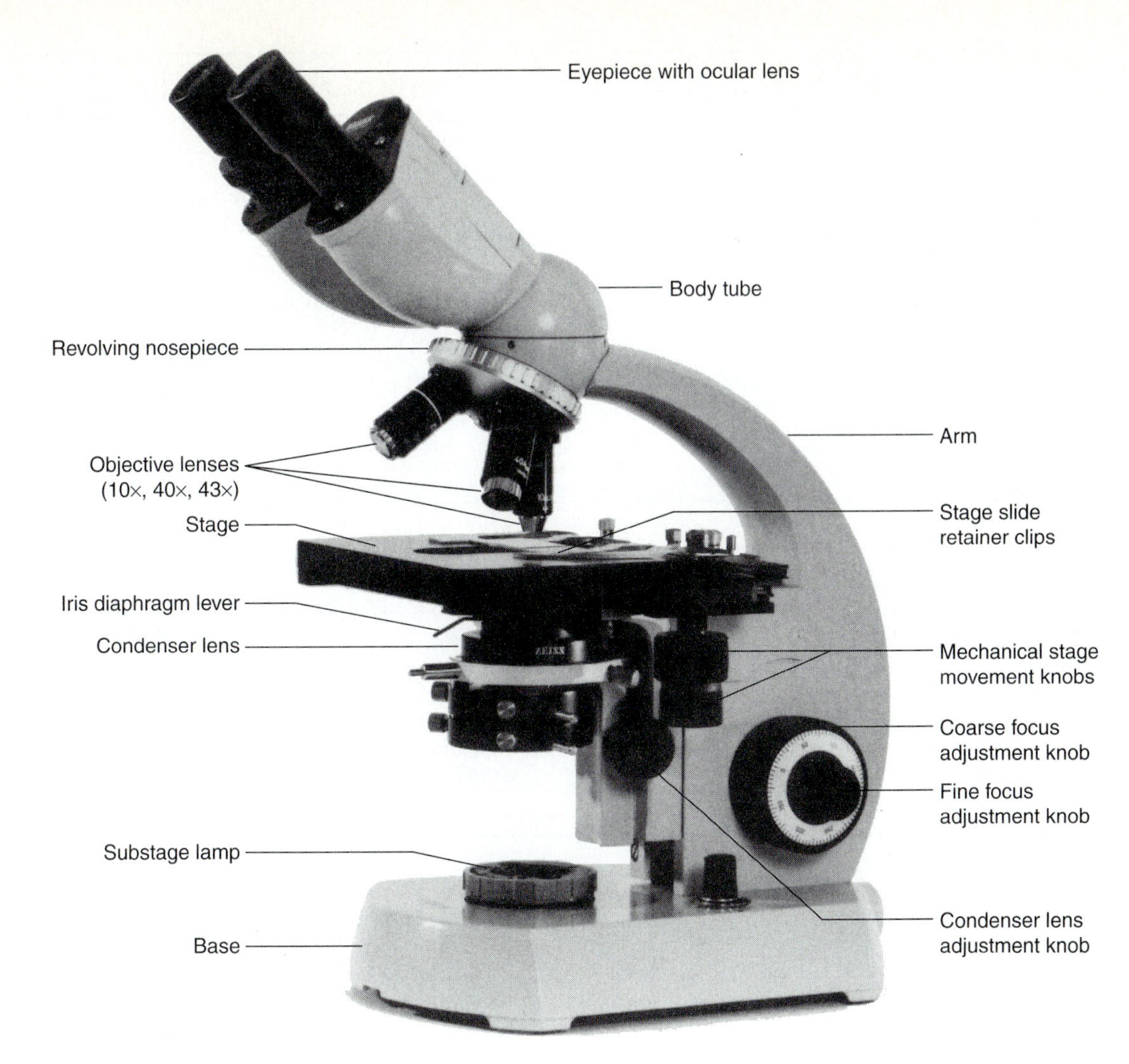


**Figure 1.1** The parts of a compound microscope.

lens until it almost touches the slide. Now, looking through the ocular lens, slowly raise the objective lens until the letter *e* comes into focus.

## PROCEDURE

1. If the visual field is dark, increase the light by adjusting the lever that opens (and closes) the iris diaphragm. If there is still not enough light, move the substage condenser lens closer to the slide by rotating its control knob. Bring the image into sharp focus using the fine focus control. Now, draw the letter *e* as it appears in the microscope.

   ___________

2. While looking through the ocular lens, rotate the mechanical stage controls so that the mechanical stage moves to the *right*. In which direction does the *e* move?

   ___________

3. While looking through the ocular lens, rotate the mechanical stage controls so that the mechanical stage moves *toward* you. In which direction does the *e* move?

   ___________

## B. The Metric System: Estimating the Size of Microscopic Objects

It is important in microscopy, as in other fields of science, that units of measure are standardized and easy to use. The **metric system** (from the Greek word *metrikos*, meaning "measure") first developed in late eighteenth-century France, is the most commonly used measurement system in scientific literature. The modern definitions of the units used in the metric system are those adopted by the General Conference on Weights and Measures, which in 1960 established the International System of Units, also known (in French) as Système International d'Unités,

## Table 1.1 International System of Metric Units, Prefixes, and Symbols

| Multiplication Factor | Prefix | Symbol | Term |
|---|---|---|---|
| $1{,}000{,}000 = 10^6$ | Mega | M | One million |
| $1{,}000 = 10^3$ | Kilo | k | One thousand |
| $100 = 10^2$ | Hecto | h | One hundred |
| $10 = 10^1$ | Deka | da | Ten |
| $1 = 10^0$ | | | |
| $0.1 = 10^{-1}$ | Deci | d | One-tenth |
| $0.01 = 10^{-2}$ | Centi | c | One-hundredth |
| $0.001 = 10^{-3}$ | Milli | m | One-thousandth |
| $0.000001 = 10^{-6}$ | Micro | μ | One-millionth |
| $0.000000001 = 10^{-9}$ | Nano | n | One-billionth |
| $0.000000000001 = 10^{-12}$ | Pico | p | One-trillionth |
| $0.000000000000001 = 10^{-15}$ | Femto | f | One-quadrillionth |

## Table 1.2 Sample Metric Conversions

| To Convert From | To | Factor | Move Decimal Point |
|---|---|---|---|
| Meter (Liter, gram) | Milli- | $\times$ 1,000 ($10^3$) | 3 places to right |
| Meter (Liter, gram) | Micro- | $\times$ 1,000,000 ($10^6$) | 6 places to right |
| Milli- | Meter (Liter, gram) | $\div$ 1,000 ($10^{-3}$) | 3 places to left |
| Micro- | Meter (Liter, gram) | $\div$ 1,000,000 ($10^{-6}$) | 6 places to left |
| Milli- | Micro- | $\times$ 1,000 ($10^3$) | 3 places to right |
| Micro- | Milli- | $\div$ 1,000 ($10^{-3}$) | 3 places to left |

and abbreviated SI (in all languages). The definitions for the metric units of *length*, *mass*, *volume*, and *temperature* are as listed here:

**meter (m)**—unit of length equal to 1,650,763.73 wavelengths in a vacuum of the orange-red line of the spectrum of krypton-86

**gram (g)**—unit of mass based on the mass of 1 cubic centimeter ($cm^3$) of water at the temperature (4° C) of its maximum density

**liter (L)**—unit of volume equal to 1 cubic decimeter ($dm^3$) or 0.001 cubic meter ($m^3$)

**Celsius (C)**—temperature scale in which 0° is the freezing point of water and 100° is the boiling point of water; this is equivalent to the centigrade scale

Conversions between different orders of magnitude in the metric system are based on powers of ten (table 1.1). Therefore, you can convert from one order of magnitude to another simply by moving the decimal point the correct number of places to the right (for multiplying by whole numbers) or to the left (for multiplying by decimal fractions). Sample conversions are illustrated in table 1.2.

### Dimensional Analysis

If you are unsure about the proper factor for making a metric conversion, you can use a technique called *dimensional analysis*. This technique is based on two principles:

1. Multiplying a number by 1 does not change the value of that number.
2. A number divided by itself is equal to 1.

These principles can be used to change the units of any measurement.

#### *Example*

Since 1 meter (m) is equivalent to 1,000 millimeters (mm),

$$\frac{1\text{ m}}{1{,}000\text{ mm}} = 1 \text{ and } \frac{1{,}000\text{ mm}}{1\text{ m}} = 1$$

Suppose you want to convert 0.032 meter to millimeters:

$$0.032\text{ m} \times \frac{1{,}000\text{ mm}}{1\text{ m}} = 32.0\text{ mm}$$

Notice that in dimensional analysis the problem is set up so that the unwanted units (meter, *m* in this example) cancel each other. This technique is particularly useful when the conversion is more complex or when some of the conversion factors are unknown.

#### *Example*

Suppose you want to convert 0.1 milliliter (mL) to microliter (μL) units. If you remember that 1 mL = 1,000 μL, you can set up the problem as shown here:

$$0.1\text{ mL} \times \frac{1{,}000\ \mu\text{L}}{1\text{ mL}} = 100\ \mu\text{L}$$

If you remember that a milliliter is one-thousandth of a liter and that a microliter is one-millionth of a liter, you can set up the problem in this way:

$$0.1\ \text{mL} \times \frac{1.0\ \text{L}}{1{,}000\ \text{mL}} \times \frac{1{,}000{,}000\ \mu\text{L}}{1.0\ \text{L}} = 100\ \mu\text{L}$$

### Visual Field and the Estimation of Microscopic Size

If the magnification power of your ocular lens is 10× and you use the 10× objective lens, the total magnification of the visual field will be 100×. At this magnification, the diameter of the visual field is approximately 1,600 micrometers (µm).

You can estimate the size of an object in the visual field by comparing it with the total diameter (line *AB*) of the visual field. Using this diagram:

How long is line AC in micrometers (µm)? ______
How long is line *AD* in micrometers (µm)? ______
How long is line *AE* in micrometers (µm)? ______

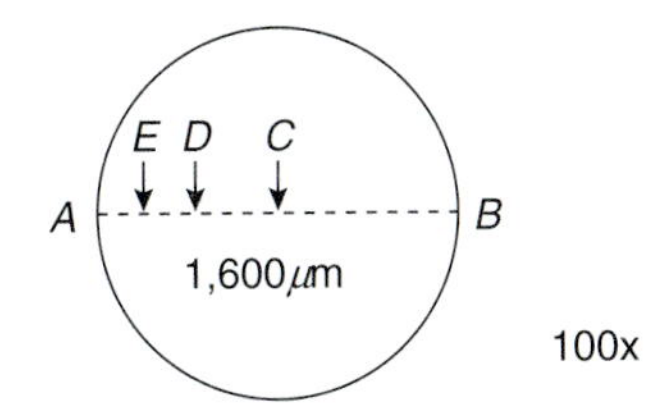


The diameter of the field of vision using the 45× objective lens (total magnification 450×) is approximately 356 micrometers. Using the diagram above and applying the same technique, answer these questions assuming use of a 45× objective lens:

How long is line AC in micrometers (µm)? ______
How long is line *AD* in nanometers (nm)? ______

## PROCEDURE

From your instructor, obtain a slide that contains a pattern of small dots and a pattern of thin lines.

1. Using the 10× objective lens:
   (a) estimate the diameter of one dot: __________ m
   (b) estimate the distance between the *nearest* edges of two adjacent dots: __________ m
2. Using the 45× objective lens:
   (a) estimate the width of one line: __________ m
   (b) estimate the distance between the *nearest* edges of two adjacent lines: __________ m

## C. Microscopic Examination of Cheek Cells

The surfaces of the body are covered and lined with *epithelial* membranes (one of the primary tissues described in exercise 1.2). In membranes that are several cell layers thick, such as the membrane lining of the cheeks, cells are continuously lost from the surface and replaced through cell division in deeper layers. In contrast to cells in the outer layer of the epidermis of the skin, which die before they are lost, the cells in the outer layer of epithelial tissue in the cheeks are still alive. You can therefore easily collect and observe living human cells by simply rubbing the inside of the cheeks.

Most living cells are difficult to observe under the microscope unless they are stained. In this exercise, the stain *methylene blue* will be used. Methylene blue is positively charged and combines with negative charges in the chromosomes to stain the nucleus blue. The cytoplasm contains a lower concentration of negatively charged organic molecules, and so appears almost clear.

## PROCEDURE

1. Rub the inside of one cheek with the cotton tip of an applicator stick (or a toothpick).
2. Press the cotton tip of the applicator stick (or the end of the toothpick) against a clean glass slide. Maintaining pressure, rotate the cotton tip against the slide and then push the cheek smear across the slide about 1/2 inch.
3. Observe the *unstained* cells under 100× and 450× total magnification.
4. Remove the slide from the microscope. Holding it over a sink or special receptacle, place a drop of methylene blue stain on the smear.
5. Place a cover slip over the stained smear and observe the stained cheek cells at 100× and 450× total magnification.
6. Using the procedure described in the previous section, estimate the size of the average cheek cell using both 100× and 450× total magnification. 100× ______ µm; 450× ______ µm.
   Are they the same? ______

## D. Cell Structure and Cell Division

Cells vary greatly in size and shape. The largest cell, an *ovum* (egg cell), can barely be seen with the unaided eye; other cells can be observed only through a microscope. Each cell has an outer *plasma membrane* (or cell membrane) and generally one *nucleus*, surrounded by a fluid matrix, or *cytoplasm*. Within the nucleus and the cytoplasm are a variety of subcellular structures, called **organelles** (fig. 1.2). The structures and principal functions of important organelles and other cellular components are listed in table 1.3.

The process of cell division, or replication, is called **mitosis** (fig. 1.3*a*). This process allows new cells to be formed to replace those that are dying and also permits body growth. Mitosis consists of a continuous sequence of four stages (table 1.4 and fig. 1.3*a*) in which both the nucleus and cytoplasm of a cell split to form two identical *daughter cells*. During mitotic cell division, the chromosomes (which had been duplicated earlier) separate, and

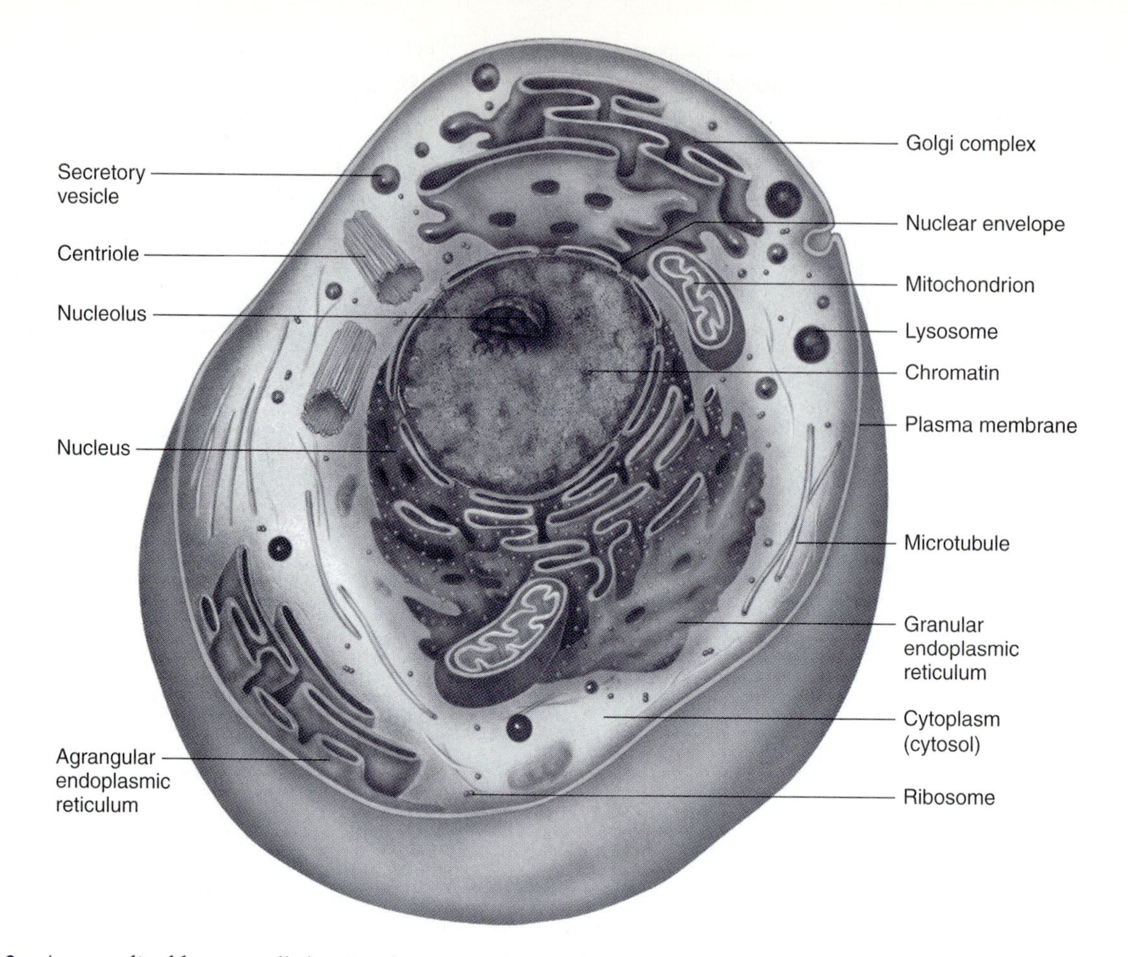


**Figure 1.2 A generalized human cell showing the principal organelles.** Since most cells of the body are highly specialized, they have structures that differ from those shown here.

**(For a full-color version of this figure, see fig. 3.1 in *Human Physiology,* eighth edition, by Stuart I. Fox.)**

one of the duplicate sets of chromosomes goes to each daughter cell. The two daughter cells therefore have the same number of chromosomes as the parent cell.

The forty-six chromosomes present in most human cells actually represent twenty-three *pairs* of chromosomes; one set of twenty-three was inherited from the mother and the other set of twenty-three from the father. A cell with forty-six chromosomes is said to be *diploid*, or *2n*.

In the process of *gamete* (sperm and ova) production in the *gonads* (testes and ovaries), specialized germinal cells undergo a type of division called **meiosis** (fig. 1.3*b*). During meiosis, each germinal cell divides twice, and the daughter cells (the gametes) get only one set of twenty-three chromosomes; they are said to be *haploid*, or *1n*. In this way the original diploid number of forty-six chromosomes can be restored when the sperm and egg unite in the process of fertilization.

## PROCEDURE

1. Study figure 1.2. Cover the labels with a blank sheet of paper and try to write them in (watch spelling!).
2. Examine a slide of a whitefish blastula (or similar early embryo) and observe the different stages of mitosis as shown in figure 1.3.

## Table 1.3 Cellular Components: Structure and Function

| Component | Structure | Function |
|---|---|---|
| Plasma (cell) membrane | Membrane composed of double layer of phospholipids in which proteins are embedded | Gives form to cell and controls passage of materials into and out of cell |
| Cytoplasm | Fluid, jellylike substance between the cell membrane and the nucleus in which organelles are suspended | Serves as matrix substance in which chemical reactions occur |
| Endoplasmic reticulum | System of interconnected membrane-forming canals and tubules | Agranular (smooth) endoplasmic reticulum metabolizes nonpolar compounds and stores $Ca^{2+}$ in striated muscle cells, granular (rough) endoplasmic reticulum assists in protein synthesis |
| Ribosomes | Granular particles composed of protein and RNA | Synthesize proteins |
| Golgi complex | Cluster of flattened membranous sacs | Synthesizes carbohydrates and packages molecules for secretion, secretes lipids and glycoproteins |
| Mitochondria | Membranous sacs with folded inner partitions | Release energy from food molecules and transform energy into usable ATP |
| Lysosomes | Membranous sacs | Digest foreign molecules and worn and damaged organelles |
| Peroxisomes | Spherical membranous vesicles | Contain enzymes that detoxify harmful molecules and break down hydrogen peroxide |
| Centrosome | Nonmembranous mass of two rodlike centrioles | Helps organize spindle fibers and distribute chromosomes during mitosis |
| Vacuoles | Membranous sacs | Store and release various substances within the cytoplasm |
| Microfilaments and microtubules | Thin, hollow tubes | Support cytoplasm and transport materials within the cytoplasm |
| Cilia and flagella | Minute cytoplasmic projections that extend from the cell surface | Move particles along cell surface or move the cell |
| Nuclear envelope | Double-layered membrane that surrounds the nucleus, composed of protein and lipid molecules | Supports nucleus and controls passage of materials between nucleus and cytoplasm |
| Nucleolus | Dense, nonmembranous mass composed of protein and RNA molecules | Produces ribosomal RNA for ribosomes |
| Chromatin | Fibrous strands composed of protein and DNA | Contains genetic code that determines which proteins (including enzymes) will be manufactured by the cell |

## Table 1.4 Major Events in Mitosis

| Stage | Major Events |
|---|---|
| Prophase | Chromosomes form from the chromatin material, centrioles migrate to opposite sides of the nucleus, the nucleolus and nuclear membrane disappear, and spindles appear and become associated with centrioles and centromeres. |
| Metaphase | Duplicated chromosomes align themselves on the equatorial plane of the cell between the centrioles, and spindle fibers become attached to duplicate parts of chromosomes. |
| Anaphase | Duplicated chromosomes separate, and spindles shorten and pull individual chromosomes toward the centrioles. |
| Telophase | Chromosomes elongate and form chromatin threads, nucleoli and nuclear membranes appear for each chromosome mass, and spindles disappear. |

## (a) Mitosis

(a) **Interphase**
- The chromosomes are in an extended form and seen as chromatin in the electron microscope.
- The nucleus is visible

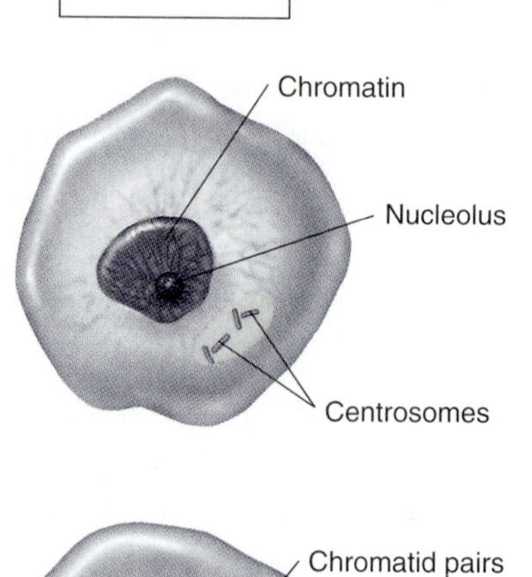


(b) **Prophase**
- The chromosomes are seen to consist of two chromatids joined by a centromere.
- The centrioles move apart toward opposite poles of the cell.
- Spindle fibers are produced and extend from each centrosome.
- The nuclear membrane starts to disappear.
- The nucleolus is no longer visible.

Chromatid pairs

Spindle fibers

(c) **Metaphase**
- The chromosomes are lined up at the equator of the cell.
- The spindle fibers from each centriole are attached to the centromeres of the chromosomes.
- The nuclear membrane has disappeared.

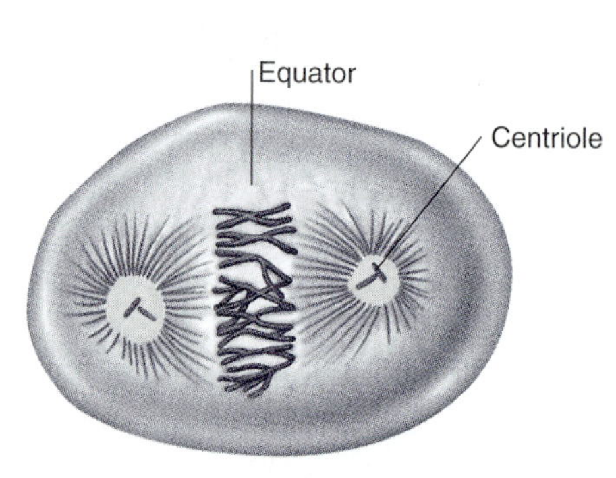


(d) **Anaphase**
- The centromere split, and the sister chromatids separate as each is pulled to an opposite pole.

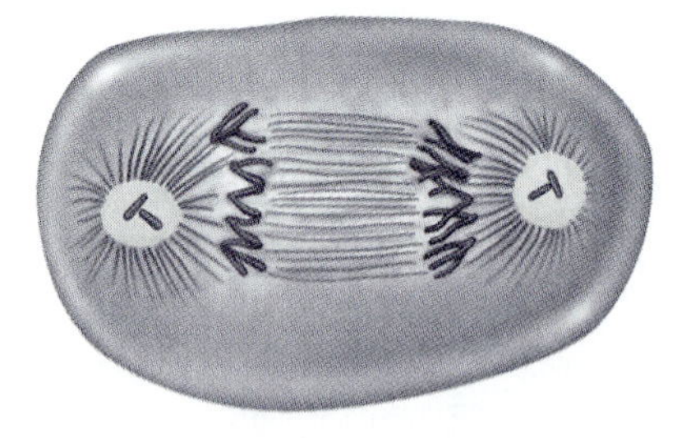

(e) **Telophase**
- The chromosomes become longer, thinner, and less distinct.
- New nuclear membranes form.
- The nucleolus reappears.
- Cell division is nearly complete.

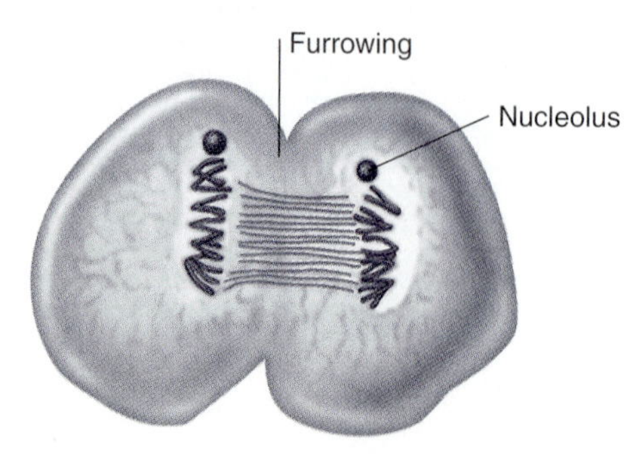


**Figure 1.3 Cell division.** (a) The stages of mitosis. (b) The stages of meiosis. Mitosis is the type of cell division that occurs when organs grow or cells within organs need to be replaced. Meiosis is the type of cell division that only occurs in the gonads for the production of gametes (sperm and ova).

**(For a full-color version of this figure, see figs. 3.29 and 3.33 in *Human Physiology*, eighth edition, by Stuart I. Fox.)**

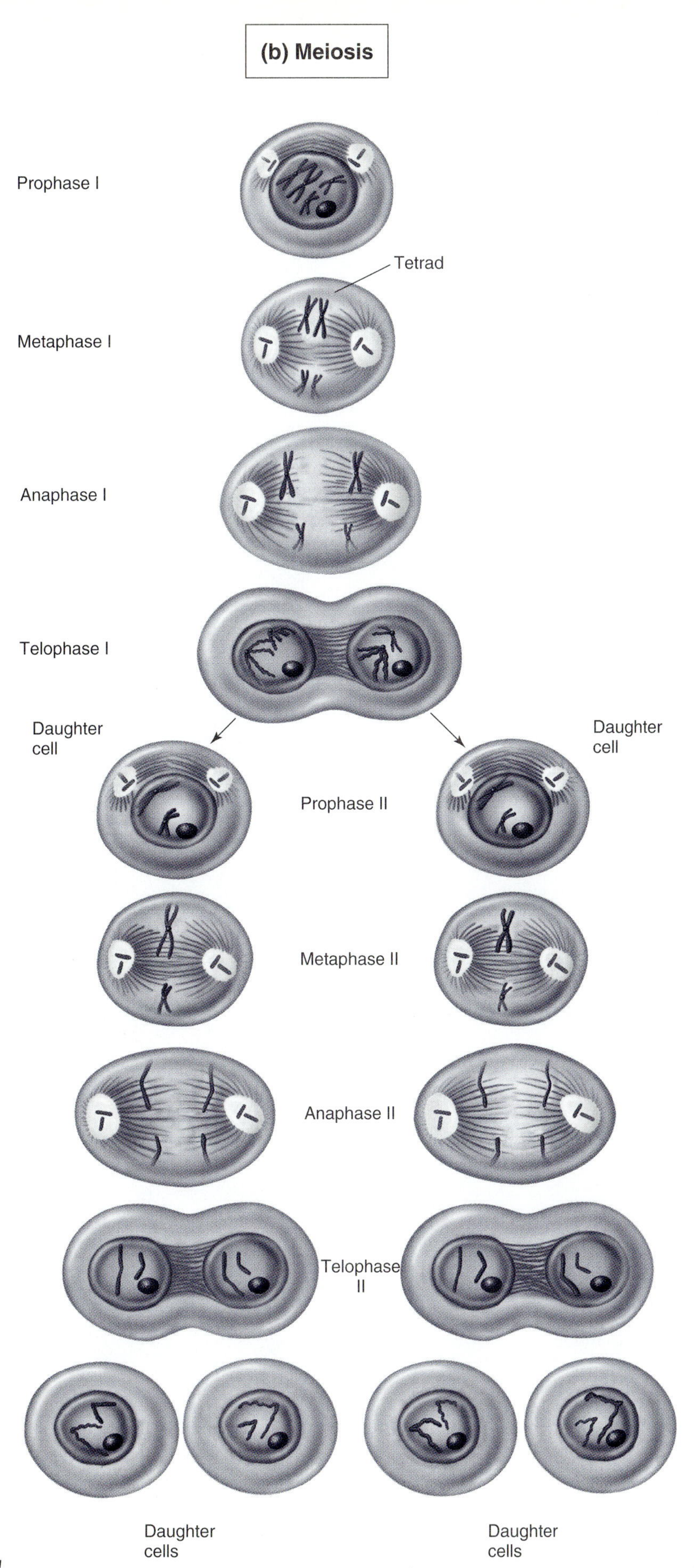


**Figure 1.3** *Continued*

# Laboratory Report 1.1

Name ______________________

Date ______________________

Section ______________________

## REVIEW ACTIVITIES FOR EXERCISE 1.1

### *Test Your Knowledge of Terms and Facts*

1. Give the *total* magnification when you use
   (a) the low-power objective lens ______________________
   (b) the high-dry power objective lens ______________________
   (c) the oil-immersion objective lens ______________________
2. Give the metric units for
   (a) the weight of 1 cubic centimeter of water at its maximum density ______________________
   (b) the temperature at which water freezes ______________________
   (c) the unit of volume equal to 0.001 cubic meter ______________________
3. Match these equivalent measurements:

| | | |
|---|---|---|
| ____ | 1. 100 mL | (a) 100 µL |
| ____ | 2. 0.10 mL | (b) 0.00001 L |
| ____ | 3. 0.0001 mL | (c) 1.0 dL |
| ____ | 4. 0.01 mL | (d) 100 nL |

4. Identify the principal organelle or cell component described here:
   (a) helps organize spindle fibers during cell division (mitosis) ______________________
   (b) the major site of energy production in the cell ______________________
   (c) a system of membranous tubules in the cytoplasm; often involved with protein synthesis ______________________
   (d) the location of genetic information ______________________
   (e) the vesicle that contains digestive enzymes ______________________
   (f) the site of protein synthesis ______________________
5. Match these events of **mitosis** with the correct name of the stage:

| | | |
|---|---|---|
| ____ | 1. the nuclear membrane disappears; spindles appear | (a) metaphase |
| ____ | 2. chromosomes line up along the equator of the cell | (b) telophase |
| ____ | 3. duplicated chromosomes separate and are pulled toward the centrioles | (c) anaphase |
| ____ | 4. chromosomes elongate into chromatin threads; nuclear membranes and nucleoli reappear | (d) prophase |

### *Test Your Understanding of Concepts*

6. Compare and contrast *mitosis* and *meiosis* in terms of where and when they occur and their end products. What are the ways that mitosis and meiosis are used in the body?

## Test Your Ability to Analyze and Apply Your Knowledge

7. In metaphase I of meiosis, the homologous chromosomes line up side by side along the equator, so that (a) crossing-over (exchange of DNA regions) can occur between the homologous pairs and (b) the homologous chromosomes can be pulled to opposite poles during anaphase I. In mitosis, by contrast, homologous chromosomes line up single-file along the equator. What benefits are derived from these two different ways that homologous chromosomes are positioned at metaphase in meiosis and mitosis?

8. Why do you think it is that scientists prefer to use the metric system over the English system of measurements? What problems might result if a country uses both systems of measurement?

# Microscopic Examination of Tissues and Organs

EXERCISE 1.2

**MATERIALS**

1. Compound microscopes
2. Lens paper
3. Prepared microscope slides of tissues

The body is composed of only four primary tissues, and each is specialized for specific functions. Most organs of the body are composed of all four primary tissues, which cooperate in determining the overall structure and function of the organ.

**OBJECTIVES**

1. Define the terms *tissue* and *organ.*
2. List the distinguishing characteristics of the four primary tissues.
3. Identify and describe the subcategories of the primary tissues.
4. In general terms, correlate the structures of the primary tissues with their function.

**Textbook Correlations**

Before performing this exercise, you should study the introductory material presented here. Further information relating to this exercise can be found in these pages of *Human Physiology,* eighth edition, by Stuart I. Fox:

- *The Primary Tissues.* Chapter 1, pp. 9–16.
- *Organs and Systems.* Chapter 1, pp. 17–19.

The trillions of cells that compose the human body have many basic features in common, but they differ considerably in size, structure, and function. Furthermore, cells neither function as isolated units nor are they haphazardly arranged in the body. An aggregation of cells that are similar in structure and that work together to perform a specialized activity is referred to as a tissue. Groups of tissues that are integrated to perform one or more common functions constitute organs. Tissues are categorized into four principal types, or **primary tissues:** (1) *epithelial,* (2) *connective,* (3) *muscular,* and (4) *nervous.*

## A. Epithelial Tissue

Epithelial tissue, or *epithelium,* functions to protect, secrete, or absorb. Epithelial membranes cover the outer surface of the body (epidermis of the skin) and the outer surfaces of internal organs; they also line the body cavities and the *lumina* (the inner hollow portions) of ducts, vessels, and tubes. All *glands* are derived from epithelial tissue. Epithelial tissues share these characteristics:

1. The cells are closely joined together and have little intercellular substance (matrix) between them.
2. There is an exposed surface either externally or internally.
3. A *basement membrane* (composed of glycoproteins) is present to anchor the epithelium to underlying connective tissue.

Epithelial tissues that are composed of a single layer of cells are called *simple;* those that contain more than one layer are known as *stratified.* Epithelial tissues may be further classified by the shape of their surface cells: *squamous* (if the cells are flat), *cuboidal,* or *columnar.* Using these criteria, one can identify these types of epithelia:

1. **Simple squamous epithelium** (fig. 1.4, *top*). This type is adapted for diffusion, absorption, filtration, and secretion—present in such places as the lining of air sacs, or *alveoli,* within the lungs (where gas exchange occurs); parts of the kidney (where blood is filtered); and the lining, or *endothelium,* of blood vessels (where exchange between blood and tissues occurs).
2. **Stratified squamous epithelium** (fig. 1.4, *middle*). This type is found in areas that receive a lot of wear and tear. The outer cells are sloughed off and replaced by new cells, produced by mitosis in the deepest layers. Stratified squamous epithelium is found in the mouth, esophagus, nasal cavity, and in

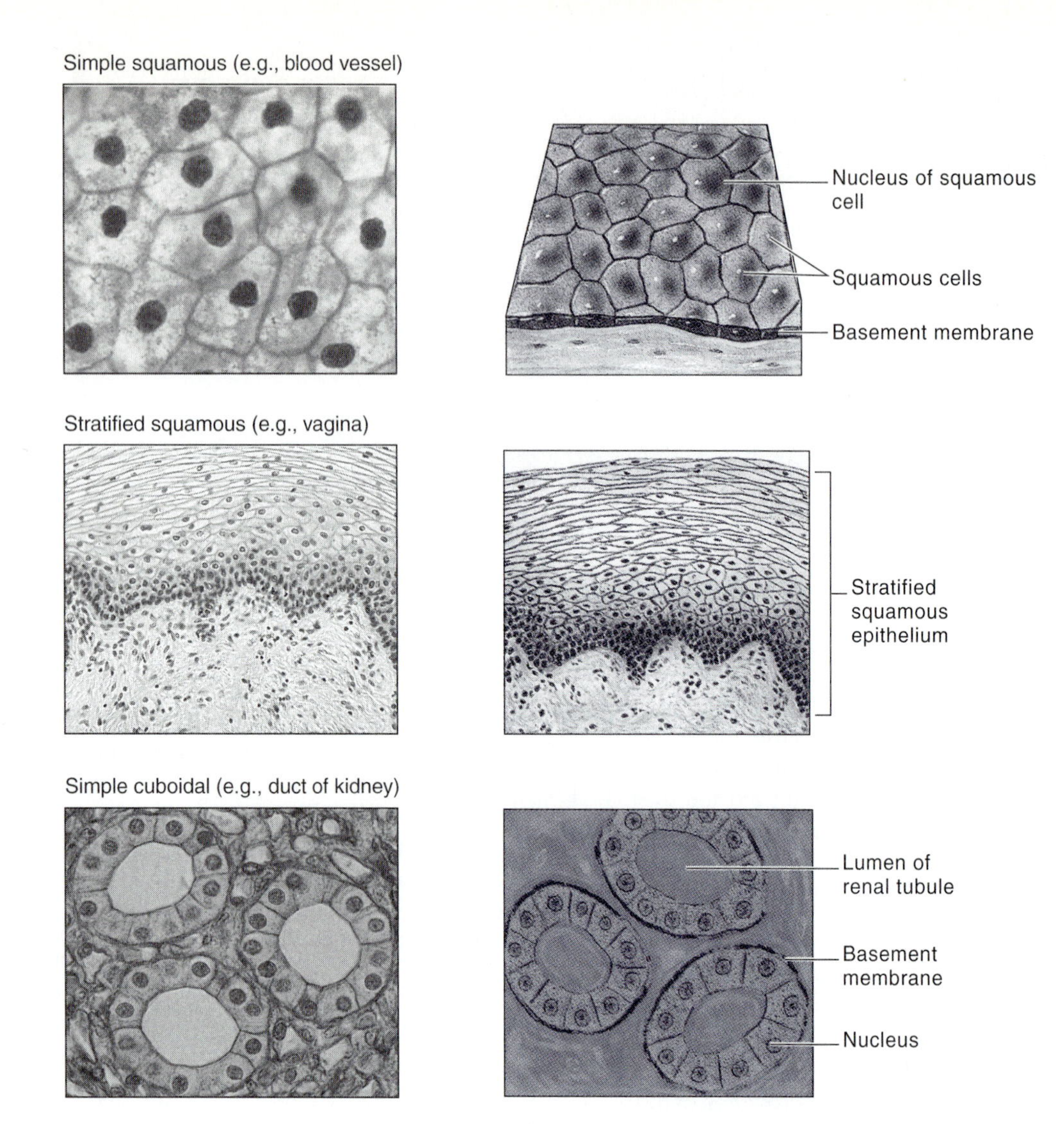


**Figure 1.4 Squamous and cuboidal epithelial membranes.** The structures shown in each photomicrograph are depicted in the accompanying diagrams.

the openings into the ears, anus, and vagina. A special *keratinized*, or *cornified*, layer of dead surface cells is found in the stratified squamous epithelium of the skin (the epidermis).

3. **Simple cuboidal epithelium** (fig. 1.4, *bottom*). This type of epithelium is usually simple and is found lining such structures as small tubules of the kidneys, and the ducts of the salivary glands or of the pancreas.
4. **Simple columnar epithelium** (fig. 1.5, *top*). This simple epithelium of tall columnar cells is found lining the lumen of the gastrointestinal tract, where it is specialized to absorb the products of digestion. It also contains mucus-secreting *goblet cells*.
5. **Simple ciliated columnar epithelium** (fig. 1.5, *upper middle*). These columnar cells support hairlike *cilia* on the exposed surface. These cilia produce wavelike movements that are characteristic along the luminal surface of female uterine tubes and the ductus deferens (vas deferens) of the male.
6. **Pseudostratified ciliated columnar epithelium** (fig. 1.5, *lower middle*). This epithelium is really simple but appears stratified because the nuclei are at different levels. Also characterized by hairlike cilia, this epithelium is found lining the respiratory passages of the trachea and bronchial tubes.
7. **Transitional epithelium** (fig. 1.5, *bottom*). This type is found only in the urinary bladder and ureters, and is uniquely stratified to permit periodic distension (stretching).

## PROCEDURE

1. Observe slides of the mesentery, esophagus, skin, pancreas, vas deferens or uterine tube, trachea, and urinary bladder.
2. Identify the type of epithelium in each of the slides.

Simple columnar (e.g., digestive tract)

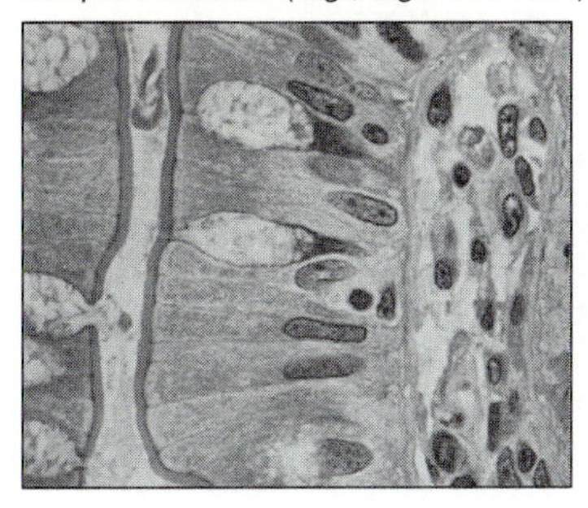

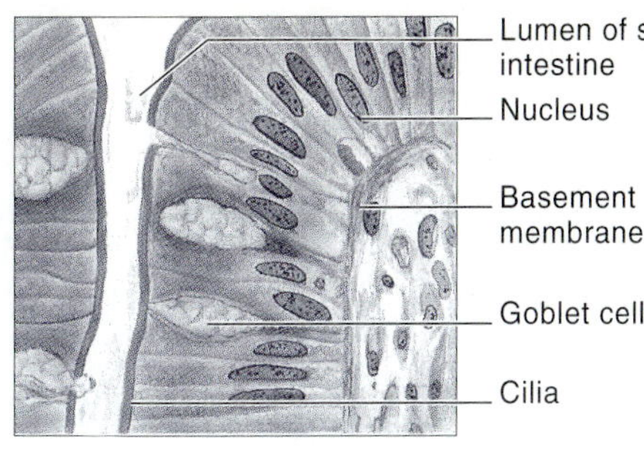


Simple ciliated columnar (e.g., uterine tube)

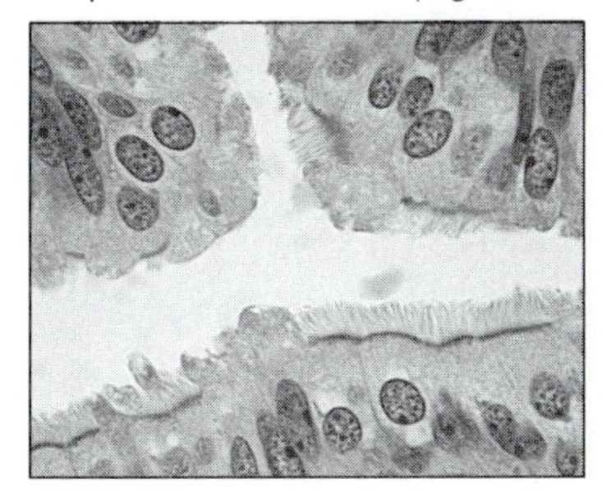

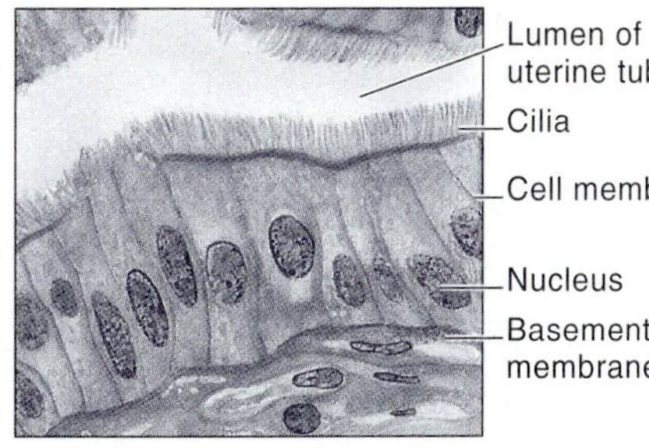


Pseudostratified ciliated columnar (e.g., lung bronchus)

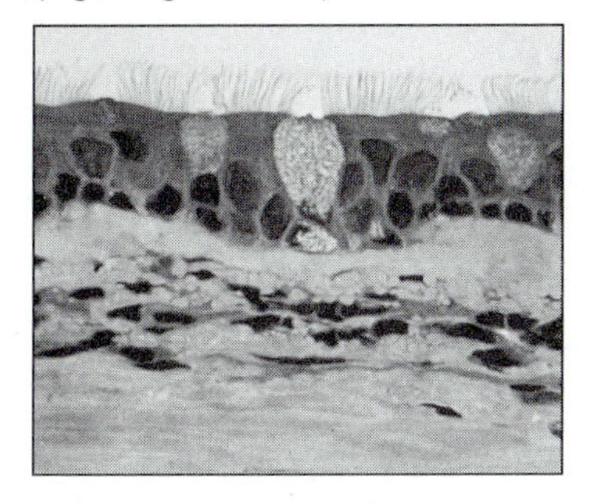

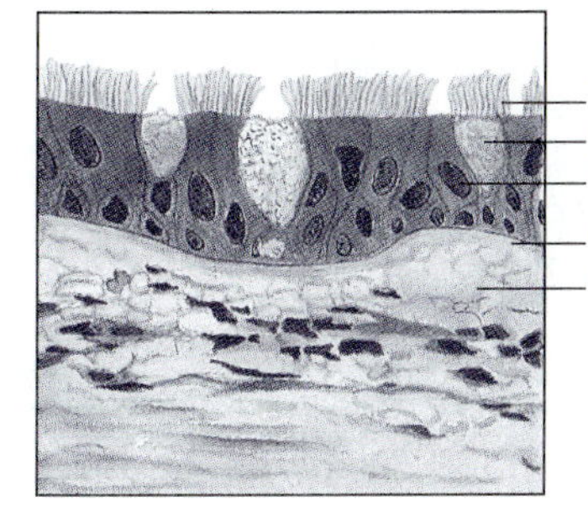


Transitional (e.g., urinary bladder)

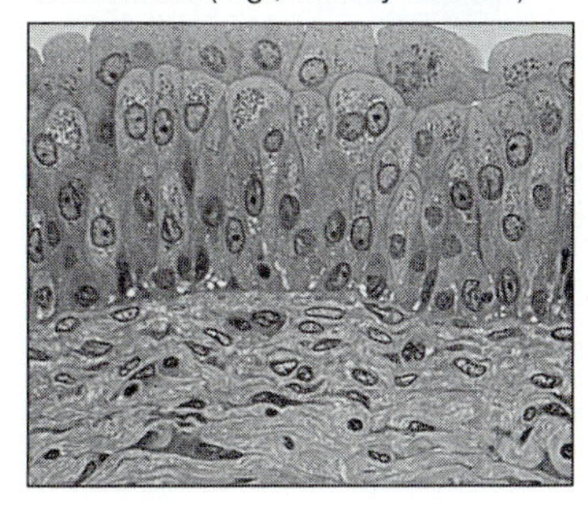

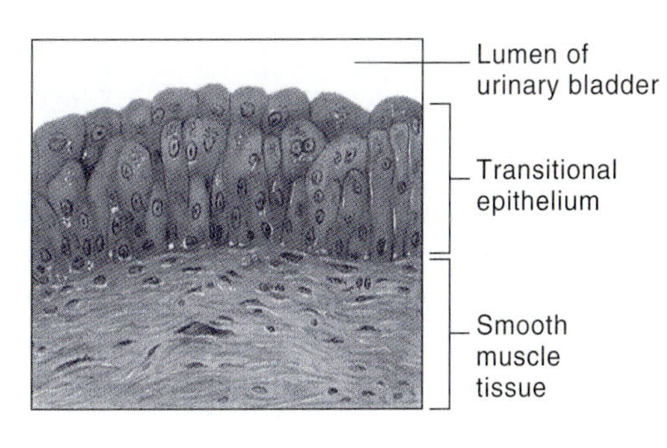


**Figure 1.5 Columnar and transitional epithelial membranes.** The structures shown in each photomicrograph are depicted in the accompanying diagrams.

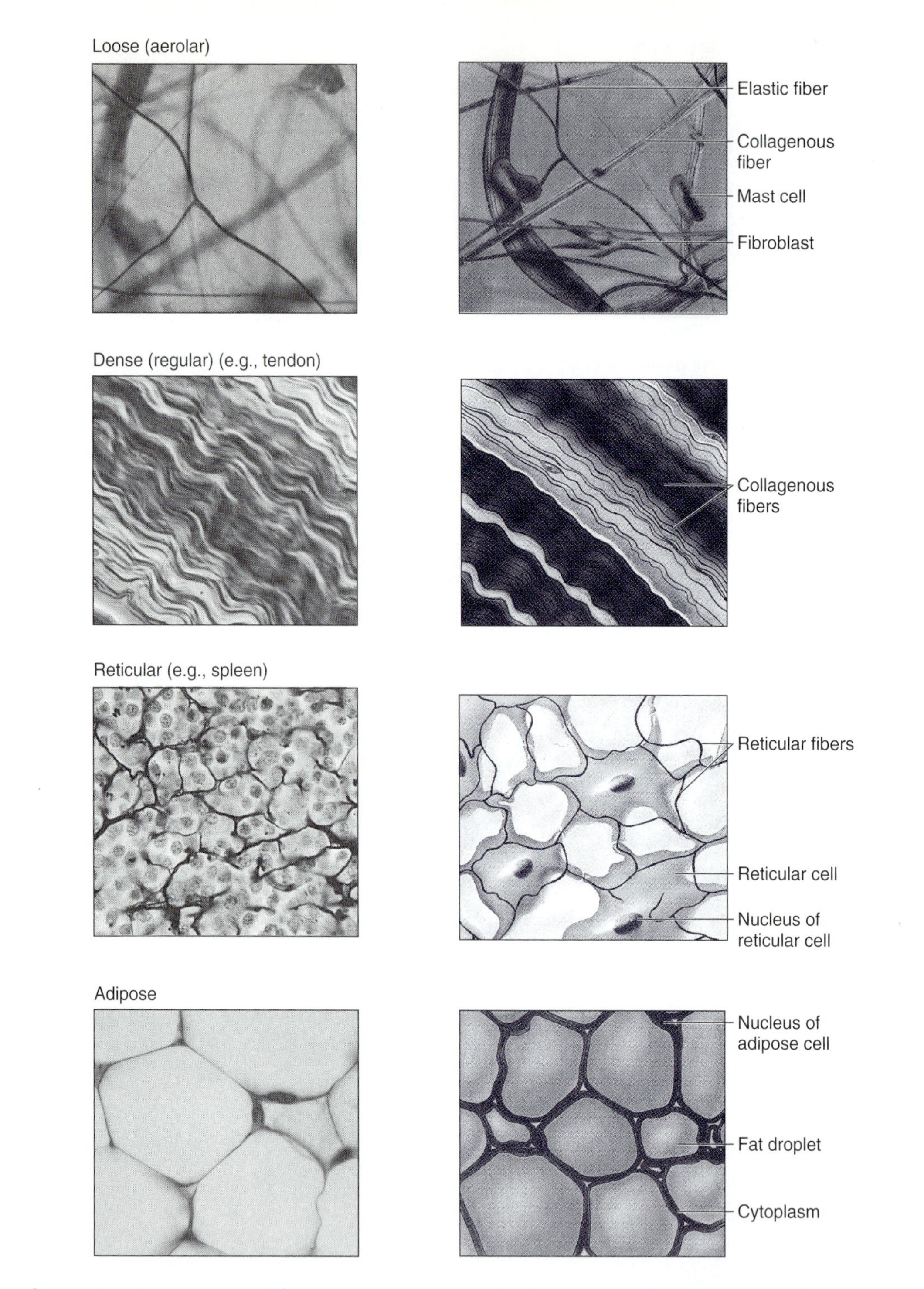


**Figure 1.6 Connective tissue proper.** The structures shown in each photomicrograph are depicted in the accompanying diagrams.

## B. Connective Tissues

Connective tissue is characterized by abundant amounts of extracellular material, or *matrix*. Unlike epithelial tissue, which is composed of tightly packed cells, the cells of connective tissue (which may be of many types) are spread out. The large extracellular spaces in connective tissue provide room for blood vessels and nerves to enter and leave organs.

There are five major types of connective tissues: (1) *mesenchyme*, an undifferentiated tissue found primarily during embryonic development; (2) *connective tissue proper*; (3) *cartilage*; (4) *bone*; and (5) *blood*.

**Connective tissue proper** (fig. 1.6) refers to a broad category of tissues with a somewhat loose, flexible matrix. This tissue may be *loose (areolar)*, which serves as a general binding and packaging material in such areas as the skin and the fascia of muscle, or *dense*, as is found in tendons

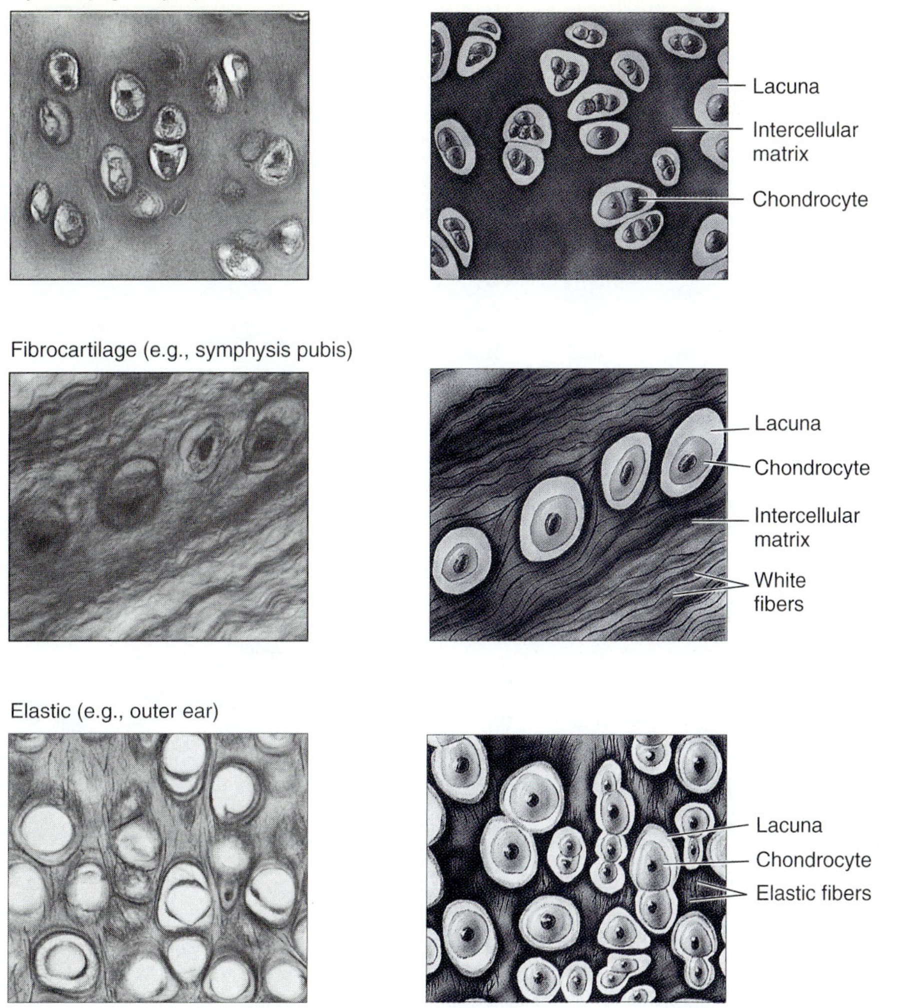


**Figure 1.7 Different forms of cartilage.** The structures shown in each photomicrograph are depicted in the accompanying diagrams.

and ligaments. The degree of denseness relates to the relative proportion of protein fibers to fluid in the matrix. These protein fibers may be made of *collagen,* which gives tensile strength to tendons and ligaments; they may be made of *elastin (elastic fibers),* which are prominent in large arteries and the lower respiratory system; or they may be *reticular fibers* providing more delicate structural support to the lymph nodes, liver, spleen, and bone marrow. *Adipose tissue* is a type of connective tissue in which the cells *(adipocytes)* are specialized to store fat.

**Cartilage** consists of cells *(chondrocytes)* and a semisolid matrix that imparts strength and elasticity to the tissue. The three types of cartilage are shown in figure 1.7. *Hyaline cartilage* has a clear matrix that stains a uniform blue. The most abundant form of cartilage, hyaline cartilage is found on the articular surfaces of bones (commonly called "gristle"), in the trachea, bronchi, nose, and the costal cartilages between the ventral ends of the first ten ribs and the sternum. *Fibrocartilage* matrix is reinforced with collagen fibers to resist compression. It is found in the symphysis pubis, where the two pelvic bones articulate, and between the vertebrae, where it forms intervertebral discs. *Elastic cartilage* contains abundant elastic fibers for flexibility. It is found in the external ear, portions of the larynx, and in the auditory canal (eustachian tube).

**Bone** (fig. 1.8*a*) contains mature cells called *osteocytes,* surrounded by an extremely hard matrix impregnated with calcium phosphate. Arranged in concentric layers, the osteocytes surround a *central canal,* containing nerves and blood vessels, and obtain nourishment via small channels in the matrix called *canaliculi.*

**Blood** (fig. 1.8*b*) is considered a unique type of connective tissue because its extracellular matrix is fluid *(plasma)* that suspends and transports blood cells (*erythrocytes, leukocytes,* and *thrombocytes*) within blood vessels. The composition of blood will be described in more detail in later exercises.

## PROCEDURE

1. Observe slides of skin, mesentery, a tendon, the spleen, cartilage, and bone.
2. Identify the types of connective tissue in each slide.

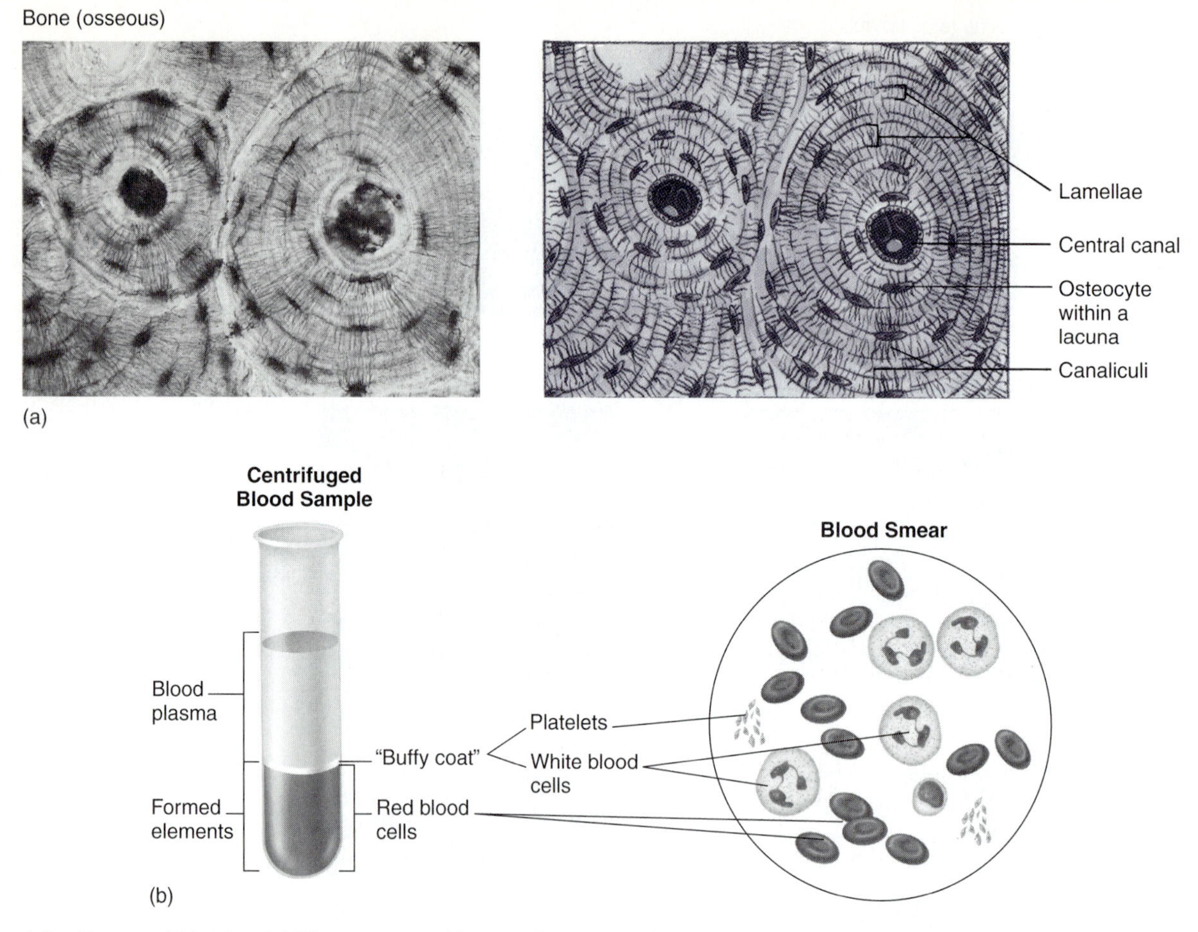


**Figure 1.8 Bone and blood.** (a) The structure of bone is shown in a photomicrograph and illustration. (b) The constituents of a centrifuged sample of blood are illustrated. When centrifuged, the solid elements—consisting of red blood cells, white blood cells, and platelets—are packed at the bottom of the tube, separating them from the fluid plasma.

**(For a full-color version of part [b] of this figure, see fig. 13.1 in *Human Physiology,* eighth edition, by Stuart I. Fox.)**

## C. Muscle Tissue

Muscles are responsible for heat production, body posture and support, and for a wide variety of movements, including locomotion. Muscle tissues, which are contractile, are composed of muscle cells, or *fibers,* that are elongated in the direction of contraction. The three types of muscle tissues—*smooth, cardiac,* and *skeletal*—are shown in figure 1.9.

**Smooth muscle** tissue is found in the digestive tract, blood vessels, respiratory passages, and the walls of the urinary and reproductive ducts. Smooth muscle fibers are long and spindle shaped, with a single nucleus near the center. **Cardiac muscle** tissue, which is found in the heart, is characterized by striated fibers that are branched and interconnected by *intercalated discs.* These interconnections allow electrical impulses to pass from one *myocardial* (heart muscle) *cell* to the next. **Skeletal muscle** tissue attaches to the skeleton, and is responsible for voluntary movements. Skeletal muscle fibers are long and thin and contain numerous nuclei. Skeletal muscle is under voluntary control, whereas cardiac and smooth muscles are classified as involuntary. This distinction relates to the type of nerves involved (innervation) and not to the characteristics of the muscles themselves. Both skeletal muscle and cardiac muscle cells are categorized as *striated muscle* because they contain cross striations.

### PROCEDURE

1. Observe prepared slides of smooth, cardiac, and skeletal muscles.
2. Identify the major distinguishing features of each type of muscle.

## D. Nervous Tissue

Nervous tissue, which forms the nervous system, consists of two major categories of cells. The nerve cell, or **neuron** (fig. 1.10), is the functional unit of the nervous system.

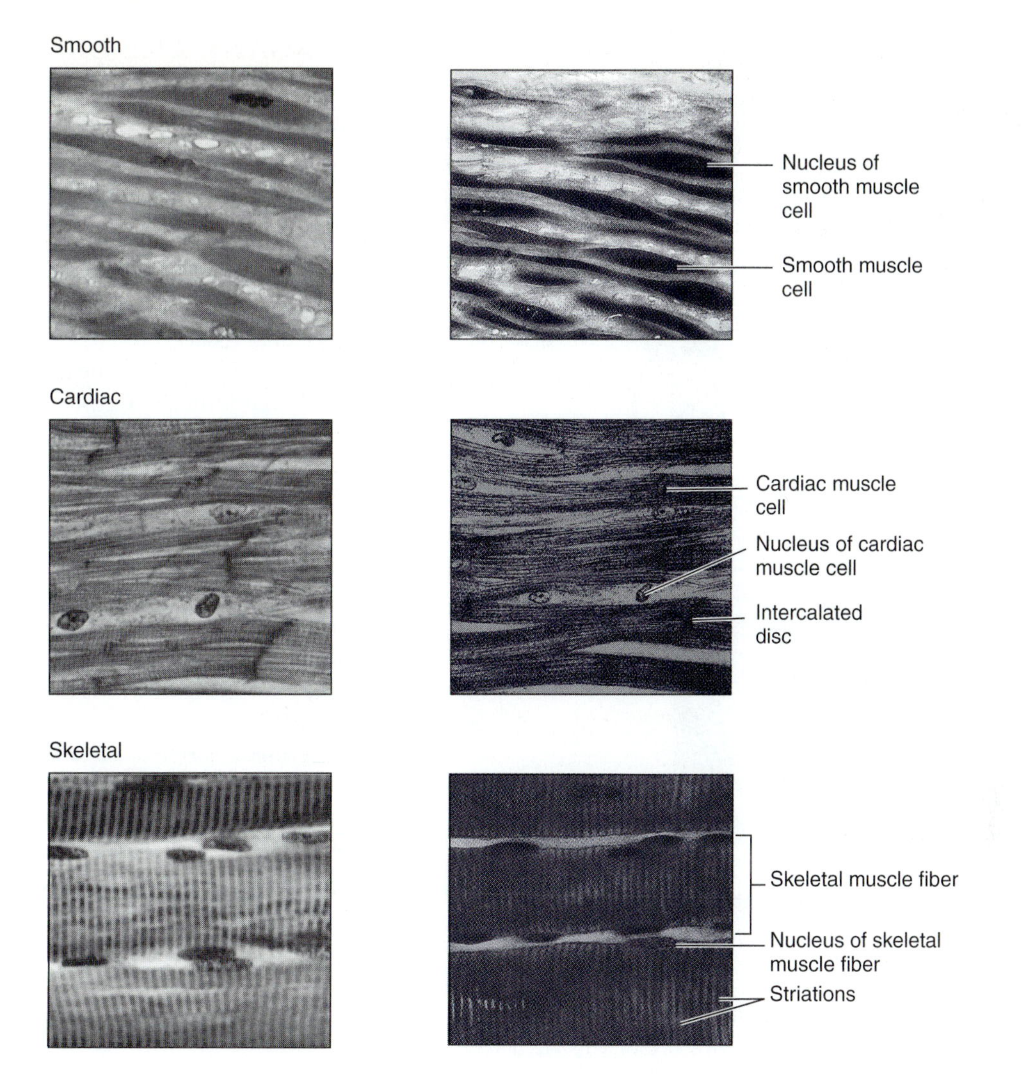


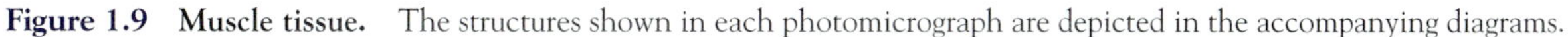

**Figure 1.9 Muscle tissue.** The structures shown in each photomicrograph are depicted in the accompanying diagrams.

The typical neuron has a *cell body* with a nucleus, smaller projections called *dendrites* branching from the cell body, and a single, long, cytoplasmic extension called an *axon*, or *nerve fiber*. The neuron is generally capable of receiving, producing, and conducting electrical impulses. Most neurons release specialized chemicals from the axon endings. A second category of cell found in the nervous system is a **neuroglial cell.** Various types of neuroglia support the neurons both structurally and functionally.

Some axons of the central nervous system (CNS) and peripheral nervous system (PNS) are surrounded by myelin sheath (are *myelinated*); others lack a myelin sheath (are *unmyelinated*). Neuroglial cells called **Schwann cells** form myelin sheaths in the PNS. When an axon in a peripheral neuron is cut, the Schwann cells form a *regeneration tube* that helps to guide the regenerating axon to its proper destination. Even a severed major nerve may be surgically reconnected, and the function of the nerve largely reestablished, if the surgery is performed before tissue death. Neuroglial cells of the CNS that form myelin sheaths are known as **oligodendrocytes.** In contrast to Schwann cells, oligodendrocytes do not form regeneration tubes. For this and other reasons that are incompletely understood, cut or severely damaged neurons of the brain and spinal cord usually result in permanent damage.

## PROCEDURE

1. Observe prepared slides of the spinal cord and the brain.
2. Identify the parts of a neuron.
3. Distinguish neurons from neuroglial cells.
4. Without referring to the caption, identify the various tissue types in the photomicrographs in figure 1.10.

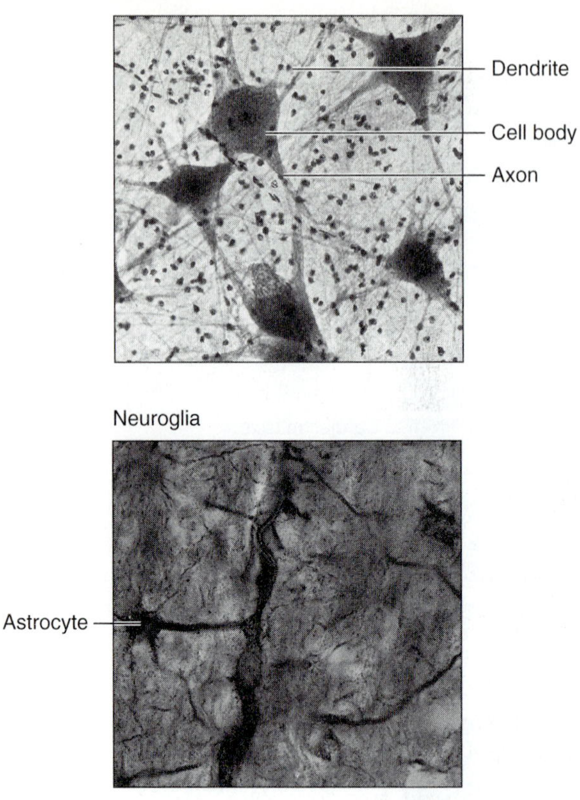


**Figure 1.10 Nervous tissue.** Photomicrographs of representative neurons and neuroglia in the CNS.

## E. An Organ: The Skin

Organs contain more than one type—and usually all four types—of primary tissue. The **skin,** the largest organ of the body, provides an excellent example.

Epithelial tissue is illustrated by the *epidermis* and the *hair follicles* (fig. 1.11). Like all glands, the oily *sebaceous glands* associated with hair follicles and the *sweat glands* are a type of epithelial tissue.

Connective tissue is seen in the *dermis*. Collagen fibers that form dense connective tissue are located in the dermis, whereas adipose connective tissue is embedded in the *hypodermis*.

Muscle tissue is represented by the *arrector pili muscle,* a smooth muscle that attaches to the hair follicle and the matrix of the dermis.

Nerve tissue is featured within skin by the sensory and motor nerves, and by *Meissner's corpuscle* (the oval structure in the dermis near the start of the sensory nerve, fig. 1.11), a sensory structure sensitive to pressure.

### PROCEDURE

1. Observe a prepared slide of the skin or scalp.
2. Identify the structures of the skin and try to find all four types of primary tissue.

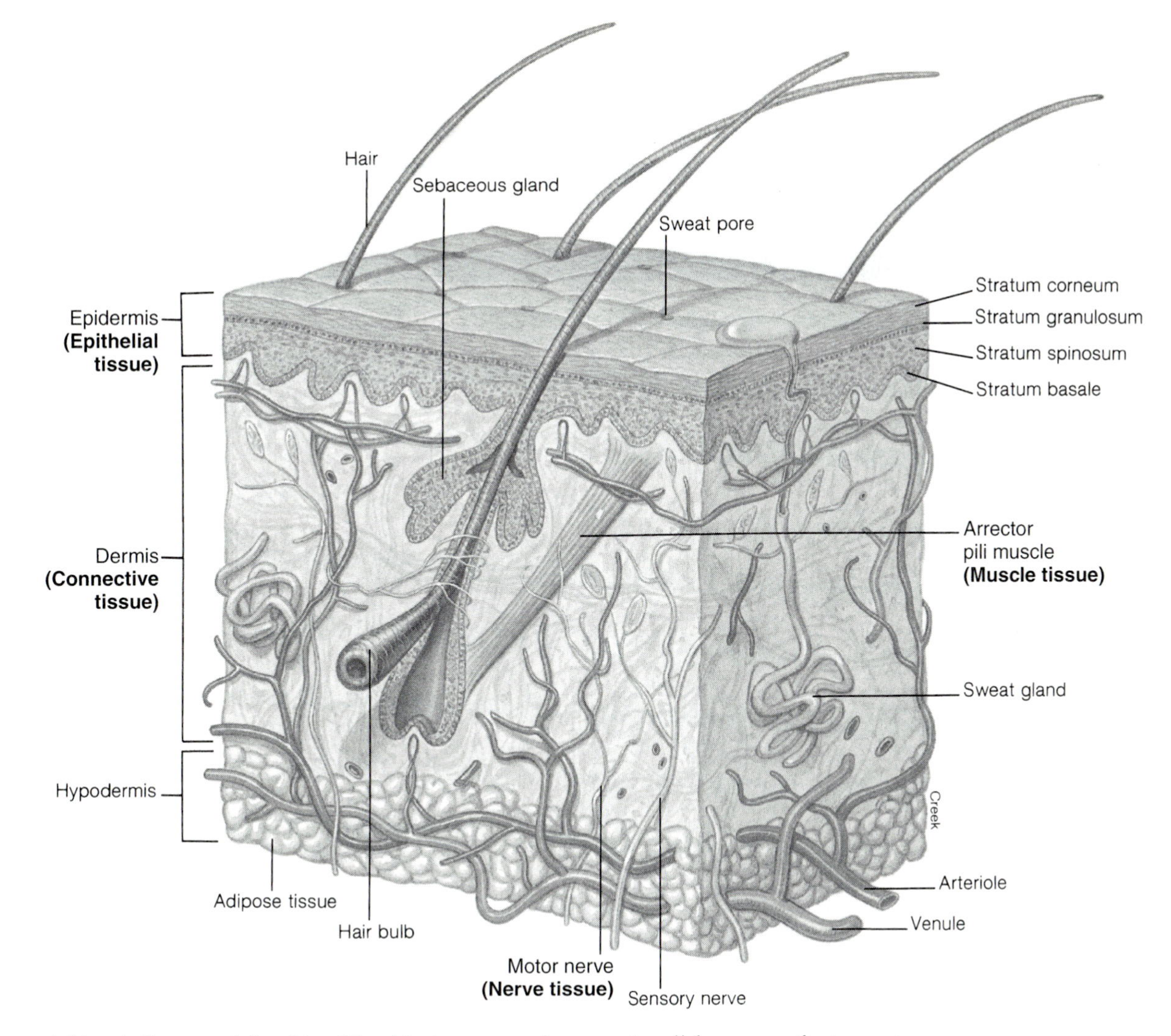


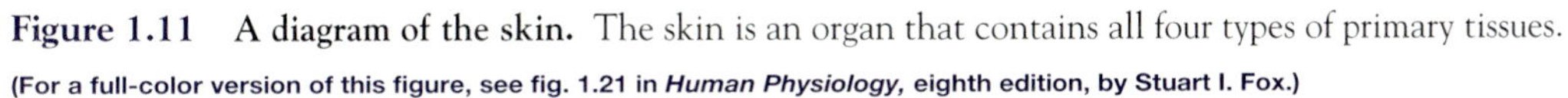

**Figure 1.11** **A diagram of the skin.** The skin is an organ that contains all four types of primary tissues.

**(For a full-color version of this figure, see fig. 1.21 in *Human Physiology,* eighth edition, by Stuart I. Fox.)**

9. Identify the distinguishing characteristic of connective tissues. Give examples of three connective tissues and describe how they fit into the connective tissue category.

## *Test Your Ability to Analyze and Apply Your Knowledge*

10. Would you expect the muscle fibers of the tongue to be striated or smooth? What about the muscle of the diaphragm? Explain your answer.

11. Blood vessels and nerves are found in connective tissues, not in epithelial membranes. Why? Would you expect to see strands of connective tissue within the pancreas and liver? Explain your answer.

# Homeostasis and Negative Feedback

EXERCISE **1.3**

MATERIALS

1. Watch or clock with a second hand
2. Hot plate; beaker; thermometer; crushed ice; constant-temperature water bath

The regulatory mechanisms of the body help to maintain a state of dynamic constancy of the internal environment known as homeostasis. Most systems of the body maintain homeostasis by operating negative feedback mechanisms that control effectors (muscles and glands).

## OBJECTIVES

1. Define the term *homeostasis.*
2. Explain how the negative feedback control of effectors helps to maintain homeostasis.
3. Explain why the internal environment is in a state of dynamic, rather than static, constancy.
4. Define the terms *set point* and *sensitivity.*
5. Explain how a normal range of values for temperature or heart rate is obtained, and discuss the significance of these values.

## Textbook Correlations*

Before performing this exercise, you should study the introductory material presented here. Further information relating to this exercise can be found in these pages of *Human Physiology,* eighth edition, by Stuart I. Fox:

- *Negative Feedback Loops.* Chapter 1, pp. 6–8.
- *Feedback Control of Hormone Secretion.* Chapter 1, p. 9.

*Multimedia Correlations (also see Appendix 3)
- *MediaPhys 2.0:* Topics 1.3–1.10

Although the structure of the body is functional, the study of body function involves much more than a study of body structure. The extent to which each organ performs the functions endowed by its genetic programming is determined by regulatory mechanisms that coordinate body functions in the service of the entire organism. The primary prerequisite for a healthy organism is the maintenance of **homeostasis,** which is the dynamic constancy of the internal environment.

When homeostasis is disturbed—for example, by an increase or decrease in body temperature from its normal value, or *set point*—a *sensor* detects the change. The sensor then activates an *effector,* which induces changes opposite to those that activated the sensor. The action action of the effector works to correct the initial disturbance, so that the initial change and its compensatory reaction result in only slight deviations from the normal value. In this way, temperature and other body parameters are maintained at a relative constancy. Homeostasis is therefore a state of *dynamic,* rather than absolute, constancy (fig. 1.12).

Since a disturbance in homeostasis initiates events that lead to changes in the opposite direction, the cause-and-effect sequence is described as a **negative feedback mechanism** (or a *negative feedback loop*). A constant-temperature water bath, for example, uses negative feedback mechanisms to maintain the temperature at which the bath is set (the **set point**). Deviations from the set point are detected by a thermostat (temperature sensor), which turns on a heating unit (the effector) when the temperature drops below the set point, and turns off the unit when the temperature rises above the set point (fig. 1.12).

By means of the negative feedback control of the heating unit, the water-bath temperature is not allowed to rise or fall too far from the set point. Keep in mind, however, that the temperature of the water is at the set point only briefly. The set point is in fact only the *average value* within a *range* (from the highest to the lowest value) of temperatures. The *sensitivity* of this negative feedback mechanism is measured by the temperature deviation from the set point required to activate the compensatory (negative feedback) response (turning the heater on or off).

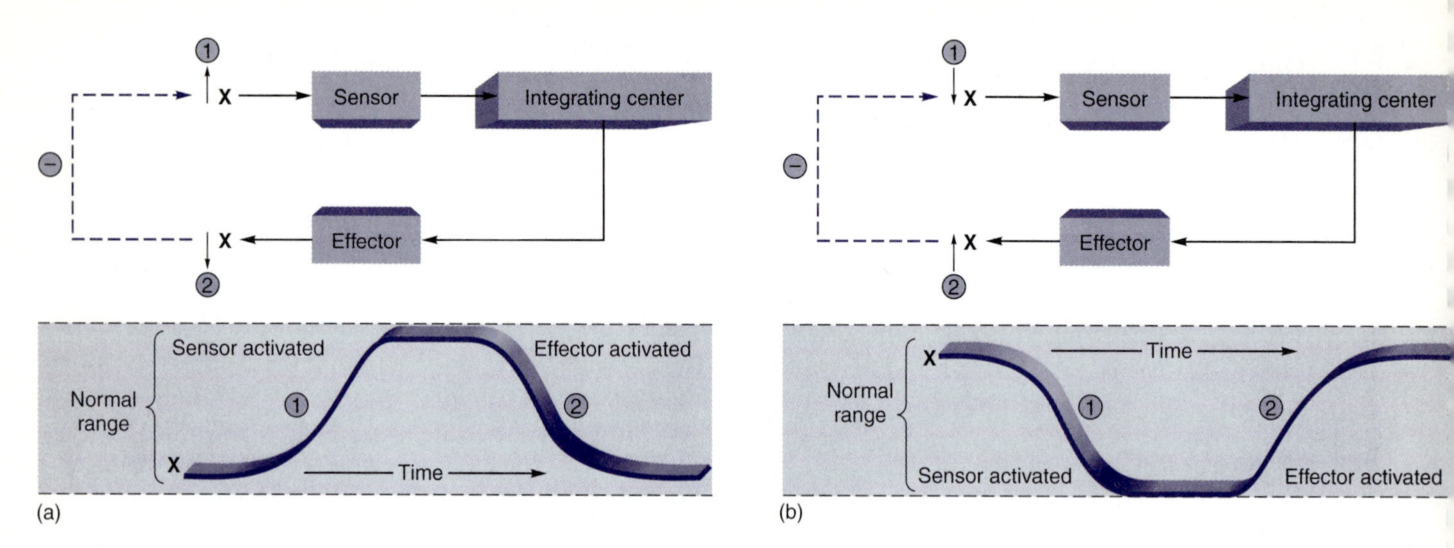


**Figure 1.12 Homeostasis is maintained by negative feedback loops.** (a) A rise in some factor in the internal environment is corrected by mechanisms that bring it back down. (b) A fall in some factor in the internal environment is corrected by mechanisms that help to raise it. In both cases, the negative feedback loops compensate for the changes to help maintain a state of dynamic constancy.

**(For a full-color version of this figure, see figs. 1.1 and 1.2 in *Human Physiology,* eighth edition, by Stuart I. Fox.)**

## A. Negative Feedback Control of Water Bath Temperature

Modern constant-temperature water baths are so sensitive that tiny deviations from the set point are quickly corrected by very efficient electronic mechanisms. This makes it difficult to observe the negative feedback corrections. However, you can observe the way that the sensor and effector (heating element) work to defend constancy of temperature by adding cold water or crushed ice to the bath. This is analogous to the way that negative feedback mechanisms in your body maintain homeostasis of body temperature when you are exposed to a cold ambient temperature.

A variation on this theme of temperature homeostasis can be performed using a hot plate, a beaker of water, and a thermometer. In this procedure, you are the sensor and the integrating center, because you adjust the effector (hot plate) in order to maintain a constant temperature of the water. You can also observe the effectiveness of **antagonistic effectors** by using some crushed ice (as well as decreased heat settings of the hot plate) to lower the water temperature. This is analogous to the antagonistic actions of sweating (which lowers body temperature) and shivering (which raises it).

### PROCEDURE

1. The temperature of the constant-temperature water bath is set by the instructor at a certain temperature, somewhere between 40° C and 60° C. A red indicator light goes on when the heating element is activated and off when the heating element is inactivated.
2. Observe the temperature of the constant-temperature water bath over the course of 5 minutes, recording its temperature once each minute in your laboratory report. Use these measurements to determine the set point of the water bath (this can be taken as the average of your measurements). Record this value in the laboratory report.
3. Add a fairly large amount of cold water or crushed ice to the constant-temperature water bath. Record the temperature of the water bath once a minute for 5 minutes after adding the cold water or crushed ice. Continue observing until the water temperature has returned to the previously established set point. In your laboratory report, record how long this takes.
4. Using a hot plate, a beaker of tap water, and a thermometer, attempt to heat the water to 37° C and then maintain this temperature by either decreasing or increasing the heat setting on the hot plate. After the 37° C temperature was first attained, record the temperature readings of the water every 5 minutes for 30 minutes.
5. Now, use some crushed ice (as well as a decreased setting on the hot plate) to lower the temperature once the water temperature exceeds 37° C. Record the temperature readings of the water every 5 minutes for another 30 minutes.

## B. Resting Pulse Rate: Negative Feedback Control and Normal Range

**Homeostasis**—the dynamic constancy of the internal environment—is maintained by negative feedback mechanisms that are far more complex than those involved in maintaining a constant-temperature water bath. In most cases, several effectors, many with antagonistic actions, are involved in maintaining homeostasis. It is as if the temperature of a water bath were determined by the antagonistic actions of both a heater and a cooling system. The cardiac rate (or pulse rate) is largely determined by the antagonistic effects of two different nerves. One of these (a *sympathetic nerve*, described in section 7) stimulates an increase in cardiac rate. A different nerve (a *parasympathetic nerve*) produces inhibitory effects that slow the cardiac rate.

The resting cardiac rate or pulse rate, measured in *beats per minute*, is maintained in a state of dynamic constancy by negative feedback loops initiated by sensors in response to changes in blood pressure and other factors. Therefore, the resting pulse rate is not absolutely constant but instead varies about a set-point value. This exercise will demonstrate that your pulse rate is in a state of dynamic constancy (implying negative feedback controls). From the data you can determine your own pulse-rate set point as the average value of the measurements.

## PROCEDURE

1. Gently press your index and middle fingers (not your thumb) against the radial artery in your wrist until you feel a pulse. Alternatively, the carotid pulse in the neck may be used for these measurements.
2. The pulse rate is usually expressed as beats per minute. However, only the number of beats per 15-second interval (quarter minute) need be measured; multiplying this by four gives the number of beats per minute. Record the number of beats per 15-second interval in the data table provided in the laboratory report.
3. Pause 15 seconds, and then count your pulse during the next 15-second interval. Repeat this procedure over a 5-minute period. Recording your count once every half minute for 5 minutes, a total of 10 measurements (expressed as beats per minute) will be obtained.
4. Using the grid provided in the laboratory report, graph your results by placing a dot at the point corresponding to the pulse rate for each measurement, and then connect the dots.

### Normal Values

Students often ask, How do my measurements compare with those of others? and Are my measurements normal? Normal values are those that healthy people have. Since healthy people differ to some degree in their particular values, what is considered normal is usually expressed as a range of values that encompasses the measurements of most healthy people. An estimate of the **normal range** is a statistical determination that is subject to statistical errors and also subject to questions about what is meant by the term *healthy*.

Healthy, in this context, means the absence of cardiovascular disease. Included in the healthy category, however, are endurance-trained athletes, who usually have lower than average cardiac rates, and relatively inactive people, who have higher than average cardiac rates. For this reason, determinations of normal ranges can vary, depending on the relative proportion of each group in the sample tested. A given class of students may therefore have an average value and a range of values that differ somewhat from those of the general population.

The concept of homeostasis is central to medical diagnostic procedures. Through the measurement of body temperature, blood pressure, concentrations of specific substances in the blood, and many other variables, the clinical examiner samples the internal environment. If a particular measurement deviates significantly from the range of normal values—that is, if that individual is not able to maintain homeostasis—the cause of the illness may be traced and proper treatment determined to bring the measurement back within the normal range.

For example, measurements of fasting blood glucose levels are commonly performed to detect **diabetes mellitus,** which is a disorder in the secretion or action of the hormone *insulin.* The negative feedback loops that maintain homeostasis involve clusters of cells (islets) in the pancreas that secrete the hormone insulin and another hormone, glucagon (fig. 1.13).

## PROCEDURE

1. Each student in the class determines his or her own average cardiac rate (pulse rate) from the previous data either by taking an arithmetic average or simply by observing the average value of the fluctuations in the previously constructed graph. Record your own average in the laboratory report.
2. Record the number of students in the class with average pulse rates in each of the rate categories shown in the laboratory report. Also, calculate the percentage of students in the class who are within each category and record this percentage in the laboratory report.
3. Divide the class into two groups: those who exercise on a regular basis (at least three times a week) and those who do not. Determine the average pulse rate and range of values for each of these groups. Enter this information in the given spaces in the laboratory report.

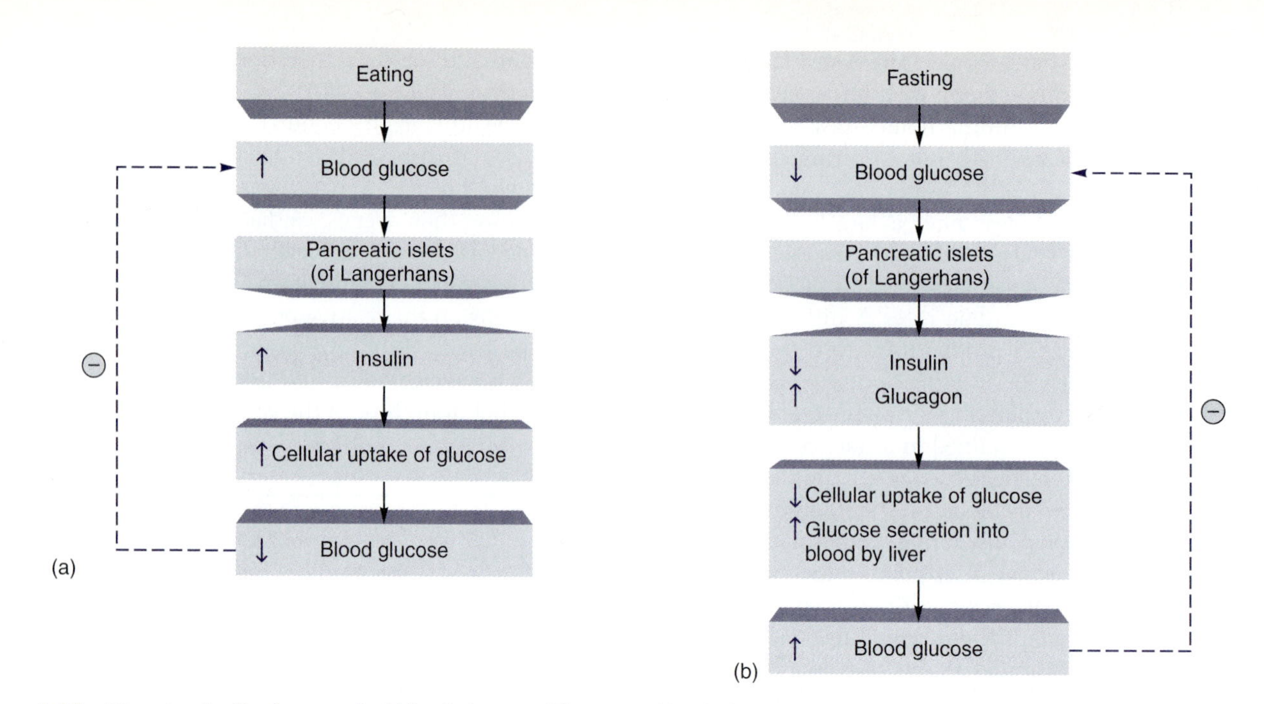


**Figure 1.13 Negative feedback control of blood glucose.** The rise in blood glucose that occurs after eating carbohydrates is corrected by the action of insulin, which is secreted in increasing amounts (a) at that time. During fasting, when blood glucose falls, insulin secretion is inhibited and the secretion of an antagonistic hormone, glucagon, is increased (b). This stimulates the liver to secrete glucose into the blood, helping to prevent blood glucose from continuing to fall. In this way, blood glucose concentrations are maintained within a homeostatic range following eating and during fasting.

**(For a full-color version of this figure, see fig. 1.6 in *Human Physiology,* eighth edition, by Stuart I. Fox.)**

# Laboratory Report 1.3

Name ________________________

Date ________________________

Section ________________________

## DATA FROM EXERCISE 1.3

**A. Negative Feedback Control of Water Bath Temperature**

1. Measurements of constant-temperature water bath temperature (once each minute for 5 minutes):
   ________; ________; ________; ________; ________.
2. Calculate the **average** temperature of the constant-temperature water bath: ________. (*Note:* Do this by adding all of your measurements and dividing by 5, the total number of measurements you made.) This is taken as the set point of the water bath.
3. What is the **range** of values in the 5 measurements? ________ to________. (*Note:* This is the lowest to the highest of your measurements.)
4. What was the temperature of the constant-temperature water bath after adding the cold water or crushed ice?
   ________
5. How long did it take for the temperature to return to the set point? ________
6. Record the temperature reading of the water in the beaker on the hot plate after first attaining 37° C (one measurement every 5 minutes for 30 minutes, adjusting only the hot plate settings): ________; ________; ________; ________; ________; ________.
   (a) Calculate the average of these measurements: ________.
   (b) Indicate the range of these measurements: ________.
7. Record the temperature reading of the water in the beaker on the hot plate (one measurement every 5 minutes for 30 minutes) when you used crushed ice as well as lower hot plate settings to lower the water temperature:
   ________; ________;________; ________; ________; ________.
   (a) Calculate the average of these measurements: ________.
   (b) Indicate the range of these measurements: ____________ to ____________.
8. Which negative feedback mechanism was more effective at maintaining constant temperature—the electronic mechanism of the constant-temperature water bath (measurements in no. 1) or the human manual methods (measurements in nos. 6 and 7)?
   ________________________________________
9. Which was more effective at maintaining a constant temperature—adjustments of one effector only (the hot plate) or adjustments of antagonistic effectors (hot plate and crushed ice)?
   ________________________________________

**B. Resting Pulse Rate: Negative Feedback Control and Normal Range**

| Measurement | 1 | 2 | 3 | 4 | 5 | 6 | 7 | 8 | 9 | 10 |
|---|---|---|---|---|---|---|---|---|---|---|
| Beats per 15 seconds | | | | | | | | | | |
| Beats per minute | | | | | | | | | | |

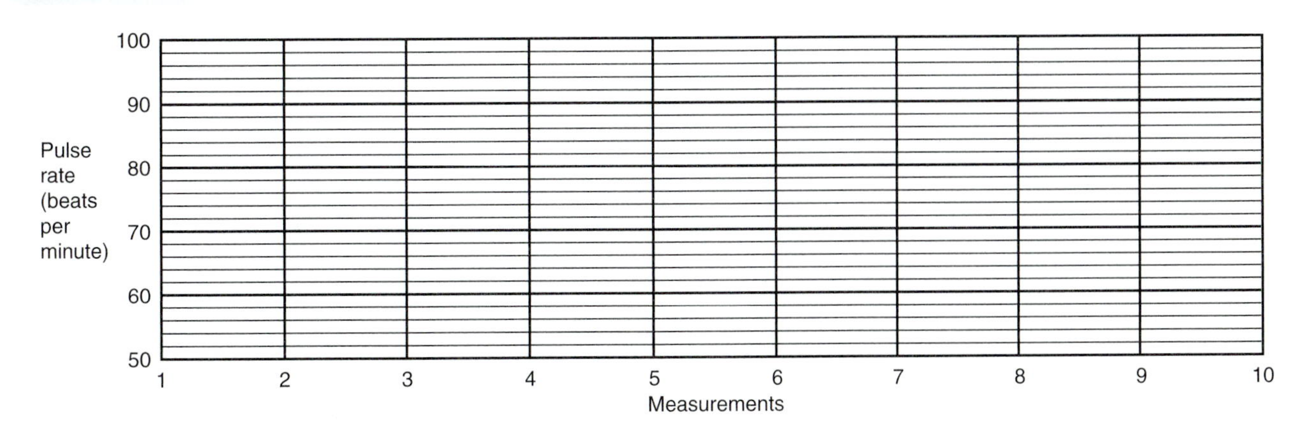

1. Calculate your **average** pulse rate: __________ pulses/minute
   (*Note:* Do this by adding all of your measurements and dividing by 10, the total number of measurements you made.)
2. What is the **range** of values in the 10 measurements? __________ pulses/minute to __________ pulses/minute
   (*Note:* This is the lowest to the highest of your measurements.)
3. What is the **sensitivity** of your negative feedback mechanism that is maintaining homeostasis of your pulse rate?
   ± __________ pulses/minute
   (*Note:* To obtain this, first take the difference between your average pulse rate and each of your individual measurements by subtracting the higher from the lower value. Then, add all of the differences together. Finally, divide this sum by 10, the number of measurements.)
4. Pulse rate averages of the class:

| Pulse Rate (beats per minute) | Number of Students | Percentage of the Total |
|---|---|---|
| Over 100 bpm | | |
| 90–100 | | |
| 80–89 | | |
| 70–79 | | |
| 60–69 | | |
| 50–59 | | |
| Under 50 bpm | | |

5. Data for the exercise and nonexercise groups:

| | Exercise Group | Nonexercise Group |
|---|---|---|
| Range of Pulse Rates | | |
| Average of Pulse Rates | | |

## REVIEW ACTIVITIES FOR EXERCISE 1.3

### *Test Your Knowledge of Terms and Facts*

1. Define the term *homeostatsis*. ____________________________________________
   ____________________________________________
2. Define the term *set point*. ____________________________________________
   ____________________________________________

### *Test Your Understanding of Concepts*

3. Explain how negative feedback mechanisms operate to maintain homeostasis. Use the terms *sensor, integrating center,* and *effector* in your answer.

4. Draw a flow diagram, illustrating cause and effect with arrows, to show how constant temperature is maintained in a water bath. (*Note:* Flow diagrams are pictorial displays of processes that occur in sequence, using arrows to indicate the direction of cause-and-effect sequences.)

5. Suppose that a constant-temperature water bath contained two antagonistic effectors: a heater and a cooler. Draw a flow diagram to show how this dual system could operate to maintain a constant temperature about some set point.

## *Test Your Ability to Analyze and Apply Your Knowledge*

6. Explain why your graph of pulse rate measurements suggests the presence of negative feedback control mechanisms.

7. Sympathetic nerves to the heart increase the rate of beat, while parasympathetic nerves decrease the rate of beat. Draw a negative feedback loop showing how sympathetic and parasympathetic nerves are affected in someone experiencing a fall in blood pressure (the initial stimulus). (*Note:* The sensor detects the fall in blood pressure.)

8. Why would there be different published values for the normal range of a particular measurement? Do these values have to be continuously updated? Why?

# Cell Function and Biochemical Measurements

# Section 2

**P**hysiological control systems maintain homeostasis of the internal chemical environment to which the organ systems are exposed. The concentrations of *glucose, protein,* and *cholesterol* in plasma (the fluid portion of the blood), for example, are maintained within certain limits despite the variety in dietary food selections and variations in our eating schedules. This regulation is necessary for health. If plasma glucose levels fall too low, for example, the brain may "starve" and a coma may result. A drop in plasma protein, as another example, may disturb the normal distribution of fluid between the blood and tissues. An abnormal rise in these values, or other abnormal changes in the chemical composition of plasma, can endanger a person's health in various ways.

Abnormal changes in the internal chemical environment, which can contribute to disease processes, are usually themselves the result of diseases that affect cell function. For example, since most plasma proteins are produced by liver cells, diseases of the liver can result in the lowering of plasma protein concentrations. Similarly, abnormal lowering of plasma glucose levels may result from excess secretion of the hormone insulin by certain cells of the pancreas. Thus, homeostasis of the internal chemical environment depends on proper cell function.

All of the molecules found in the body's internal environment, aside from those few obtained directly from food, are produced within the cells. Some molecules remain within the cells; others are secreted into the tissue fluids and blood. Almost all of these molecules are produced by chemical reactions catalyzed by special proteins known as **enzymes.** All enzymes in the body are produced within tissue cells according to information contained in the **DNA** (genes). In this way, the overall metabolism of carbohydrates, lipids, proteins, and other molecules in the cell is regulated largely by genes. Defects in these genes can result in the production of defective enzymes, which result in impaired metabolism. Thus, the study of organ system physiology is intertwined with the study of cell function and biochemistry, as well as with the study of genetics.

Proper cell function also depends upon the integrity of the plasma (cell) membrane. Composed primarily of two semifluid phospholipid layers, cell membranes can regulate the passive transport of molecules moving from higher to lower concentration by diffusion. Special membrane proteins can serve as channels for the passage of larger or more polar molecules, whereas other membrane proteins serve as carriers that require the expenditure of energy to "pump" molecules across the membrane "uphill" from lower to higher concentrations (a process called active transport).

# Measurements of Plasma Glucose, Cholesterol, and Protein

EXERCISE 2.1

## Materials

1. Pyrex (or Kimax) test tubes, mechanical pipettors for 40 μL, 50 μL, 100 μL, and 5.0 mL volumes; and corresponding pipettes (0.10 mL and 5.0 mL total volume—see fig. 2.1)
2. Constant-temperature water bath, set at 37° C
3. Colorimeter and cuvettes
4. Glucose kit ("Glucose LiquiColor Test," Stanbio Laboratory, Inc.)
5. Cholesterol kit ("Cholesterol Liquicolor Test," Stanbio Laboratory, Inc.)
6. Total Protein Standard (10 g/dL from Stanbio Laboratory, Inc.); the following concentrations: 2, 4, 6, 8 g/dL can be prepared by dilution.
7. Biuret reagent. To a 1.0-L volumetric flask, add 45 g of sodium potassium tartrate and 15 g of $CuSO_4 \cdot 5\,H_2O$. Fill 2/3s full with 0.2N NaOH and shake to dissolve. Add 5 g of potassium iodide and fill to 1.0 L volume with 0.2N NaOH.
8. Serum (Artificial "Normal" and "Abnormal Control" sera can be purchased from Stanbio Laboratory, Inc.)

The concentrations of glucose, protein, and cholesterol in plasma (or serum) can be measured using colorimetric techniques in the laboratory. Abnormal concentrations of these molecules are associated with specific disease states.

## OBJECTIVES

1. Describe how Beer's law can be used to determine the concentration of molecules in solution.
2. Use the formula method and graphic method to determine the concentration of molecules in plasma (serum) samples.
3. Explain the physiological roles of glucose, protein, and cholesterol in the blood.
4. Explain why abnormal measurements of plasma glucose, protein, and cholesterol are clinically significant.

## Textbook Correlations

Before performing this exercise, you should study the introductory material presented here. Further information relating to this exercise can be found in these pages of *Human Physiology,* eighth edition, by Stuart I. Fox:

- *Carbohydrates and Lipids.* Chapter 2, pp. 31–38.
- *Proteins.* Chapter 2, pp. 38–41.
- *Exchange of Fluid between Capillaries and Tissues.* Chapter 14, pp. 413–415.

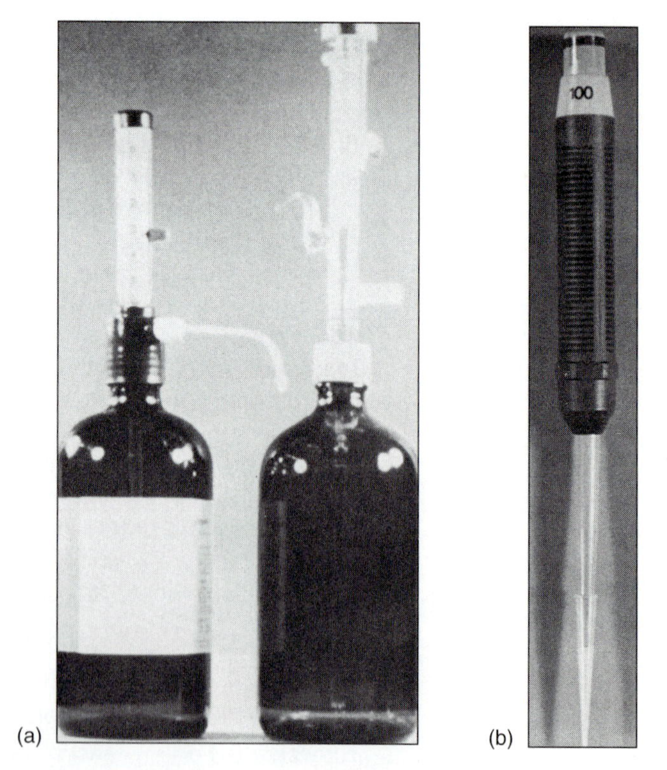


**Figure 2.1 Automatic devices for dispensing fluids.** (a) Device to dispense milliliters (such as 5.0 mL) of reagent. (b) An automatic microliter pipettor (Eppendorf) for dispensing 100 μL (0.10 mL) of solution, or similar volumes.

## Table 2.1 Examples of Monomers and Polymers

| Monomer | Examples | Polymer | Examples |
|---|---|---|---|
| Monosaccharides | Glucose, fructose | Polysaccharides | Starch, glycogen |
| Amino acids | Glycine, phenylalanine | Proteins | Hemoglobin, albumin |
| Fatty acids and glycerol | | Triglycerides | Fats, oils |
| Ribonucleotides and deoxyribonucleotides | | Nucleic acids | DNA and RNA |

*Organic* molecules found in the body contain the atoms carbon (C), hydrogen (H), and oxygen (O) in various ratios, and some of these molecules also contain the atoms nitrogen (N), phosphorus (P), and sulfur (S). Many organic molecules are very large. They consist of smaller repeating subunits that are chemically bonded to each other. The term *monomer* refers to the individual subunits; the term *polymer* refers to the long chain formed from these repeating subunits.

When two monomers are bonded together, a molecule of water (HOH) is released. This reaction is called **condensation,** or **dehydration synthesis.**

$$\mathbf{A}\text{—OH} + \mathit{HO}\text{—}\mathbf{B} \rightarrow \mathbf{A}\text{—}\mathbf{B} + \mathit{HOH}$$

The new molecule (**A**—**B**) formed from the two monomers (**A** and **B**) is called a *dimer*. This dimer may participate in a condensation reaction with a third monomer to form a *trimer*. The stepwise addition of new monomers to the growing chain by condensation reactions will result in the elongation of the chain and the formation of the full polymer. Examples of monomers and polymers are given in table 2.1.

When the chemical bond between monomers is broken, a molecule of water is consumed. This **hydrolysis reaction** is the reverse of a condensation reaction.

$$\mathbf{A}\text{—}\mathbf{B} + \mathit{HOH} \rightarrow \mathbf{A}\text{—}\mathit{OH} + \mathbf{B}\text{—}\mathit{OH}$$

Ingested foods are usually polymers—mainly proteins, carbohydrates, and triglycerides. In the stomach and small intestine, these polymers are hydrolyzed (in the process of *digestion*) into their respective monomers: amino acids, monosaccharides, fatty acids, and glycerol. These monomers are then moved across the wall of the small intestine into the blood of the capillaries (a process called *absorption*). The vascular system transports them primarily to the liver and then to all the other tissues of the body.

Once inside the cells of the body, the monomers can be either hydrolyzed into smaller molecules, by a process that yields energy for the cell, or condensed to form new, larger polymers in the cytoplasm. Some of these new polymers are released into the blood (e.g., hormones and the plasma proteins), whereas others remain inside the cell and contribute to its structure and function. In turn, some of the new polymers of the cell can eventually be hydrolyzed to form new monomers, which may be used by the cell or released into the blood for use by other cells in the body.

In the healthy person, the concentrations of the different classes of monomers and polymers in the blood plasma are held remarkably constant and vary only within narrow limits. When the concentration of one of these molecules in the blood deviates from the normal range, specific compensatory mechanisms are activated that bring the concentration back to normal (negative feedback). Homeostasis is thus maintained.

When the concentration of any of the monomers or polymers in the blood remains consistently above or below normal, the health of the person may be threatened. Abnormal concentrations of different molecules in the blood are characteristic of different diseases and aid in their diagnosis. The disease *diabetes mellitus*, for example, is characterized by a high blood glucose concentration. Therefore, accurate measurement of the concentrations of different molecules in the blood is extremely important in physiology and clinical laboratories.

### The Colorimeter

The colorimeter is a device used in physiology and clinical laboratories to measure the concentration of a substance in a solution. This is accomplished by the application of **Beer's law,** which states that the concentration of a substance in a solution is directly proportional to the amount of light absorbed (*Absorbance, A*) by the solution and inversely proportional to the logarithm of the amount of light transmitted (*Percent Transmittance, %T*) by the solution.

Absorbance (*A*)

Concentration

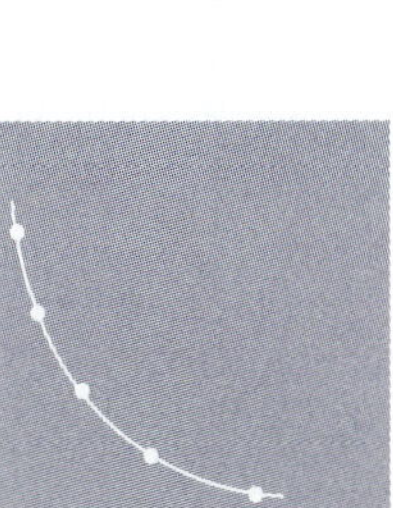

Percent Transmittance (%*T*)

Concentration

Beer's law will apply only if the incident light (the light entering the solution) is monochromatic—that is, light composed of a single wavelength. White light is a mixture of many different wavelengths between 380 and 750 nanometers (nm), or millimicrons (mμ). The rods and cones within the eyes respond to the light waves, and the brain interprets these different wavelengths as different colors.

| | |
|---|---|
| Violet | 380–435 nanometers (nm) |
| Blue | 436–480 nanometers (nm) |
| Green | 481–580 nanometers (nm) |
| Yellow | 581–595 nanometers (nm) |
| Orange | 596–610 nanometers (nm) |
| Red | 611–750 nanometers (nm) |

By means of a prism or diffraction grating, the colorimeter can separate white light into its component wavelengths. The operator of this device can select incident light of any wavelength by simply turning the appropriate dial to that wavelength. This light enters a specific tube, the *cuvette*, which contains the test solution. A given fraction of the incident light is absorbed by the solution and the remainder of the light passes through the cuvette. The transmitted light generates an electric current by means of a photoelectric cell, and the amount of this current is registered on a galvanometer scale.

The colorimeter scale indicates the percent transmittance (%). Since the amount of light that goes into the solution and the amount of light that leaves the solution are known, a ratio of the two indicates the light absorbance (A) of that solution. The colorimeter also includes an absorbance scale. In the following exercises, the absorbance scale will be used rather than the percent transmittance scale because absorbance and concentration are directly proportional to each other. This relationship can be described in a simple formula, where 1 and 2 represent different solutions:

$$\frac{\text{Concentration}_1}{\text{Absorbance}_1} = \frac{\text{Concentration}_2}{\text{Absorbance}_2}$$

One solution might be a sample of plasma whose concentration (e.g., of glucose) is unknown. The second solution might be a *standard*, which contains a known concentration of the test substance (such as glucose). When the absorbances of both solutions are recorded from the colorimeter, the concentration of the test substance in plasma (i.e., the unknown) can easily be calculated:

$$C_x = C_{std} \frac{A_x}{A_{std}}$$

where

$x$ = the unknown plasma
$std$ = the standard solution
A = the absorbance value
C = the concentration

Suppose there are four standards. Standard 1 has a concentration of 30 mg per 100 mL (or mg per deciliter, dL). Standards 2, 3, and 4 have concentrations of 50 mg/dL, 60 mg/dL, and 70 mg/dL, respectively. Since standard 3 has twice the concentration of standard 1, it should (according to Beer's law) have twice the absorbance. The second standard (at 50 mg/dL), similarly, should have an absorbance value midway between that of the first and the fourth standard, since its concentration is midway between 30 and 70 mg/dL. Experimental errors, however, make this unlikely. Therefore, it is necessary to average the answers obtained for the unknown concentration when different standards are used. This can be done either arithmetically by applying the previous formula, or by means of the graph below.

A graph plotting the four standard data points, including a straight line of "best fit" drawn closest to these points, is called a **standard curve.**

| Standard | Concentration (mg/dL) | Absorbance |
|---|---|---|
| 1 | 30 mg/100 mL | 0.25 |
| 2 | 50 mg/100 mL | 0.38 |
| 3 | 60 mg/100 mL | 0.41 |
| 4 | 70 mg/100 mL | 0.57 |
| **Unknown** | **??** | **0.35** |

Now suppose that a solution of unknown concentration has an absorbance of 0.35. The standard curve graph can be used to determine its concentration.

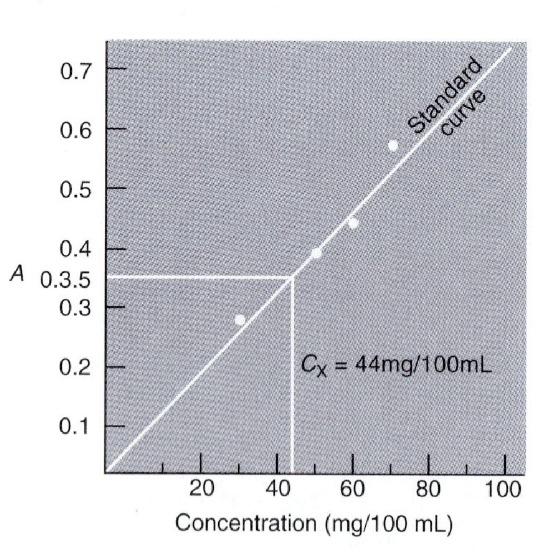


### *Standardizing the Colorimeter*

This procedure is intended specifically for the Spectronic 20 (Bausch & Lomb) colorimeter (fig. 2.2*a*). Although the general procedure is similar for all colorimeters, specific details may vary between different models.

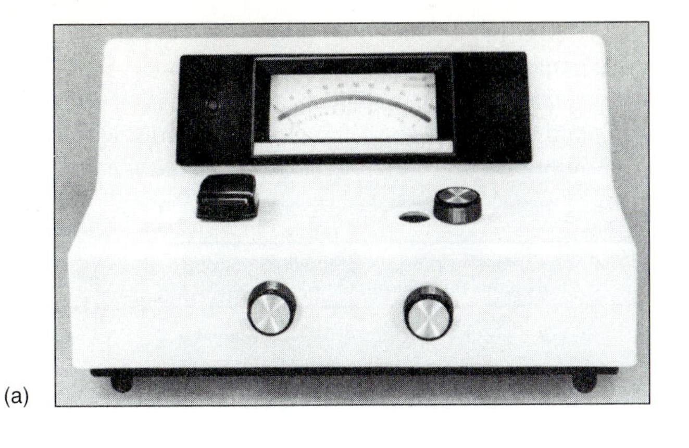

(a)

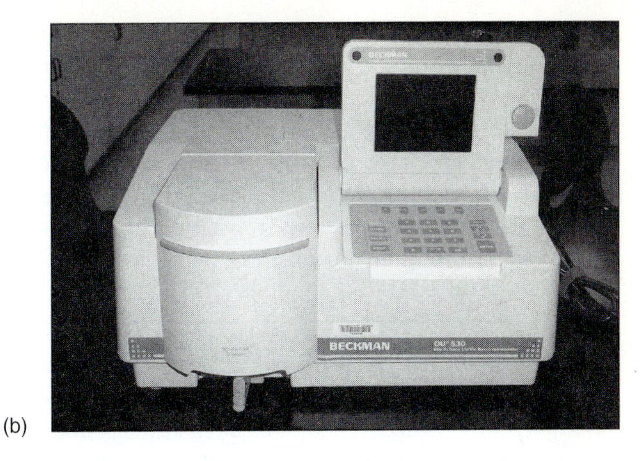


(b)

**Figure 2.2 Devices used to measure the concentrations of solutions by their light absorbances.** (a) A colorimeter, used to measure light absorbance in the visible spectrum only. (b) A spectrophotometer that can measure the abosorbance of light in the ultraviolet as well as the visible region.

## PROCEDURE

1. Set the wavelength dial so that the correct wavelength in nanometers (provided in each exercise) is lined up with the indicator in the window adjacent to this dial.
2. When there is no cuvette in the cuvette holder, the light source is blocked. The pointer should read zero transmittance or infinite absorbance at the left end of the scale. Turn knob to align the pointer with the *left* end of the scale.
3. Place the cuvette, which contains all the reagents *except* the test solution (e.g., glucose), into the cuvette holder. This tube is called the **blank** because it has a concentration of test substance equal to zero. It should therefore have an absorbance of zero (or a transmittance of 100%) that is read at the right end of the scale. Turn knob to align the pointer with the *right* end of the scale.
4. Repeat steps 2 and 3 to confirm settings.
5. Place the other cuvettes, which contain the standard solutions and the unknown, in the cuvette holder. Close the hatch and read the absorbance value of each solution.

**Note:** *Before placing each cuvette in the chamber, wipe it with a lint-free, soft paper towel. If the cuvette has a white indicator line, place the cuvette so that this line is even with the line at the front of the cuvette holder.*

## A. Carbohydrates: Measurement of Plasma Glucose Concentration

The monomers of the carbohydrates are the **monosaccharides,** or simple sugars. The general formula for these molecules is $C_nH_{2n}O_n$, where *n* can be any number. *Glucose,* for example, has the formula $C_6H_{12}O_6$. The monosaccharide *fructose* has the same formula but differs from glucose in the arrangement of the atoms (glucose and fructose are *isomers*).

Two monosaccharides can join together by means of a dehydration synthesis (condensation) reaction to form a **disaccharide.** *Sucrose* (common table sugar), for example, is a disaccharide of glucose and fructose, whereas *maltose* is a disaccharide of two glucose subunits.

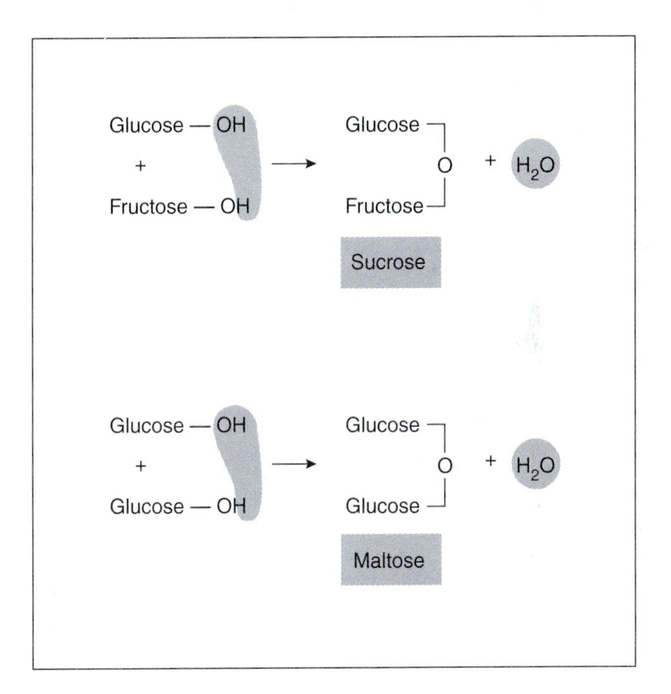


The continued addition of glucose subunits to maltose will result in the production of a long, branched chain of repeating glucose subunits, forming the **polysaccharide** *glycogen* (or animal starch). This polysaccharide is formed inside muscle and liver cells and serves as an efficient storage form of glucose. When the blood glucose level drops below normal, the liver cells can hydrolyze stored glycogen and release glucose into the blood. Conversely, when the blood sugar level rises above normal, the liver cells can take glucose from the blood and store it as glycogen for later use. In this way, the equilibrium between blood glucose and liver glycogen helps to maintain constancy *(homeostasis)* of the blood sugar level. This process is regulated by hormones that include epinephrine (adrenaline), insulin, hydrocortisone, and glucagon.

The most important regulator of the blood glucose level is the hormone *insulin,* produced by the islets of Langerhans in the pancreas. This hormone stimulates the transport of blood glucose into the cells of the body; hence, it lowers the blood glucose level (see fig. 1.13). An elevated blood glucose level, *hyperglycemia,* results from insufficient insulin secretion. This disease is called **diabetes mellitus.** Low blood glucose, *hypoglycemia,* is clinically rare, but can result from the excessive secretion (or injection) of insulin. In addition, slight hypoglycemia may be associated with arthritis, renal disease, and the late stages of pregnancy.

## PROCEDURE

### Measurement of Plasma Glucose Concentration

1. Obtain three test tubes, and label them *U* (unknown), *S* (standard), *B* (blank).
2. Using a mechanical pipettor, pipette 5.0 mL of the glucose reagent into *each* tube.
3. Use a microliter pipettor to add 40 μL (0.04 mL) of the following solutions into each of the indicated test tubes to avoid contamination. Use different pipette tips for adding each solution.

| Tube | Serum | Standard | Water | Reagent |
|---|---|---|---|---|
| Unknown *(U)* | 40 μL | — | — | 5.0 mL |
| Standard *(S)* (100 mg/dL) | — | 40 μL | — | 5.0 mL |
| Blank *(B)* | — | — | 40 μL | 5.0 mL |

**Note:** *All tubes must contain equal volumes (5.04 mL) of solution: 1 deciliter (dL) = 100 milliliters (mL).*

4. Gently tap each tube to mix the contents and allow the tubes to stand at room temperature for 10 minutes.
5. Set the monochromator (wavelength) dial at 500 nm and standardize the colorimeter, using solution *B* as the blank.
6. Record the absorbance values of solutions *U* and *S* in the chart in the laboratory report.
7. Using Beer's law formula, calculate the concentration of glucose in the unknown plasma sample and enter the value in the laboratory report.
8. Using graph paper that follows this exercise, draw a graph of absorbance versus glucose concentration (mg/dL). Plot the standard *(S)* absorbance value and draw a standard curve. Then determine the unknown glucose concentration from the graph.

The normal fasting range of glucose in the plasma is 70–100 mg per 100 mL (or 70–100 mg/dL).

## B. Lipids: Measurement of Plasma Cholesterol Concentration

The lipids are an extremely diverse family of molecules that share the common property of being soluble (dissolvable) in organic solvents such as benzene, ether, chloroform, and carbon tetrachloride, but *are not soluble in water or plasma.* The lipids found in blood can be classified as **free fatty acids (FFA),** also known as nonesterified fatty acids (NEFA); **triglycerides** (or neutral fats); **phospholipids;** and **steroids.** As found in carbohydrate molecules, carbon, hydrogen, and oxygen form the basic structure of lipids; however, these elements are not present in the same predictable ratio.

Fatty acids are long chains, ranging from sixteen to twenty-four carbons in length. When adjacent carbons are linked by single bonds, the fatty acid is said to be *saturated;* when adjacent carbons are linked by double bonds, the fatty acid is said to be *unsaturated.*

**Saturated:** $—CH_2—CH_2—CH_2—CH_2—CH_2—$
**Unsaturated:** $=CH—CH_2—CH=CH—CH_2—$

The triglycerides consist of three fatty acids bonded to a molecule of the alcohol *glycerol.* Triglycerides with few sites of unsaturation and that are solid at room temperature are called *fats,* whereas those with many sites of unsaturation and that are liquid at room temperature are called *oils.*

Like the triglycerides, the phospholipids consist of two fatty acids bonded to a glycerol molecule. However, as their name implies, each phospholipid also contains the element phosphorus (in the form of phosphate, $PO_4^{3-}$) bonded to the third position on the glycerol molecule. Phospholipids are important components of cell membranes.

The steroids are characterized by a structure consisting of four rings. One of the most important steroids in the body is cholesterol.

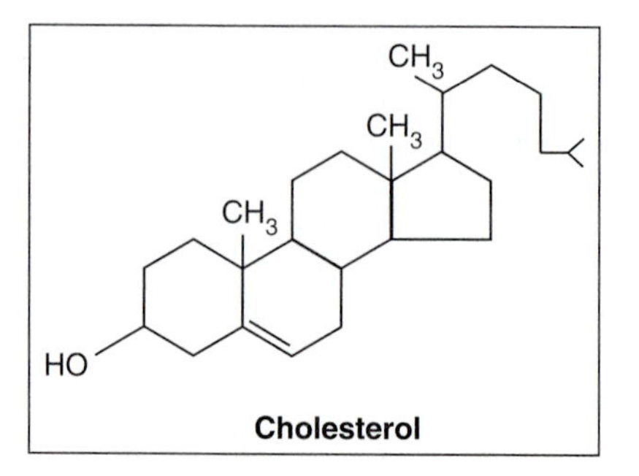

Cholesterol

There is evidence that high blood cholesterol, together with other risk factors, such as hypertension and cigarette smoking, contributes to **atherosclerosis.** In atherosclerosis, deposits of cholesterol and other lipids, calcium salts, and smooth muscle cells build up in the walls of arteries and reduce blood flow. These deposits—called *atheromas*—also serve as sites for the production of *thrombi* (blood clots), which further occlude blood flow. The reduction in blood flow through the artery may result in heart disease or cerebrovascular accident (stroke). It is generally believed that blood cholesterol levels, and the risk of atherosclerosis, may be significantly lowered by a diet low in cholesterol and saturated fats.

## PROCEDURE

### Measurement of Plasma Cholesterol Concentration

1. Obtain three test tubes, and label them *U* (unknown), *S* (standard), *B* (blank).
2. Using a mechanical pipettor, pipette 5.0 mL of the cholesterol reagent into each tube.
3. Use a microliter pipettor to add 50 μL (0.05 mL) of the following solutions into each of the indicated test tubes. Use different pipette tips for adding each solution to avoid contamination.

| Tube | Serum | Standard | Water | Reagent |
|---|---|---|---|---|
| Unknown *(U)* | 50 μL | — | — | 5.0 mL |
| Standard *(S)* (200 mg/dL) | — | 50 μL | — | 5.0 mL |
| Blank *(B)* | — | — | 50 μL | 5.0 mL |

**Note:** *All tubes must contain equal volumes (5.05 mL) of solution: 1 deciliter (dL) = 100 milliliters (mL).*

4. Gently tap each tube to mix the contents and allow the tubes to stand at room temperature for 10 minutes.
5. Transfer the solutions to three cuvettes. Standardize the spectrophotometer at 500 nm, using solution *B* as the blank.
6. Record the absorbance values of solutions *U* and *S* in the laboratory report.
7. Using Beer's law formula, calculate the concentration of cholesterol in the unknown plasma sample. Enter this value in the laboratory report.
8. Using graph paper that follows this exercise, draw a graph of absorbance versus cholesterol concentration (mg/dL). Plot the standard (*S*) absorbance value and draw a standard curve. Then determine the unknown cholesterol concentration from the graph.

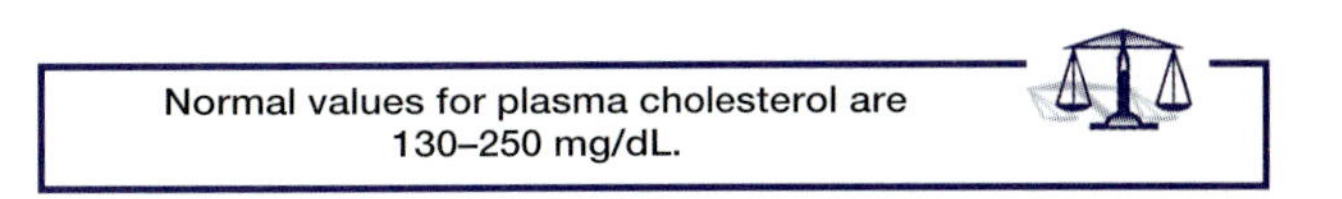

$$C_{plasma} = \frac{A_{plasma}}{A_{standard}} \times C_{standard}$$

Normal values for plasma cholesterol are 130–250 mg/dL.

## C. Proteins: Measurement of Plasma Protein Concentration

Proteins are long chains of amino acids bonded to one another by condensation reactions. Each amino acid has an amino (—$NH_2$) end and a carboxyl (—COOH) end, as shown by the general formula:

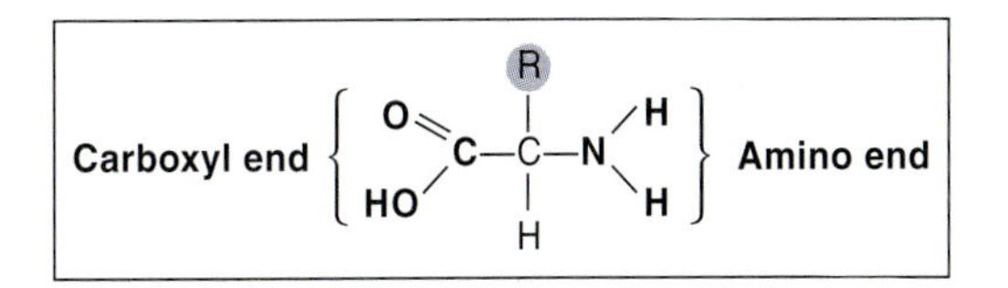


When amino acids bond (through a **peptide bond**) to form a protein, one end of the protein will have a free amino group and the other end will have a free carboxyl group.

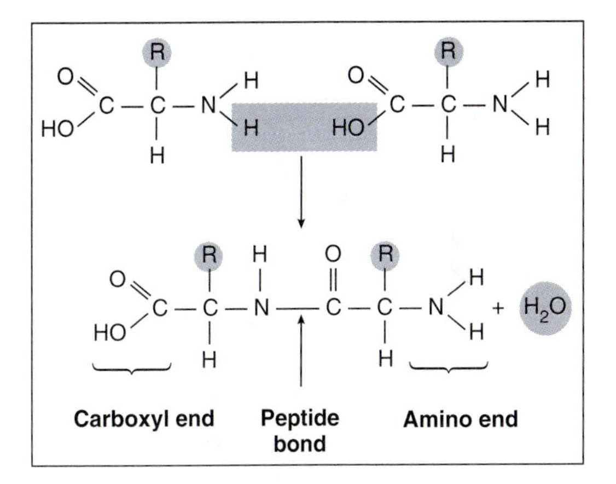


There are more than twenty-two different amino acids in nature, each differing from the others with respect to the combination of atoms in the *R group*, sometimes known as the *functional group*. The amino acid **glycine,** for example, has a hydrogen atom (H) in the *R* position, whereas the amino acid **alanine** has a methyl group ($CH_3$) in the *R* position.

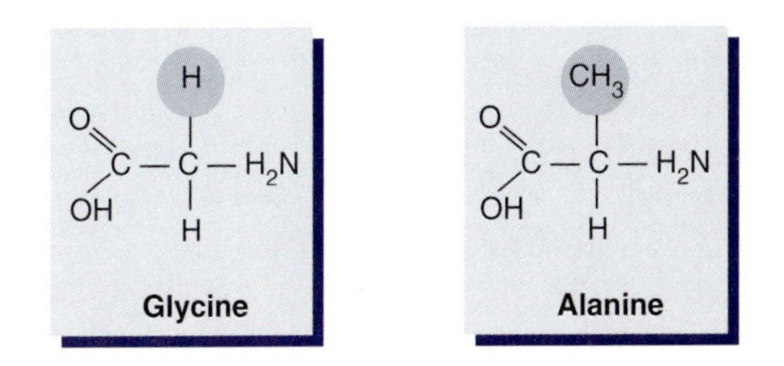

An abnormally low concentration of total blood protein *(hypoproteinemia)* may be due to an inadequate production of protein by the liver caused by liver disease (such as cirrhosis or hepatitis), or to the loss of protein in the urine *(albuminuria)* caused by kidney diseases. Hypoproteinemia decreases the colloid osmotic pressure of the blood and may lead to accumulation of excess fluid in the tissue spaces, a condition called **edema.**

An abnormally high concentration of total plasma protein *(hyperproteinemia)* may be due to dehydration or to an increased production of the plasma proteins. An increased production of gamma globulins (antibodies), for example, is characteristic of pneumonia and many other infections, and of parasitic diseases such as malaria.

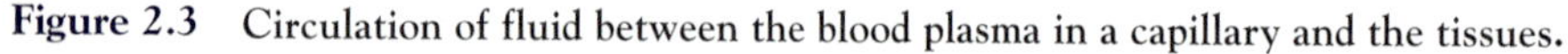

Proteins in the plasma serve a variety of functions. Some proteins may be active as enzymes, hormones, or carrier molecules (transporting lipids, iron, or steroid hormones in the blood), while others have an immune function (antibodies). The **plasma proteins** are classified according to their behavior during biochemical separation procedures. These classes include the *albumins*, the *alpha* and *beta globulins* (synthesized mainly in the liver from amino acids absorbed by the intestine), and *gamma globulins* (antibodies produced by the lymphoid tissue).

In addition to the separate functions of the different plasma proteins, the total concentration of proteins in the plasma is physiologically important. The plasma proteins exert an osmotic pressure, the **colloid osmotic** (or **oncotic**) **pressure,** which pulls fluid from the tissue spaces into the capillary blood. This force compensates for the continuous filtration of fluid from the capillaries into the tissue spaces produced by the *hydrostatic pressure* of the blood (fig. 2.3).

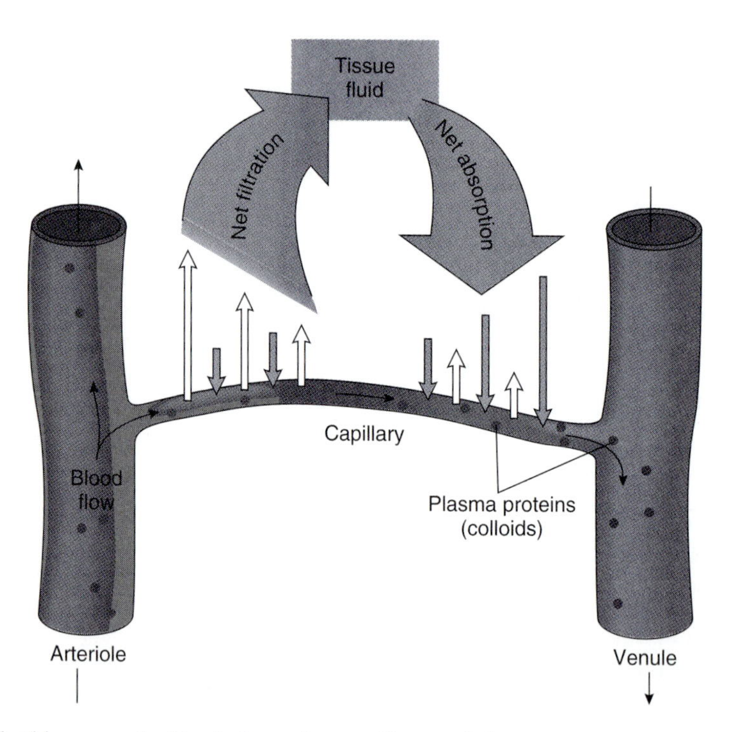


**Figure 2.3 Circulation of fluid between the blood plasma in a capillary and the tissues.**
Arrows pointing away from the capillary indicate the force exerted by the blood pressure, whereas arrows pointing toward the capillary indicate the force exerted by the colloid osmotic pressure of the plasma proteins. The arrow labeled "blood flow" indicates the direction of flow along the capillary from arteriole to venule.

**(For a full-color version of this figure, see fig. 14.9 in *Human Physiology*, eighth edition, by Stuart I. Fox.)**

| Tube Number | | Serum | Standard | Water | Reagent |
|---|---|---|---|---|---|
| 1 Blank *(B)* | | — | — | 50 μL | 5.0 mL |
| 2 Standard (S) | (2.0 g/dL) | — | 50 μL | — | 5.0 mL |
| 3 | (4.0 g/dL) | — | 50 μL | — | 5.0 mL |
| 4 | (6.0 g/dL) | — | 50 μL | — | 5.0 mL |
| 5 | (8.0 g/dL) | — | 50 μL | — | 5.0 mL |
| 6 | (10.0 g/dL) | — | 50 μL | — | 5.0 mL |
| 7 Unknown *(U)* | | 50 μL | — | — | 5.0 mL |

## PROCEDURE

### Measurement of Total Plasma Protein Concentration

1. Obtain seven test tubes, and label them 1–7.
2. Using a mechanical pipettor, pipette 5.0 mL of biuret reagent into each tube.
3. Use a microliter pipettor to add 50 μL (0.05 mL) of the following solutions into each of the indicated test tubes. Use different pipette tips for adding each solution to avoid contamination.

**Note:** *All tubes must contain equal volumes (5.05 mL) of solution. (The expression g/dL is equivalent to g per 100 mL or g%.)*

4. Gently tap each tube to mix the contents, and allow the tubes to stand at room temperature for at least 5 minutes.
5. Transfer the solutions to seven cuvettes. Standardize the spectrophotometer at 550 nm, using solution *B* as the blank.
6. Record the absorbance values of the unknown (*U*) and the five standard solutions (*S* tubes 2–6) in the laboratory report.
7. Using Beer's law formula, calculate the concentration of total protein in the unknown plasma sample. Enter this value in the laboratory report.
8. Using graph paper that follows this exercise, draw a graph of absorbance versus total protein concentration (g/dL). Plot the standard (*S* tubes 2–6) absorbance values and draw a standard curve. Then determine the unknown total protein concentration from the graph.

**The normal fasting total protein level is 6.0–8.0 g/dL.**

# Laboratory Report 2.1

Name ____________________

Date ____________________

Section ____________________

## DATA FROM EXERCISE 2.1

### A. CARBOHYDRATES: Measurement of Plasma Glucose Concentration

1. Record the absorbance values of solutions *U* and *S*.
   Absorbance of solution *U*: __________
   Absorbance of solution *S*: __________
2. Enter the unknown serum glucose concentration derived from the equation for Beer's law: __________ mg/dL.
   How does this value for the unknown compare with that derived from the graph? Explain.

3. Is the glucose concentration of the serum sample normal or abnormal? Explain.

### B. LIPIDS: Measurement of Plasma Cholesterol Concentration

1. Record the absorbance values of solutions *U* (unknown) and *S* (standard).
   Absorbance of solution *U*: __________
   Absorbance of solution *S*: __________
2. Enter the unknown serum cholesterol concentration derived from the equation for Beer's law: __________ mg/dL.
   How does this value for the unknown compare with that derived from the graph? Explain.

3. Is the cholesterol concentration of the serum sample normal or abnormal? Explain.

## C. PROTEINS: Measurement of Plasma Protein Concentration

1. Record your absorbance values in this data table.

| Tube Number | Protein Concentration (g/dL) | Absorbance |
|---|---|---|
| 1 | 0 (blank) | 0 |
| 2 | 2.0 | |
| 3 | 4.0 | |
| 4 | 6.0 | |
| 5 | 8.0 | |
| 6 | 10.0 | |
| 7 | **Unknown** (serum sample) | |

2. Use the graph paper on page 42 or 43 to plot a standard curve.
3. Derive an estimated protein concentration for the unknown serum sample.
   Unknown protein concentration from the curve: __________ g/dL.
4. Use the equation for Beer's law to determine the protein concentration of the unknown serum.
   Unknown protein concentration from Beer's equation: __________ g/dL.
5. How does this value for the unknown compare with that derived from the graph? Explain.

6. Is the total protein concentration of the unknown serum sample normal or abnormal? Explain: What clinical condition(s) might have caused an abnormal result?

# REVIEW ACTIVITIES FOR EXERCISE 2.1

## *Test Your Knowledge of Terms and Facts*

1. The concentration of a solution is ______________________________ proportional to its absorbance.
2. The above relationship is described by ______________________________ law.
3. Hyperglycemia is characteristic of the disease ______________________________.
4. Hyperglycemia may be caused by a deficiency of the hormone ______________________________.
5. Cholesterol belongs to the general category of molecules known as ______________________________ and to the specific category of molecules known as ______________________________.
6. High blood cholesterol, along with other risk factors, is a contributing factor in the disease______________________.
7. Most of the plasma proteins are produced by the ______________________________ (organ).
8. Low plasma protein concentration is described clinically as ______________________________, and can produce a physical condition called ______________________________.
9. The colloid osmotic pressure of the blood is related to the plasma concentration of ______________________________.
10. "All fats are lipids, but not all lipids are fats." Explain.

## *Test Your Understanding of Concepts*

11. Describe the functions of the plasma proteins. Where do these proteins originate?

12. What does the blank tube contain, and what is its function in a colorimetric assay?

## *Test Your Ability to Analyze and Apply Your Knowledge*

13. Do any of the proteins in your plasma come from food proteins? Does the starch (glycogen) in your liver come from food starch? Explain your answers.

14. Why do you draw a linear (straight line) graph of absorbance vs. concentration even though your experimental values deviated slightly from a straight line? Why must your line intersect the origin of the graph (zero concentration equals zero absorbance)? Explain.

# Thin-Layer Chromatography of Amino Acids

EXERCISE **2.2**

## MATERIALS

1. Silica gel plates (F-254 rapid, adhered to plastic or glass); capillary tubes; chromatography (or hair) dryers; rulers

**Note:** *Chromatography paper can be used as an alternative*

2. Developing chambers or oven set at about 60° C
3. Amino acid solutions: arginine, cysteine, aspartic acid, phenylalanine—1.0 mg/mL of each dissolved in 0.1N HCl:isopropyl alcohol (9:1); "unknown" solution of amino acids, containing two of these amino acids in the same solution
4. Developing solvent: 20 mL of 17% $NH_4OH$ (dilute concentrated $NH_4OH$ with an equal amount of water), 40 mL of ethyl acetate, and 40 mL of methanol per developing chamber
5. Ninhydrin spray

Amino acids, the subunits of protein structure and function, are divisible into approximately twenty chemically unique molecules. The distinctive properties of each amino acid provide the basis for its separation from the others and its identification. This information can be clinically useful in the diagnosis of genetic diseases that involve amino acid metabolism.

## OBJECTIVES

1. Using the general formula for amino acids, explain how one amino acid differs from another.
2. Explain how thin-layer chromatography can separate different amino acids that are present together in a single solution.
3. Explain what the $R_f$ value signifies. Calculate the $R_f$ values for different spots and use this information to identify unknown amino acids.

## Textbook Correlations

Before performing this exercise, you should study the introductory material presented here. Further information relating to this exercise can be found in these pages of *Human Physiology,* eighth edition, by Stuart I. Fox:

- *Proteins.* Chapter 2, pp. 38–41.
- *Inborn Errors of Metabolism.* Chapter 4, pp. 91–92.

In this exercise, you will attempt to identify two unknown amino acids that are present in the same solution. To accomplish this task, you must (1) *separate* the two amino acids, and (2) *identify* these amino acids by comparing their behavior with that of known amino acids.

Since each amino acid has a chemically different *R* group, each will dissolve in a given solvent to a different degree. These differences will be used to separate and identify the amino acids on a **thin-layer plate.**

The thin-layer plate consists of a thin layer of porous material (in this procedure, silica gel) that is coated on one side of a plastic, glass, or aluminum plate. The solutions of amino acids are applied to different spots on the plate (a procedure called *spotting*) and allowed to dry. The plate is then placed on edge in a solvent bath with the spots above the solvent.

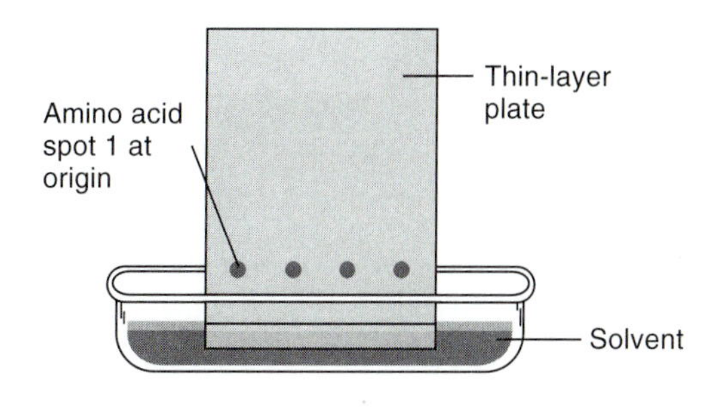

As the solvent creeps up the plate by capillary action, it will wash the amino acids off their original spots (or *origins*) and carry them upward toward the other end of the plate. Since the solubility of each amino acid is different, the ability of the solvent to dissolve and carry a particular amino acid will vary with the properties of that amino acid. The most soluble amino acids in the given solvent will wash and carry farther than those that are less soluble. Since the process is halted shortly before the solvent front reaches the top of the plate, some amino acids will have migrated farther from the origin than others.

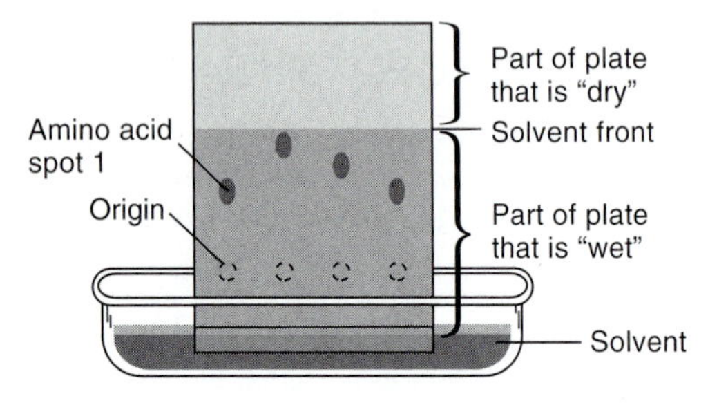


If this chromatography were repeated using the same amino acids and the same solvent, the final pattern *(chromatogram)* would be the same as that obtained previously. In other words, the distance that a given amino acid migrates in a given solvent, relative to the *solvent front*, can be used as an identifying characteristic of that amino acid. A numerical value (the $R_f$ value) can be assigned to this characteristic by calculating the distance the amino acid traveled relative to that traveled by the solvent front, as:

$$R_f = \frac{\text{distance from origin to spot}}{\text{distance from origin to solvent front}}$$

An unknown amino acid can be identified by comparing its $R_f$ value in a given solvent with the $R_f$ values of known amino acids in the same solvent.

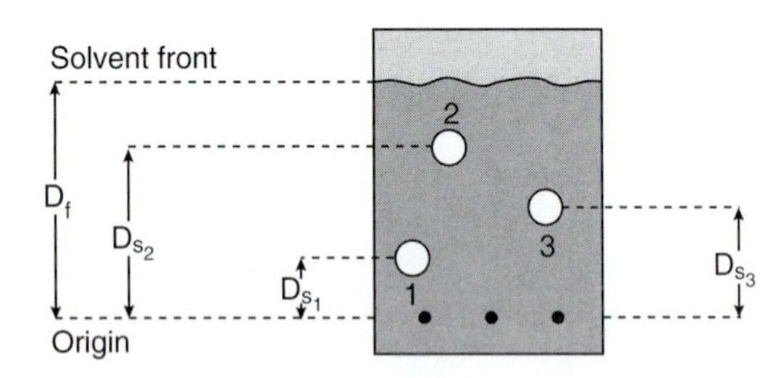


## PROCEDURE

1. Fill a capillary tube with amino acid solution 1 (arginine). Gently touch it to the first spot on the plate 1 1/2 inches from the bottom and 1/2 inch from the left-hand edge. Dry the spot with a hair dryer. Repeat the spotting and drying procedure until you have made *five* applications of the *same* amino acid to the *same* spot.

An abnormally high concentration of certain amino acids or their metabolites in the blood frequently results in deterioration of the central nervous system and mental retardation. These abnormal concentrations are usually due to defective enzymes that are involved in the degradation of the amino acids.

In the disease **phenylketonuria (PKU),** for example, the enzyme that converts the amino acid phenylalanine to tyrosine is defective, leaving phenylalanine and its other metabolites to accumulate in the body. This condition, which affects one baby in every 10,000 to 20,000, results in severe mental retardation. Other diseases of similar etiology (cause) include *homocystinuria, alkaptonuria,* and *maple syrup disease* (the name refers to a characteristic odor of the urine). The defective enzymes are synthesized by defective genes; hence, the diseases of amino acid metabolism that have this etiology are referred to as **inborn errors of metabolism** (see exercise 2.5).

**Note:** *(1) Be careful to apply the amino acid solution to exactly the same spot each time (use a spotting guide, or make X's lightly with pencil before spotting); (2) dry the spot thoroughly between applications; and (3) spot gently so that you do not gouge out the silica gel.*

2. Repeat the spotting procedure for amino acid 2 (cysteine), using a new capillary tube and applying the spot about 1/2 inch to the right of the first amino acid.
3. Repeat the spotting procedure for amino acids 3 (aspartic acid) and 4 (phenylalanine), applying them about 1/2 inch to the right of the preceding spot.
4. Repeat the spotting procedure with the solution containing two unknown amino acids, applying it about 1/2 inch to the right of amino acid 4.
5. Carefully place the thin-layer plate in a chromatography developing chamber that has been previously filled with solvent. Cover the chamber and allow the solvent to migrate up the plate for 1 hour.
6. Remove the plate from the developing chamber, dry it, and then spray it with ninhydrin in a well-ventilated area.

**Note:** *Since amino acids are colorless, it is necessary to react them with ninhydrin, a reagent that combines with the amino acids to produce a blue-colored complex.*

7. Heat the plate in an oven set at approximately 60° C for *10–15* minutes.
8. Remove the plate and measure the distance, in centimeters, from the origin to the solvent front and from the origin to the center of each amino acid spot. Record these values and calculate $R_f$ values. Enter your data in the laboratory report.

# Laboratory Report 2.2

Name ______________________________

Date ______________________________

Section ______________________________

## DATA FROM EXERCISE 2.2

1. Record your data in this table and calculate the $R_f$ value for each spot.

| **Amino Acid** | $D_s$ | $D_f$ | $R_1$ |
|---|---|---|---|
| Arginine | | | |
| Cysteine | | | |
| Aspartic acid | | | |
| Phenylalanine | | | |
| *Unknown 1* | | | |
| *Unknown 2* | | | |

2. The unknown solution contained the two amino acids ________________ and ________________.

## REVIEW ACTIVITIES FOR EXERCISE 2.2

### *Test Your Knowledge of Terms and Facts*

1. Which part of an amino acid distinguishes it and grants it chemical specificity? ______________________________

2. Inherited defects in the ability to convert one amino acid into another are in a class of disorders called ______________________________

3. Define the term $R_f$. ______________________________

### *Test Your Understanding of Concepts*

4. Why do different amino acids have different $R_f$ values?

5. Suppose an amino acid is 8 cm from the origin, and the solvent front is 12 cm from the origin. What is the $R_f$ value for this amino acid?

6. What is the maxmum $R_f$ value that a spot can have? Explain.

## *Test Your Ability to Analyze and Apply Your Knowledge*

7. Suppose amino acid "A" has a higher $R_f$ value than amino acid "B" in a solvent system "1" and the order is reverse in solvent system "2." Further, suppose that solvent system "1" has a higher ratio of methanol to ethyl acetate (and is thus more polar) than solvent system "2." What can you conclude about the structure of amino acid "A" compared to the structure of amino acid "B"?

# Electrophoresis of Serum Proteins

EXERCISE **2.3**

## Materials

1. Plastic troughs, buffer chamber, sample applicator, power supply, Sepharose strips (the Sepra Tek System, Gelman), forceps
2. High-resolution buffer, Ponceau S stain (Gelman), 5% acetic acid (v/v)
3. Sterile lancets, 70% ethanol, disposal receptacle for all blood-contaminated objects (Alternatively, test tubes containing previously prepared serum samples may be used.)
4. Unheparinized capillary tubes, clay sealant, and microhematocrit centrifuge

Plasma contains classes of proteins that differ in structure and function. These classes can be separated from one another and identified by electrophoresis.

## OBJECTIVES

1. Explain what is meant by the term *amphoteric* and describe how amphoteric molecules can be separated by electrophoresis.
2. Identify the different bands of serum proteins in an electrophoresis pattern.
3. Describe the origin and functions of the different classes of plasma proteins.

## Textbook Correlations

Before performing this exercise, you should study the introductory material presented here. Further information relating to this exercise can be found in these pages of *Human Physiology,* eighth edition, by Stuart I. Fox:

- *Proteins,* Chapter 2, pp. 38–41.
- *Plasma Proteins,* Chapter 13, p. 368.
- *Antibodies,* Chapter 15, pp. 453–456.

The unique structure and physiological role of each type of protein is determined by the specific number, type, and sequence of its component amino acids. Proteins differ in size, range in shape from elliptical (globular) to fibrous, and contain different numbers of positive and negative charges.

Under acidic conditions, the amino group (—$NH_2$) of an amino acid tends to gain an $H^+$ and become positively charged (—$NH_3^+$), whereas under basic conditions, the carboxyl group (—COOH) loses an $H^+$ and becomes negatively charged (—$COO^-$). Since amino acids can have either polarity, depending on the pH, they are said to be **amphoteric.**

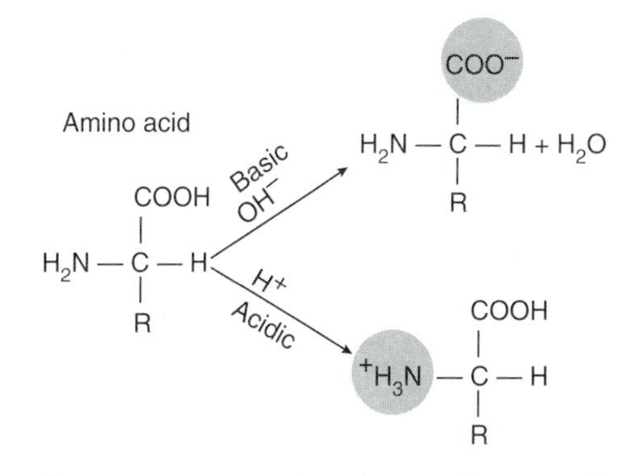


When an amino acid is electrically neutral, its amphoteric nature can be shown by the *zwitterion formula,* in which neutrality is indicated by a balance between positive and negative charges.

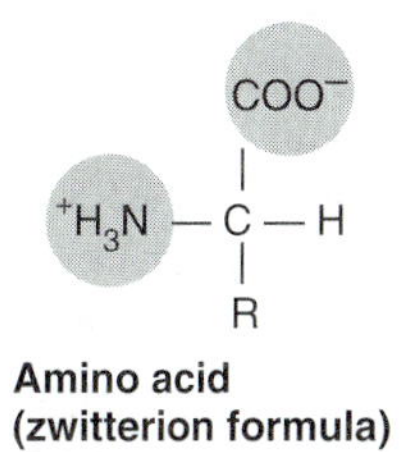


**Amino acid (zwitterion formula)**

In addition to the amino and carboxyl ends of a protein, the functional groups (*R* groups) of many amino acids have either an amino-containing functional group (such as lysine or arginine) or a carboxyl-containing functional group (such as aspartic acid or glutamic acid). Since each type of protein has a characteristic ratio of these two types of amino acids, each protein will have a characteristic net charge at a given pH.

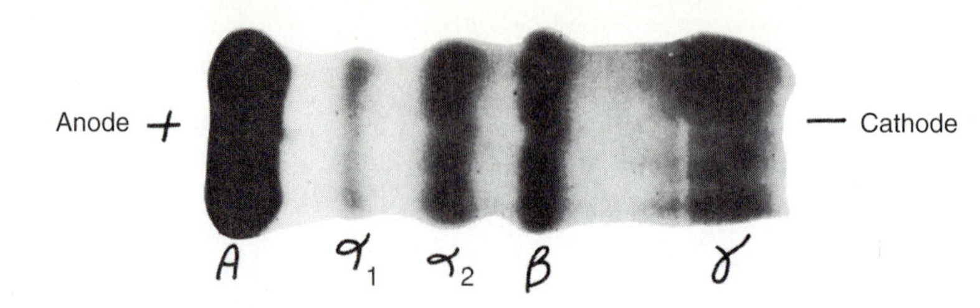


**Figure 2.4 Electrophoresis pattern of normal serum.** The bands include albumin (A) and the $\alpha_1$, $\alpha_2$, $\beta$, and $\gamma$ globulins.

In many instances, the diagnosis of a disease can be aided by an analysis of the electrophoresis pattern made by migrating plasma proteins. Although direct observation is useful, more reliable information can be gained by a quantitative measurement of the proteins in each band. These measurements are made by a *densitometer,* a device that optically scans the electrophoresis pattern and graphically records the absorbance of different regions of the strip.

Diseases that can be accurately diagnosed through electrophoresis include acute inflammations (elevated alpha-2 proteins), viral hepatitis (change in gamma globulin and albumin), and cirrhosis of the liver (broad gamma globulin band). In addition, electrophoresis is valuable in the diagnosis of nephrotic syndrome, malignant tumors, and many other diseases.

At a pH of 8.8 (slightly basic), each of the variety of proteins found in plasma will have a different degree of net negative charge. When plasma proteins are placed in an electric field, each protein will migrate away from the negative pole (cathode) and toward the positive pole (anode) at different rates. The rates at which they travel will also be influenced by their size and shape. This technique, known as **electrophoresis,** can be used to separate and identify the different classes of plasma proteins.

There are two main types of plasma proteins: albumin and the globulins (see section C, exercise 2.1). The latter type is composed of four primary subclasses: *alpha-1* ($\alpha_1$), *alpha-2* ($\alpha_2$), beta ($\beta$), and *gamma* ($\gamma$) *globulins* (fig. 2.4).

The fluid part of the blood as it circulates in the vessels is **plasma.** When blood clots, a soluble protein in the plasma *(fibrinogen)* is converted into an insoluble thread-like protein called *fibrin*. The strands of fibrin intertwine to form the meshwork of the blood clot. **Serum,** which is the fluid that remains after the clot has formed, does not contain fibrinogen and is incapable of further clotting.

*Albumins* are the most abundant of the serum proteins, serving as carrier molecules for hormones, lipids, and bile pigment; and they are responsible for most of the colloid osmotic pressure exerted by the blood. The *alpha* and *beta globulins* serve a variety of functions and, like albumin and fibrinogen, are synthesized by the liver. The *gamma globulins* are **antibodies,** which are produced by white blood cells known as lymphocytes.

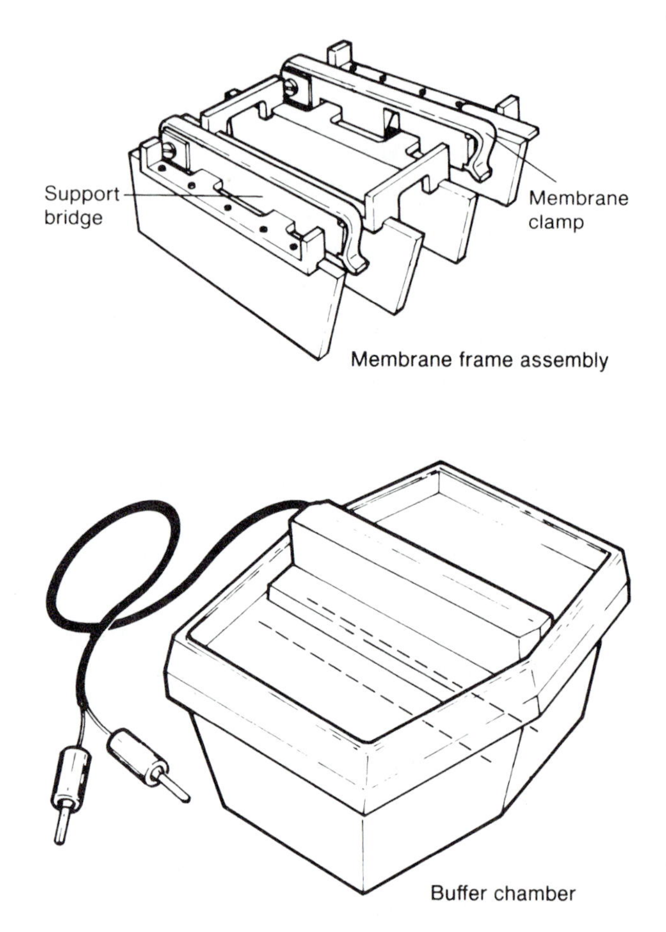


**Figure 2.5 An electrophoresis system.** A membrane frame assembly and buffer chamber for the Gelman Sepra Tek electrophoresis system.

## PROCEDURE

1. Float a strip of cellulose acetate in buffer for 1 minute; then immerse it in the buffer for *10 minutes*.
2. Using forceps, remove the cellulose acetate strip from the buffer and blot it with filter paper that has been premoistened with buffer.
3. Raise the tension latch of the frame assembly, placing the movable support bridge in the vertical position (figs. 2.5 and 2.6*a*).*
4. Center the cellulose acetate strip on the support bridges and fasten it with the membrane clamps. Tension the membrane by releasing the latch (fig. 2.6*b*,*c*).

*For Gelman Sepra Tek System.

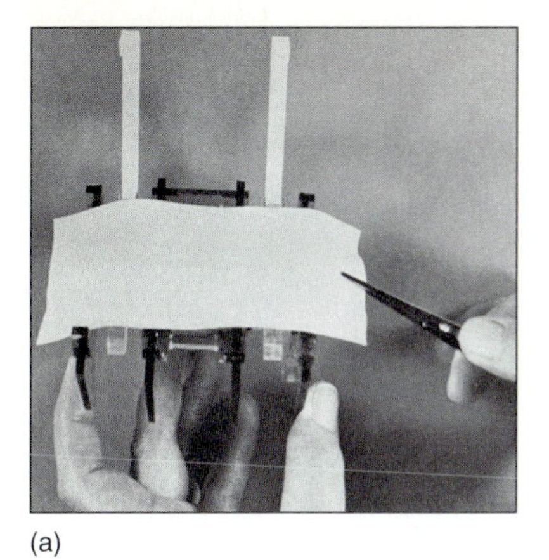
(a)

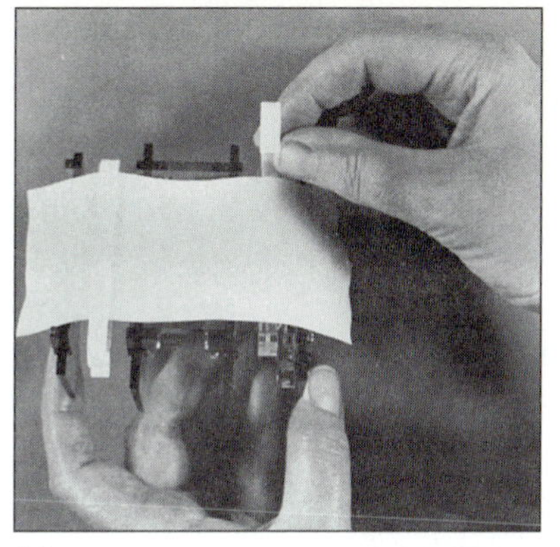
(b)

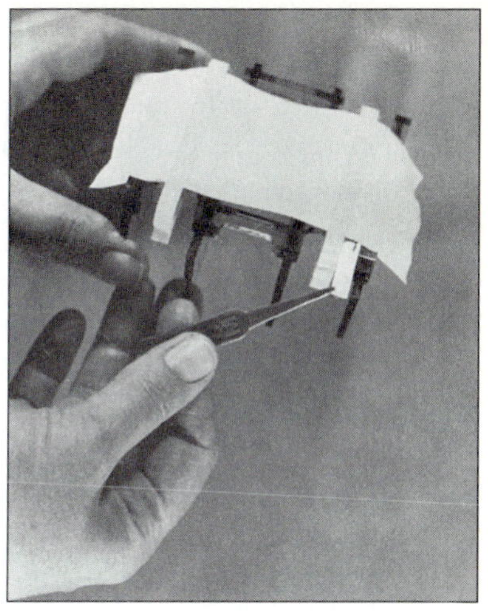
(c)

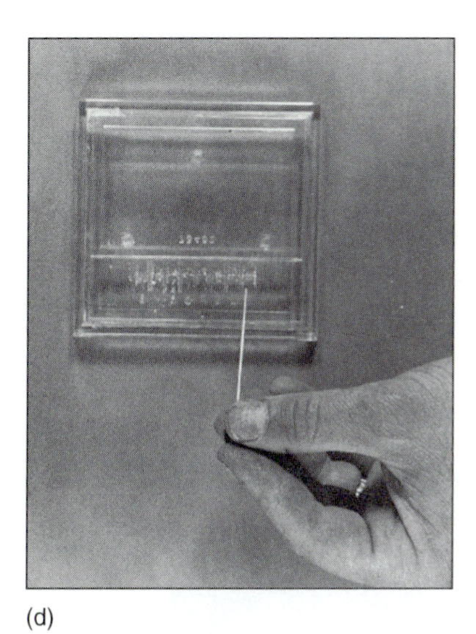
(d)

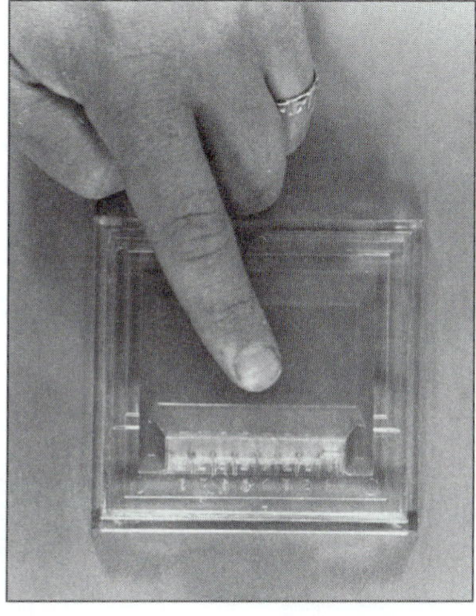
(e)

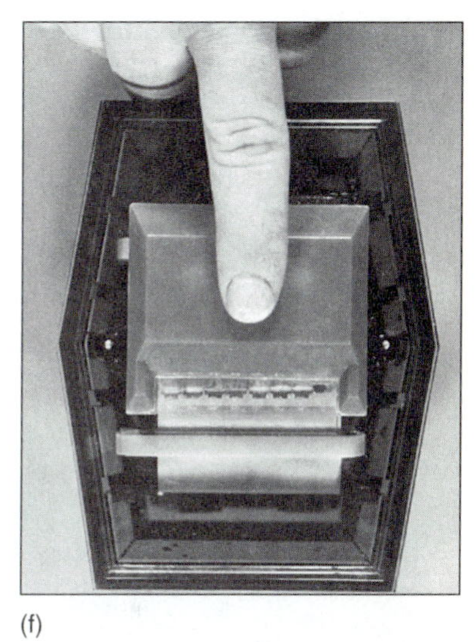
(f)

(g)

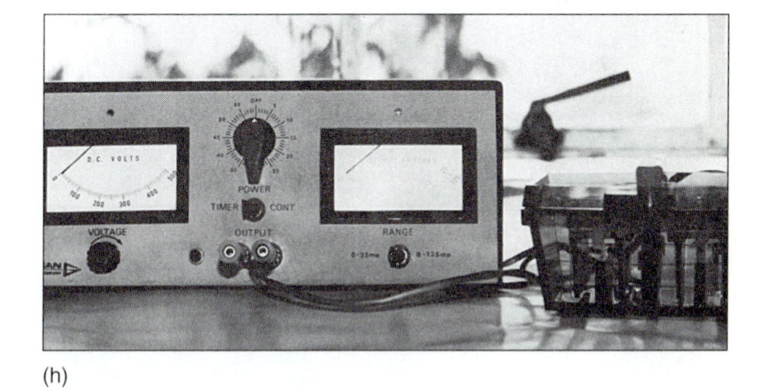

(h)

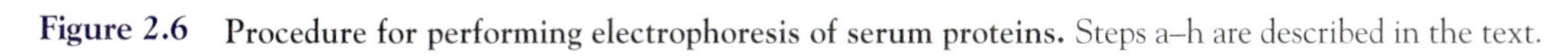
**Figure 2.6 Procedure for performing electrophoresis of serum proteins.** Steps a–h are described in the text.

5. Place the membrane frame assembly in the chamber, bringing the strip ends into contact with the buffer, and position the cover on the chamber.
6. Cleanse the tip of a finger with 70% ethanol and puncture it with a sterile lancet.

**Note:** *Extreme caution must be exercised when handling blood to guard against contracting infectious agents. Handle only your own blood and discard all objects containing blood into the receptacles provided by the instructor.*

7. Quickly fill an *unheparinized* capillary tube with blood, seal one end, and immediately centrifuge for 3 minutes. (Do *not* use heparinized capillary tubes—the anticoagulant heparin is a protein and will interfere with the test.)
8. Break the capillary tube at the junction of the packed red blood cells and the pale yellow serum. Place a drop of serum on the sample well of the applicator block by lightly touching it with the capillary tube. Similarly, a sample of serum from six different students can be placed in each of six wells (fig. 2.6*d*).
9. Fill a new capillary tube with the serum provided by the instructor and place it on sample wells 7 and 8.
10. Position the applicator on the applicator block, and load it by depressing the button for 4 seconds (fig. 2.6*e*).
11. Position the loaded applicator on the chamber cover. Depress the button for 4 seconds to place the serum on the cellulose acetate strip (fig. 2.6*f*).
12. Connect the chamber to the power supply and electrophorese for *20 minutes* at 200 V (figs. 2.6*g,h* and 2.7).

**Note:** *During this time, clean the applicator by placing it on filter paper moistened with the buffer. Rinse with tap water and distilled water.*

13. When the voltage is off, open the chamber and remove the membrane frame assembly. Raise the tension latch and strip clamps and remove the cellulose acetate strip with forceps.
14. Float the strip on Ponceau S stain for *1 minute;* then immerse it completely in the stain for *10 minutes*.
15. Remove the strip with forceps and rinse it in two successive baths of 5% acetic acid. Tape the cellulose acetate strip in the space provided (or draw a facsimile) and label the bands.

**Note:** *The albumin band (nearest the anode) is the darkest band; the gamma globulin band (nearest the cathode) is the widest band.*

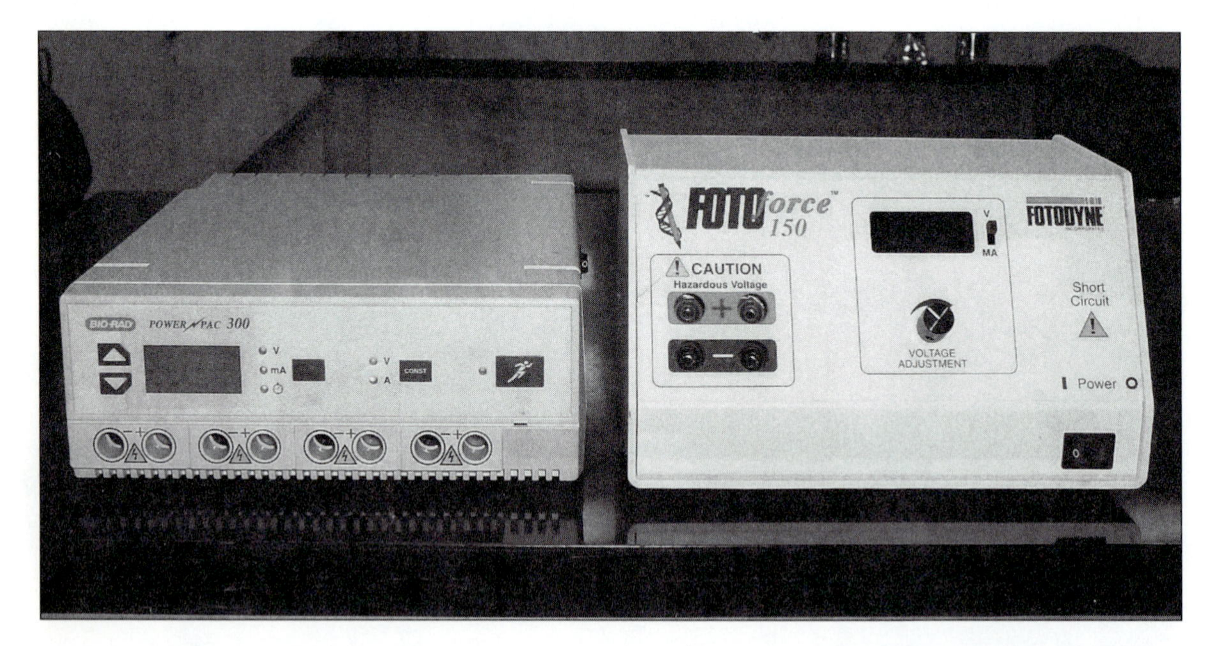


**Figure 2.7 Electophoresis power supplies.** The models shown here are examples of equipment with digital displays.

# Laboratory Report 2.3

Name ____________________

Date ____________________

Section ____________________

## DATA FROM EXERCISE 2.3

1. In this space, tape your electrophoresis strip (or draw a facsimile).

2. Label the protein bands in your strip.

## REVIEW ACTIVITIES FOR EXERCISE 2.3

### *Test Your Knowledge of Terms and Facts*

1. Describe the property of an amphoteric molecule. ____________________
____________________

2. The amphoteric nature of an amino acid that is electrically neutral can best be shown by its ____________________ formula. Draw one here.

3. Name two amino acids that have
   (a) an extra amino group: ____________________ and ____________________.
   (b) an extra carboxyl group: ____________________ and ____________________.
4. Proteins have a net ____________________ charge in acidic solutions and a net ____________________ charge in basic solutions.
5. The soluble protein in plasma that aids in the formation of blood clots is ____________________.
6. The insoluble protein that aids in the formation of blood clots is ____________________.
7. The most abundant class of plasma proteins is the ____________________.
8. Albumin and many other plasma proteins are made by the ____________________.
9. The class of plasma proteins containing antibodies is the ____________________.

## *Test Your Understanding of Concepts*

10. What is the difference between serum and plasma? Explain.

11. What factors contribute to the distance that a protein will migrate during electrophoresis?

## *Test Your Ability to Analyze and Apply Your Knowledge*

12. Why is the albumin band the darkest of the bands? What does the width of the gamma globulin band reveal about the structure of antibodies?

13. How might liver disease affect the density of the bands? Which bands would be the most affected? What effects might these changes in plasma proteins have on the body?

# Measurements of Enzyme Activity

EXERCISE 2.4

## Materials

1. Beakers, rusty nails, chicken liver
2. Hydrogen peroxide
3. Test tubes, mechanical pipettors, automatic microliter pipettes
4. Constant-temperature water bath set at 37° C
5. Cuvettes and spectrophotometer (colorimeter)
6. Serum or reconstituted normal and abnormal serum (available, for example, from Bio-Analytic Laboratories, Inc., or Stanbio Laboratory, Inc.).
7. Alkaline Phosphatase Test and Lactate Dehydrogenase Test (available, for example, from Biotron Diagnostics, Inc. or from Bio-Analytic Laboratories, Inc.)

The presence of a specific enzyme can be detected by the reaction it catalyzes, and the enzyme concentration can be measured by the amount of product it forms in a given period of time. Enzyme activity is affected by pH, temperature, and the availability of substrates and coenzymes.

## OBJECTIVES

1. Describe the lock-and-key model of enzyme activity and use this model to explain enzyme specificity.
2. Describe the effects of pH and temperature on enzyme activity.
3. Describe how enzyme concentration is measured.

Enzymes are biological **catalysts;** that is, enzymes are substances that increase the rate of chemical reaction without changing the nature of the reaction and without being altered by the reaction. The catalytic process occurs in two stages: (1) the reactants (hereafter referred to as the *substrates* of the enzyme) bond in a specific manner to the enzyme, forming an *enzyme-substrate complex,* and (2) the enzyme-substrate complex dissociates into the free, unaltered enzyme and *products*.

## Textbook Correlations

Before performing this exercise, you should study the introductory material presented here. Further information relating to this exercise can be found in these pages of *Human Physiology,* eighth edition, by Stuart I. Fox:

- *Enzymes as Catalysts.* Chapter 4, p. 86.
- *Effects of Temperature and pH.* Chapter 4, p. 89.
- *Cofactors and Coenzymes.* Chapter 4, p. 90.

All enzymes are proteins (although not all proteins are enzymes). The polypeptide chain of each enzyme bends and folds in a unique way to produce a characteristic three-dimensional structure. The substrates interact with a specific part of this structure, the **active site,** which is complementary in shape to the substrate molecules. This is most easily visualized by the *lock-and-key model* of enzyme action (fig. 2.8).

The shape of the active site is determined by the amino acid sequence of the protein; this shape is different for different enzymes. Enzymes are *relatively specific,* interacting only with specific substrates and catalyzing selective reactions.

The relative specificity of an enzyme can often be deduced from its name. Thus, the enzyme *lactate dehydrogenase* removes a hydrogen from lactic acid (the suffix *-ase* denotes an enzyme), whereas *phosphatase* enzymes hydrolyze the phosphate group from a wide variety of organic compounds. These guidelines do not apply to enzymes that were discovered before a systematic terminology was developed, such as the digestive enzymes *pepsin* and *trypsin* that hydrolyze peptide bonds.

Since enzymatic activity is dependent on the delicate bending and folding of polypeptide chains, changes in pH and temperature, which affect the three-dimensional structure of proteins, also affect enzymatic activity. Although most enzymes display maximum activity at pH 7 and at 40° C to 45° C, the **pH optimum** and **temperature optimum** of one enzyme may be significantly different from those of another enzyme. Enzymatic activity diminishes

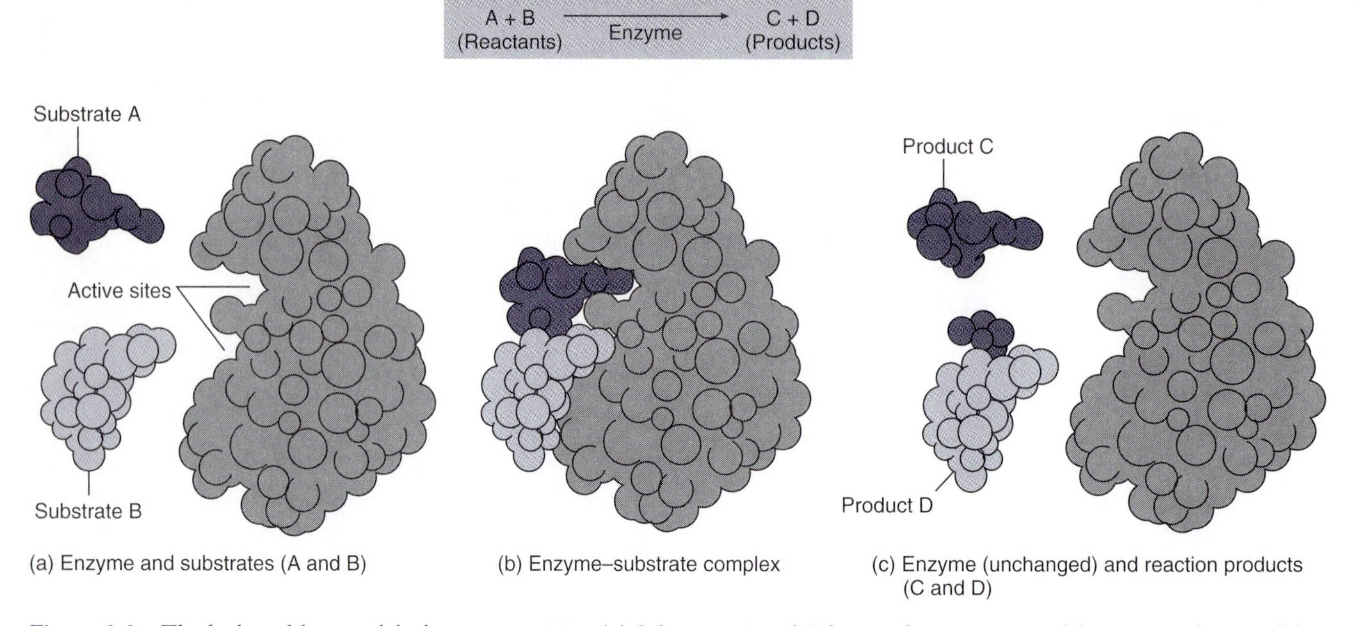


**Figure 2.8 The lock-and-key model of enzyme activity.** (a) Substrates A and B fit into the active sites of the enzyme, forming (b) an enzyme-substrate complex. This complex then dissociates (c) to yield free enzyme and the products of the reaction.

**(For a full-color version of this figure, see fig. 4.2 in *Human Physiology*, eighth edition, by Stuart I. Fox.)**

when the pH or temperature varies from the optimal value for that enzyme.

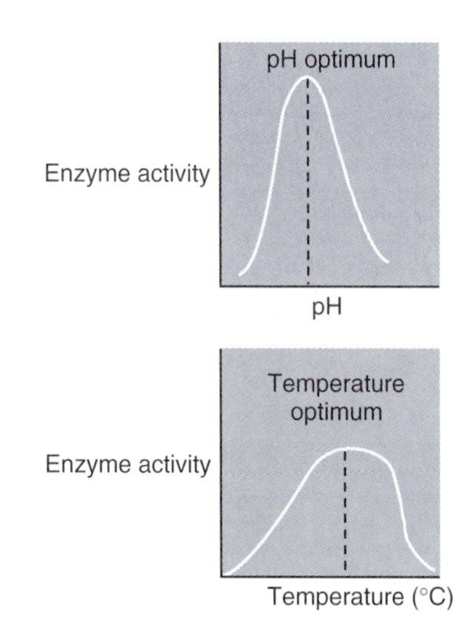


The activity of many enzymes is absolutely dependent on the presence of specific, smaller, nonprotein molecules called *cofactors* or *coenzymes*. Cofactors are specific inorganic ions (e.g., $Mg^{2+}$) required for enzyme activity, and coenzymes are organic compounds derived from the water-soluble vitamins that play a similar role.

To assay (measure) the enzyme activity in an unknown sample, the appropriate cofactors or coenzymes must be provided, and the pH and temperature must be standardized. In addition, the concentration of substrate should be very large relative to the concentration of enzyme, so that the availability of substrate does not limit the rate of the reaction. Under these conditions, the amount of product doubles when the concentration of enzyme doubles or when the reaction time doubles. If the reaction time is held constant, the amount of product formed is *directly proportional* to the enzyme concentration.

The concentration of the enzyme in a sample is usually measured in **units of activity,** determined under conditions of specified pH, temperature, reaction time, and other controlled factors. A unit of enzyme activity is a measure of the reaction rate and may be determined by either the quantity of substrate consumed or the quantity of product formed in a given time interval. Specifically, one **international unit (IU)** is defined as *the quantity of enzyme required to convert one micromole of substrate per minute into products,* under specified conditions of pH, temperature, and other controlled conditions. For example, the normal concentration of the enzyme alkaline phosphatase in serum ranges from 9 to 35 international units per liter (IU/L).

In these exercises, you will assay the activity of three enzymes found in serum. These enzymes are not normally active in serum but are enzymes released into the blood from ruptured cells of damaged tissues. These enzymatic assays are often valuable in the diagnosis of certain diseases.

## A. Catalase in Liver

**Catalase** is an enzyme that converts hydrogen peroxide ($H_2O_2$)to water and oxygen gas. Found in many tissues, catalase is one of the most rapidly acting enzymes in the body.

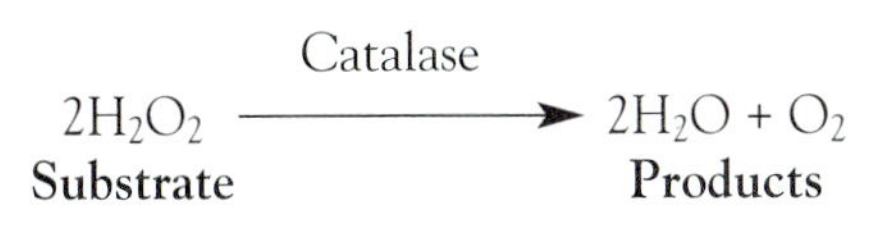


### PROCEDURE

1. Fill two small 100-mL beakers to the halfway mark with hydrogen peroxide.
2. Immerse a rusty nail or similar object in the solution in one beaker and then gently stir. Observe the effect of an inorganic catalyst and record your observations in the laboratory report.
3. Mince a fresh chicken liver with scissors and add it to the hydrogen peroxide in the second beaker. Stir the solution and record your observations in the laboratory report.

## B. Measurement of Alkaline Phosphatase in Serum

Plasma contains a number of enzymes that are released from damaged tissue cells. Assays of these enzymes can thus be useful in clinical diagnosis. In this exercise, you will determine the concentration of plasma **alkaline phosphatase** by measuring the increase in the absorbance of a solution caused by the accumulation of the product of the enzymatic reaction over time. The increase in absorbance of the solution with time is due to the formation of the yellow product, p-nitrophenol, which occurs when alkaline phosphatase hydrolyzes the phosphate from the substrate p-nitrophenylphosphate at a high pH.

> There are two enzymes in serum that display phosphatase activity (remove phosphate from organic compounds). One of these has a pH optimum of 4.9 *(acid phosphatase);* the other has a pH optimum of 9.8 *(alkaline phosphatase).* Abnormally high levels of serum acid phosphatase activity may be noted in patients with cancer of the prostate, whereas elevated alkaline phosphatase activity is primarily associated with various liver and bone diseases.

### PROCEDURE

1. Obtain three cuvettes. Label one *U* (for unknown serum), one *S* (for standard reference), and one *B* (for the water blank).
2. Pipette 0.5 mL of substrate reagent into each cuvette and place in a 37° C water bath. Let the solutions sit in the water bath for approximately 5 minutes.
3. Pipette 50 μL (0.05 mL) of the unknown serum to the cuvette marked *U*. Wait 30 seconds, and pipette 50 μL of the enzyme standard (containing 25 IU/L) to the cuvette marked *S*. Wait 30 seconds and pipette 50 μL of distilled water to the cuvette marked *B*. Be sure to return the cuvettes to the water bath immediately.
4. Ten minutes after having added the serum to cuvette *U*, add 2.5 mL of color developer to this cuvette, cap, and mix. Repeat this procedure 30 seconds later with cuvette *S*, and then with cuvette *B*. Note that each cuvette, in this way, will have incubated for exactly 10 minutes before you terminated the reaction by adding color developer.
5. Set the colorimeter at a wavelength of 590 nm. Standardize the colorimeter using the blank cuvette *(B)* and read the absorbance values of cuvettes *U* and *S*. Record these values in your laboratory report.
6. Calculate the concentration of alkaline phosphatase in the unknown serum sample (cuvette *U*) using the Beer's law formula below and enter your value in the laboratory report.

#### *Example*

Suppose the enzyme reference standard (25 IU/L) had an absorbance of 0.28, and the unknown serum had an absorbance of 0.34. The concentration of the unknown could be calculated as:

$$\frac{0.34}{0.28} \times 25 \text{ IU/L} = 30 \text{ IU/L}$$

> The normal range for alkaline phosphatase concentrations in plasma is 9–35 IU/L.

## C. Measurement of Lactate Dehydrogenase in Serum

An enzyme of great importance in physiology and medicine is **lactate dehydrogenase (LDH).** In skeletal muscles during heavy exercise, this enzyme catalyzes the conversion of pyruvic acid to lactic acid. In the process of this reaction, the coenzyme NADH + $H^+$ is oxidized to NAD:

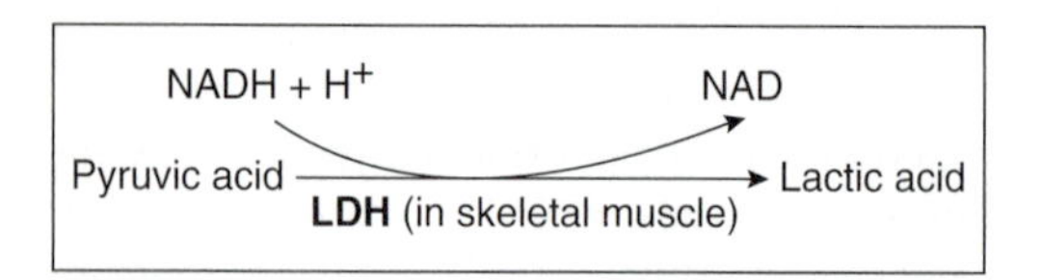


The same enzyme also catalyzes the reverse reaction, whereby lactic acid is converted into pyruvic acid. This reverse reaction occurs in normal heart tissue. It also occurs in the liver, which can utilize lactic acid for energy production (cell respiration) or in the formation of glucose. This reaction is accompanied by the reduction of NAD to NADH + $H^+$ as:

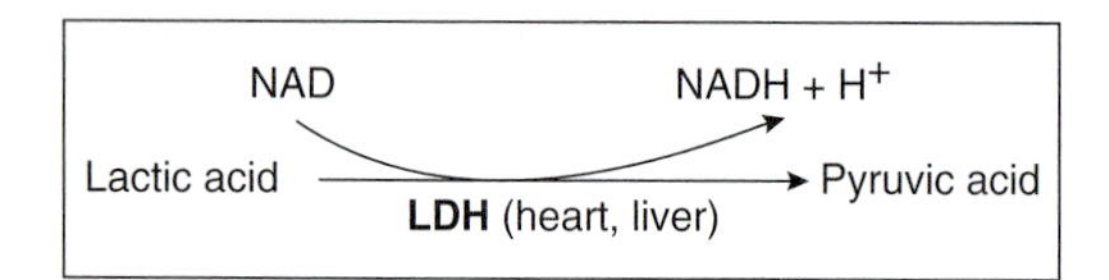

Serum **lactate dehydrogenase (LDH)** concentrations are elevated following a myocardial infarction (MI or "heart attack"). Measurements of these levels, in conjunction with other tests, can help diagnose heart attacks. Serum LDH levels are also elevated in some renal (kidney) diseases, cirrhosis of the liver, and hepatitis.

## PROCEDURE

1. Using a mechanical pipettor (do not pipette by mouth), pipette 1.0 mL of working reagent (containing lactate and NAD) into a cuvette.
2. Place the cuvette in a 37° C water bath for 5 minutes. This is a preincubation, to bring the substrate to the proper temperature.
3. While waiting, set the spectrophotometer to a wavelength of 340 nm.
4. While waiting, fill a cuvette with water, and using this as the blank, set the spectrophotometer to read zero absorbance.
5. Pipette 0.05 ml (50 μL) of the serum sample into the cuvette that was prewarmed in the water bath. Immediately return it to the water bath.
6. Wait *30 seconds*, and then quickly remove the cuvette from the water bath, insert it into the spectrophotometer, and take an absorbance reading.
7. Return the cuvette to the water bath, and then, *1 minute* after your first reading, take a second absorbance reading. The second absorbance reading should be higher, because the conversion of NAD to NADH in the reaction results in an increase in absorbance.
8. Subtract the first absorbance reading from the second absorbance reading to obtain the change in absorbance per minute: __________.
9. In order to obtain the enzyme concentration in IU/L, multiply the change in absorbance above by 3,376 (a conversion factor). Enter your measurement in the laboratory report.

### *Example*

Suppose the first absorbance reading was 0.10 and the second absorbance reading was 0.15. The difference in absorbance is 0.05. Then,

$$0.05 \times 3376 = 169 \text{ IU/L}$$

**The normal LDH concentration for healthy adults is 100–225 IU/L.**

# Laboratory Report 2.4

Name ____________________

Date ____________________

Section ____________________

## DATA FROM EXERCISE 2.4

### A. Catalase in Liver

1. Describe what occurred when the rusty nail was placed in the beaker of hydrogen peroxide. Write a simple equation that might describe this reaction.

2. Describe what occurred when the chicken liver was placed in the beaker of hydrogen peroxide. Explain what occurred and write a simple equation that might describe this reaction.

### B. Measurement of Alkaline Phosphatase in Serum

1. Record your absorbance values for cuvettes *U* and *S* in these spaces.
   Absorbance of *U*: ____________________
   Absorbance of *S*: ____________________
   Calculate the alkaline phosphatase concentration of the unknown serum: ____________________ IU/L.
2. Was the value you obtained in the normal range? What might an abnormally high value indicate?

**C. Measurement of Lactate Dehydrogenase in Serum**

1. Calculate the lactate dehydrogenase concentration of the unknown serum: ____________ IU/L.
2. Was the value you obtained in the normal range? What might an abnormally high value indicate?

## REVIEW ACTIVITIES FOR EXERCISE 2.4

### *Test Your Knowledge of Terms and Facts*

1. Define an *international unit (IU)* of enzyme activity. ____________
2. Define the *pH optimum* of an enzyme. ____________
3. Define the *temperature optimum* of an enzyme. ____________
4. What are cofactors and coenzymes? ____________
5. Judging from its name, describe the activity of these enzymes:
   (a) phosphatase ____________
   (b) glycogen synthetase ____________
   (c) lactate dehydrogenase ____________
   (d) DNA polymerase ____________

### *Test Your Understanding of Concepts*

6. What must you add to a sample of plasma in order to measure the activity of a particular enzyme in plasma? How might you measure the activity of a different enzyme in the same tube of plasma?

7. What happens to the structure of an enzyme when the pH and temperature are changed from the optima for that enzyme?

8. Describe the reactions catalyzed by lactate dehydrogenase in the skeletal muscles, liver, and heart. What is the metabolic significance of these reactions? What happens to NAD and NADH in these reactions?

## *Test Your Ability to Analyze and Apply Your Knowledge*

9. Why were you so careful to have each tube incubate for exactly 10 minutes? What might have occurred if you allowed the tubes to incubate overnight before finally stopping the reaction? Would this be an accurate test? Explain.

10. Pancreatic amylase is an enzyme normally secreted by the pancreas into the small intestine, where it catalyzes the hydrolysis of starch. Should it be in the blood? What might explain its presence in the blood? Would it have any activity as it circulates in the blood plasma? Explain.

# Genetic Control of Metabolism

EXERCISE 2.5

## Materials

1. Test tubes, Pasteur pipettes (droppers), urine collection cups
2. Phenistix (Ames Laboratories), silver nitrate (3 g per 100 mL), 10% ammonium hydroxide, 40% sodium hydroxide, lead acetate (saturated)

Since a different gene codes for the production of each enzyme, a defective gene can result in a specific metabolic disorder. These inborn errors of metabolism can be detected by tests for specific enzyme products.

## OBJECTIVES

1. Define the terms *genotype, phenotype, transcription,* and *translation.*
2. Describe how genes regulate metabolic pathways.
3. Describe how inborn errors of metabolism are produced.
4. Explain the etiology (cause) of phenylketonuria (PKU).

## Textbook Correlations

Before performing this exercise, you should study the introductory material presented here. Further information relating to this exercise can also be found in these pages of *Human Physiology,* eighth edition, by Stuart I. Fox:

- *Cell Nucleus and Gene Expression.* Chapter 3, pp. 61–64.
- *Metabolic Pathways.* Chapter 4, pp. 91–93.

The physical appearance of an individual (**phenotype**) is largely determined by the individual's genetic endowment (**genotype**). The control of the phenotype by the genotype is achieved by means of the genetic regulation of cellular metabolism.

All of the chemical reactions of cellular metabolism are catalyzed by specific enzymes, and the information for the synthesis of each specific enzyme is coded by a specific gene. That is, one gene makes one enzyme, which is capable of catalyzing one type of reaction (changing A to B, for example).

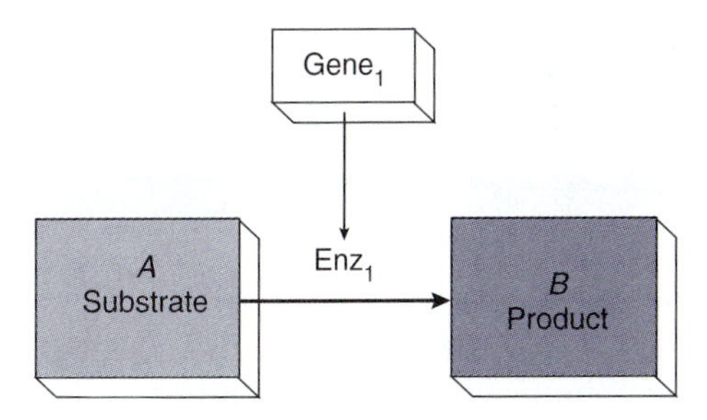


The product of this reaction may become the substrate of a second enzyme, made by a second gene, which converts B into a new product, C. A third enzyme, made by a third gene, may then convert C into D. Thus, a **metabolic pathway** is formed, where the product of one enzyme becomes the substrate of the next. In a metabolic pathway, some initial substrate, A, is converted into a final product (such as D) through a number of *intermediates* (B and C, for example).

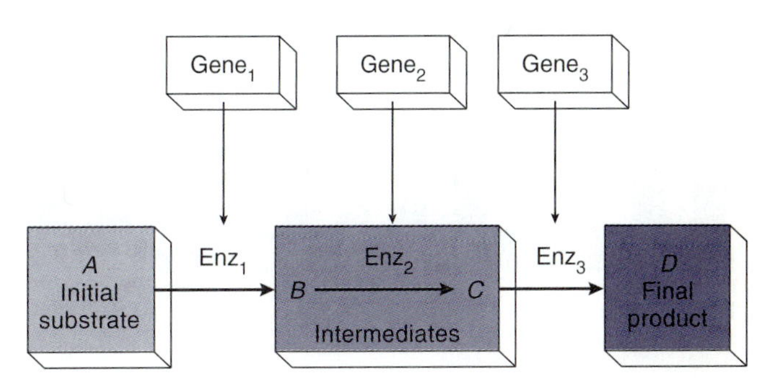

Genes contain the information for the synthesis of all proteins, not only those with enzymatic activity. The genetic code is based on the sequence of **DNA** components known as **nucleotide bases** *(adenine, guanine, cytosine, thymine)*. Thus, the sequence of these bases is different in different genes.

The sequence of bases on the DNA that composes one gene is used as a template for the synthesis of one **RNA** molecule. The RNA molecule consists of a linear sequence of nucleotide bases (*uracil*, cytosine, guanine, and adenine), which is precisely complementary to the sequence of bases on the region of the DNA (gene) on which it was made. This complementarity is ensured by the fact that only a specific base on the RNA can bind to a specific base on the DNA.

A part of the genetic code is thus transcribed by the synthesis of a specific RNA molecule—a process called **transcription.** This RNA contains a part of the genetic message and is called *messenger RNA (mRNA)*.

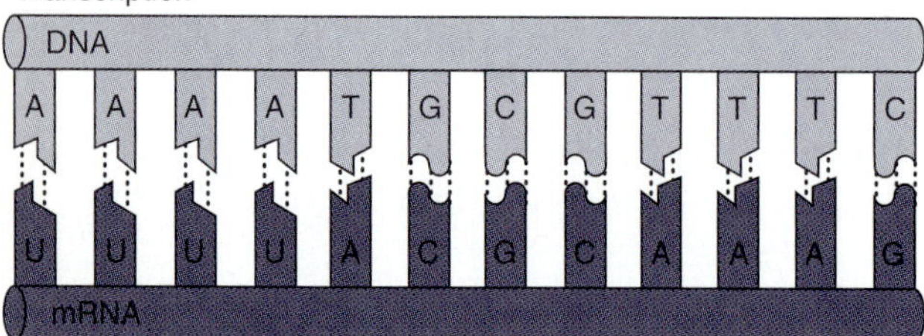


The messenger RNA, in association with ribosomes, forms the template for **protein synthesis,** where the sequence of amino acids in the protein is specified by the sequence of bases in the mRNA. The genetic code is based on the fact that a sequence of three mRNA bases (a *base triplet*) can bind only to a specific amino acid through an intermediate compound called *transfer RNA (tRNA)*.

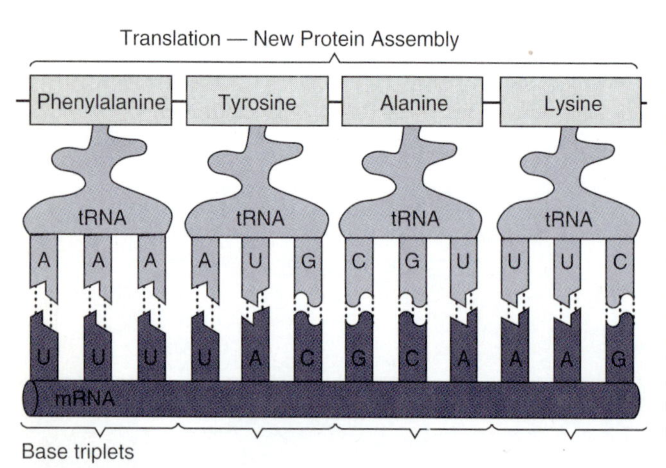


In this way, the sequence of bases in the mRNA determines the sequence of amino acids in the protein. This process is called **translation.** The sequence of amino acids, in turn, determines how the protein will fold (i.e., its three-dimensional structure). The three-dimensional structure of a protein directly determines its function.

## Inborn Errors of Metabolism

When a gene is missing or defective, the enzyme that it makes will also be missing or defective. This will result in a hereditary metabolic disorder in which there will be a decrease in the intermediates formed *after* the step normally catalyzed by that enzyme. The intermediates formed *before* this step will accumulate in the blood and body tissues and will be excreted in the urine.

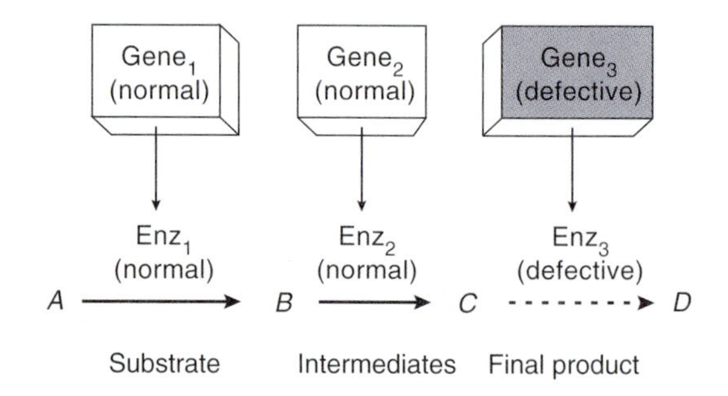


Thus, intermediate *D* decreases, while intermediates *A*, *B*, and *C* increase.

There are a number of genetic defects associated with the metabolism of the amino acids *phenylalanine* and *tyrosine*. The phenotypic (expressed) effects of these **inborn errors of metabolism** depend on which enzymes are defective and therefore on which intermediate products either are absent or accumulate abnormally in the body tissues.

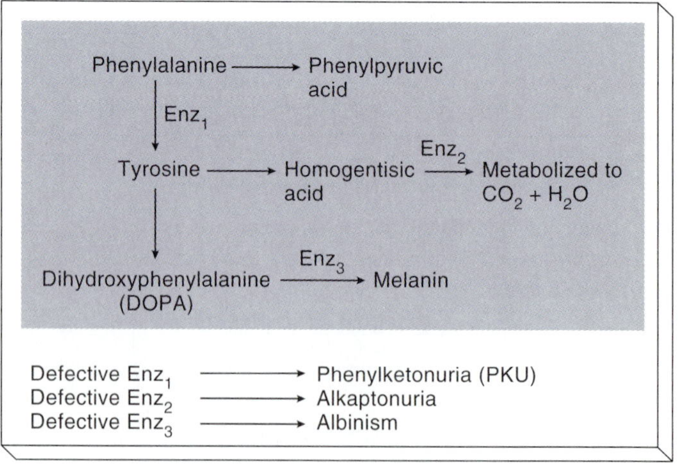

Probably the best known phenotypic effect of this group of hereditary disorders is the lack of *melanin* pigment characteristic of **albinism. Phenylketonuria (PKU)** is the most clinically serious disorder of this group, since the accumulation of phenylpyruvic acid can affect the developing central nervous system and produce mental retardation. The odious effects of PKU can be avoided only by placing the affected child on a special low phenylalanine diet.

Defective genes cannot be replaced with "good" genes, although the first such treatment has recently been reported. At present, therefore, treatment of those with genetic disorders depends upon early diagnosis and preventive measures. Newborn babies are routinely tested for PKU; and those with the enzyme defect can be placed on low phenylalanine diets. Unfortunately, many other inborn errors of metabolism cannot be treated by dietary restrictions. The only way to prevent some of these diseases is through genetic counseling for prospective parents who are carriers of the disease.

When there is a defective enzyme in a metabolic pathway, the molecule that is the substrate of that enzyme accumulates in particular tissues of the body. Such inborn errors of metabolism are not restricted to pathways of amino acid metabolism. In *Tay-Sachs disease,* for example, there is a defect in the enzyme (hexosaminidase A) that breaks down a complex type of lipid known as a ganglioside, resulting in lipid accumulation in the brain and retina. This disease, which occurs primarily in certain families of European Jewish origin (specifically, Ashkenazic Jews), is invariably fatal. Inborn errors also occur in carbohydrate metabolism. In *glycogen storage disease,* for example, the enzyme that breaks down glycogen may be defective, resulting in the excessive accumulation of glycogen that causes liver damage.

## A. Phenylketonuria

A person with PKU excretes large amounts of phenylpyruvic acid because of the inability to convert the amino acid phenylalanine into the amino acid tyrosine.

### PROCEDURE

1. Dip the test end of a Phenistix strip into a sample of urine.

**Note:** *Use care when handling all body fluids, including urine. Clean spills and dispose of urine containers properly, as described by your instructor.*

2. Compare the color of the strip with the color chart provided.

**Note:** *Federal law now mandates the Guthrie test using a spot of blood from newborn babies for diagnostic screening for PKU. Therefore, this procedural kit may not be readily available.*

## B. Alkaptonuria

A person with **alkaptonuria** excretes large amounts of homogentisic acid, which reacts with silver to form a black precipitate. This condition does not have immediate adverse effects on health, but may cause the development of a characteristic type of joint degeneration later in life.

### PROCEDURE

1. Add 10 drops of urine to a test tube containing 5 drops of 3% silver nitrate ($AgNO_3$).
2. Add 5 drops of 10% ammonium hydroxide ($NH_4OH$) to the tube and mix.
3. A positive test for homogentisic acid is indicated by the presence of a black precipitate.

## C. Cystinuria

The renal tubules can normally reabsorb all of the amino acids filtered in the glomerulus. People with the recessive trait known as **cystinuria,** however, have an impaired ability to reabsorb the amino acid *cystine* and the related amino acids lysine, arginine, and ornithine.

Cystine is the least soluble amino acid and may precipitate in the urinary tract to form stones. This condition accounts for 1% of the cases of renal stones in the United States. (About 10% are uric acid stones, and the remainder are formed from calcium salts, primarily calcium oxalate.)

### PROCEDURE

1. Add 1.0 mL of 40% NaOH to a test tube containing 5.0 mL of urine. Allow the tube to cool.
2. Add 3.0 mL of lead acetate. A brown-to-black precipitate indicates the presence of cystine in the urine.

## *Test Your Ability to Analyze and Apply Your Knowledge*

5. Propose reasons why it would be difficult to correct an inborn error of metabolism. Explain why a simple pill or even an injection probably wouldn't be effective.

6. Speculate as to how the information obtained from the sequencing of the human genome might be used to someday eliminate inborn errors of metabolism.

# Diffusion, Osmosis, and Tonicity

EXERCISE 2.6

## Materials

1. Test tubes, thistle tubes, dialysis tubing
2. Beakers, ring stands, burette clamp
3. Lancets and alcohol swabs; or animal (dog, cat, or rat) blood from veterinary or other appropriate source. Biohazard receptacle for all blood contaminated items.
4. Microscopes, slides, cover slips, and transfer pipettes
5. Make two dilutions of molasses, 200 mL molasses (a 12-oz. jar) diluted with water to 1 L (20% solution); and 250 mL molasses diluted with water to make 1 L (25% solution).
6. Sodium chloride solutions in dropper bottles at these concentrations: 0.20 g per 100 mL; 0.45 g per 100 mL; 0.85 g per 100 mL; 3.5 g per 100 mL; and 10 g per 100 mL.
7. Toluene, potassium permanganate crystals, vegetable oil, laboratory detergent

Osmosis is the net diffusion of water (solvent) through a membrane that separates two solutions. Osmosis occurs passively when the two solutions have different total concentrations of molecules (solutes) to which the membrane is relatively impermeable. If there is no osmosis when a membrane separates two solutions, those solutions are said to be isotonic to each other.

## OBJECTIVES

1. Distinguish between the terms *solute, solvent,* and *solution.*
2. Define the terms *passive transport, diffusion,* and *active transport.*
3. Define the terms *osmosis, osmotic pressure,* and *osmolality.*
4. Define the terms *isotonic, hypotonic,* and *hypertonic.*
5. Calculate the osmolality of solutions when the concentration of solute (in g/L) and the molecular weight of a solute are known.
6. Describe how red blood cells (RBCs) are affected when they are placed in isotonic, hypotonic, and hypertonic solutions.

## Textbook Correlations*

Before performing this exercise, you should study the introductory material presented here. Further information relating to this exercise can be found in these pages of *Human Physiology,* eighth edition, by Stuart I. Fox:

- *Osmosis.* Chapter 6, pp. 130–133.
- *Regulation of Blood Osmolality,* Chapter 6, pp. 133–134.

If you were to drop a pinch of sugar (the *solute*) into a beaker of water (the *solvent*), the resulting *solution* would, after a time, have a uniform sweetness. The uniform sweetness would result from the constant state of motion of all of the solute and solvent molecules in the solution, producing a net movement of solute molecules from regions of higher concentration to regions of lower concentration. This net movement of *solute* molecules is known as **diffusion.**

The rate of diffusion is proportional to the concentration differences that exist in the solution. The diffusion rate will steadily decrease as the solute becomes evenly distributed in the solvent, and diffusion will cease entirely when the solution becomes uniform.

A molecule may move into or out of a cell by diffusion if (1) a difference in the concentration of that molecule *(concentration gradient)* exists between the intracellular and extracellular compartments, and (2) the cell membrane will allow the passage of that molecule.

The movement of a molecule across the plasma (cell) membrane by diffusion is called **passive transport.** The term *passive* is used because the cell need not expend energy in the process. By contrast, cells often must move molecules across the plasma membrane from lower to higher concentrations; that is, cells must "fight" diffusion in the attempt to maintain a concentration difference across the membrane. However, to move molecules "uphill" against their concentration gradients, the cells must expend energy. This process is called **active transport.**

*Multimedia Correlations (also see Appendix 3)
- *MediaPhys 2.0:* Topics 3.9–3.24

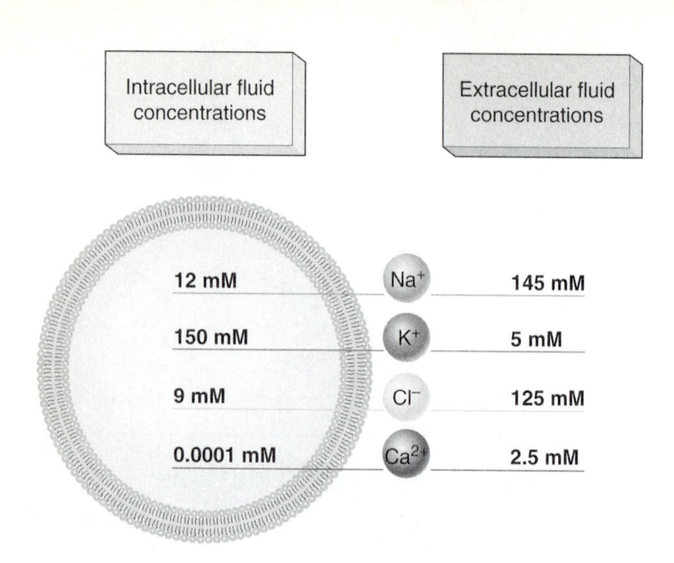


**Figure 2.9 Concentrations of ions in the intracellular and extracellular fluids.** This distribution of ions, and the different permeabilities of the plasma membrane to these ions, affects the membrane potential and other physiological processes. (Molarity, symbolized by M, is a unit of concentration based on the atomic weight.)

**(For a full-color version of this figure, see fig. 6.23 in *Human Physiology*, eighth edition, by Stuart I. Fox.)**

Sodium, for example, is maintained at a higher concentration outside the cell than inside the cell, whereas potassium is maintained at a higher concentration inside the cell than outside the cell (fig. 2.9).

The *permeability* of a membrane refers to the ease with which substances can pass through (permeate) it. A membrane that is completely permeable to all molecules is not a barrier to diffusion, whereas a membrane that is completely impermeable to all molecules essentially divides the solutions into two noncommunicating compartments. Since a living cell must selectively interact with its environment, taking in raw materials and excreting waste products, the cell is surrounded by a membrane that is **semi-** (or **selectively**) **permeable.** A semipermeable membrane is completely permeable to some molecules, slightly permeable to others, and completely impermeable to still others.

The **plasma (cell) membrane** is composed primarily of two semifluid phospholipid layers with proteins. Some proteins are partially submerged; others span the complete thickness of the membrane. In this way, the membrane is not continuous but behaves as if tiny protein channels were serving as waterways for diffusion, allowing the passage of ions and smaller molecules while excluding the passage of molecules larger than the channels. (fig. 2.10).

## A. Solubility of Compounds in Polar and Nonpolar Solvents

Most of the molecules that a cell encounters are water soluble (easily dissolved in water). Such molecules have charged groups and are said to be *polar*. The lipids of the cell membrane are *nonpolar* and serve as a barrier to the passage of polar molecules across the membrane. Small polar molecules may pass through the pores in the lipid barrier, but large polar molecules, such as proteins and polysaccharides, are restricted by the pore size.

Many organic solvents (benzene, or toluene, for example) are nonpolar; that is, they are soluble in lipids but not in water. Such nonpolar molecules are not limited in their passage by the membrane pores and can rapidly diffuse into the cells by passing through the lipid layers of the membrane.

### PROCEDURE

1. Pour about 2.0 mL of water and 2.0 mL of toluene into a test tube.
2. Shake the tube and record your observations in the laboratory report.
3. Using forceps, drop 2–3 crystals of potassium permanganate ($KMnO_4$) into the tube. Shake the tube and record your observations in the laboratory report.
4. Add about 1.0 mL of yellow vegetable oil to the tube. Shake the tube and record your observations in the laboratory report.
5. Add a pinch of laboratory detergent to the tube. Shake the tube and record your observations in the laboratory report.

**Note:** *One end of the detergent molecule is polar (charged); the other end is nonpolar. The detergent can thus act as a bridge between the two phases.*

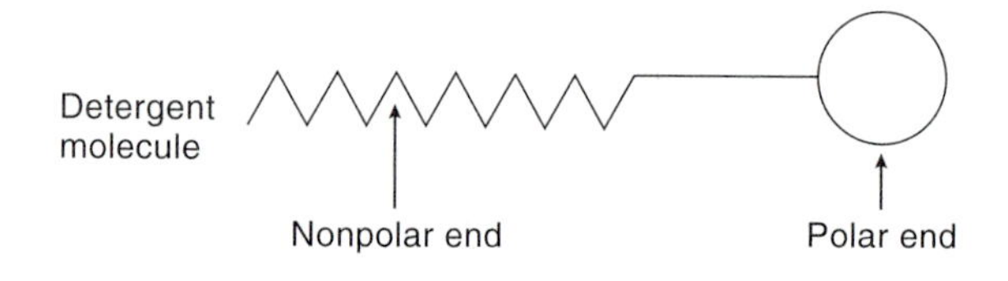


## B. Osmosis across an Artificial Semipermeable Membrane

Imagine a solution divided into two compartments by a membrane. If the membrane is completely permeable to solute and solvent molecules, these molecules will be able to diffuse across it, so that the solute/solvent ratio (concentration) will be the same on both sides of the membrane. Suppose, however, that the membrane is permeable to the solvent but not to the solute. If the solvent is water, the water will diffuse from the region where the solute/solvent ratio is *lower* (relatively more water) to the region where the solute/solvent ratio is *higher* (relatively less water), until the solute/solvent ratio (concentration) is the same on both sides of the membrane. The net diffusion of water across a membrane is called **osmosis** (fig. 2.11).

In osmosis, water diffuses into the more concentrated (greater solute/solvent ratio) solution from the less concentrated solution. The more highly concentrated solution is

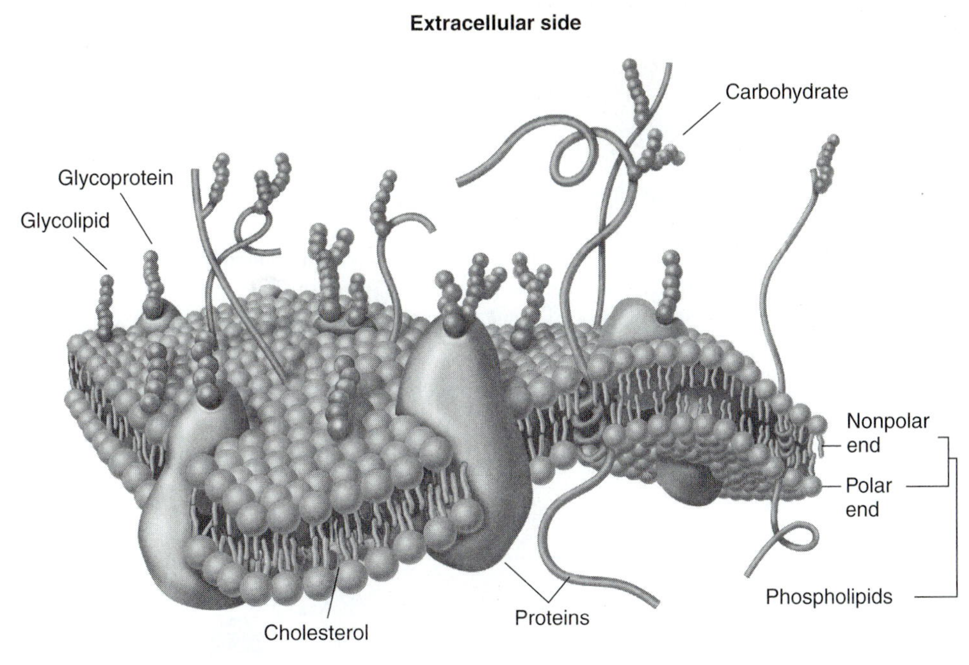

**Figure 2.10 Structure of the plasma (cell) membrane.** The plasma membrane consists of a double layer of phospholipids, with the phosphate-containing polar ends (spheres) oriented outward, and the nonpolar portions of the molecules (wavy lines) oriented towards the center. Proteins are interposed between the phospholipids, and carbohydrates are often bound to the external surface of these proteins.
**(For a full-color version of this figure, see fig. 3.2 in *Human Physiology,* eighth edition, by Stuart I. Fox.)**

said to have a greater **osmotic pressure** than the less concentrated solution. The osmotic pressure is a measure of the ability of a solution to "pull in" water from another solution that is separated from it by a semipermeable membrane. Keep in mind, however, that the "pulling" is a metaphor; since osmosis is the simple diffusion of water through a membrane, water moves into the more concentrated solution as a result of the higher-to-lower *water* concentration gradient. Since the osmotic pressure of a solution is proportional to its solute concentration, the osmotic pressure of distilled water is zero.

## PROCEDURE

1. Cut a 2 1/2-inch piece of dialysis tubing. Soak this piece in tap water until the layers separate. (Speed this process by rotating the tubing between two fingers.) Slide one blade of the scissors inside the tube and cut lengthwise, producing a single rectangular sheet of dialysis membrane.

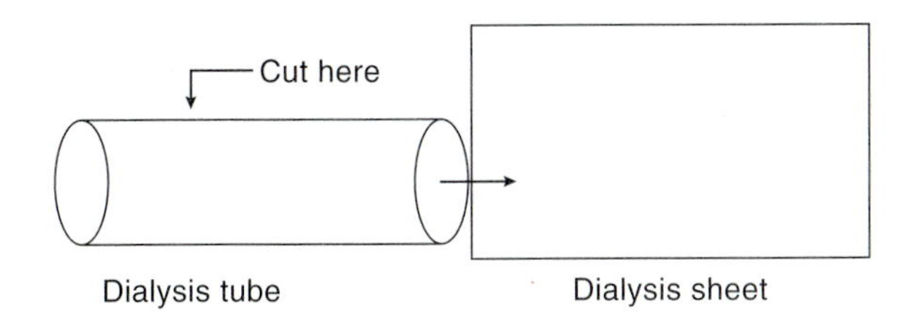


**Note:** *Dialysis tubing is a plastic porous material used to separate molecules on the basis of their size. It is an artificial semipermeable membrane. Molecules that are larger than the pore size remain inside the tubing, whereas smaller molecules (including water) can move through the membrane by diffusion. This technique of physical separation is called* ***dialysis.***

2. Divide the class into two molasses groups (20% and 25%). Hold a thistle tube vertically with the mouth upward. Have one person hold a finger over the lower opening while another pours a 20% or 25% molasses solution into the mouth of the tube until the solution is about to overflow the tube.
3. Place the rectangular piece of dialysis tubing tightly over the mouth of the thistle tube so that no air is trapped between the dialysis tubing and the molasses solution. Keeping the dialysis tubing taut, secure it to the thistle tube with several wrappings of a rubber band.
4. Invert the thistle tube, rinse with water, and check for leaks. If leaks are observed, remove the dialysis tubing and repeat step 3.
5. With the thistle tube inverted, immerse it in a large beaker of water (fig. 2.12). Secure the inverted thistle tube using a ring stand and a burette clamp. The narrow part of the thistle tube is more safely secured with a folded wad of paper towel in the clamp.

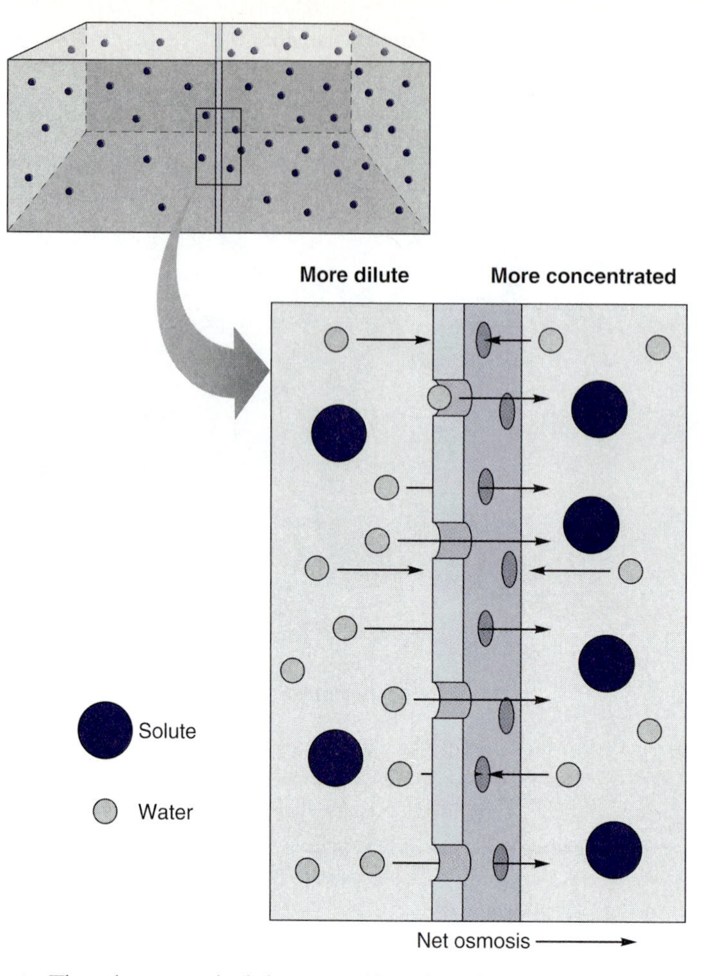


**Figure 2.11 A model of osmosis.** The solution on the left is more dilute than the one on the right, causing water to diffuse from left to right by osmosis across the semipermeable membrane.

**(For a full-color version of this figure, see fig. 6.5 in *Human Physiology,* eighth edition, by Stuart I. Fox.)**

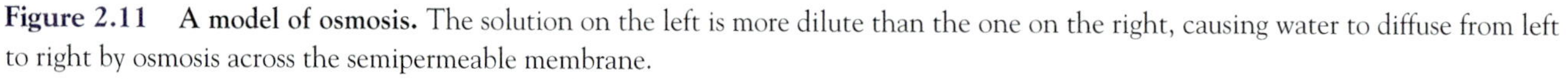

6. Mark the meniscus of the molasses solution on the thistle tube with a grease pencil. Every 15 minutes for a 1-hour period, record the change in the level of this meniscus (in centimeters) in the laboratory report.

## C. Concentration and Tonicity

Osmosis, the net diffusion of water across a membrane, requires that the membrane be completely permeable to water but only partially permeable or completely impermeable to solute molecules. In this case, the nonpenetrating solutes are said to be **osmotically active.** Osmosis is then caused by a difference in the concentration of osmotically active solutes, and the osmotic pressure of a solution is proportional to the concentration of osmotically active solutes. Solutes that are as freely permeable as water are not osmotically active.

The simplest way to express the concentration of a solution is the weight of the solute, in grams or milligrams, per 100 mL of solution. For example, 150 mg in 100 mL may be expressed as 150 mg per 100 mL—or as 150 mg%, or 150 mg/dL.

It is frequently more useful to express concentration in terms of *molarity* or *molality*. These measurements take into account the different molecular weights of the solutes; for example, a one-molar (1 M) or a one-molal (1 *m*) solution of sodium chloride (NaCl) would require you to weigh out a different amount of solute than you would for a 1 M or a 1 *m* solution of glucose ($C_6H_{12}O_6$). The common unit of weight in a 1 M NaCl solution and a 1 M glucose solution is the **mole.** The significance of the mole value is that *solutions with equal molarities have*

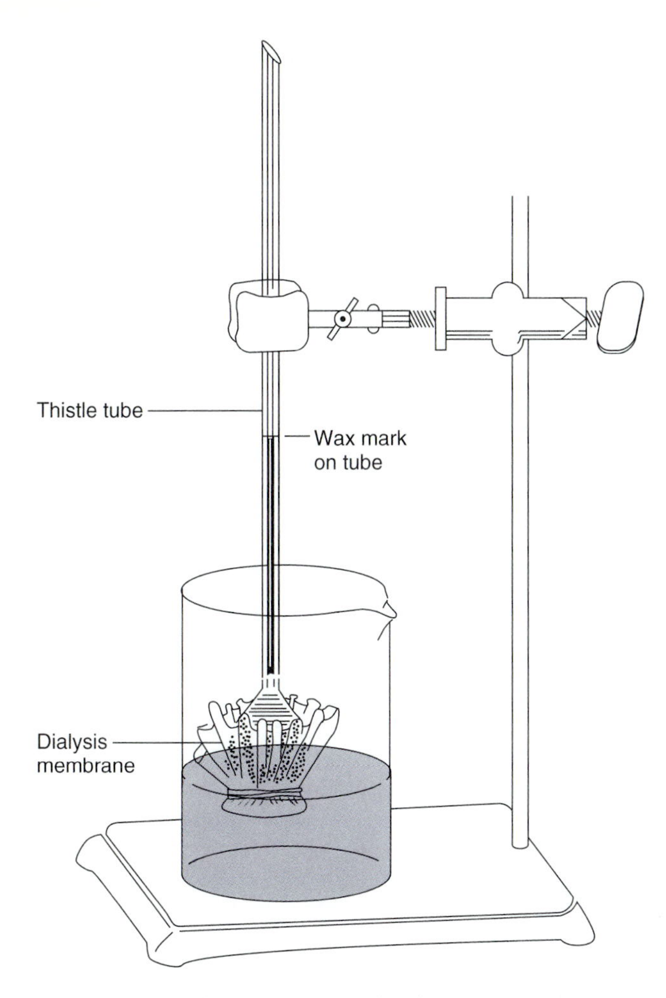


**Figure 2.12** **A thistle tube setup for the osmosis exercise.**

*equal numbers of molecules*. Although they weigh different amounts, a mole of NaCl contains the same number of molecules (Avogadro's number—$6.02 \times 10^{23}$) as a mole of glucose.

One mole is equal to the molecular weight of the solute in grams. The molecular weight is obtained by adding the atomic weights of each element in the molecule.

| | *Sodium chloride* (*NaCl*) | *Glucose* ($C_6H_{12}O_6$) |
|---|---|---|
| Atomic weights: | Na = 23.0 | $C_6$ 12 × 6 = 72 |
| | Cl = 35.5 | $H_{12}$ 1 × 12 = 12 |
| | | $O_6$ 16 × 6 = 96 |
| Molecular weights: | 58.5 | 180 |

A one-molar (1 M) solution contains 1 mole of solute in 1 L of solution. Thus, 1 M NaCl contains 58.5 g of NaCl per liter, whereas 1 M glucose contains 180 g of glucose per liter. A one-molal (1 *m*) solution of NaCl contains 1 mole of NaCl (58.5 g) dissolved in 1,000 g of solvent. If water is the solvent, 1,000 g equals 1,000 mL (at maximum density, 4° C). A 1 M solution has a final volume of 1,000 mL, whereas a 1 *m* solution has a final volume that usually exceeds 1,000 mL.

Notice that decreasing the value of both the solute and the solvent by the same proportion does not change the concentration of the solution. A 1 M glucose solution can be made by dissolving 90 g of glucose in a final volume of 500 mL or by dissolving 180 mg of glucose in 1 mL of solvent.

If the concentration of osmotically active solute is the *same* on both sides of a membrane (if the osmotic pressures are equal), osmosis will not occur. These two solutions are said to be **isotonic** to each other (iso means "same"). If the concentration of a third solution is *less* than that of the first two solutions, it is said to be **hypotonic** to the first two solutions (*hypo* means "below"). Water will diffuse from the third solution into the first two solutions if these solutions are separated by a semipermeable membrane. If the concentration of a fourth solution is *greater* than that of the first two solutions, it is said to be **hypertonic** (*hyper* means "above") to the first two solutions. Water will diffuse out of the first two solutions and into the fourth if these solutions are separated by a semipermeable membrane.

In all cases, water diffuses from the solution of lower osmotic pressure (lower solute concentration) to the solution of greater osmotic pressure (greater solute concentration). Notice that the osmotic pressure of a solution is proportional to the number of solute molecules in solution. A one-molal solution of glucose, for example, has 1 mole of glucose molecules in solution. This solution would have the same osmotic pressure as a 1 *m* solution of sucrose or a 1 *m* solution of urea. These solutions are said to have the same **osmolality** (1.0 osmole/kg water, or 1.0 Osm). A solution containing 1 mole of glucose plus 1 mole of urea would have an osmolality of 2 (2.0 Osm). The osmolality of a solution is determined by *the sum of all the moles of solute in a solution*.

Some molecules *dissociate* (come apart) when they are dissolved in solution. Common table salt (NaCl), for example, completely dissociates in solution to $Na^+$ and $Cl^-$ ions. Thus, a one-molal solution of NaCl has a total osmolality of 2 (one mole of $Na^+$ plus one mole of $Cl^-$), written 2.0 Osm. This one-molal solution of NaCl is isotonic to a two-molal solution of glucose, a molecule that does not dissociate when dissolved in solution.

It is frequently convenient to express concentration in terms of *milliosmolality* (*mOsm*). A 0.1 *m* solution of NaCl, for example, has an osmolal concentration of 200 mOsm, whereas a 0.1 *m* solution of glucose has an osmolal concentration of 100 mOsm.

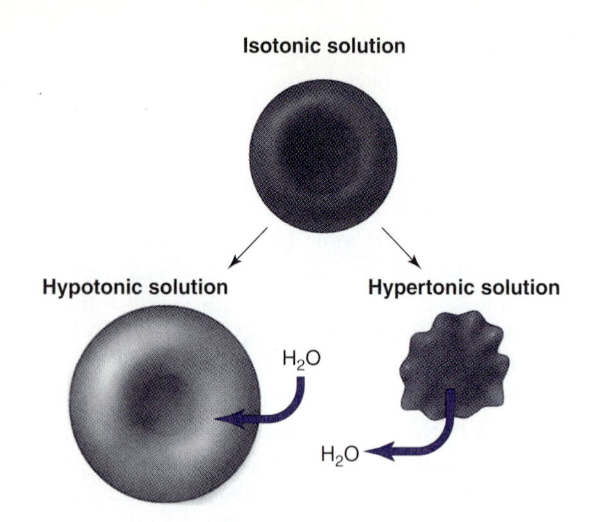


**Figure 2.13 Red blood cells in isotonic, hypotonic, and hypertonic solutions.** In each case, the external solution has an equal, lower, or higher osmotic pressure, respectively, than the intracellular fluid. As a result, water moves by osmosis into the red blood cells placed in hypotonic solutions, causing them to swell and even to burst. Similarly, water moves out of red blood cells placed in a hypertonic solution, causing them to shrink and become crenated.

**(For a full-color version of this figure, see fig. 6.11 in *Human Physiology*, eighth edition, by Stuart I. Fox.)**

### Tonicity of Saline Solutions Using Red Blood Cells as Osmometers

The **red blood cell (RBC)** has the same osmolality and the same osmotic pressure as plasma. When a red blood cell is placed in a hypotonic solution, it will expand or perhaps even burst (a process called *hemolysis*) as a result of the influx of water, extruding its hemoglobin into the solution. When placed in a hypertonic solution, a red blood cell will shrink (a process called *crenation*, fig. 2.13) as a result of the efflux of water.

Red blood cells can thus be used as *osmometers* to determine the osmolality of plasma, since RBCs will neither expand nor shrink in an isotonic solution.

Osmolality determines the distribution of water between the intracellular and extracellular fluid compartments of the body. Greater osmolality within the muscle fibers of a body builder is accompanied by greater fluid volume and cell enlargement (hypertrophy). Similarly, an accumulation of extracellular fluid in the tissues (edema), for example, can result when the osmolality of the tissue spaces increases due to an abnormal accumulation of proteins. Intravenous infusions for the purpose of maintaining blood volume and pressure must be isotonic to prevent the expansion or crenation of the body cells. **Normal saline** (0.9% NaCl) and **dextrose 5% in water (D5W)** are examples of such isotonic solutions.

## PROCEDURE

1. Measure 2.0 mL of the solutions indicated in part C of the laboratory report into each of five numbered test tubes.
2. Wipe the tip of a finger with alcohol and, using a sterile lancet, prick the finger to draw a small drop of blood. Alternatively, dog or cat blood may be provided by the instructor.

**Note:** *Caution must be exercised when handling blood to guard against contracting infectious agents. Handle only your own blood and discard all objects containing blood into the receptacles provided by the instructor.*

3. Allow the drop of blood to drain down the side of test tube 1. Mix the blood with the saline (salt) solution by inverting the test tube a few times.
4. Repeat the above procedure for test tubes 2–5. Additional drops of blood can be obtained by milking the finger.
5. Using a transfer pipette, place a drop of solution 1 on a slide, and cover it with a cover slip. Observe the cells using the 45× objective.
6. Repeat step 5 for the other solutions and record your observations in the laboratory report.

# Laboratory Report 2.6

Name ______________________

Date ______________________

Section ______________________

## DATA FROM EXERCISE 2.6

### A. Solubility of Compounds in Polar and Nonpolar Solvents

1. Describe the appearance of the solutions after completing steps 2 and 3. What is your conclusion regarding solubility and solvents?

2. Describe the appearance of the solutions after completing steps 4 and 5. What is your conclusion regarding solubility, solvents, and the detergent?

### B. Osmosis across an Artificial Semipermeable Membrane

1. Enter your data in this table.
   Time when meniscus level was marked: __________

| Time | 20% Molasses Solution, Distance Meniscus Moved | 25% Molasses Solution, Distance Meniscus Moved |
|---|---|---|
| 15 minutes | | |
| 30 minutes | | |
| 45 minutes | | |
| 60 minutes | | |

2. Was there any change in the solution level with time? Explain the forces involved.

3. Compare the movement of the 20% molasses solution to that of the 25% solution. Was the distance traveled by the two solutions predictable? Explain.

### C. Concentration and Tonicity

1. Enter your data in this table.

| Tube and Contents | Molality | Milliosmolality | Visual Appearance of RBCs | Estimated RBC Diameter (μm) |
|---|---|---|---|---|
| 1 10 g/dL NaCl | | | | |
| 2 3.5 g/dL NaCl* | | | | |
| 3 0.85 g/dL NaCl | 0.145 m | 290 mOsm | | |
| 4 0.45 g/dL NaCl | | | | |
| 5 0.20 g/dL NaCl | | | | |

*Approximately the concentration of seawater.
**Note:** *dL = 100 mL.*

2. Which solution is isotonic?______________________________

3. Which solutions are hypotonic? ______________________________

4. Which solutions are hypertonic? ______________________________

# REVIEW ACTIVITIES FOR EXERCISE 2.6

## *Test Your Knowledge of Terms and Facts*

1. Define the term *osmosis*. ______________________________

2. Describe what is meant by the term *osmotic pressure*. ______________________________

3. Define the term *isotonic*. ______________________________

4. Red blood cells in a hypertonic solution will ______________________________

5. A 0.10 M NaCl solution is ______________ (iso/hypo/hypertonic) to a 0.10 M glucose solution.

## *Test Your Understanding of Concepts*

6. Suppose a salt and a glucose solution are separated by a membrane that is permeable to water but not to the solutes. The NaCl solution has a concentration of 1.95 g per 250 mL (molecular weight = 58.5). The glucose solution has a concentration of 9.0 g per 250 mL (molecular weight = 180).

   Calculate the molality, millimolality, and milliosmolality of both solutions. State whether or not osmosis will occur and, if it will, in which direction. Explain your answer.

7. When the body needs to conserve water, the kidneys excrete a hypertonic urine. What do the terms *isotonic* and *hypertonic* mean? Since the fluid that is to become urine begins as plasma, an isotonic solution, what must happen to change it to a hypertonic urine? Explain.

8. What component of the molasses solution was osmotically active? Explain why this is true.

## Test Your Ability to Analyze and Apply Your Knowledge

9. The receptors for thirst are located in a part of the brain called the *hypothalamus*. These receptors, called *osmoreceptors*, are stimulated by an increase in blood osmolality. Imagine a man who has just landed on a desert island. Trace the course of events leading to his sensation of thirst. Can he satisfy his thirst by drinking seawater? Explain.

10. Before the invention of refrigerators, pioneers preserved meat by salting it. Explain how meat can be preserved by this procedure. (Hint: Think about what salting the meat would do to decomposer organisms, such as bacteria and fungi.)

# The Nervous System and Sensory Physiology

# Section 3

Despite changes in the external temperature, the availability of foods, the presence of toxic and threatening agents, and other influences, the internal environment of the body remains remarkably constant. The science of physiology is largely a study of the regulatory mechanisms that maintain this internal constancy *(homeostasis).* This regulation is largely a function of the nervous system and endocrine system.

To maintain homeostasis, an organism must be able to recognize specific environmental factors and make appropriate responses. At its simplest level, recognition is achieved through the stimulation of specific types of **sensory receptors,** such as cutaneous (skin) receptors. A given receptor will usually be responsive to only one *modality* (specific type) of stimulus. The rod and cone photoreceptors of the eye, for example, are stimulated by light. Each receptor transduces the particular environmental stimulus into electrical nerve impulses that then go to the specific part of the brain where the sensation is identified. This principle is known as the *law of specific nerve energies.*

The appropriate response to the environmental stimulus occurs through the neural activation of **effector organs,** which are *muscles* and *glands.* The voluntary activation of skeletal muscles by somatic motor nerves emerging from the spinal cord may produce a simple reflex action or involve more complex nervous system interaction. Autonomic motor nerves are involuntary and stimulate cardiac muscles, smooth muscles, exocrine glands (such as sweat glands and gastric glands), and some endocrine glands (the adrenal medulla, for example). Many other endocrine glands are indirectly regulated by the hypothalamus through its control of the anterior pituitary, thus wedding the endocrine system to the nervous system. Interposed between these *afferent* (sensory) and *efferent* (motor) pathways are millions of association neurons within the brain and spinal cord that integrate these activities while promoting learning and memory. The brain is the ultimate interpretation center for all sensations, including touch, pain, vision, hearing, equilibrium, and taste.

# Recording the Nerve Action Potential

EXERCISE 3.1

## Materials

1. Frogs
2. Dissecting equipment and trays, glass probes, thread
3. Oscilloscope and nerve chamber
4. Frog Ringer's solution (see Materials, exercise 5.1)

The potential difference across axon membranes undergoes changes during the production of action potentials. As the axons of a nerve produce action potentials, the surface of the nerve at this region has a difference in potential in relation to the surface of an unstimulated region. The polarity of this potential difference and its magnitude in millivolts can be seen in an oscilloscope.

### OBJECTIVES

1. Describe the resting membrane potential and the distribution of $Na^+$ and $K^+$ across the axon membrane.
2. Describe the events that occur during the production of an action potential.
3. Demonstrate the recording of action potentials in the sciatic nerve of a frog and explain how this recording is produced.

A **neuron** (nerve cell) consists of three regions that are specialized for different functions: (1) the *dendrites* receive input from sensory receptors or from other neurons; (2) the *cell body* contains the nucleus and serves as the metabolic center of the cell; and (3) the *axon* conducts the nerve impulse to other neurons or to effector organs (fig. 3.1). A bundle of axons that leaves the brain or spinal cord is known as a **peripheral nerve.**

## Membrane Potentials

If one lead of a voltmeter is placed on the surface of an axon and the other lead is placed inside the cytoplasm, a *potential difference* (or voltage) will be measured across the axon membrane. The inside of the cell about 70 millivolts negative (–70 mV) with respect to the outside. (The surface of the axon is positive with respect to the cytoplasm.) This **resting membrane potential** is maintained by the unequal distribution of ions on the two sides of the membrane. $Na^+$ is present in higher concentrations outside the cell than inside, whereas $K^+$ is more concentrated inside the cell. These differences are maintained, in part, by active transport processes (the sodium-potassium pump).

When a neuron is appropriately stimulated, the barriers to $Na^+$ are lifted and the positively charged $Na^+$ is allowed to diffuse into the cell along its concentration gradient. The flow of positive ions into the cell first eliminates the potential difference across the membrane and then continues until the polarity is actually reversed and the inside of the cell is positive with respect to the outside (about +30 mV). This phase is called **depolarization.** At this point, the barriers to $K^+$ are lifted, and the flow of this positively charged ion out of the cell along its concentration gradient helps to reestablish the resting potential (**repolarization**). During the time of repolarization, the diffusion of $Na^+$ into the cell is blocked and the neuron is refractory (not capable of responding) to further stimulation; it is in a *refractory period*. The momentary

## Textbook Correlations*

Before performing this exercise, you should study the introductory material presented here. Further information relating to this exercise can be found in these pages of *Human Physiology,* eighth edition, by Stuart I. Fox:

- *Action Potentials.* Chapter 7, pp. 161–165.
- *Conduction of Nerve Impulses.* Chapter 7, pp. 165–167.

*Multimedia Correlations (also see Appendix 3)
- A.D.A.M. *InterActive Physiology* (Nervous System I): The Action Potential (orientation, anatomy review)
- *MediaPhys 2.0*: Topics 3.27–3.34; Topics 4.4–4.22

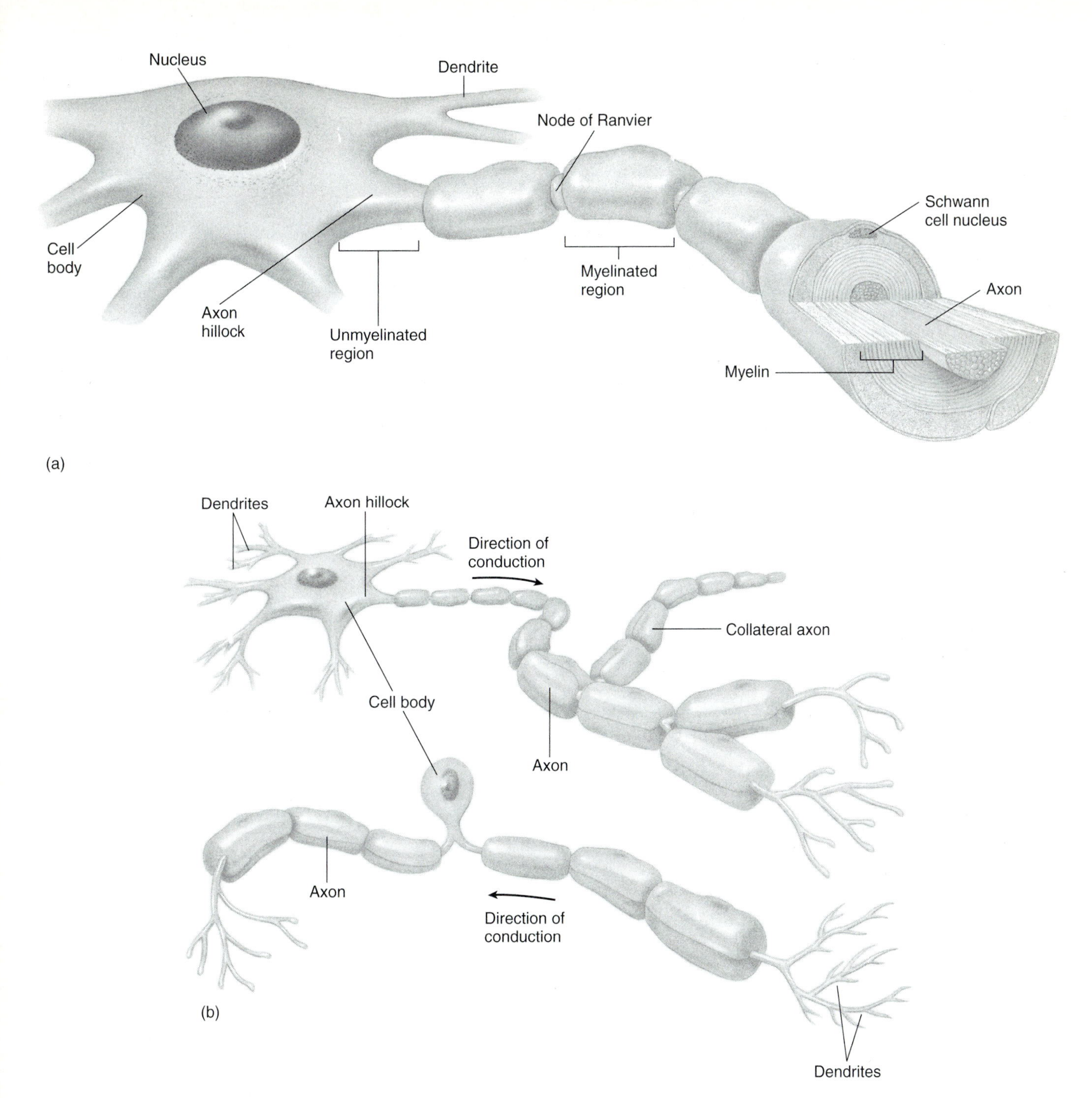


**Figure 3.1** **Neuron structure.** (a) The components of a neuron. (b) A motor and a sensory neuron, showing the pathwa conduction (arrows).

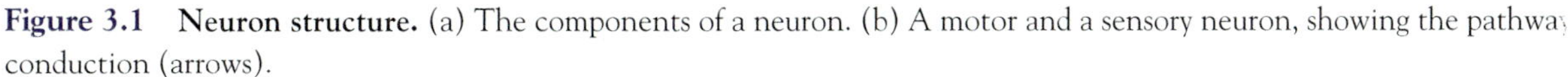

**(For a full-color version of this figure, see figs. 7.1 and 7.2 in *Human Physiology,* eighth edition, by Stuart I. Fox.)**

reversal and reestablishment of the resting potential is known as the **action potential** (fig. 3.2).

In this exercise, a frog's sciatic nerve will be dissected and placed on two pairs of electrodes. One pair of electrodes (the stimulating electrodes) will deliver a measured pulse of electricity to one point on the nerve; the other pair (the recording electrodes) will be connected to a cathode-ray tube of an oscilloscope that is adjusted to sweep an electron beam horizontally across a screen when the nerve is unstimulated. The small pulse of electricity through the stim trodes produces a small vertical deflection ning of the horizontal sweep. This ini deflection (the *stimulus artifact*) increases (height) as the strength of the stimulating creased (fig. 3.3).

When the stimulating voltage reaches level (the *threshold potential*), the region of th

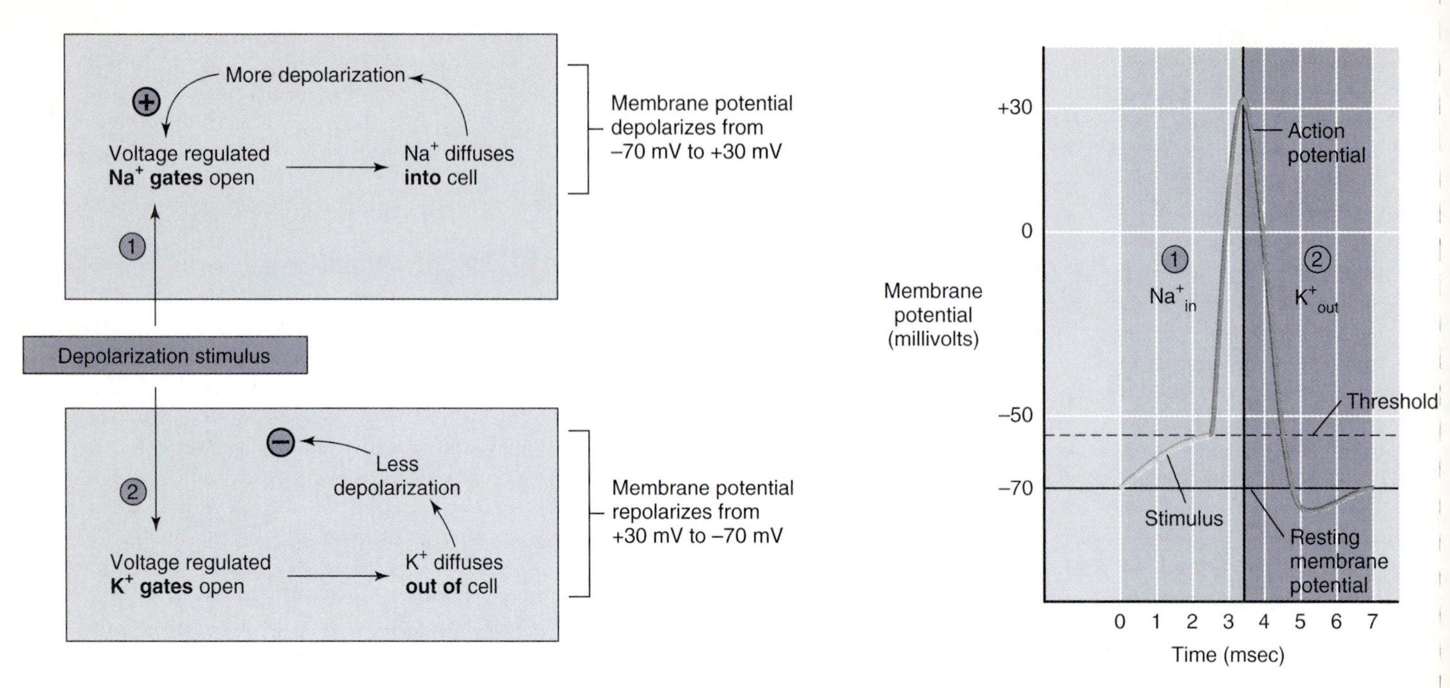


**Figure 3.2 Depolarization of an axon affects $Na^+$ and $K^+$ diffusion in sequence.** (1) $Na^+$ gates open and $Na^+$ diffuses into the cell. (2) After a brief period, $K^+$ gates open and $K^+$ diffuses out to the cell. An inward diffusion of $Na^+$ causes further depolarization, which in turn causes further opening of $Na^+$ gates in a positive feedback (+) fashion. The opening of $K^+$ gates and outward diffusion of $K^+$ makes the inside of the cell more negative, and thus has a negative feedback effect (–) on the initial depolarization.

**(For a full-color version of this figure, see fig. 7.13 in *Human Physiology,* eighth edition, by Stuart I. Fox.)**

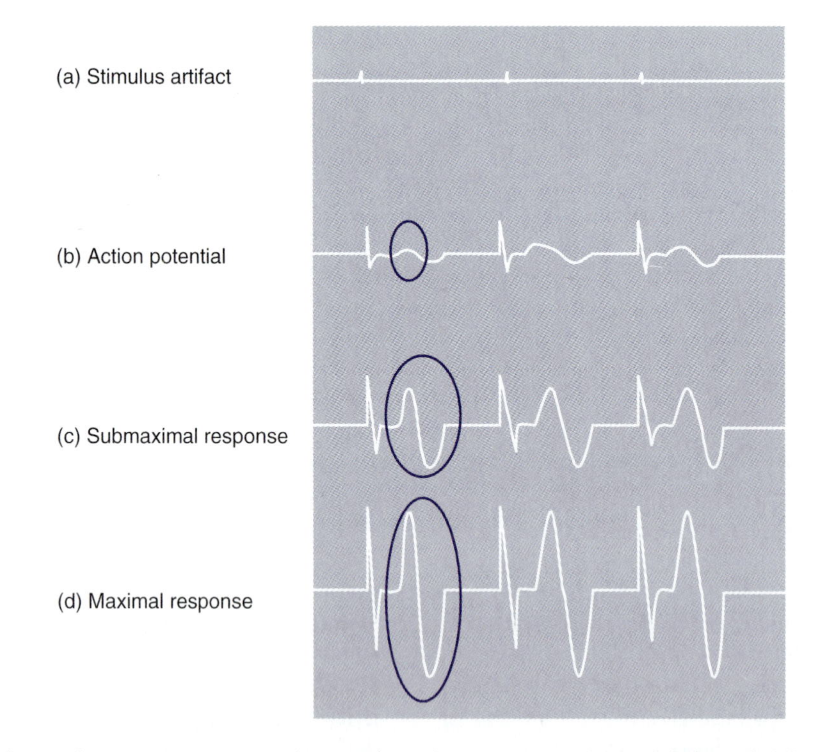


**Figure 3.3 Recording from a frog sciatic nerve.** As the stimulus voltage is increased from (a) through (d), the amplitude of the stimulus artifact and nerve action potential (circled areas) are also increased, but only to a maximum value, shown in (d). This effect is due to the fact that the sciatic nerve contains hundreds of nerve fibers. The number of individual nerve fibers stimulated to produce all-or-none action potentials increases with increasing stimulus intensity because some fibers are located closer to the stimulating electrodes than others.

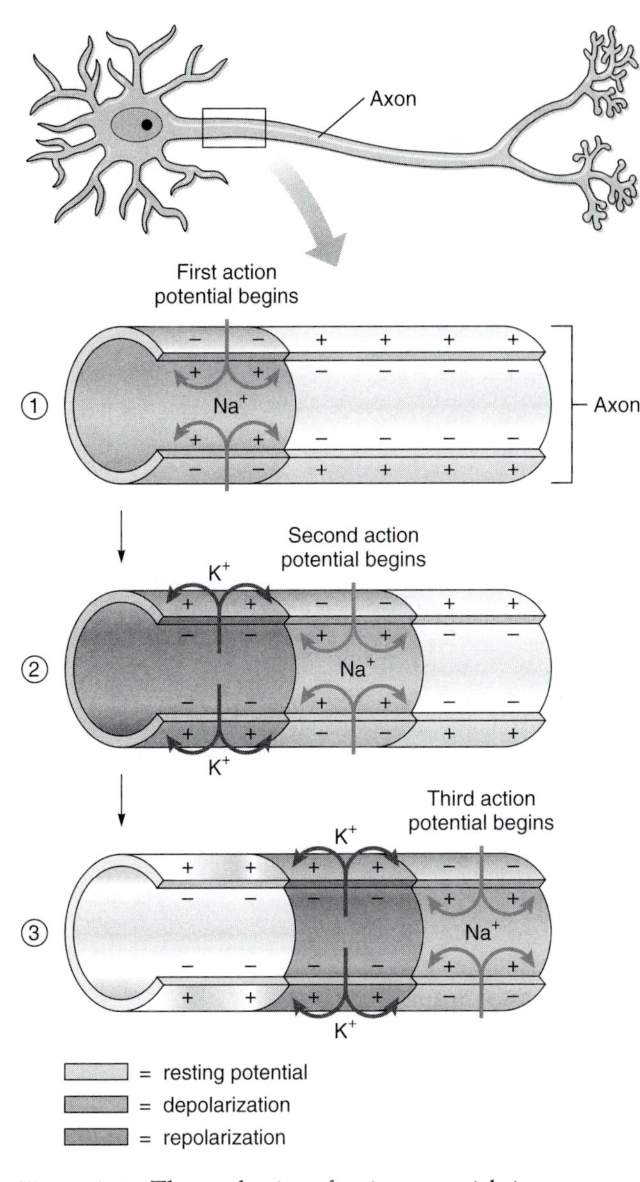


**Figure 3.4 The conduction of action potentials in an unmyelinated axon.** Each action potential "injects" positive charges that spread to adjacent regions. The region that has just produced an action potential is refractory. The next region, not having been stimulated previously, is partially depolarized. As a result, its voltage-regulated $Na^+$ gates open and the process is repeated. Successive segments of the axon thereby regenerate, or "conduct," the action potential.

**(For a full-color version of this figure, see fig. 7.16 in *Human Physiology*, eighth edition, by Stuart I. Fox.)**

to the stimulating electrodes becomes depolarized (the outside of the nerve becomes negative with respect to the inside). By the creation of "minicircuits," this region depolarizes the adjacent region of the nerve, while it (the region nearest the stimulating electrodes) is being repolarized (fig. 3.4). In this manner, a wave of "surface negativity" is conducted from the stimulating electrodes toward the two recording electrodes. When this wave reaches the first recording electrode, this electrode becomes electrically negative with respect to the second recording electrode (since both are on the surface of the nerve and the depolarization wave has not yet reached this second electrode). The potential difference between these two recording electrodes produces a vertical deflection of the electron beam a few milliseconds (msec) after the stimulus artifact. This second event is the action potential of the nerve.

The electrical activity of the nerve can be intensified by increasing either the frequency or the strength of stimulation. Increasing the frequency of stimulation will increase the number of impulses conducted by the nerve in a given time, up to a maximum amount. This maximum (impulses about 2 msec apart) is due to the refractory period of the nerve and ensures that the action potentials will remain separate events even at high frequencies of stimulation. Similarly, increasing the strength of each stimulus will increase the amplitude of each action potential, up to a maximum level. This is because the action potential recorded from a nerve is the sum of the action potentials of all stimulated axons in that nerve. As the strength (voltage) of the stimulus is increased, the number of axons that are depolarized increases, increasing the amplitude of the overall nerve response (fig. 3.3).

When impulses are recorded from individual axons, the amplitude of the action potentials does not increase with increasing strength of stimulation. A neuron either does not "fire" (to any subthreshold stimulus) or it "fires" maximally (to any suprathreshold stimulus). This is the **all-or-none law** of nerve physiology. The stronger the stimulus, the greater the number (not the size) of action potentials carried by a nerve fiber in a given time. The strength of a stimulus, therefore, is coded in the nervous system by the *frequency* (not the amplitude) of action potentials.

## Use of the Oscilloscope

Electrical activity in nerves is frequently observed by using an oscilloscope. In an oscilloscope, electrons from a cathode-ray "gun" are sprayed across a fluorescent screen, producing a line of light. Changes in the potential difference between the two recording electrodes cause this line to deflect. Movement of the line upward or downward is proportional to the incoming voltage from the nerve preparation. The oscilloscope can thus function as a fast-responding voltmeter, displaying voltage changes with time as vertical line deflections (fig. 3.3).

The electron beam is made to sweep from left to right across the screen at a particular rate. The image on the screen is a plot of voltage (*y* axis) against time (*x* axis). If the sweep of the electron beam is triggered by a stimulus to the nerve preparation, the electrical response of the nerve will always appear at the same time after the sweep has begun and at the same location on the screen. Since the phosphor in the screen continues to emit light long after it has been struck with electrons, an action potential produced in response to a second stimulus will be superimposed on the one produced in response to the previous stimulus. Therefore, an observer will see an

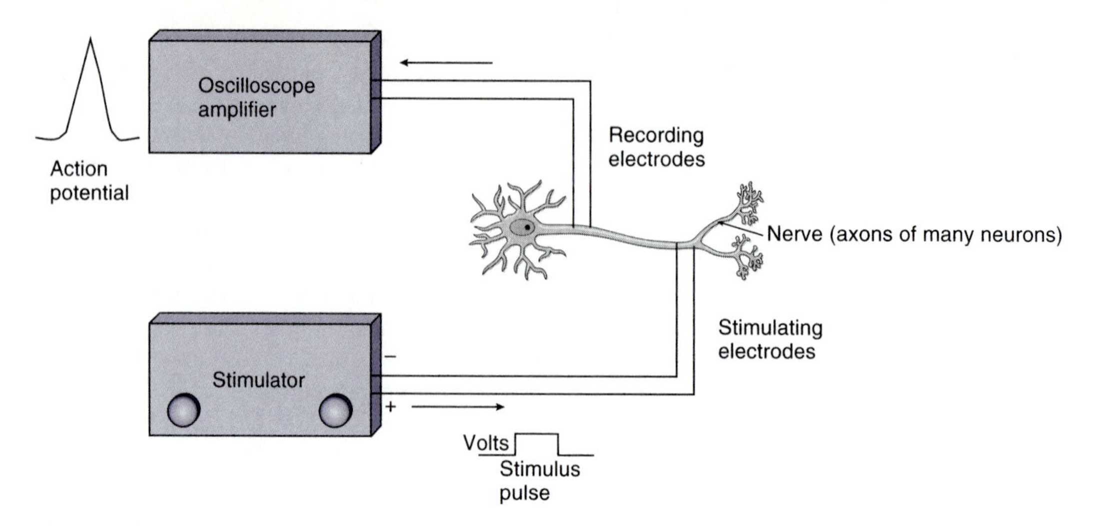


**Figure 3.5 Setup for recording from an isolated nerve.** A nerve is laid across a pair of stimulating electrodes and a pair of recording electrodes. A stimulator delivers a square-wave pulse of a given voltage and duration to the nerve via the stimulating electrodes. This can be seen at the recording electrodes as the stimulus artifact (not shown). The action potential observed after the stimulus artifact represents the response of the nerve to the stimulus.

apparently stable image of an action potential, even though each action potential lasts for only about 3 msec.

Vertical deflections of the electron beam are produced when a **potential difference,** or voltage, develops between the two *recording electrodes* that touch the nerve some distance away from a pair of *stimulating electrodes* (fig. 3.5). The first action potential, produced near the stimulating electrodes, results from a depolarizing current whose voltage, duration, and frequency (number of "shocks" per second) can be varied by the operator adjusting the oscilloscope stimulator module. Conducted by the nerve, the action potential is recreated in the region of the nerve in contact with the recording electrodes, and a vertical line deflection is observed on the oscilloscope.

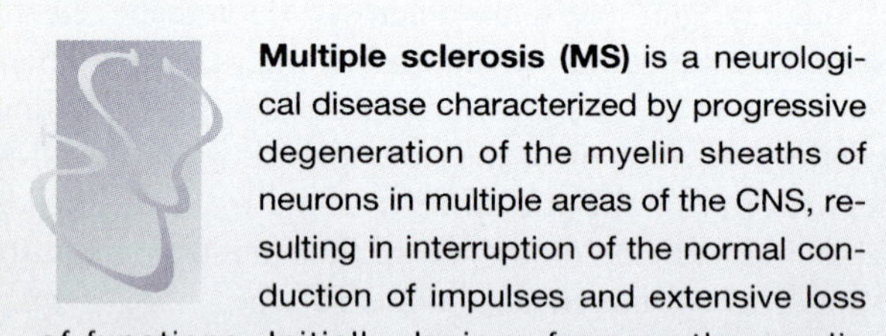

**Multiple sclerosis (MS)** is a neurological disease characterized by progressive degeneration of the myelin sheaths of neurons in multiple areas of the CNS, resulting in interruption of the normal conduction of impulses and extensive loss of functions. Initially, lesions form on the myelin sheaths and soon develop into hardened *scleroses,* or scars, effectively interfering with the formation of action potentials along the affected axons. Because myelin degeneration is widespread, MS has a wider variety of symptoms than most neurological diseases. This multiplicity of symptoms combined with a characteristic pattern of remission and relapse makes MS a disease that is exceedingly difficult to diagnose and treat.

## PROCEDURE

1. Decapitate or double-pith a frog (see section 5, pages 175–176 and figs. 5.6 and 5.7, for this procedure), skin its legs and the lower portion of its back, and place it in a prone position in a dissecting tray (fig. 3.6).
2. Make a 1-inch incision on both sides of its spine from the anal region toward the head. Lift the portion of spine free from surrounding muscle, and excise it to expose the right and left sciatic nerves, which run lateral and parallel to the spine.

**Note:** *Be careful not to touch the nerve with your fingers or metal tools; and not to let the nerve touch the surface of the frog's skin or cut muscle. Keep the nerve moist with Ringer's solution.*

3. Lift one of the sciatic nerves with a glass probe, tie it with a few inches of thread, and cut the nerve closer to the head, beyond the tie.
4. Separate the large posterior muscles of the thigh to expose the distal portion of the sciatic nerve, and lift it with a glass probe.

**Note:** *The sciatic nerve is easily identified because it runs in the same connective tissue sheath as the sciatic artery. Tie the nerve with a few inches of thread, and cut the nerve distally beyond the tie.*

5. Lift the two ends of the nerve with the two lengths of thread (fig. 3.6) and carefully free it from attached muscle and fascia.
6. Lay the nerve across the stimulating and recording electrodes of the nerve chamber, always keeping the nerve moist with Ringer's solution.

**Figure 3.6 Exposing the sciatic nerve of a frog.** Two glass probes are used to handle the nerve. The two ends have been tied with thread just beyond where they will be cut.

7. Connect the stimulating electrodes to the stimulator and the *recording* electrodes to the preamplifier of the oscilloscope.
8. Adjust the horizontal sweep; set the stimulus frequency, duration, and amplitude at their *lowest* values.
9. Slowly increase the stimulus strength until the *threshold* voltage is obtained (the lowest stimulus that produces an observed action potential). Threshold stimulus: _____ V

**Note:** *The first deflection on the screen is the stimulus artifact; the deflection to the right of the stimulus artifact is the action potential.*

10. Slowly increase the *strength* of the stimulating voltage until the action potential is at its maximum amplitude. Stimulus producing maximum response: _____ V
11. With the stimulating voltage set at a slightly suprathreshold level, gradually increase the frequency of stimulation and note this effect on the amplitude of the action potential.

# Laboratory Report 3.1

Name ______________________

Date ______________________

Section ______________________

## REVIEW ACTIVITIES FOR EXERCISE 3.1

### *Test Your Knowledge of Terms and Facts*

1. The parts of a neuron that receive input from other neurons or from sensory receptors are the ____________.
2. The voltage across the membrane of a particular axon may be –70mV. This is known as its ____________
   ____________.
3. During the action potential, the membrane polarity momentarily reverses.
   (a) This process is called ____________.
   (b) It is caused by the diffusion of ____________ into the axon.
4. The number of action potentials produced per unit time is known at its ____________;
   this is proportional to the ____________ of the stimulus.
5. What is the all-or-none law regarding action potentials? ____________
   ____________

### *Test Your Understanding of Concepts*

6. Both recording electrodes in this exercise were placed on the surface of the nerve—that is, both were extracellular. What must be done to record the resting membrane potential of an axon?

7. Describe the permeability properties of the axon membrane when the nerve is at rest, and the permeability properties during the production of an action potential. Describe the direction and nature of the ion movements during the production of an action potential.

## *Test Your Ability to Analyze and Apply Your Knowledge*

8. How is the strength of a sensory stimulus coded by a single axon? What additional way could the strength of a stimulus be coded by a collection of axons (a nerve) in the intact nervous system?

9. In this exercise, the amplitude of the action potential produced by the frog sciatic nerve increased when the stimulating voltage increased. How is this possible if the action potentials of individual axons obey the all-or-none law?

# Electroencephalogram (EEG)

EXERCISE **3.2**

## Materials

1. Oscilloscope and EEG selector box (Phipps and Bird), or physiograph and high-gain coupler (Narco), or electroencephalograph recorder (Lafayette Instrument Company)
2. EEG electrodes and surface electrode
3. Long ECG elastic band and ECG electrolyte gel
4. Alternatively, the Biopac equipment can be used for visualizing the EEG on a computer (Biopac lessons 3 and 4), available in their basic, advanced, and ultimate systems.

## Textbook Correlations

Before performing this exercise, you should study the introductory material presented here. Further information relating to this exercise can be found in these pages of *Human Physiology*, eighth edition, by Stuart I. Fox.

- *The Synapse.* Chapter 7, pp. 167–170.
- *Acetylcholine as a Neurotransmitter.* Chapter 7, pp. 170–175.
- *Cerebral Cortex.* Chapter 8, pp. 193–197.

Chemical neurotransmitters released by presynaptic nerve fibers produce excitatory or inhibitory postsynaptic potentials. These postsynaptic potentials account for most of the electrical activity of the brain and contribute to the electroencephalogram recorded by surface electrodes positioned over the brain.

## OBJECTIVES

1. Describe the structure of a chemical synapse.
2. Describe EPSPs and IPSPs and explain their significance.
3. Demonstrate the recording of an electroencephalogram and explain how an EEG is produced.

The basic unit of neural integration is the **synapse,** the functional connection between the axon of one neuron and the dendrites or cell body (occasionally even the axon) of another neuron. A train of action potentials travels down the axon of the first neuron and stimulates the release of packets (*vesicles*) of chemical transmitter substances (fig. 3.7). These chemical transmitters, such as **acetylcholine (ACh)**, diffuse across the small space separating the two neurons (the *synaptic cleft*) and reach the membrane of the second neuron (the *postsynaptic membrane*).

The interaction of these chemical transmitters with their specific receptors may stimulate a depolarization in the postsynaptic membrane. Unlike the all-or-none action potential, this **excitatory postsynaptic potential,** or **EPSP,** is a *graded* response—the larger the number of transmitter vesicles released, the greater the depolarization. When the EPSP reaches a critical level of depolarization (the threshold) it generates an action potential, which can then be conducted down the axon of the second neuron to the next synapse. Chemical transmitters released by other neurons that synapse with the second cell may produce the opposite response—a *hyperpolarization* of the postsynaptic membrane (the inside of the cell becomes even more negative with respect to the outside). This is called an **inhibitory postsynaptic potential,** or **IPSP.** The production of action potentials by the second cell, as well as their frequency, will be determined by the algebraic sum of these EPSPs and IPSPs produced by the convergence (fig. 3.8) of multiple neurons on the second cell.

These electrical activities—action potentials, EPSPs and IPSPs—are not equal in two different regions of the brain at the same time. Therefore, an extracellular potential difference (which fluctuates between 50 and 100 millionths of a volt) can be measured by placing two electrode leads on the scalp over two different regions of the brain. The recording of these "brain waves" is known as an **electroencephalogram (EEG).**

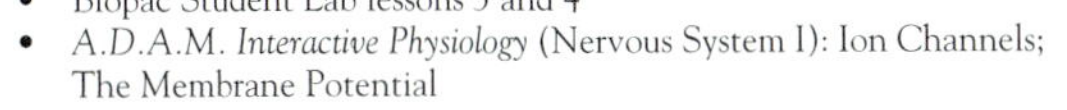

*Multimedia Correlations (also see Appendix 3)
- Biopac Student Lab lessons 3 and 4
- A.D.A.M. *Interactive Physiology* (Nervous System I): Ion Channels; The Membrane Potential

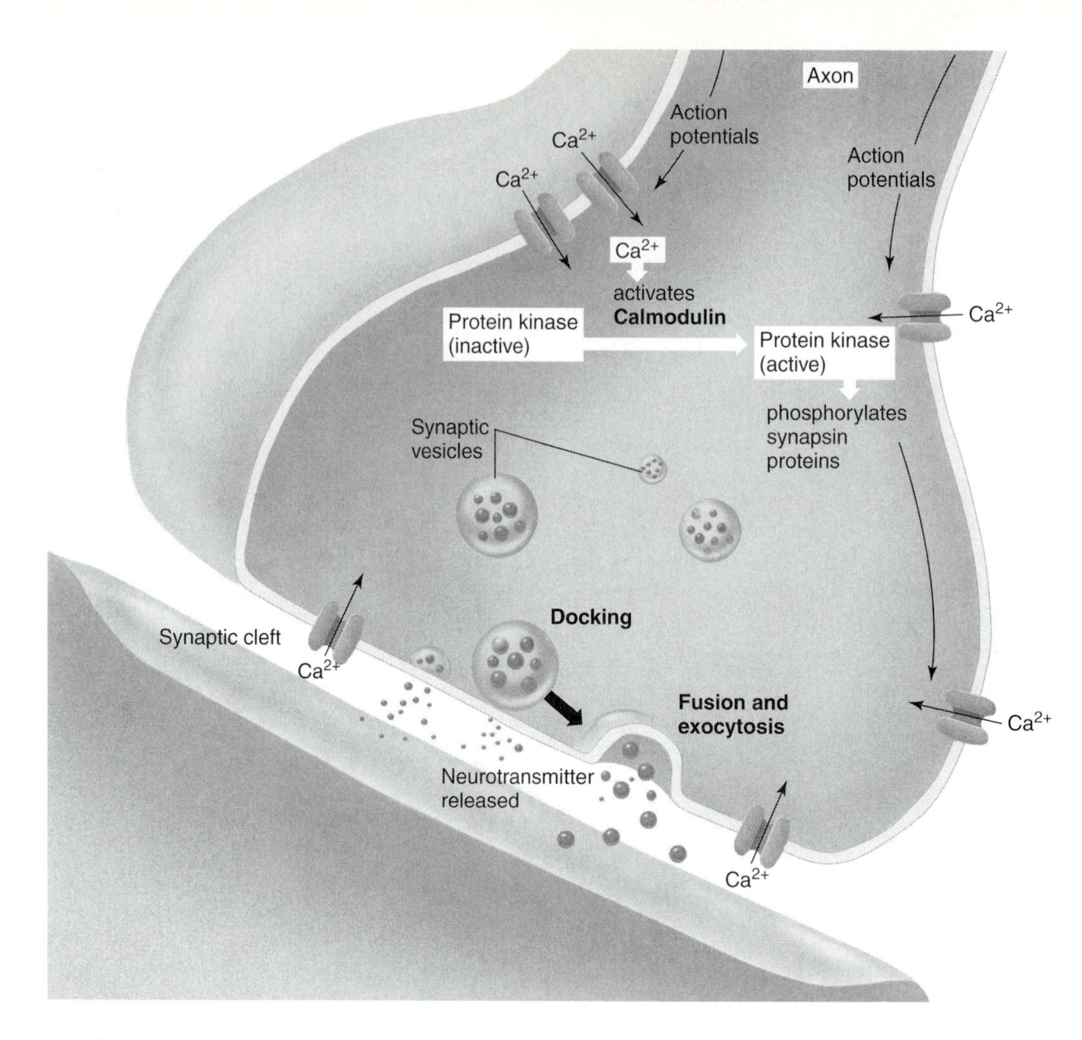


**Figure 3.7 The release of neurotransmitter.** Action potentials, by opening $Ca^{2+}$ channels, stimulate the fusion of docked synaptic vesicles with the cell membrane of the axon terminals. This leads to exocytosis and the release of neurotransmitter. The activation of protein kinase by $Ca^{2+}$ may also contribute to this process.

**(For a full-color version of this figure, see fig. 7.21 in *Human Physiology,* eighth edition, by Stuart I. Fox.)**

In actual practice, nineteen electrodes are placed at various standard positions on the scalp, with each pair of electrodes connected to a different recording pen. The record obtained reflects periodic waxing and waning of synchronous neuronal activity, producing complex waveforms that are characteristic of the regions of the brain sampled and the state of the subject (fig. 3.9).

There are four characteristic types of EEG wave patterns. **Alpha waves,** consisting of rhythmic oscillations with a frequency of 8 to 12 cycles per second (cps), were the first patterns to be characterized (fig. 3.9). These waves, which can be seen with a single pair of electrodes, are produced by the visual association areas of the parietal and occipital lobes and predominate when the subject is relaxed and awake but has eyes closed. Alpha waves can be suppressed by opening the eyes or by doing mental arithmetic, and are normally absent in a significant number of people. The alpha rhythm of children under the age of 8 occurs at a lower frequency (4–7 cps).

**Beta waves** (13–25 cps) are strongest from the frontal lobes and reflect the evoked activity produced by visual stimuli and mental activity; they are enhanced by barbiturate drugs. **Theta waves** (5–8 cps) are emitted from the temporal and occipital lobes and are common in newborn infants. In adults, theta wave recordings generally indicate severe emotional stress and can be a forewarning of a nervous breakdown. **Delta waves** (1–5 cps) seem to be emitted from the cerebral cortex. Delta waves are seen in awake infants. Common in adults during deep sleep, the presence of delta waves in an awake adult indicates brain damage. During a *petit mal epileptic seizure,* the EEG pattern may show regular spikes and waves at 3 cps. Abnormal patterns not observed in the "resting" EEG can often be revealed by stimulation of the subject through hyperventilation or flashing of a strobe light at different frequencies.

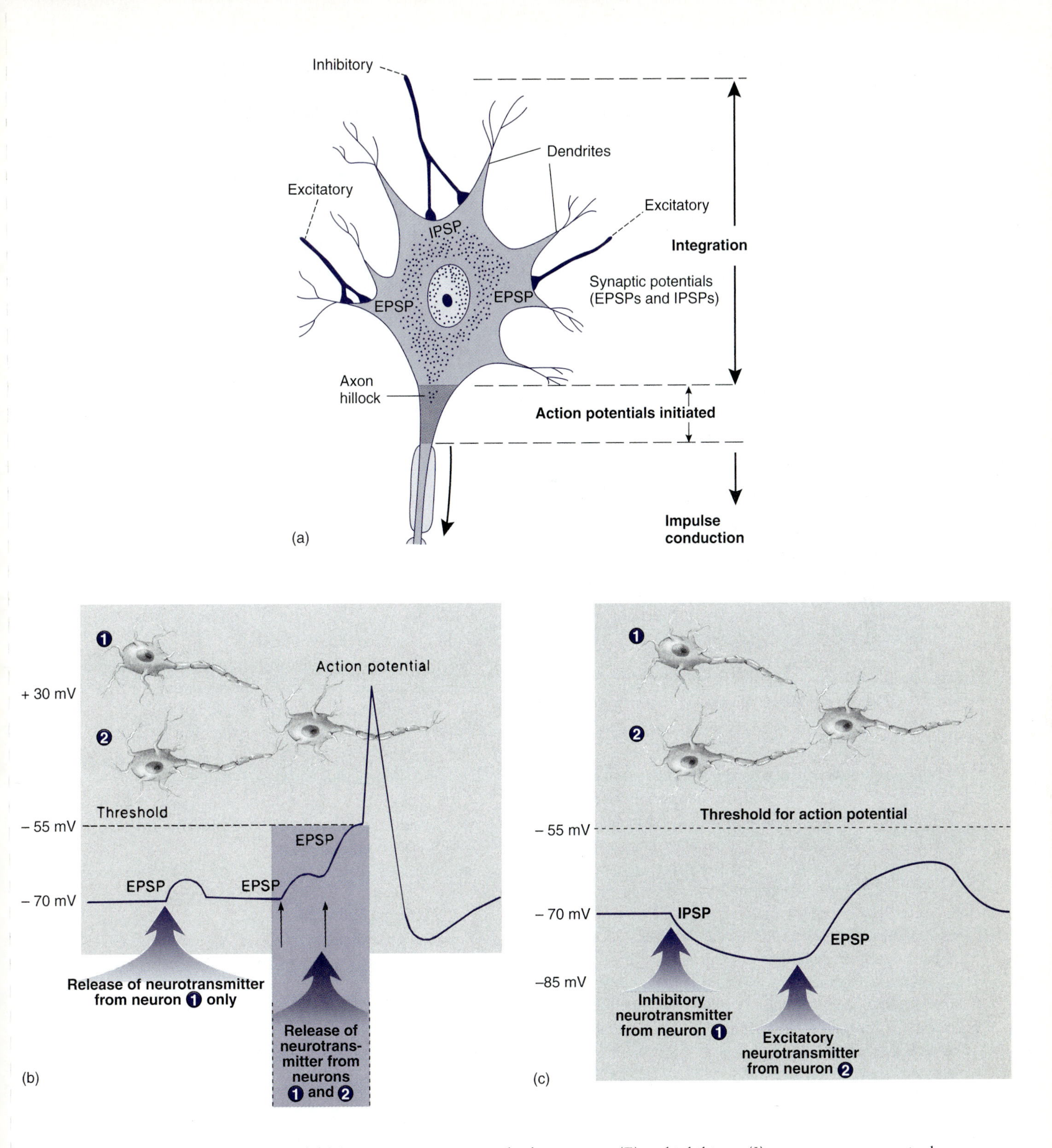


**Figure 3.8** **Synaptic integration.** (a) Many presynaptic inputs, both excitatory (E) and inhibitory (I), can converge on a single neuron. (b) Excitatory input—the depolarizations of EPSPs—from different presynaptic neurons can summate in the postsynaptic neuron. (c) Inhibitory input— the hyperpolarization of an IPSP—can also summate with EPSPs, taking the membrane potential farther from the threshold required for action potentials.

**(For a full-color version of this figure, see figs. 7.29 and 7.30 in *Human Physiology*, eighth edition, by Stuart I. Fox.)**

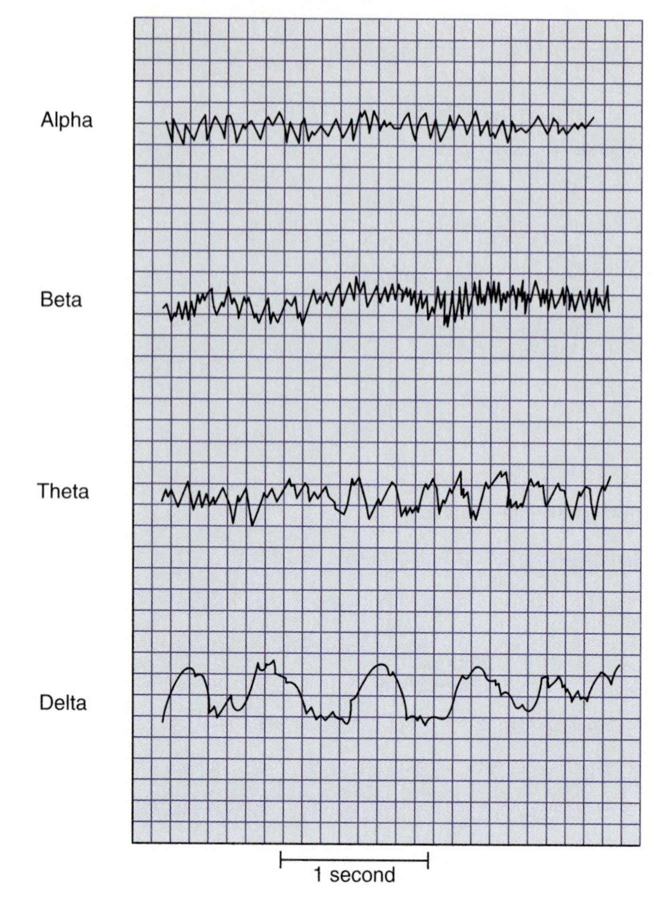


**Figure 3.9** **Electroencephalograph (EEG) rhythms.**

Use of the *electroencephalograph* may help to diagnose a number of brain lesions, including epilepsy, intracranial infections, and encephalitis. It has been discovered that meditation, as performed by Zen monks and yogis, results in the production of slow alpha waves (7 cps) of increased amplitude and regularity (even with the eyes partially open), and that this change is associated with a decrease in metabolism and sympathetic nerve activity. Many people, through biofeedback techniques, attempt to enhance their ability to produce alpha rhythms so they can relax and lower their sympathetic nerve effects.

## PROCEDURE

1. If an **oscilloscope** will be used,
   (a) connect the EEG selector box to the preamplifier of the oscilloscope, and adjust the horizontal sweep and sensitivity;
   (b) plug the lead for the forehead into the EEG selector box outlet labeled *L. Frontal* and the lead for the earlobe into the outlet for the *ground*.

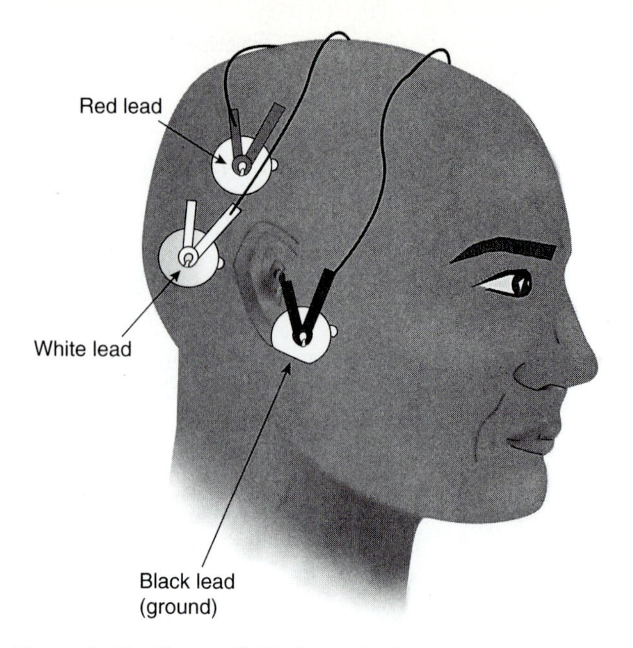


**Figure 3.10** **Biopac EEG electrode placement.** Be sure that the EEG leads are in firm contact with the scalp.

2. If a **physiograph** will be used,
   (a) insert a *high-gain coupler* into the physiograph;
   (b) set the *gain* on ×100, the *sensitivity* on 2 or 5, and the *time constant* on 0.03 or 0.3;
   (c) connect two *EEG electrodes* and one *surface electrode* to the high-gain coupler.
3. If an **electroencephalograph** will be used,
   (a) insert the cable from the electrode box/lead selector into the EEG amplifier within the recorder;
   (b) insert the electrodes into the selector box, and follow the instructions for the particular equipment model used.
4. Obtain a long elastic ECG strap and tie it around the forehead (with the knot at the back of the head) to form a snug headband.
   (a) If an **oscilloscope** will be used for recording, dab a little electrolyte gel onto the single electrode from the left frontal outlet and place it under the headband in the middle of the forehead.
   (b) If a **physiograph** will be used for recording, dab electrolyte gel onto both EEG electrodes and place them under the headband on the right and left sides of the forehead.
5. Dab a little electrolyte gel onto the ground electrode (the surface electrode for the physiograph), and with your fingers, press this electrode against the skin behind the ear.
6. If the **Biopac** system is used (fig 3.10), follow the procedural steps provided on the computer screen. You can complete their lab exercise 3 and 4 once the electrodes are in place.

**Note:** *Alpha, beta, delta, and theta waves will be recorded and stored in your student folder.*

7. With the subject in a relaxed position (no muscular movements), with his or her eyes closed, observe the EEG pattern, checking particularly for the presence of alpha waves. Since many people do not produce alpha waves in this situation, test a number of subjects.

**Note:** *Interference from room electricity at 60 cps sometimes occurs. This will appear as regular, fast, low-amplitude waves usually superimposed on the slower, more irregular brain waves of larger amplitude.*

8. Observe the effect of opening the eyes, doing mental arithmetic, and hyperventilating on the production of alpha waves.

# Laboratory Report 3.2

Name ______________________________

Date ______________________________

Section ______________________________

## REVIEW ACTIVITIES FOR EXERCISE 3.2

### Text Your Knowledge of Terms and Facts

1. A chemical released by an axon is known as a ______________________________.
2. Binding of the above-named chemical to its receptor in the postsynaptic membrane may produce a depolarization called a(n) ______________________________.
3. Different chemicals released by axons may produce a hyperpolarization of the postsynaptic membrane; such a hyperpolarization is called a(n) ______________________________.
4. Action potentials are all-or-none; synaptic potentials, by contrast, are ______________________________.

Match these terms:

____ 5. alpha rhythm (a) common in awake, tense subjects
____ 6. beta rhythm (b) observed in some relaxed subjects
____ 7. theta rhythm (c) observed in deep sleep
____ 8. delta rhythm (d) seen in some children

### Test Your Understanding of Concepts

9. Describe where an EPSP is produced, and explain how it is produced. What is its significance?

10. Compare the properties of an EPSP with those of an action potential.

11. What is the significance of IPSPs? How are they produced?

## *Test Your Ability to Analyze and Apply Your Knowledge*

12. Propose a rationale by which the observation of a person's brain waves might be used to help that person lower his or her blood pressure.

13. Which type of neural activity, action potentials or synaptic potentials, produces the EEG? Explain your answer. Speculate as to why the EEG may be influenced by hyperventilation.

# Reflex Arc

EXERCISE **3.3**

## Materials

1. Rubber mallets
2. Blunt probes

**Textbook Correlations***

Before performing this exercise, you should study the introductory material presented here. Further information relating to this exercise can be found in these pages of *Human Physiology,* eighth edition, by Stuart I. Fox:

- *Spinal Cord Tracts.* Chapter 8, pp. 209–211.
- *Cranial and Spinal Nerves.* Chapter 8, pp. 212–214.
- *Neural Control of Skeletal Muscles.* Chapter 12, pp. 347–353.

In a reflex, specific sensory stimuli evoke characteristic motor responses very rapidly because few synapses are involved. Since a specific simple reflex arc occurs at a specific spinal cord segment and involves particular nerves, tests for simple reflex arcs are very useful in diagnosing neurological disorders.

## OBJECTIVES

1. Describe the neurological pathways involved in a simple reflex arc.
2. Describe the structure and function of muscle spindles.
3. Demonstrate muscle stretch reflexes and explain the clinical significance of these tests.
4. Demonstrate a Babinski reflex (Babinski's sign) and explain the clinical significance of this test.

The speed of a motor response to an environmental stimulus depends, in part, on the number of synapses to be crossed between the afferent flow of impulses and the activation of efferent nerves. A **reflex** is a relatively simple motor response that is made without the involvement of large numbers of association neurons.

The simplest reflex requires only one synapse between the sensory and motor neurons (the **knee-jerk,** or **patellar reflex,** for example). Impulses traveling on the sensory axons enter the CNS in the *dorsal root* of the peripheral nerve, make a single synapse with a motor neuron (an alpha motoneuron) within the central gray matter, and then leave the CNS in the *ventral root* of the spinal nerve (fig. 3.11). In more complicated reflexes, the sensory impulses may travel longitudinally and transversely within the gray matter, stimulating other motor neurons. This may lead to the contraction of other flexor muscles on the same side (*ipsilateral* muscles) and the contraction of extensor muscles on the opposite side (*contralateral* muscles), while inhibiting the contraction of antagonistic muscles (ipsilateral extensors and contralateral flexors).

## A. Tests for Spinal Nerve Stretch Reflexes

In this exercise, a number of reflex arcs will be tested that are initiated by distinctive *stretch receptors* within muscles. These receptors, called **muscle spindles,** are embedded within the connective tissue of the muscle and consist of specialized thin muscle fibers *(intrafusal fibers)* that are innervated by sensory neurons (fig. 3.11). The intrafusal fibers are arranged in parallel with the normal muscle cells *(extrafusal fibers)*, so that stretch of the muscle also places tension on the intrafusal fibers. Located within the spindles, the intrafusal fibers respond to the tension by causing the stimulation (depolarization) of the sensory neuron. The sensory neuron arising from the intrafusal fiber synapses with the motor neuron in the spinal cord that, in turn, innervates the extrafusal fibers. The resultant

*Multimedia Correlations (also see Appendix 3)
- Intelitool: Flexicomp

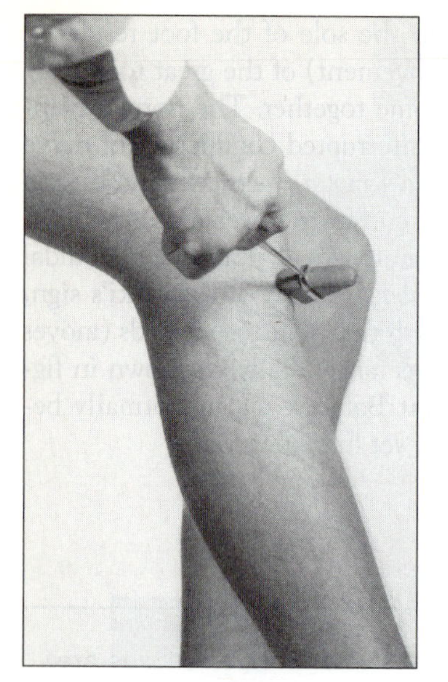

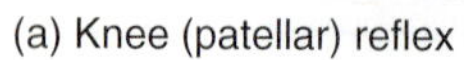

(a) Knee (patellar) reflex

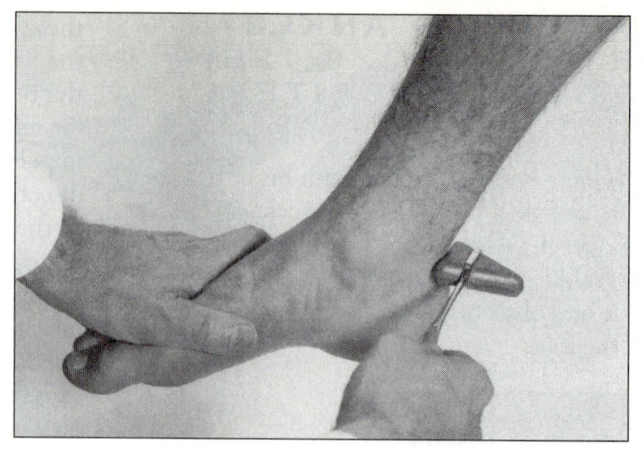

(b) Ankle (Achilles) reflex

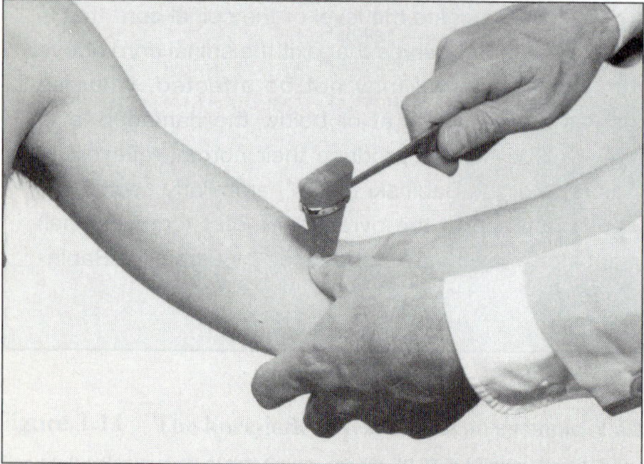

(c) Biceps reflex

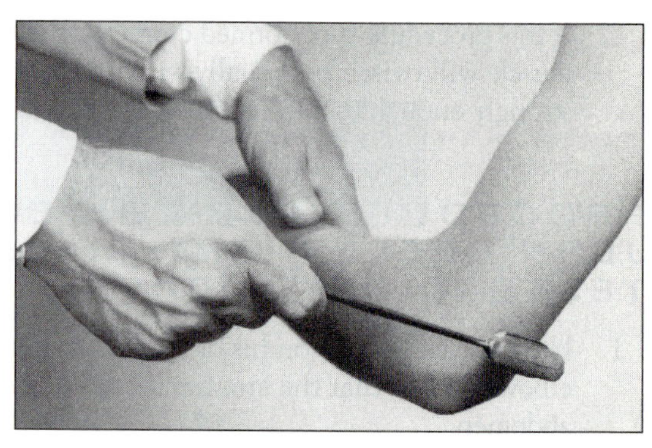

(d) Triceps reflex

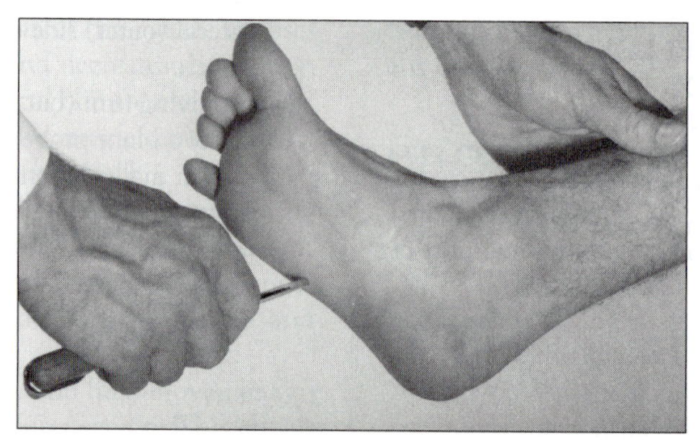

(e) Plantar reflex

**Figure 3.12** Some reflexes of clinical importance.

# Laboratory Report 3.3

Name ______________________________

Date ______________________________

Section ______________________________

## REVIEW ACTIVITIES FOR EXERCISE 3.3

### *Test Your Knowledge of Terms and Facts*

1. Afferent neurons are ______________________________.
2. Efferent neurons are ______________________________.
3. Muscle spindles are receptors sensitive to ______________________________.
4. A muscle spindle is composed of several ______________________________.
5. How many synapses are crossed in a single reflex arc during a muscle stretch reflex? ____________
6. The ventral root of spinal nerves contain ______________________ neurons, whereas the dorsal root contains ______________________ neurons.
7. The test you performed that involves the stimulation of ascending and descending spinal cord tracts: ______________________________.

Match these tests with the nerve involved:

____ 8. biceps jerk (a) femoral nerve
____ 9. triceps jerk (b) musculocutaneous nerve
____10. knee jerk (c) medial popliteal nerve
____11. ankle jerk (d) radial nerve

### *Test Your Understanding of Concepts*

12. Describe the sequence of events that occurs from the time the patellar tendon is stretched to the time the leg is extended (knee-jerk reflex).

13. Compare the neural pathway involved in a muscle stretch reflex with that of the plantar (Babinski) reflex.

## Table 3.1 Cutaneous Receptors

| Receptor | Structure | Sensation | Location |
| --- | --- | --- | --- |
| Free nerve endings | Unmyelinated dendrites of sensory neurons | Light touch; hot; cold; nociception (pain) | Around hair follicles; throughout skin |
| Merkel's discs | Expanded dendritic endings | Sustained touch and pressure | Base of epidermis (stratum basale) |
| Ruffini corpuscle (endings) | Enlarged dendritic endings within open, elongated capsule | Sustained pressure | Deep in dermis and hypodermis |
| Meissner's corpuscles | Dendrites encapsulated in connective tissue | Changes in texture; slow vibrations | Upper dermis (papillary layer) |
| Pacinian corpuscles | Dendrites encapsulated by concentric lamellae of connective tissue structures | Deep pressure; fast vibrations | Deep in dermis |

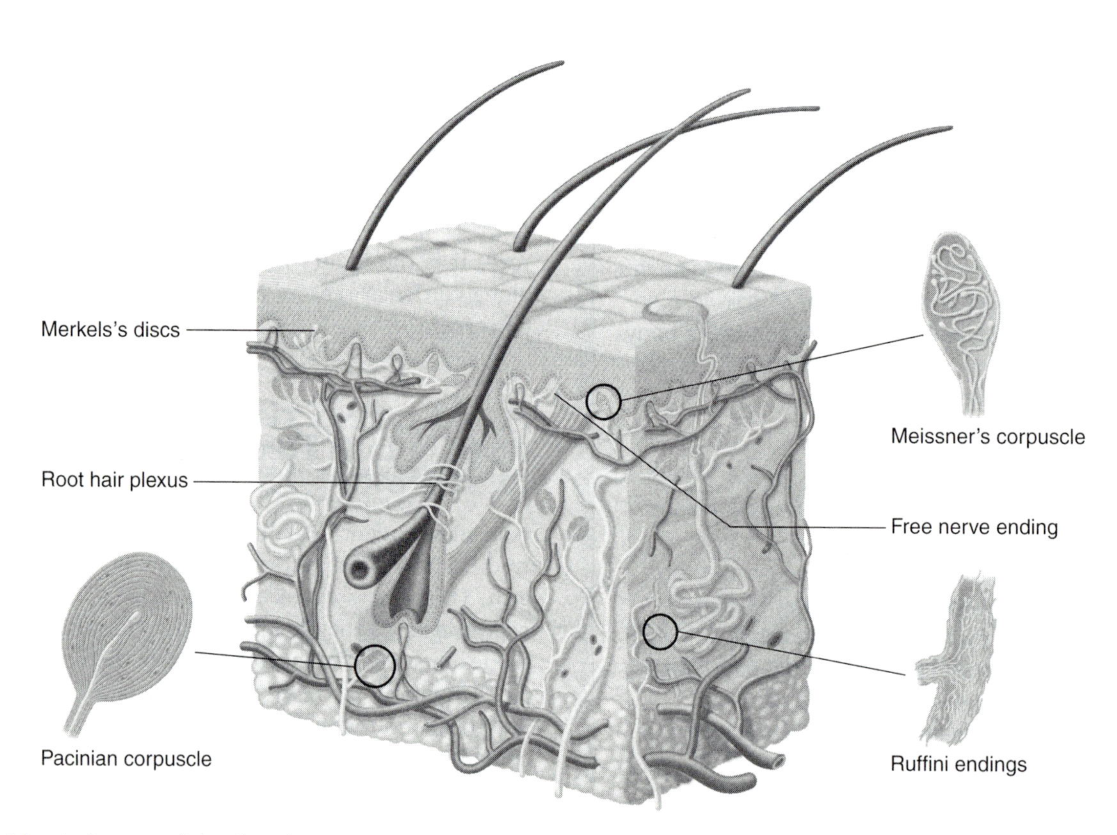


**Figure 3.13** A diagram of the skin showing cutaneous receptors.

(For a full-color version of this figure, see fig. 10.4 in *Human Physiology,* eighth edition, by Stuart I. Fox.)

## PROCEDURE

1. With a ballpoint pen, draw a square (2 cm per side) on the ventral surface of the subject's forearm. Alternatively, a square ink stamp may be used.
2. With the subject's eyes closed, gently touch a dry, ice-cold, metal rod to different points in the square. Mark the points of cold sensation with a dark dot.
3. With the subject's eyes closed, gently touch a dry, warm, metal rod (heated to about 45° C in a water bath) to different points in the square. Mark the points of warm sensation with an open circle.
4. Gently touch a thin bristle to different areas of the square and indicate the points of touch sensation with small *x*'s.
5. Reproduce this map in your laboratory report.

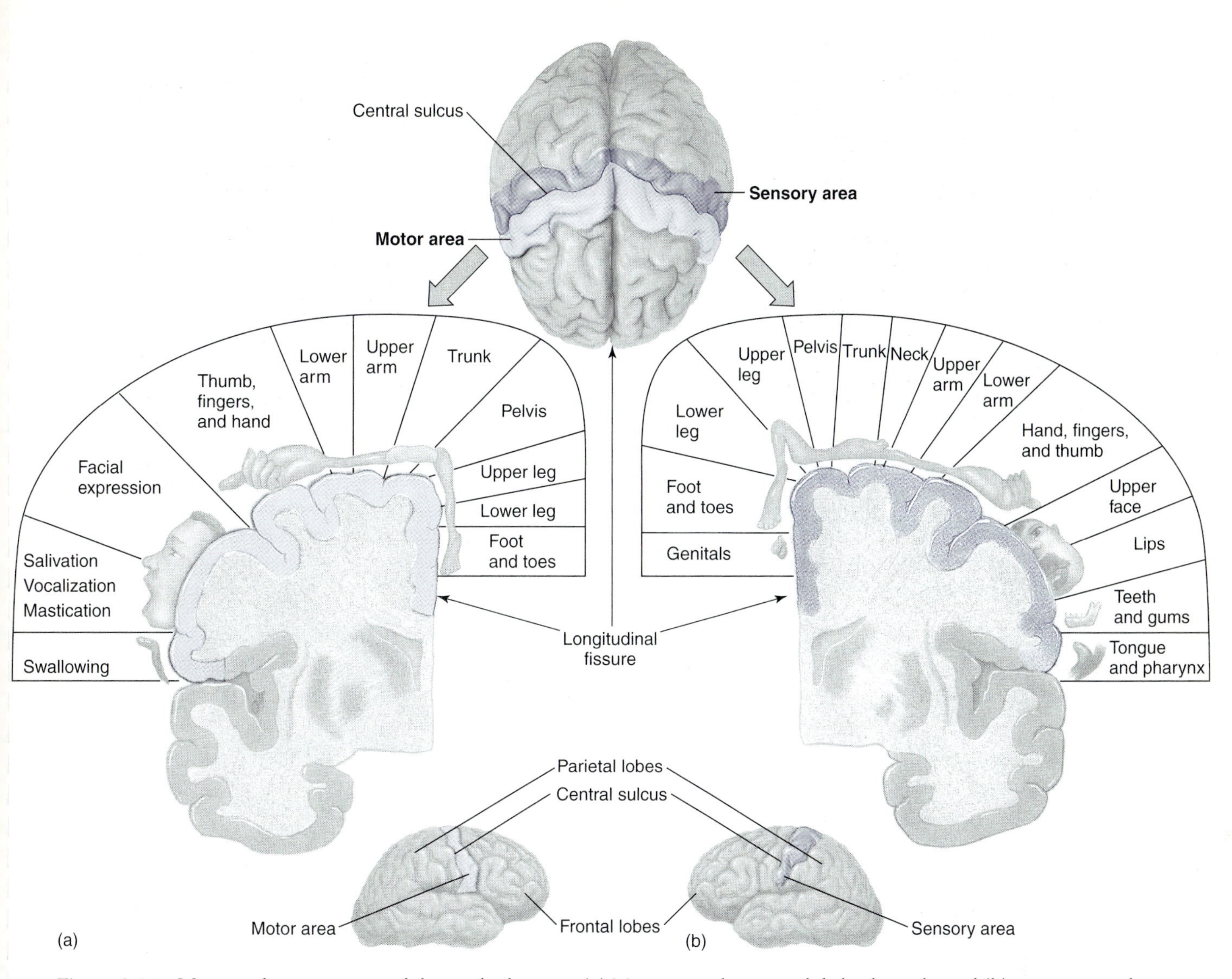


**Figure 3.14 Motor and sensory areas of the cerebral cortex.** (a) Motor areas that control skeletal muscles and (b) sensory areas that receive somatesthetic sensations.

**(For a full-color version of this figure, see fig. 8.7 in *Human Physiology,* eighth edition, by Stuart I. Fox.)**

## B. The Two-Point Threshold in Touch Perception

The density of touch receptors in some parts of the body is greater than in other parts. Therefore, the areas of the **sensory cortex** (*postcentral gyrus* of the central fissure, fig. 3.14*b*) that correspond to different regions of the body are of different sizes. Those areas of the body that have the largest density of touch receptors also receive the greatest motor innervation; the areas of the **motor cortex** (*precentral gyrus* of the central fissure, fig. 3.14*a*) that serve these regions are correspondingly larger than other areas. Therefore, a map of the sensory and motor areas of the brain reveals that large areas are devoted to the touch perception and motor activity of the face (particularly the tongue and lips) and hands, whereas relatively small areas are devoted to the trunk, hips, and legs.

The density of touch receptors is measured by the **two-point threshold test.** The two points of a pair of adjustable calipers are simultaneously placed on the subject's skin with equal pressure, and the subject is asked whether two separate points of contact are felt. If the answer is yes, the points of the caliper are brought closer together, and the test is repeated until only one point of contact is felt. The minimum distance at which two points of contact can be felt is the two-point threshold.

Sensory information from the cutaneous receptors projects to the **postcentral gyrus** of the cerebral cortex. Therefore, direct electrical stimulation of the post-central gyrus produces the same sensations as those felt when the cutaneous receptors are stimulated. Much of this information has been gained by the electrical stimulation of the brain of awake patients undergoing brain surgery; the surgeon must often map the areas of the brain in order to locate the site of the lesion and avoid damage to healthy tissue. Since the cutaneous receptors are more densely arranged in the face, tongue, and hands than on the back and thighs, larger areas of the brain are involved in analyzing information from the former areas than from the latter. Consequently, the areas of the brain map representing the face, tongue, and hands are larger than those representing the back and thighs. The map is also upside down, with the feet represented near the superior surface and the head represented more inferiorly and laterally in the cortex (fig. 3.14).

### PROCEDURE

1. Starting with the calipers wide apart and the subject's eyes closed, determine the two-point threshold on the back of the hand. (Randomly alternate the two-point touch with one-point contacts, so that the subject cannot anticipate you.)
2. Repeat this procedure with the palm of the hand, fingertip, and back of the neck.
3. Write the minimum distance (in mm) in the data table provided in your laboratory report.

## C. Adaptation of Temperature Receptors

Many of our sense receptors respond strongly to acute changes in our environment and then stop responding when these stimuli become constant. This phenomenon is known as **sensory adaptation.** Our sense of smell, for example, quickly adapts to the odors of the laboratory; and our touch receptors soon cease to inform us of our clothing, until these stimuli change. Sensations of pain, by contrast, adapt little if at all.

When one hand is placed in warm water and another in cold water, the strength of stimulation gradually diminishes until both types of temperature receptors have adapted to their new environmental temperature. If the two hands are then placed in water at an intermediate temperature, the hand that was in the cold water will feel warm, and the hand that was in the warm water will feel cold. The "baseline," or "zero," of the receptors has obviously changed. The sensations of temperature are therefore not absolute but relative to the baseline previously established by sensory adaptation.

### PROCEDURE

1. Place one hand in warm water (about 40° C) and the other in cold water, and leave them in the water for a minute or two (remove them if the water becomes too uncomfortable).
2. Now place both hands in lukewarm water (about 22° C) and record your observations and conclusions regarding your sensations in the laboratory report.

## D. Referred Pain

Receptor organs are sensory transducers, changing environmental stimuli into afferent nerve impulses (action potentials). Since the action potentials in one nerve are the same as those in another, the perception of the sensation is determined entirely by the area of the brain stimulated, which is different for each sensory nerve. Although a given sensory nerve is normally stimulated by a specific receptor, trauma to the nerve along the afferent pathway may also evoke action potentials, and this will be interpreted by the brain as the normal sensation. An example of traumatic stimulation would be seeing flashes of light or "stars" when punched in the eye.

Amputees frequently report feelings of pain in their missing limbs as if they were still there; this is part of the **phantom limb phenomenon.** The source of nerve stimulation is trauma to the cut nerve fibers, yet the brain perceives the pain as coming from the amputated region of the body that had originally produced the action potentials along these nerves. This is a **referred pain** because the source of nerve stimulation is different from the perceived location of the stimulus.

Referred pains are important clinically, particularly for deep visceral (organs within the abdominal or thoracic cavities) pain that is characteristically dull and poorly localized. In ischemic heart disease, for example, the pain is referred to the left pectoral region and left arm and shoulder areas (fig. 3.15*a*); this is called **angina pectoris.** In many patients with stomach ulcers, the pain is referred to the region between the scapulae of the back. In general, the deep pain is referred to a surface location served by nerves from the same segmental level of the spinal cord (fig. 3.15*b*).

### PROCEDURE

1. Gently tap the ulnar nerve where it crosses the median epicondyle of the elbow.
2. Describe the locations where you perceive tingling or pain.

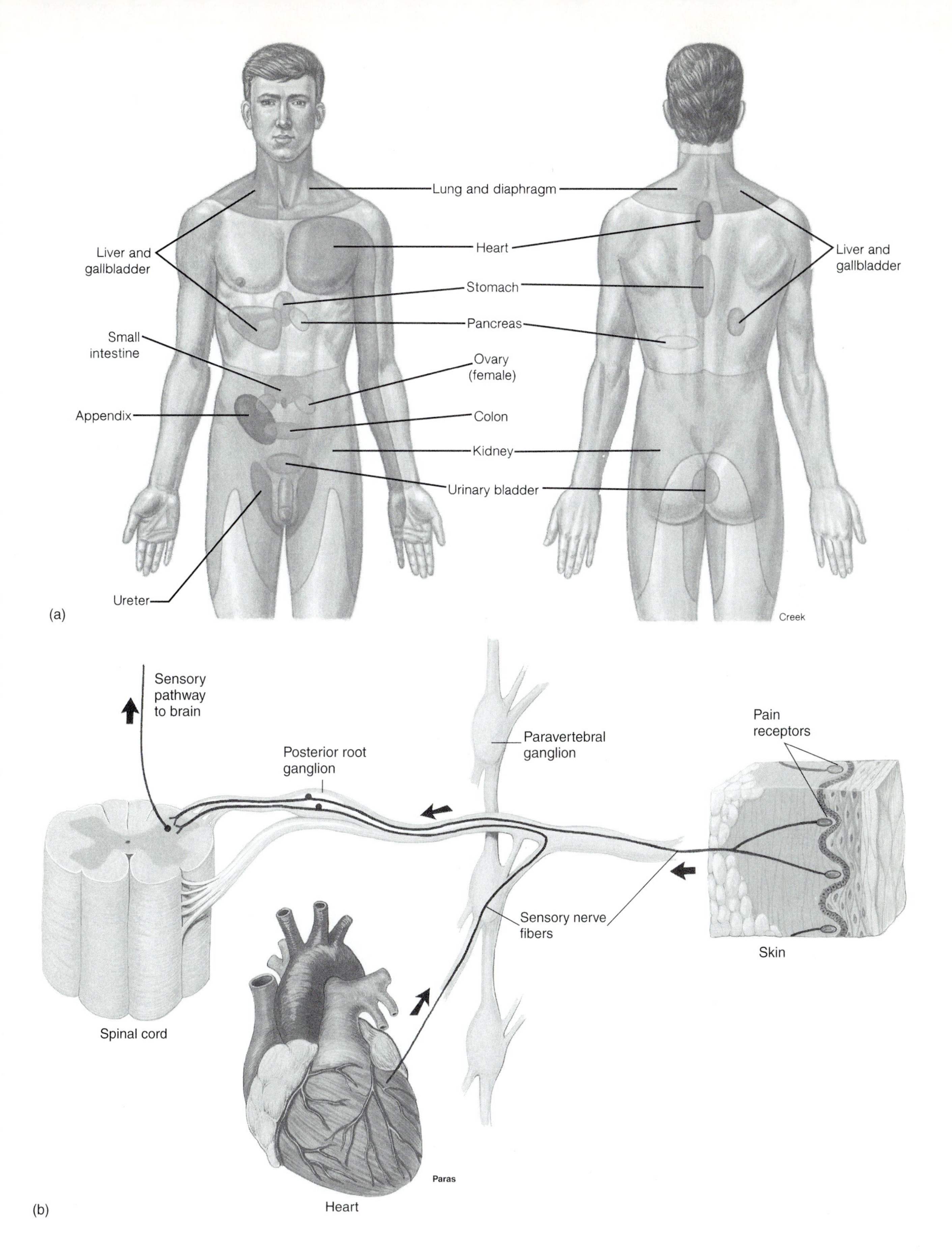


**Figure 3.15 Referred pain.** (a) Sites of referred pain are perceived cutaneously but actually originate from specific visceral organs. (b) One explanation why pain from a visceral organ, such as the heart, might be perceived over a particular area of skin. In this scenario, sensations from the two regions share a common nerve pathway.

# Laboratory Report 3.4

Name ______________________

Date ______________________

Section ______________________

## DATA FROM EXERCISE 3.4

### A. Mapping the Temperature and Touch Receptors of the Skin

1. In this box, reproduce the map of the hot, cold, and touch receptors from the square on your forearm.

### B. The Two-Point Threshold in Touch Perception

1. Write your results in this data table.

| Location | Two-Point Threshold (mm) |
|---|---|
| Back of hand | |
| Palm of hand | |
| Fingertip | |
| Back of the neck | |

2. Compare your results to the average values provided in chapter 10, table 10.3 in the textbook *Human Physiology*, eighth edition, by Stuart I. Fox.

### C. Adaptation of Temperature Receptors

1. Record your observations and conclusions here.

### D. Referred Pain

1. Describe the locations where tingling or pain was felt. Was this feeling perceived to be in a different location than where the mallet was struck? If so, where?

# REVIEW ACTIVITIES FOR EXERCISE 3.4

## Test Your Knowledge of Terms and Facts

Match the receptor with the sensation with which it is most associated:

____ 1. free nerve endings (a) light touch; hot and cold

_____2. Ruffini corpuscle (b) sustained touch and pressure

_____3. Pacinian corpuscle (c) sustained pressure

_____4. Meissner's corpuscle (d) changes in texture; slow vibrations

_____5. Merkel's discs (e) deep pressure; fast vibrations

6. The motor cortex is the ____________________ gyrus of the cerebral cortex; the sensory cortex is the ____________________ gyrus.
7. Define *sensory adaptation*. ____________________________________________
8. Name a sensory modality that adapts quickly: ____________________;
   name one that adapts slowly, if at all: ____________________.
9. Angina pectoris is an example of a(n) ____________________ pain.
10. Pain that is perceived in a limb that has been amputated is known as the ________________________________
    ____________________________________________.

## Test Your Understanding of Concepts

11. Which parts of your body have the highest density of touch receptors? What benefits may be derived from that fact?

12. What does the map of the sensory cortex reveal about the density of touch receptors? Explain.

13. How did your right and left hands feel when placed in the same lukewarm water bath? Explain how this occurred.

14. Describe the importance of referred pain in the diagnosis of deep visceral pain and give examples.

## *Test Your Ability to Analyze and Apply Your Knowledge*

15. Describe the map of the motor cortex. How does it compare with the map of the sensory cortex? What does the map of the motor cortex reveal about motor control?

16. "Our perceptions of the external world are created by our brains." Discuss this concept, using the phantom limb phenomenon to support your argument.

# Eyes and Vision

EXERCISE **3.5**

## Materials

1. Snellen eye chart and astigmatism chart
2. Wire screen and meter stick
3. Ophthalmoscope
4. Lamp
5. Red, blue, and yellow squares on larger sheets of black paper or cardboard
6. Ishihara color blindness cards

### Textbook Correlations*

Before performing this exercise, you should study the introductory material presented here. Further information relating to this exercise can be found in these pages of *Human Physiology,* eighth edition, by Stuart I. Fox:

- *The Eyes and Vision.* Chapter 10, pp. 261–268.
- *Retina.* Chapter 10, pp. 268–275.

The elastic properties of the eye lens allow its refractive power to be varied so that the image of an object from almost any distance can be focused properly on the retina. Photoreceptors—rods and cones—are located in the retina. The refractive abilities of the eye and the functions of its inner structures are routinely tested in eye examinations.

### OBJECTIVES

1. Describe the structure of the eye and the functions of its component parts.
2. Test for visual acuity and accommodation and describe common refractive problems.
3. Identify the extrinsic eye muscles and describe their functions.
4. Describe the optic disc and fovea centralis and explain their significance.
5. Demonstrate the presence of a blind spot and explain why light focused on this spot cannot be seen.

You should be familiar with the gross structure of the eye (fig. 3.16). The eye has three walls, or tunics, that form an *outer fibrous layer* (the **sclera** and **cornea**), a middle vascular layer (the **choroid**), and an inner layer (the **retina**). The **lens,** suspended by a *suspensory ligament* attached to the **ciliary body,** divides the eye into anterior and posterior *cavities*. Filled with *aqueous humor*, the anterior cavity is further divided into anterior and posterior *chambers* by the colored **iris.** The iris is a muscular diaphragm that regulates the entry of light into the eye through its aperture *(pupil)*. The semigelatinous *vitreous humor* (or *vitreous body*) occupies the posterior chamber and lends structural support to the eye.

## A. Refraction: Test for Visual Acuity and Astigmatism

Light rays are bent *(refracted)* when they pass from air to a medium of greater density, where their rate of transmission is slower. The light rays that diverge from an object in the visual field are refracted by the cornea, aqueous humor, lens, and vitreous humor of the eye so that the rays converge (are focused) on the retina and form an inverted image, reversed from left to right (fig. 3.17).

The refractive power of the cornea and vitreous humor is constant. The strength of the lens (i.e., its ability to refract light) can be varied by making it more or less convex. The greater the degree of convexity, the greater the strength of the lens (i.e., the greater the ability to bring parallel rays of light to a focus). A lens that brings light to a focus 0.25 m from its center is stronger (more convex) than a lens that brings light to a focus 1 m from its center. The strength of a lens is expressed in **diopters.**

$$\text{Strength (diopters)} = \frac{1}{\text{focal length (meters)}}$$

*Multimedia Correlations (also see Appendix 3)
- *MediaPhys 2.0*: Topics 6.42–6.47

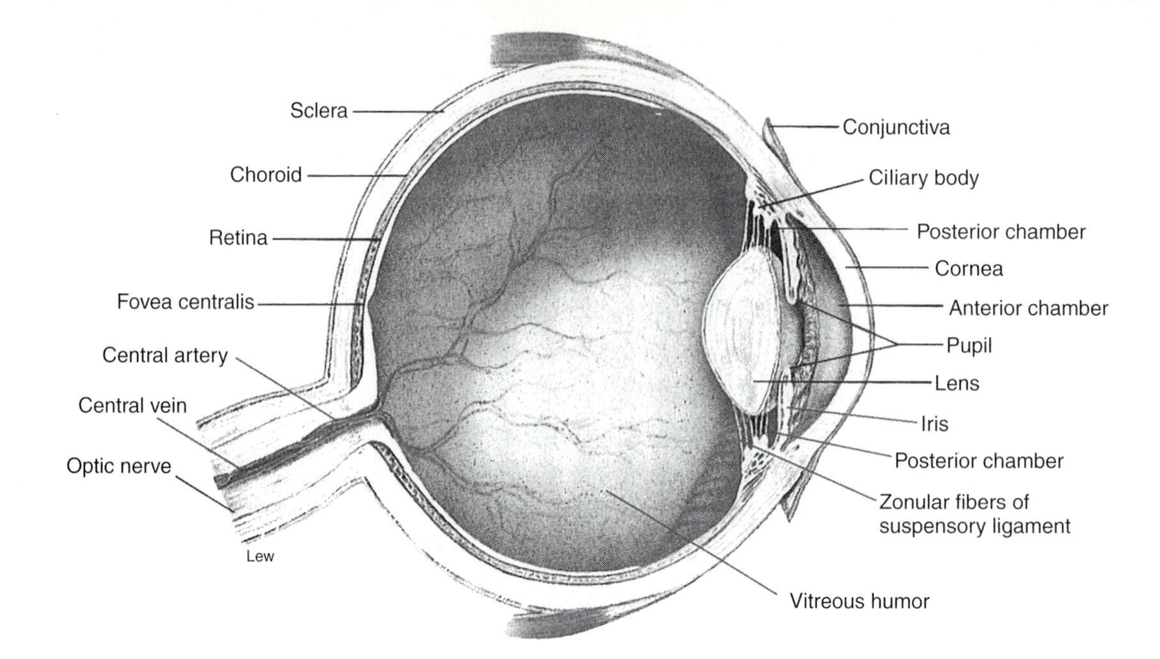


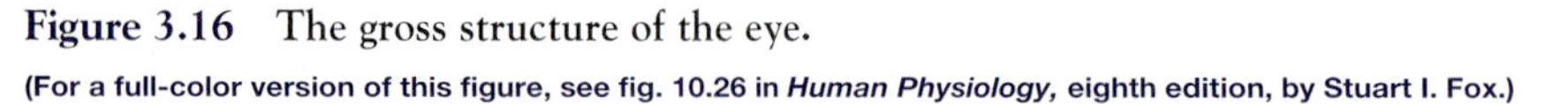

**Figure 3.16** The gross structure of the eye.

**(For a full-color version of this figure, see fig. 10.26 in *Human Physiology,* eighth edition, by Stuart I. Fox.)**

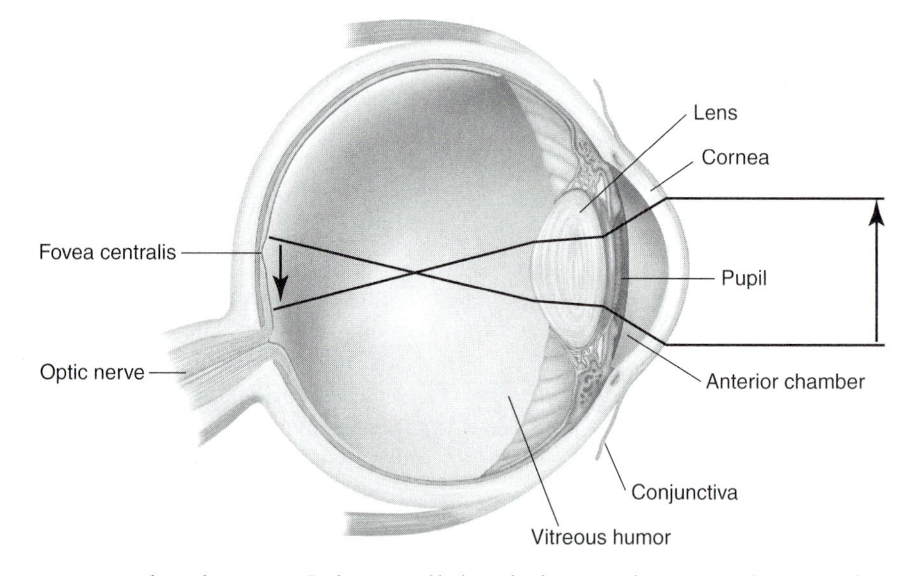


**Figure 3.17 The image is inverted on the retina.** Refraction of light, which causes the image to be inverted, occurs to the greatest degree at the air-cornea interface. Changes in the curvature of the lens, however, provide the required fine focusing adjustments.

**(For a full-color version of this figure, see fig. 10.30 in *Human Physiology,* eighth edition, by Stuart I. Fox.)**

The lens that brings parallel waves of light to a focus 0.25 m from its center has a *focal length* of 0.25 m and a strength of 4 diopters, whereas the lens that brings light to a focus 2 m from its center has a focal length of 2 m and a strength of 0.5 diopters. The refractive power of the normal eye when an object is 20 feet or more away is 67 diopters.

When the light rays that diverge from two adjacent points in the visual field are each brought to a perfect focus on the retina, two points will clearly be perceived. If, however, the light rays converge on a point in front of or behind the retina, the two points in the visual field will be perceived as one fuzzy or blurred point. To correct this defect in **visual acuity** (sharpness of vision), the individual must either adjust the distance between the eye and the object or wear corrective lenses that change the degree of refraction.

When a distant object (20 feet or more) is brought to a focus in front of the retina, the individual is said to have **myopia** *(nearsightedness)*. Myopia is usually due to an elongated eyeball (excessive distance from lens to retina) and is corrected by a concave lens. In the opposite condition, **hyperopia** (hypermetropia, or *farsightedness*), the image is brought to a focus behind the retina. This condition is usually due to an eyeball that is too short. In this case, an increase in refractive power is needed, and a convex lens is used. Normal visual acuity is called **emmetropia.**

Visual acuity is frequently tested by means of the **Snellen eye chart.** A person with normal visual acuity can read the line marked 20/20 from a distance of 20 feet. An individual with 20/40 visual acuity must stand 20 feet away from a line that a normal person can read at 40 feet. An individual with 20/15 visual acuity can read a line at a distance of 20 feet that the average, normal young adult could not read at a distance greater than 15 feet. The person with 20/40 vision has myopia, but the person with 20/15 vision does not necessarily have hyperopia. The farsighted person has a decreased ability to see near objects but cannot see distant objects any better than a person with normal vision.

An **astigmatism** is a visual defect produced by an abnormal curvature of the cornea or lens, or by an irregularity in their surface. Because of this abnormality, the refraction of light rays in the horizontal plane is different from the refraction in the vertical plane. At a given distance from an astigmatism chart, therefore, lines in the visual field oriented in one plane will be clear, while lines oriented in the other plane will be blurred. Astigmatism is corrected by means of a cylindrical lens.

The strength of corrective lenses prescribed is given in diopters, preceded by either a plus sign (convex lens for hyperopia; e.g., +14 diopters) or a minus sign (concave for myopia; e.g., –5 diopters). The correction for astigmatism indicates both the strength of the cylindrical lens (e.g., +2) and the axis of the defect (90° for vertical plane, 180° for horizontal plane). A correction for both myopia and astigmatism may be indicated, for example, as –3 + 2 axis 180°.

## PROCEDURE

1. Stand 20 feet (6 m) from the *Snellen eye chart*. Covering one eye, attempt to read the line with the smallest letters you can see (with glasses off, if applicable). Walk up to the chart and determine the visual acuity of that eye.
2. Repeat this procedure using the other eye (with glasses off, if applicable).
3. Repeat this procedure for each eye with glasses on (if applicable).
4. Stand about 20 feet away from an *astigmatism chart* and cover one eye (glasses off). This chart consists of a number of dark lines radiating from a central point, like spokes on a wheel. If astigmatism is present, some of the spokes will appear sharp and dark, whereas others will appear blurred and lighter because they are coming to a focus either in front of or behind the retina. Still covering the same eye, slowly walk up to the chart while observing the spokes.
5. Repeat this procedure using the other eye.
6. Repeat the test for astigmatism for both eyes with glasses on (if applicable).
7. To verify that astigmatism has been corrected with glasses, hold the glasses in front of your face while standing 10 feet from the chart and rotate the glasses 90°. The shape of the wheel should change when the glasses are rotated.

## B. Accommodation

If the refractive power (strength) of a lens is constant, the distance between the lens and the point of focus (focal length) will increase as an object moves closer to the lens. For example, if the image of an object that is 20 feet away is in focus on the retina (or on the photosensitive film of a camera), the image of an object 10 feet away will be focused *behind* the retina (or the camera film) and will appear blurred. A camera can be adjusted to focus on an object 10 feet away by moving the lens outward until the focal length of the image equals the distance between the lens and the film. The object 10 feet away will now be in focus, but the object 20 feet away will be blurred because its image will now come to a focus in front of the film.

The human eye differs from the camera in that the distance between the retina and the lens of the eye cannot be changed to bring objects into focus. Since the human lens is elastic, however, its degree of convexity (and therefore its refractive power) can be altered by changing the tension placed on it by the suspensory ligament; this, in turn, is regulated by the degree of contraction of the ciliary muscle. When the ciliary muscle is relaxed, the suspensory ligament pulls on the lens, thereby decreasing its convexity and power; distant objects (more than 20 feet away) are thus brought to a focus on the retina. Near objects are brought to a focus on the retina by contraction of the ciliary muscle. The contraction reduces the tension on the suspensory ligament, allowing the lens to assume a more convex shape. This ability of the eye to focus the images of objects that are at different distances from the lens is called **accommodation.**

The convexity of the normal lens can be adjusted to give it a range of power from 67 diopters (for distant vision; least convex) to 79 diopters (for near vision; most convex). The elasticity of the lens and the degree of convexity it can assume for near vision decreases with age, a condition called **presbyopia,** or *old eyes*. Lens elasticity can be tested by measuring the **near point of vision** (the closest an object can be brought to the eyes while still maintaining visual acuity). The near point of vision changes dramatically with age, averaging about 8 cm at age 10 and 100 cm at age 70. Presbyopia is corrected with *bifocals*, which contain two lenses of different refractive strengths.

In addition to tests of refraction, measurements of intraocular (eyeball) pressure are frequently performed with a device known as a *tonometer.* About 6 mL of aqueous humor is formed per day by the ciliary body. This fluid is drained by the *canal of Schlemm* (or *venous sinus*). If the drainage of aqueous humor is blocked, the intraocular pressure may rise; this is a condition known as **glaucoma,** which may cause damage to the optic nerve and blindness. Glaucoma may also damage the cornea, resulting in replacement of the normally transparent tissue with opaque scar tissue. When this happens the cornea can be surgically removed and replaced with either a contact lens or a grafted cornea. Because the cornea is avascular, corneal grafts can be performed with less risk of the transplanted tissue being rejected by the immune system.

### PROCEDURE

1. Place a square of wire screen about 10 inches in front of your eyes, and observe a distant object through the screen.
2. After closing your eyes momentarily, open them and note whether the screen or the distant object is in focus.
3. Repeat this procedure, this time focusing the eyes on the screen before closing and re-opening them.
4. To measure the near point of vision, place one end of a meter stick under one eye and extend it outward. Holding a pin at arm's length, gradually bring the pin toward the eye.
5. Record the distance at which the pin first appears blurred or doubled.
   Near point of vision __________ cm
6. Repeat this procedure, determining the near point of vision for the other eye.
   Near point of vision _______ cm

**Note:** *If the average near point of vision at age 10 is 8 cm, and at age 70 is 100 cm, what is your expected near point of vision?*

## C. Extrinsic Muscles of the Eye and Nystagmus

The six extrinsic muscles of the eye are shown in figure 3.18. The cranial nerve innervations and the actions of these muscles are summarized in table 3.2. These muscles allow the eyes to follow a moving object by maintaining the image on the same location of the retina of each eye, the *fovea centralis*, which provides maximum visual acuity. These muscles also allow the visual field of each eye to maintain the correct amount of central overlap. (The medial regions of each visual field overlap, while the more lateral regions are different for each eye; this **retinal disparity** helps in three-dimensional vision and depth perception.) When an object is brought closer, the correct amount of overlap and retinal disparity is maintained by the medial movement, or *convergence*, of the eyes.

The actions of antagonistic ocular muscles normally maintain the eyes in a midline position. If the tone of one muscle is weak as a result of muscle or nerve damage, the eyes will drift slowly in one direction followed by a rapid movement back to the correct position. This phenomenon is known as **nystagmus.**

In a typical examination of the ocular muscles, the subject is asked to follow an object (such as a pencil) with his or her eyes as it is moved up and down, right and left. Continued oscillations of the eye (slow phase in one direction, fast phase in the opposite direction) indicate the presence of nystagmus. Inability to move the eye outward indicates damage either to the abducens (sixth cranial) nerve or the lateral rectus muscle (table 3.2). Inability to move the eye downward when it is moved inward indicates damage to the trochlear (fourth cranial) nerve or the superior oblique muscle. All other defects in eye movement may be due either to damage of the oculomotor (third cranial) nerve or to damage of the specific muscles involved.

### PROCEDURE

1. Observe retinal disparity by holding a pencil in front of your face with one eye closed and then quickly changing eyes and noting the apparent position of the pencil.
2. Observe convergence by asking a subject to focus on the tip of a pencil as it is slowly brought from a distance of 2 feet in front of the face to the bridge of the nose. Notice the change in the diameter of the subject's pupil during this procedure.
3. Hold a pencil about 2 feet away from the bridge of the subject's nose. Then move the pencil to the left, to the right, and up and down, leaving the pencil in each position at least 10 seconds. Observe the movement of the eyes and note the presence or absence of nystagmus.

## D. Pupillary Reflex

The correct amount of light is admitted into the eye through an adjustable aperture, the pupil, surrounded by the iris. The iris consists of two groups of smooth muscles with opposing actions. Operating like sphincters, the *circular muscles* constrict the pupil in bright light, whereas the *radial muscles* work to dilate the pupil in dim light. These responses are mediated by the autonomic nervous system. Sympathetic nerves stimulate the radial muscles to dilate the pupil, and parasympathetic nerves stimulate the circular muscles to constrict the pupil.

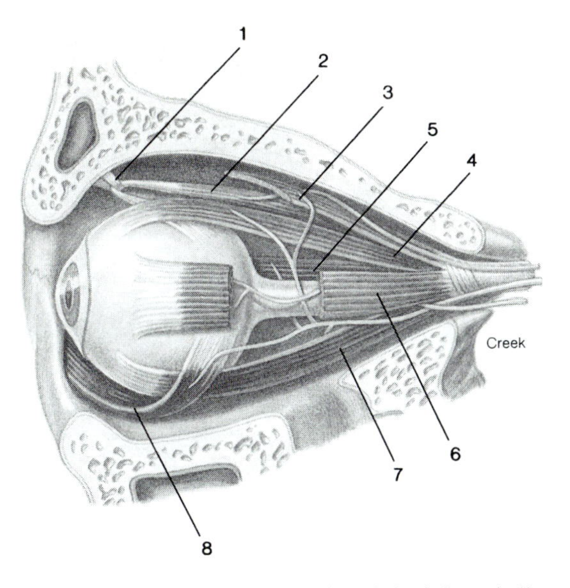


1. Trochlea
2. Superior oblique
3. Levator palpebrae superioris (cut)
4. Superior rectus
5. Medial rectus
6. Lateral rectus
7. Inferior rectus
8. Inferior oblique

**Figure 3.18** A lateral view of ocular muscles of the left eyeball.

## Table 3.2 The Ocular Muscles

| Muscle | Cranial Nerve Innervation | Movement of Eyeball |
|---|---|---|
| Lateral rectus | Abducens | Lateral |
| Medial rectus | Oculomotor | Medial |
| Superior rectus | Oculomotor | Superior and medial |
| Inferior rectus | Oculomotor | Inferior and medial |
| Inferior oblique | Oculomotor | Superior and lateral |
| Superior oblique | Trochlear | Inferior and lateral |

### PROCEDURE

1. Stay with the subject in a darkened room for at least 1 minute, allowing his or her eyes to adjust to the dim light. This adjustment is known as *dark adaptation*.
2. Shine a narrow beam of light (from a pen flashlight or an ophthalmoscope, for example) from the right side into the subject's right eye. Observe the pupillary reflex in the right eye and also in the left eye. The pupillary reflex in the other (left) eye is called the *consensual reaction*.
3. Repeat this procedure (first dark-adapting the eyes again) from the left side with the left eye.

## E. Examination of the Eye with an Ophthalmoscope

An **ophthalmoscope** is a device used to observe the posterior inner part of the eye (the *fundus*). A mirror positioned at the top of the instrument deflects light at a right angle into the eye, enabling an observer to see the interior of the eye through a small slit in the mirror. Different depths of focus are attained by changing the lenses that are positioned in the slit. The lenses are carried on a wheel in regular order according to their focal lengths. The strength of each lens is given in diopters preceded by a plus (+) for a convex lens or a minus (–) for a concave lens, with 0 indicating no lens.

In this exercise, an opthalmoscope will be used to observe the arteries and veins of the fundus and two regions of the retina: the **optic disc** (the region where the optic nerve exits the eye, otherwise known as the *blind spot*) and the **macula lutea** (fig. 3.19). The macula lutea is a yellowish region containing a central pit, the **fovea centralis,** where is found the highest concentration of the photoreceptors responsible for visual acuity (the cones). When the eyes are looking directly at an object, the image is focused on the fovea.

Clinical examination of the fundus **(ophthalmoscopy)** can aid the diagnosis of a number of ocular and systemic (body) diseases. The features noted in these examinations include the condition of the blood vessels; the color and shape of the disc; the presence of particles, exudates, or hemorrhage; the presence of edema and inflammation of the optic nerve *(papilledema);* and myopia and hyperopia.

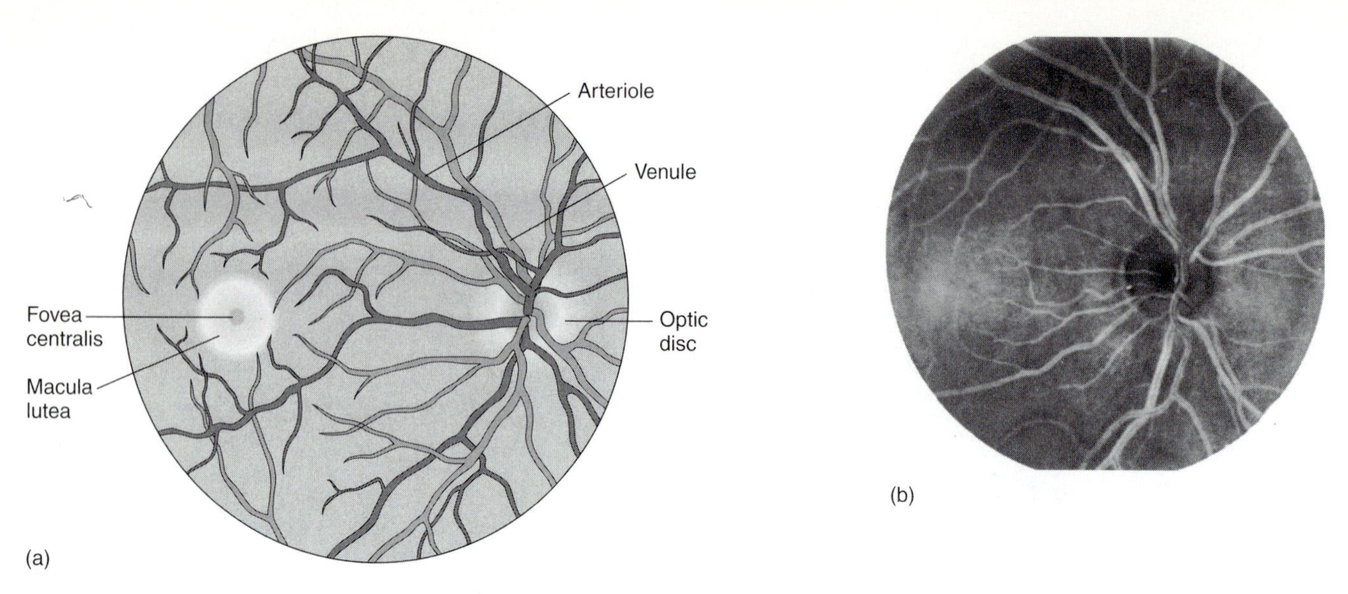


**Figure 3.19 A view of the retina as seen with an ophthalmoscope.** Optic nerve fibers leave the eyeball at the optic disc to form the optic nerve. (Note the blood vessels that can be seen entering the eyeball at the optic disc.)

**(For a full-color version of this figure, see fig. 10.29 in *Human Physiology*, eighth edition, by Stuart I. Fox.)**

### PROCEDURE

1. Have the subject sit in a darkened room and look at a distant object (blinking as needed).
2. Position your chair so that you are close to and facing the subject. Hold the ophthalmoscope with your right hand and use your right eye when observing the subject's right eye. (The situation is reversed when viewing the left eye.)
3. With your forefinger on the lens adjustment wheel and your eye as close as possible to the small hole in the ophthalmoscope (glasses off if applicable), bring the instrument as close as possible to the subject's eye. (Steady your hand by resting it on the subject's cheek.)
4. Examine the subject's eye from the front to the back. Looking slightly from the side of the eye (not directly in front), examine the iris and lens using a +20 to +15 lens. The lens selected will vary among examiners who normally wear glasses. If both your eyes and the subject's eyes are normal, you will be able to see clearly without a lens. (On the 0 setting, the refractive strength of the subject's eye will be sufficient to focus the light on the eye.)
5. Rotate the wheel counterclockwise to examine the fundus.
   (a) If a positive (convex) lens is necessary to focus on the fundus, and your eyes are normal, the subject has hyperopia (hypermetropia).
   (b) If a negative (concave) lens is necessary to focus on the fundus, and your eyes are normal, the subject has myopia.
6. Observe the arteries and veins of the fundus and follow them to their point of convergence. This will enable you to see the optic disc (fig. 3.19).
7. Finally, at the end of the examination, observe the macula lutea by asking the subject to look directly into the light of the ophthalmoscope.

## F. The Blind Spot

The retina contains two types of *photoreceptors*, **rods** and **cones.** Rods and cones synapse with other cells *(bipolar neurons)*, which in turn synapse with *ganglion cells* whose axons form the optic nerve transmitting sensory information out of the eye to the brain. In the fovea, only one cone will synapse with one bipolar cell, whereas several rods may converge on a given bipolar cell (fig. 3.20). Thus, the rods are more sensitive to low levels of illumination, whereas the cones require more light but provide greater visual acuity. The rods, therefore, are responsible for night *(scotopic)* vision, when sensitivity is most important. The cones are responsible for day *(photopic)* vision, when visual acuity is most important. The cones also provide color vision—colors are seen during the day, whereas night vision is in black and white.

The axons of all ganglion cells in the retina gather together to become the optic nerve that exits the eye at the optic disc. This is also called the **blind spot** because there are no rods or cones in the optic disc, so an object whose image is focused here will not be seen.

### PROCEDURE

1. Hold the drawing of the circle and the cross (fig. 3.21) about 20 inches from your face with the left eye covered or closed. Focus on the circle; this is most easily done if the circle is positioned in line with the right eye.
2. Keeping the right eye focused on the circle, slowly bring the drawing closer to your face until the cross

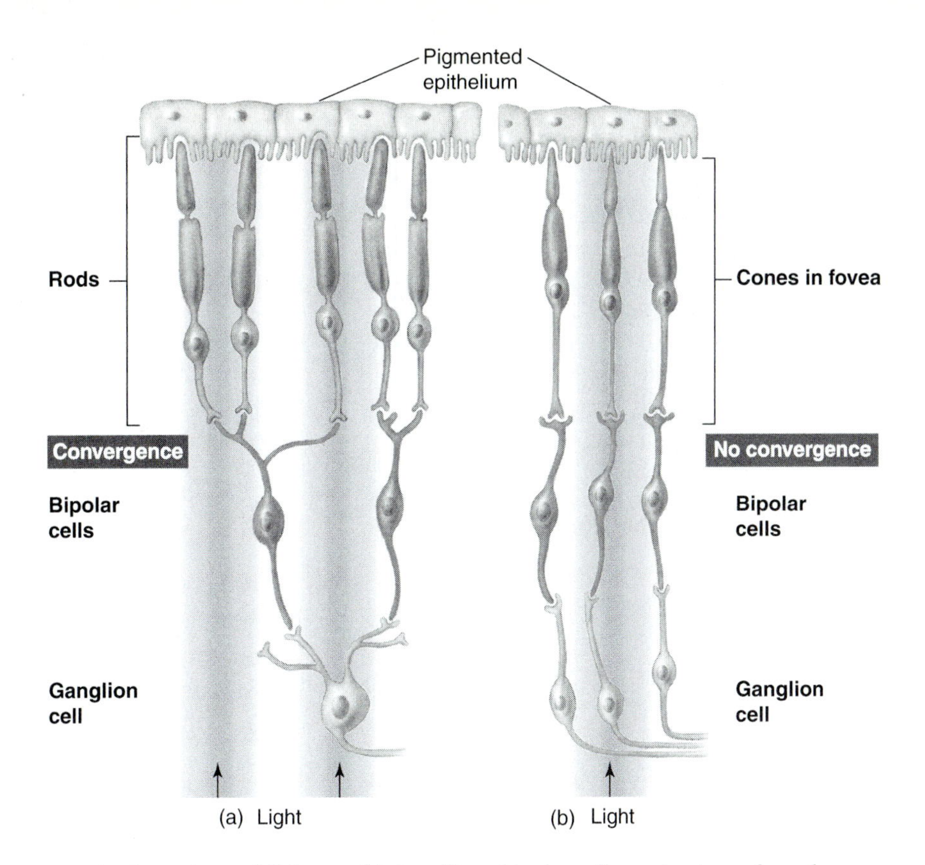


**Figure 3.20 Convergence in the retina and light sensitivity.** Since bipolar cells receive input from the convergence of many rods (a), and since a number of such bipolar cells converge on a single ganglion cell, rods maximize sensitivity to low levels of light at the expense of visual acuity. By contrast, the 1:1:1 ratio of cones to bipolar cells to ganglion cells in the fovea (b) provides high visual acuity, but sensitivity to light is reduced.

**(For a full-color version of this figure, see fig. 10.40 in *Human Physiology,* eighth edition, by Stuart I. Fox.)**

**Figure 3.21 A diagram for demonstrating the blind spot.**

disappears. Continue to move the drawing slowly toward your face until the cross reappears.

3. Repeat this procedure with the right eye closed or covered and the left eye focused on the cross. Observe the disappearance of the circle as the drawing is brought closer to your face.

## G. The Afterimage

The light that strikes the receptors of the eye stimulates a photochemical reaction in which the pigment **rhodopsin** (within the rods) dissociates to form the pigment *retinene* and the protein *opsin*. This chemical dissociation produces electrical changes in the photoreceptors, which trigger a train of action potentials in the axons of the optic nerve. These events cannot be repeated in a given rod receptor until the rhodopsin is regenerated. This requires a series of chemical reactions in which one isomer of retinene is converted to another through the intermediate compound *vitamin* $A_1$. In other words, after the rhodopsin visual pigment in the rod has been "bleached" or dissociated by light from an object, a certain period of time is required before that receptor can again be stimulated.

When an eye that has adapted to a bright light, such as a lightbulb, is closed or quickly turned towards a wall, the bright image of the lightbulb will still be seen. This is called a **positive afterimage** and is caused by the continued "firing" of the photoreceptors. After a short period, the dark image of the lightbulb, called the **negative afterimage,** will appear against a lighter background due to the "bleaching" of the visual pigment of the affected receptors.

According to the **Young-Helmholtz** theory of color vision, there are three systems of cones that respond respectively to *red*, *green*, and *blue* (or violet) light, and all other colors are seen by the brain's interpretation of mixtures of impulses from these three systems. Color discrimination will of course be impaired if one system of cones is defective (color blindness), or if one system of cones has been "bleached" by the continued viewing of an object. In the latter case, the positive afterimage of the object will appear in the complementary color.

It is important to remember that the eye is a receptor, transducing light into electrical nerve impulses. We actually see with our brain. Impulses from the retina pass, via the *lateral geniculate bodies,* to the *visual cortex* of the occipital lobe, where the patterns of impulses are integrated to produce an image. The importance of the visual cortex in vision is illustrated by **strabismus,** a condition in which weak extrinsic eye muscles prevent the two eyes from converging on an object and fusing the images. To avoid confusion, the cortical cells eventually stop responding to information from one eye, making that eye functionally blind. Visual information is integrated with input from the other senses in the cortex of the *inferior temporal lobe.* If this area is damaged (the **Klüver-Bucy syndrome**), visual recognition is impaired. Although the image is seen, it lacks meaning and emotional content.

### PROCEDURE

1. After staring at a lightbulb, suddenly shift your gaze to a blank wall. Observe the appearance of the *negative afterimage*.
2. For 1 minute, stare at a dot made in the center of a small red square that has been pasted on a larger sheet of black paper.
3. Suddenly shift your gaze to a sheet of white paper and note the color of the *positive afterimage*. Positive afterimage (red) is______________________________.
4. Repeat this procedure using blue squares and yellow squares.
   Positive afterimage (blue) is ____________________.
   Positive afterimage (yellow) is ___________________.

## H. Color Vision and Color Blindness

Humans have **trichromatic color vision.** As stated in exercise 3.5G, color vision is provided by the stimulation of three types of cones, designated *blue, green,* and *red.* These terms do not refer to the color of the cones; rather, they refer to the region of the visible spectrum in which each cone's pigment absorbs light best (fig. 3.22). This is the cone's *absorption maximum.* On this basis, the blue cones are also called *S cones* (for short wavelength); the green cones are also called *M cones* (for medium wavelength); and the red cones are called *L cones* (for long wavelength). Each of these cones has the same retinene pigment present in the rods, but each has a different protein (known as the *photopsins*) associated with the retinene to give the cone its distinct absorption maximum.

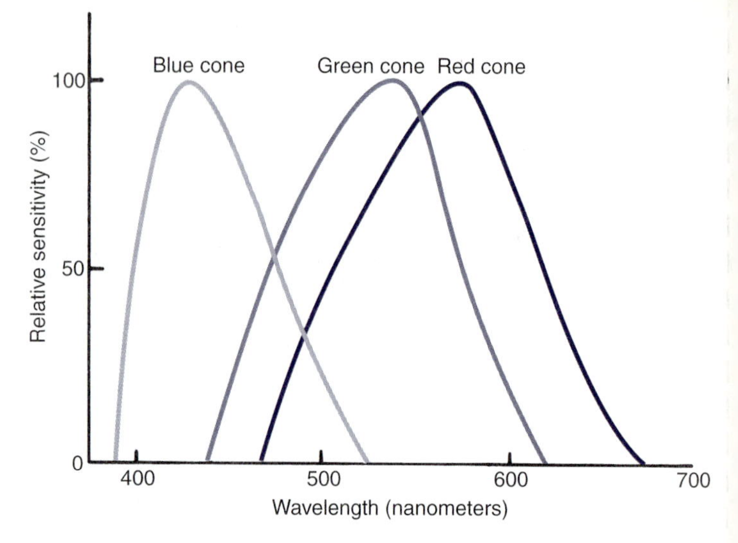


**Figure 3.22 The three types of cones.** Each type contains retinene, but the protein with which the retinene is combined is different in each case. Thus, each different pigment absorbs light maximally at a different wavelength. Color vision is produced by the activity of these blue cones, green cones, and red cones.

**(For a full-color version of this figure, see fig. 10.40 in *Human Physiology,* eighth edition, by Stuart I. Fox.)**

**Color blindness** is caused by an inherited lack of one or more type of cones (see exercise 11.3C), usually the M (green) or L (red) cones. These people have only two functioning types of cones, and their vision is thus dichromatic. Since the photopsins of the M and L cone pigments are coded on the X chromosome, and since men have only one X chromosome (and so cannot carry the trait in a recessive state), such red-green color blindness is far more common in men than in women.

### PROCEDURE

You can test for red-green color blindness using the *Ishihara test.* In this test, colored dots are arranged in a series of circles in such a way that a person with normal vision can see a number embedded within each circle. By contrast, a color-blind person will see only an apparently random array of colored dots.

# Laboratory Report 3.5

Name ______________________________

Date ______________________________

Section ______________________________

## REVIEW ACTIVITIES FOR EXERCISE 3.5

### *Test Your Knowledge of Terms and Facts*

1. The photoreceptors responsible for color vision are the __________________; of these, there are __________________ different kinds.
2. The region of the retina in which there are no photoreceptors is called the ____________________; this is also known as the ____________________.
3. Retinene (retinaldehyde) is derived from which vitamin? __________________
4. When light enters the retina, it first passes through the __________________ cell layer, then the __________________ cell layer, before reaching the photoreceptors.
5. The axons of __________________ cells gather together to produce the optic nerve.
6. Define these terms:
   (a) *visual acuity* ______________________________________________
   (b) *accommodation* ______________________________________________

Match these terms with the appropriate description:

_____ 7. myopia (a) abnormal curvature of the cornea or lens
_____ 8. hyperopia (b) eye too long
_____ 9. presbyopia (c) abnormally high intraocular pressure
_____10. astigmatism (d) eye too short
_____11. glaucoma (e) loss of lens elasticity

### *Test Your Understanding of Concepts*

12. Describe the muscle layers of the iris and the innervation to these muscles. Use this information to explain how pupils constrict in bright light and dilate in dim light.

13. Describe how the curvature of the lens is regulated, and use this information to explain how images are kept in focus on the retina as a distant object is brought closer to the eyes.

14. Explain the reason for blurred vision in a person with myopia, and describe how this person's vision is improved by the lenses in a pair of glasses.

## *Test Your Ability to Analyze and Apply Your Knowledge*

15. Carrots contain large amounts of *carotene*, a precursor of vitamin A. Can eating carrots help improve blurred vision from myopia or hyperopia? Explain. Also, explain how eating carrots may influence scotopic (night) vision.

16. "You see with your brain, not with your eyes." Defend this statement with reference to the Young-Helmholtz theory of color vision and to the optic disc.

# Ears: Cochlea and Hearing

EXERCISE 3.6

**MATERIALS**

1. Tuning forks
2. Rubber mallets

Sound is conducted by the middle ear to the inner ear, where events within the cochlea result in the production of nerve impulses. Clinical tests of middle ear (conductive) and inner ear (sensory) function aid in the diagnosis of hearing disorders.

**OBJECTIVES**

1. Describe the structure of the middle ear and explain how the ossicles function.
2. Describe the structure of the inner ear and explain how the cochlea functions.
3. Demonstrate Rinne's test and Weber's test and explain the significance of each.
4. Explain how the source of a sound is localized.

**Textbook Correlations***

Before performing this exercise, you should study the introductory material presented here. Further information relating to this exercise can be found in these pages of *Human Physiology,* eighth edition, by Stuart I. Fox:

- *The Ears and Hearing.* Chapter 10, pp. 255–261.

*Multimedia Correlations (also see Appendix 3)
- *MediaPhys 2.0:* Topics 6.49–6.57

You should be familiar with the gross structure of the ear (fig. 3.23). Sound waves are conducted through the **outer ear** (the pinna and the external auditory meatus) to the tympanic membrane (eardrum), causing it to vibrate. The vibration of the tympanic membrane causes the three ossicles of the **middle ear**—the *malleus* (hammer), *incus* (anvil), and *stapes* (stirrup)—to vibrate, thus pushing the footplate of the stapes against a flexible membrane, the *oval window.* Vibration of the oval window produces compression waves in the fluid-filled cochlea of the **inner ear.**

The compression waves of cochlear fluid flow over a thin, flexible membrane within the cochlea called the *basilar membrane,* causing it to vibrate. Within the **organ of Corti,** the basilar membrane is coated with sensory *hair cells,* which are displaced upward by this vibration into a stiff overhanging structure called the *tectorial membrane* (fig. 3.24). The distortion of the hair cells produced by this action stimulates a train of action potentials that travels along the cochlear branch of the vestibulocochlear (eighth cranial) nerve to the brain. Here, the action potentials are interpreted as the sound of a specific **pitch,** which is determined by the location of the stimulated hair cells on the basilar membrane, and of a specific **loudness,** which is coded by the frequency of action potentials.

## A. Conduction of Sound Waves through Bone: Rinne's and Weber's Tests

Although hearing is normally produced by the vibration of the oval window in response to sound waves conducted through the movements of the middle-ear ossicles, the *endolymph* fluid of the cochlea can also be made to vibrate in response to sound waves conducted through the skull bones directly, thereby bypassing the middle ear. This makes it possible to differentiate between deafness resulting from middle-ear damage (**conduction deafness,** such as from damage to the ossicles in *otitis media* or immobilization of the stapes in *otosclerosis*) and deafness resulting from damage to the cochlea or vestibulocochlear nerve (**sensory deafness,** such as from infections, streptomycin toxicity, or prolonged exposure to loud sounds).

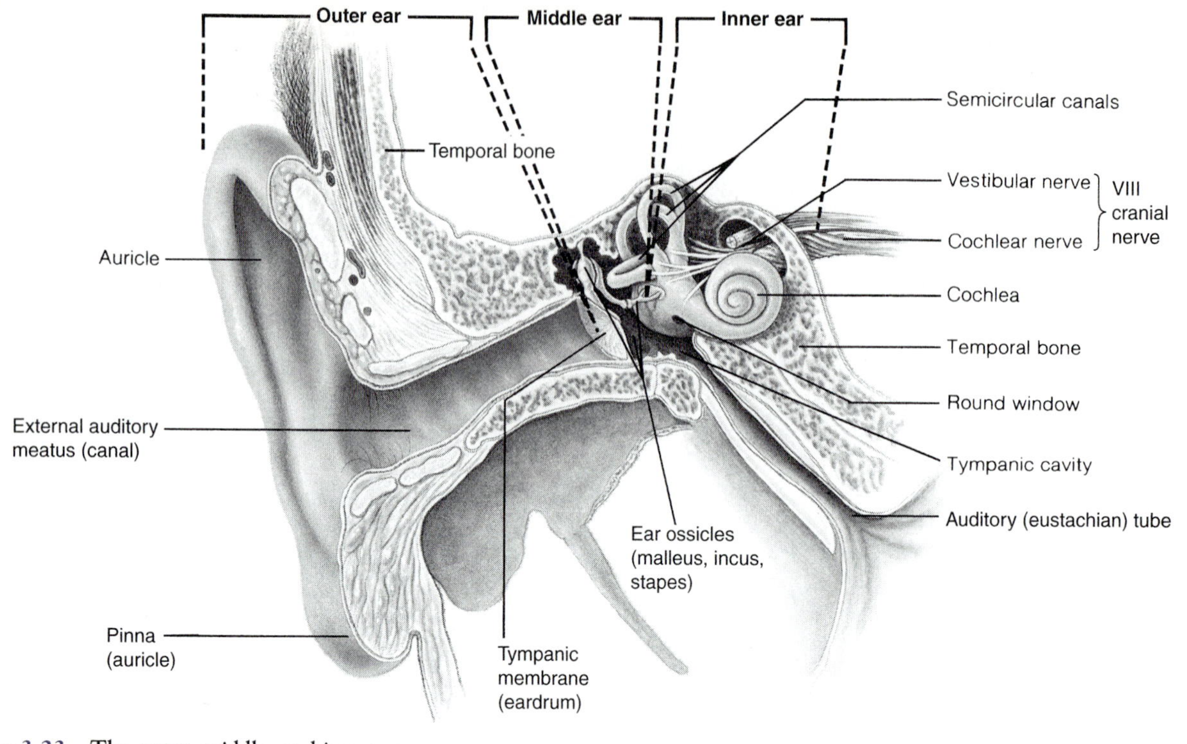


**Figure 3.23** The outer, middle, and inner ear.

**(For a full-color version of this figure, see fig. 10.17 in *Human Physiology,* eighth edition, by Stuart I. Fox.)**

Conduction deafness may be caused by infections of the middle ear **(otitis media)**, infections of the tympanic membrane **(tympanitis)**, or an excessive accumulation of ear wax *(cerumen).* People with conduction deafness often wear hearing aids over the mastoid process of the temporal bone. These devices amplify sounds and transmit them by bone conduction to the cochlea. Hearing aids do not help in cases of complete sensory (nerve) deafness. The first cochlear implant device received FDA approval in 1984. Designed to bypass damaged or destroyed hair cells, cochlear implants stimulate the auditory nerve directly. Although this procedure has met with limited success, recent technological breakthroughs may offer future remedies for deafness.

## PROCEDURE

### *Rinne's Test*

1. Strike a tuning fork with a rubber mallet to produce vibrations.
2. Perform Rinne's test by placing the handle of the vibrating tuning fork against the mastoid process of the temporal bone (the bony prominence behind the ear), with the tuning fork pointed down and behind the ear. When the sound has almost died away, move the tuning fork (by the handle) near the external auditory meatus. If there is no damage to the middle ear, the sound will reappear.
3. Simulate conduction deafness by repeating Rinne's test with a plug of cotton in the ear. Notice that in conductive deafness, conduction by bone (via the mastoid process) is more effective than conduction by air.

### *Weber's Test*

1. Perform Weber's test by placing the handle of the vibrating tuning fork on the midsagittal line of the head, and listen. In conduction deafness, the sound will seem louder in the affected ear (room noise is excluded but bone conduction continues), whereas in sensory deafness, the cochlea is defective and the sound will be louder in the normal ear.
2. Repeat Weber's test with one ear plugged with your finger. The sound will appear louder in the plugged ear because external room noise is excluded.

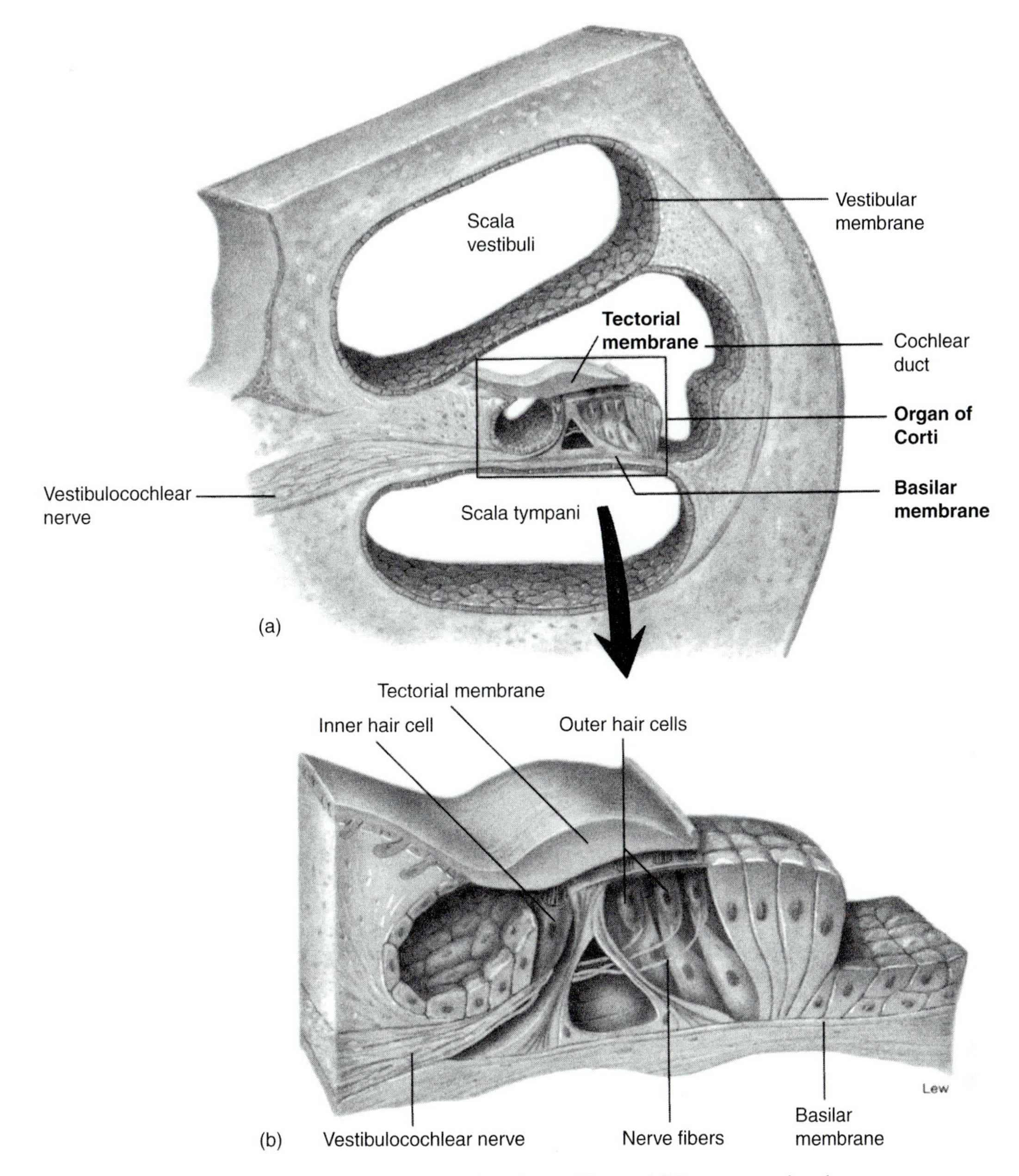

**Figure 3.24 The organ of Corti.** (a) The organ of Corti within the cochlea and (b) in greater detail.

**(For a full-color version of this figure, see fig. 10.22 in *Human Physiology,* eighth edition, by Stuart I. Fox.)**

## B. Binaural Localization of Sound

Just as binocular vision provides valuable clues for viewing scenes in three dimensions, binaural hearing helps to localize sounds (people have stereoscopic vision and stereophonic hearing). The ability to localize the source of a sound depends partly on the difference in loudness of the sound that reaches the two ears and partly on the difference in the time of arrival of the sound at the two ears. The difference in loudness is more important for high-pitched sounds, where the sound waves are blocked by the head. The difference in the time of arrival is more important for low-pitched sounds, whose wavelengths are large enough to bend around the head.

### PROCEDURE

1. With both eyes closed, the subject is asked to locate the source of a sound (e.g., a vibrating tuning fork).
2. The vibrating tuning fork is placed at various positions (front, back, and sides about a foot from the subject's head), and the subject is asked to describe the location of the tuning fork.
3. Repeat the above procedures with one of the subject's ears plugged.

# Laboratory Report 3.6

Name ______________________________

Date ______________________________

Section ______________________________

## REVIEW ACTIVITIES FOR EXERCISE 3.6

### *Test Your Knowledge of Terms and Facts*

1. List the three ossicles of the middle ear in sequence, from the outermost to the innermost.
   Outermost ______________________________
   Middle ______________________________
   Innermost ______________________________
2. The scientific name for the eardrum is the ______________________________.
3. The middle chamber of the cochlea is the ______________________________.
4. Sensory hair cells in the cochlea are located on a membrane called the ______________________________.
5. The innermost middle ear ossicle presses against a flexible membrane called the ______________________________.
6. The sensory structure of the inner ear responsible for transducing vibrations into nerve impulses is known as the ______________________________.

### *Test Your Understanding of Concepts*

7. In sequence, describe the events that occur between the arrival of sound waves at the tympanic membrane and the production of nerve impulses.

8. Explain how different pitches of sound affect the basilar membrane, and how this relates to the information sent to the brain.

## *Test Your Ability to Analyze and Apply Your Knowledge*

9. Explain the result that might be obtained by performing the Rinne's and Weber's test on a patient with otosclerosis. How might these results compare with those obtained from a patient with sensory deafness? Explain.

10. Hair cells are mechanoreceptors that are employed in the inner ear of mammals and also in the lateral line organs on the body of fish. How might the bending of hair cell processes lead to depolarization? Propose a general description of the function of hair cells in the animal kingdom.

# Ears: Vestibular Apparatus—Balance and Equilibrium

EXERCISE **3.7**

## Materials

1. Swivel chair

The vestibular apparatus provides a sense of balance and equilibrium. As a result of inertia acting on the structures within the vestibular apparatus, changes in the position of the head result in the production of afferent nerve impulses that are conducted to the brain along the eighth cranial nerve. This information results in eye movements and other motor activities that help to orient the body in space.

## OBJECTIVES

1. Describe the structure of the semicircular canals and explain how movements of the head result in the production of nerve impulses.
2. Describe vestibular nystagmus and explain how it is produced.

## Texbook Correlations*

Before performing this exercise, you should study the introductory material presented here. Further information relating to this exercise can also be found in these pages of *Human Physiology,* eighth edition, by Stuart I. Fox:

- *Vestibular Apparatus and Equilibrium.* Chapter 10, pp. 251–255.

*Multimedia Correlations (also see Appendix 3)
- *MediaPhys 2.0*: Topics 6.59–6.64

The **vestibular apparatus,** located in the inner ear above the cochlea, consists of three *semicircular canals* (oriented in three planes), the *utricle*, and the *saccule* (fig. 3.25). The utricle and saccule are together called *otolith organs*, and provide a sense of linear acceleration. These structures, like the cochlea, are filled with *endolymph* and contain sensory cells activated by bending. These sensory hair cells of the semicircular canals support numerous hairlike extensions, which are embedded in a gelatinous "sail" (the *cupula*) that projects into the endolymph (fig. 3.26). Movement of the endolymph fluid, induced by acceleration or deceleration, bends the extensions of the hair cells. This sends a train of impulses to the brain along the *vestibulocochlear* (*eighth* cranial) *nerve*. The sensory hair cells of the utricle and the saccule serve to orient the head with respect to the gravitational pull of the earth.

Afferent impulses from the vestibular apparatus help make us aware of our position in space and affect a variety of efferent somatic motor nerves (such as those regulating the voluntary extrinsic eye muscles). Under intense vestibular activity, autonomic motor nerves may also become stimulated, producing involuntary responses that can include vomiting, perspiration, and hypotension. This exercise will test the effect of vestibular activity on the extrinsic muscles of the eye by producing involuntary eye oscillations known as **vestibular nystagmus.**

The semicircular canals can be stimulated by rotating a subject in a chair. Stimulation of the lateral canal can occur when the head is flexed 30° forward, almost touching the chest. When the subject is first rotated to the right, the cupula will be bent to the left because of the inertial lag of the endolymph. This will cause nystagmus in which the eyes drift slowly to the left followed by a quick movement to the right (midline position). Nystagmus will continue until the inertia of the endolymph has been overcome and the cupula returns to its initial unbent position. At this time, the endolymph and the cupula are moving in the same direction and at the same speed. When the rotation of the subject is abruptly stopped, the greater inertia of the endolymph will cause bending of the cupula in the previous direction of spin (to the right), producing nystagmus with a slow drift phase to the right and a rapid phase to the left. This activation of

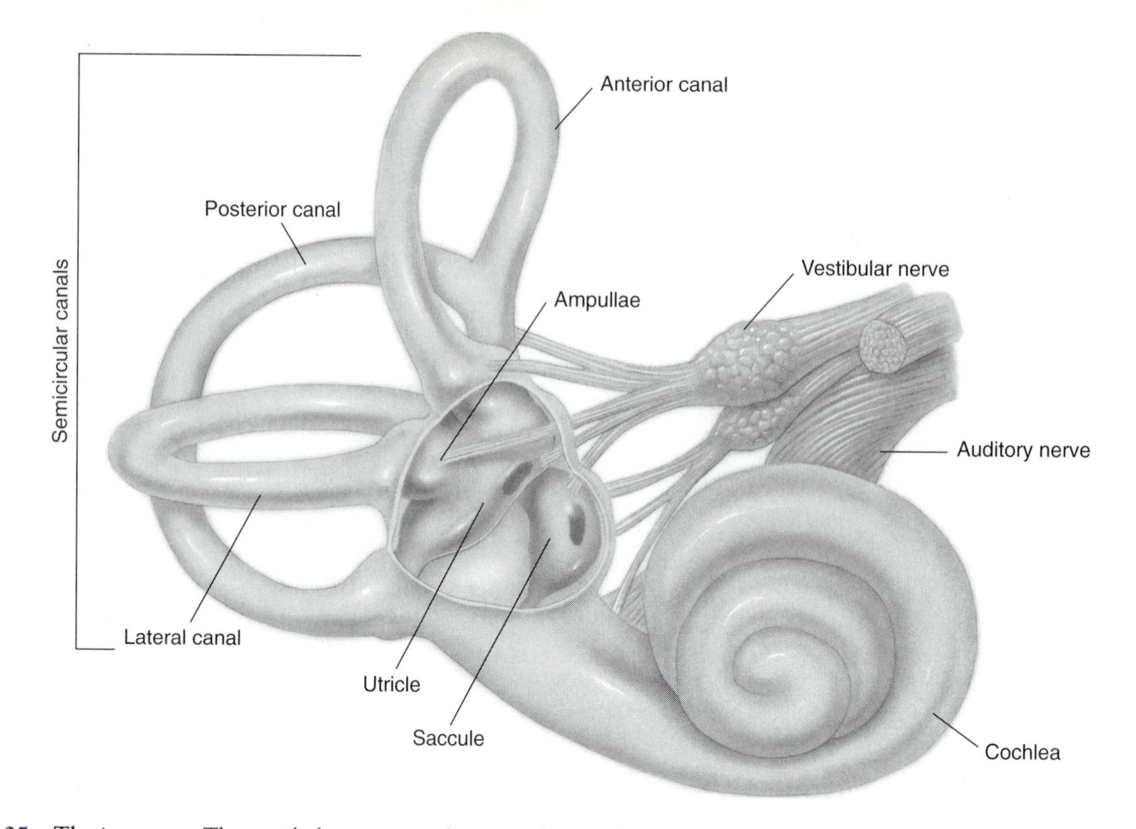


**Figure 3.25 The inner ear.** The vestibular apparatus (semicircular canals, utricle, and saccule), required for a sense of equilibrium, and the cochlea, required for hearing, make up the inner ear.

**(For a full-color version of this figure, see fig. 10.11 in *Human Physiology,* eighth edition, by Stuart I. Fox.)**

the vestibular apparatus is often accompanied by **vertigo** (an illusion of movement, or spinning) and a tendency to fall to the right.

> Vertigo may be accompanied by dizziness, but the two sensations are not the same. A person may be dizzy without experiencing the more severe effects of vertigo. In the procedure described below, activation of the vestibular apparatus by rotation of the subject produces oscillatory eye movements (nystagmus). The eyes, in turn, can activate the vestibular apparatus and produce vertigo, as occurs in motion sickness (seasickness or car sickness, for example). Vertigo can also accompany diseases unrelated to the special senses, such as cardiovascular disease. Many of the unpleasant symptoms associated with vertigo, such as nausea and vomiting, are the result of activation of the autonomic motor system. Drugs taken for motion sickness (e.g., Dramamine or scopolamine) act by suppressing these autonomic responses.

## PROCEDURE

1. Have the subject sit in a swivel chair with the eyes open and the head flexed 30° forward (chin almost touching the chest). Rotate the chair quickly to the right for 20 seconds (about 10 revolutions). After noting the initial nystagmus, have the subject close his or her eyes.

**Note:** *Only subjects who are not subject to motion sickness should be used. The exercise should be stopped immediately if the subject feels sick.*

2. Abruptly stop the chair and have the subject open his or her eyes as wide as possible. Note the direction of nystagmus (left to right or right to left, circle one).
3. Repeat this procedure (using different subjects), alternating with the head resting on the right shoulder and the left shoulder (this stimulates the vertical canals), and note the direction of the postrotational nystagmus.

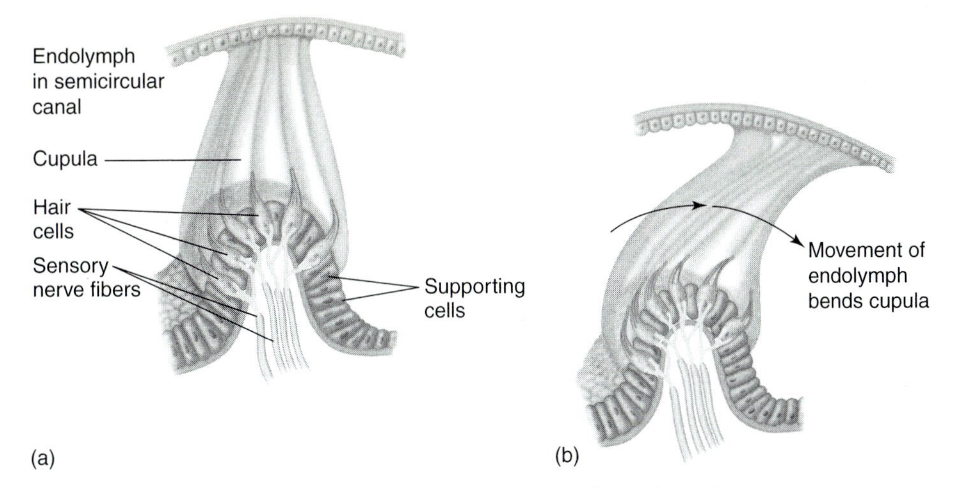


**Figure 3.26 The cupula and hair cells within the semicircular canals.** (a) Shown here, the structures are at rest or at a constant velocity. (b) Here, movement of the endolymph during rotation causes the cupula to bend, thus stimulating the hair cells.

**(For a full-color version of this figure, see fig. 10.15 in *Human Physiology,* eighth edition, by Stuart I. Fox.)**

# Laboratory Report 3.7

Name ____________________

Date ____________________

Section ____________________

## REVIEW ACTIVITIES FOR EXERCISE 3.7

### *Test Your Knowledge of Terms and Facts*

1. The organ of equilibrium is called the ____________________.
2. The structures sensitive to angular acceleration in three planes are the ____________________.
3. The structures sensitive to linear acceleration are the ____________________ and ____________________. Together, they are called the ____________________ organs.
4. The fluid within the organs of equilibrium is known as ____________________.
5. The sense of equilibrium is transmitted by the ____________________ (its number) cranial nerve, also known as the ____________________ (its name) nerve.
6. An illusion of movement or spinning is called ____________________.

### *Test Your Understanding of Concepts*

7. Describe the meaning of the term *inertia* and explain how inertia affects the hair cell processes of the inner ear during acceleration and deceleration.

8. Including all of the relevant anatomical structures (and ending with the activation of the cranial nerve), explain how we sense spinning in a chair. What happens when the chair is suddenly stopped?

9. Which structures of the inner ear would be activated by a somersault? By a cartwheel? By fast acceleration in a race car?

## *Test Your Ability to Analyze and Apply Your Knowledge*

10. Does nystagmus occur after a person in a rotating chair has achieved constant velocity? Explain?

11. Propose an explanation of the causes of vertigo and nausea in a seasick person.

# Taste Perception

EXERCISE

## MATERIALS

1. Cotton-tipped applicator sticks
2. Solutions of 5% sucrose, 1% acetic acid, 5% NaCl, and 0.5% quinine sulfate
3. Fruit (apples, grapes, or others)

There are four modalities of taste perception: sweet, sour, bitter, and salty. Taste buds sensitive to each of these modalities have a characteristic distribution on the tongue. Therefore, each of these taste modalities is perceived most acutely in a particular tongue region.

## OBJECTIVES

1. Describe the structure of a taste bud and the location of taste buds on the tongue.
2. List the *four* primary taste modalities, and describe how they are stimulated.

## Textbook Correlations

Before performing this exercise, you should study the introductory material presented here. Further information relating to this exercise can be found in these pages of *Human Physiology,* eighth edition, by Stuart I. Fox:

- *Taste and Smell.* Chapter 10, pp. 248–250.

The **taste buds** consist of specialized epithelial cells arranged in the form of barrel-shaped receptors (fig. 3.27*c*), associated with sensory (afferent) nerves. Long microvilli extend through a pore at the external surface of the taste bud and are bathed in saliva. Although not considered neurons, taste cells can be depolarized to release chemical neurotransmitters that, in turn, stimulate associated sensory neurons leading to the brain. In adults, these receptors are located primarily on the surface of the tongue, with a lesser number on the soft palate and epiglottis. In children the sense of taste is more diffuse, with additional receptors located on the inside of the cheeks.

The specialized epithelial cells of the taste bud are known as **taste cells.** The different categories of taste are produced by different chemicals that come into contact with the microvilli of these cells. Four different categories of taste are traditionally recognized: *salty, sour, sweet,* and *bitter.* There may also be a fifth category of taste, termed *umami* (a Japanese term related to a meaty flavor), for the amino acid glutamate (and stimulated by the flavor-enhancer monosodium glutamate). Although scientists long believed that different regions of the tongue were specialized for different tastes, this is no longer believed to be true. Indeed, it seems that each taste bud contains taste cells responsive to each of the different taste categories! It also appears that a given sensory neuron may be stimulated by more than one taste cell in a number of different taste buds, and so one sensory fiber may not transmit information specific for only one category of taste. The brain interprets the pattern of stimulation of these sensory neurons, together with the nuances provided by the sense of smell, as the complex tastes that we are capable of perceiving.

The *sour* taste of solutions is due to the presence of acid or hydrogen ions ($H^+$), and the *salty* taste is due to the presence of $Na^+$ ions (fig. 3.28), but modified by the anion—sodium chloride tastes saltier than other sodium salts, such as sodium acetate. The chemical basis for *bitter* and *sweet* taste is largely unknown, since these can be produced by a variety of seemingly unrelated compounds. (Fructose tastes the sweetest, followed by sucrose, and then by glucose. The artificial, nonsugar sweeteners, however, taste sweeter than any of these.)

The afferent pathway from the taste buds to the brain involves primarily two cranial nerves (fig. 3.29). Taste buds on the anterior two-thirds of the tongue are served by the *facial* (seventh cranial) *nerve;* whereas those on the posterior third of the tongue have a sensory pathway along the *glossopharyngeal* (ninth cranial) *nerve.* The vagus (tenth cranial) nerve also has limited sensory

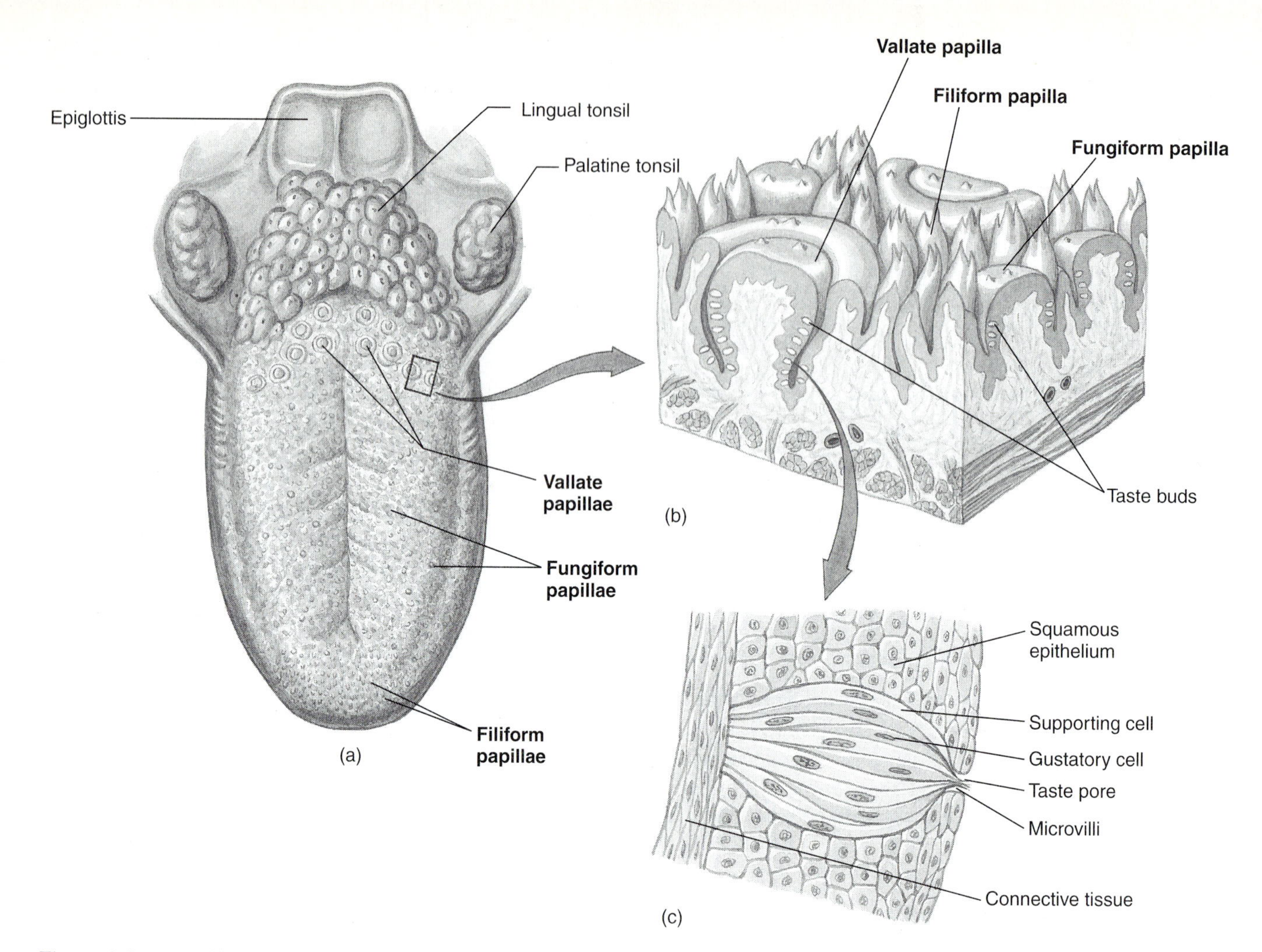


**Figure 3.27 Papillae of the tongue and taste buds.** (a,b) The structure of the tongue, with its papillae. (c) The structure of a taste bud. Taste buds are found throughout the tongue within the vallate and fungiform papillae.

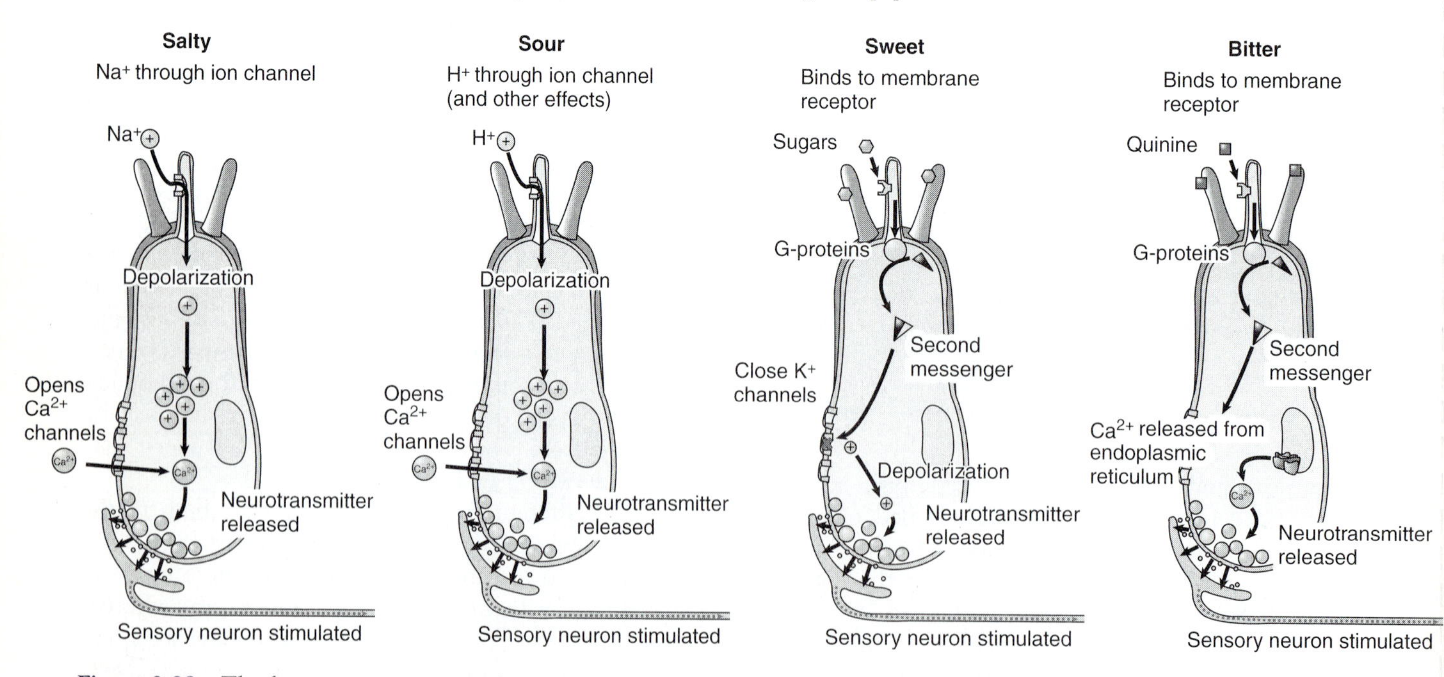


**Figure 3.28 The four major categories of taste.** Each category of taste activates specific taste cells by different means. Notice that taste cells for salty and sour are depolarized by ions ($Na^+$ and $H^+$, respectively) in the food, whereas taste cells for sweet and bitter are depolarized by sugars and quinine, respectively, by means of G-protein-coupled receptors and the actions of second messengers.

**(For a full-color version of this figure, see fig. 10.8 in *Human Physiology*, eighth edition, by Stuart I. Fox.)**

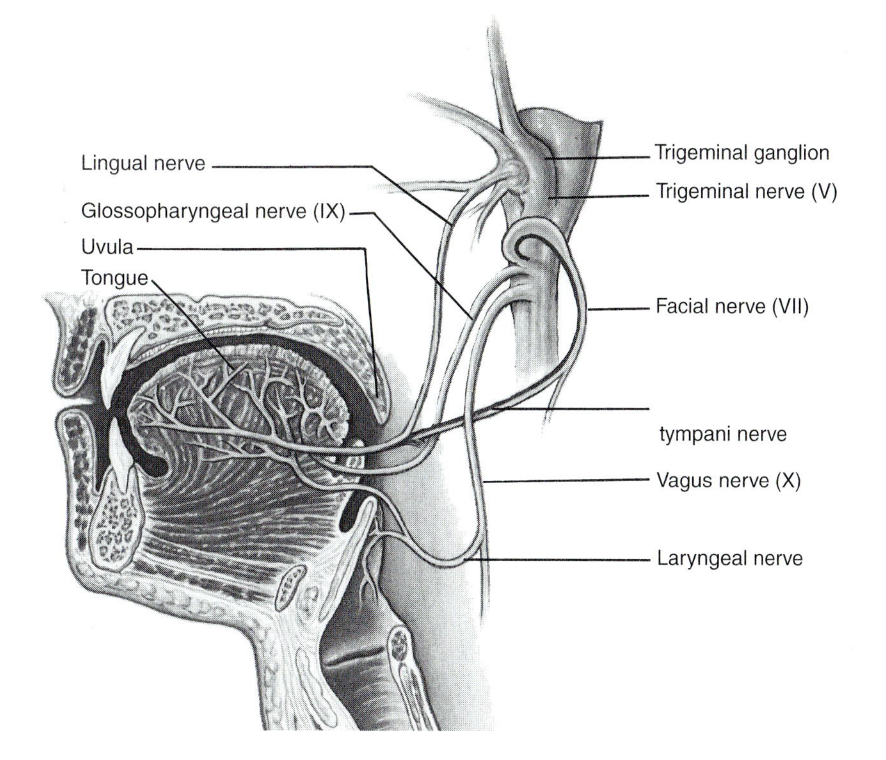


**Figure 3.29** The innervation of the tongue.

innervation in the epiglottis area. Originating from stimulated taste buds, impulses in these nerves are conducted through the thalamus to the postcentral gyrus of the cerebral cortex (ending in the region represented by the tongue), where they are interpreted. The afferent pathway from taste buds to brain is ipsilateral (traveling along the same side of the head), as opposed to most sensory tracts, which cross over.

## PROCEDURE

1. Dry the tongue with a paper towel and, using an applicator stick, apply a dab of 5% sucrose solution to the tip, sides, and back of the tongue.
2. Repeat this procedure using 1% acetic acid, 5% NaCl, and 0.5% quinine sulfate, being sure to rinse the mouth and dry the tongue between applications.

**Note:** *Apply quinine sulfate last; the effect is dominant and often lingering.*

3. Using the sketch provided in the laboratory report, record the location where you tasted each solution. Use the symbols *sw* for sweet, *sl* for salty, *sr* for sour, and *b* for bitter.
4. Pinch your nostrils closed with your fingers and bite into an apple, grape, or other fruit. Chew the food and try to describe its flavor.
5. Now, remove your fingers from your nostrils so that you can also smell the food, and continue to bite and chew the fruit. Describe how its taste compares to what you described in step 4.

The sensations of *gustation* (taste) and *olfaction* (smell) are often grouped together in a single category—the *chemical senses.* The molecular basis of taste and smell is complex. Apparently, both sweet and bitter tastes (as well as perception of particular odorant molecules) are mediated by receptors that are coupled to membrane G-proteins. Dissociation of the G-protein subunit activates second-messenger systems that lead to depolarization of the receptor cell (see fig. 3.28). Together, these chemoreceptors function to provide the proper nuances of taste, which are extremely well developed in some people, such as wine tasters. Humans can distinguish up to 10,000 different odors, with a sensitivity capable of detecting a billionth of an ounce of perfume in the air. Since the sense of smell is so important in tasting, a stuffy nose from a cold or allergy can greatly affect the taste of foods.

# Laboratory Report 3.8

Name ______________________________

Date ______________________________

Section ______________________________

## DATA FROM EXERCISE 3.8

1. Map the areas of the tongue that seem to be sensitive to sweet, salty, bitter, and sour.

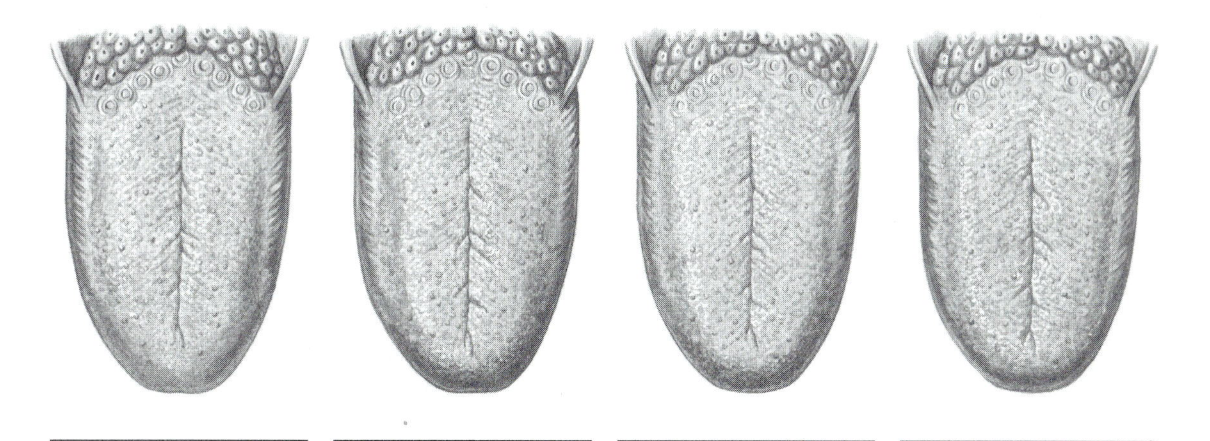

____________ ____________ ____________ ____________

## REVIEW ACTIVITIES FOR EXERCISE 3.8

### *Test Your Knowledge of Terms and Facts*

1. What aspect of a solution causes it to taste sour? ______________________________
2. What aspect of a solution causes it to taste salty? ______________________________
3. Were tastes of sweet, salty, sour or bitter restricted to any particular area of the tongue? Did any area of the tongue seem particularly sensitive to one modality of taste? If so, describe this distribution.

### *Test Your Understanding of Concepts*

4. Describe the common mechanism of action of molecules that taste sweet or bitter.

5. What does the term *chemical senses* mean? Explain how the sense of olfaction and gustation interact.

6. Did your map of taste perception support the traditional concept that different tastes could be mapped to specific locations on the tongue? Propose scientific tests to resolve this issue.

7. Humans are said to be able to distinguish 10,000 different odors. Therefore, do you think we have 10,000 different receptors, each specialized for a different odor? Why or why not? Propose another mechanism that might account for the ability to distinguish so many different odors.

# The Endocrine System

# Section 4

**G***lands* are clusters of secretory cells derived from glandular epithelial membranes that invaginate into the underlying connective tissues (fig. 4.1). When the invagination persists, a duct is formed leading from the secretory cells to the epithelial membrane surface and to the outside of the body. Ducts may thus lead directly to an external body surface or indirectly to the luminal lining of tubes within the digestive, respiratory, urinary, or reproductive tract, which ultimately lead to the outside. These glands are called **exocrine glands,** and include the sebaceous (oil) glands, sweat glands, mammary glands, salivary glands, and the pancreas. In contrast, if the invagination disappears, the chemical product of the gland, a **hormone,** is secreted internally into the blood capillaries. Ductless glands producing hormones are called **endocrine glands.**

Hormones secreted by endocrine glands regulate the activities of other organs (table 4.1). This regulation complements that of the nervous system and serves to direct the metabolism of the hormone's target organs and cells along paths that benefit the body as a whole. By regulating the activity of enzymes within their target cells, hormones primarily regulate total body metabolism and the function of the reproductive system.

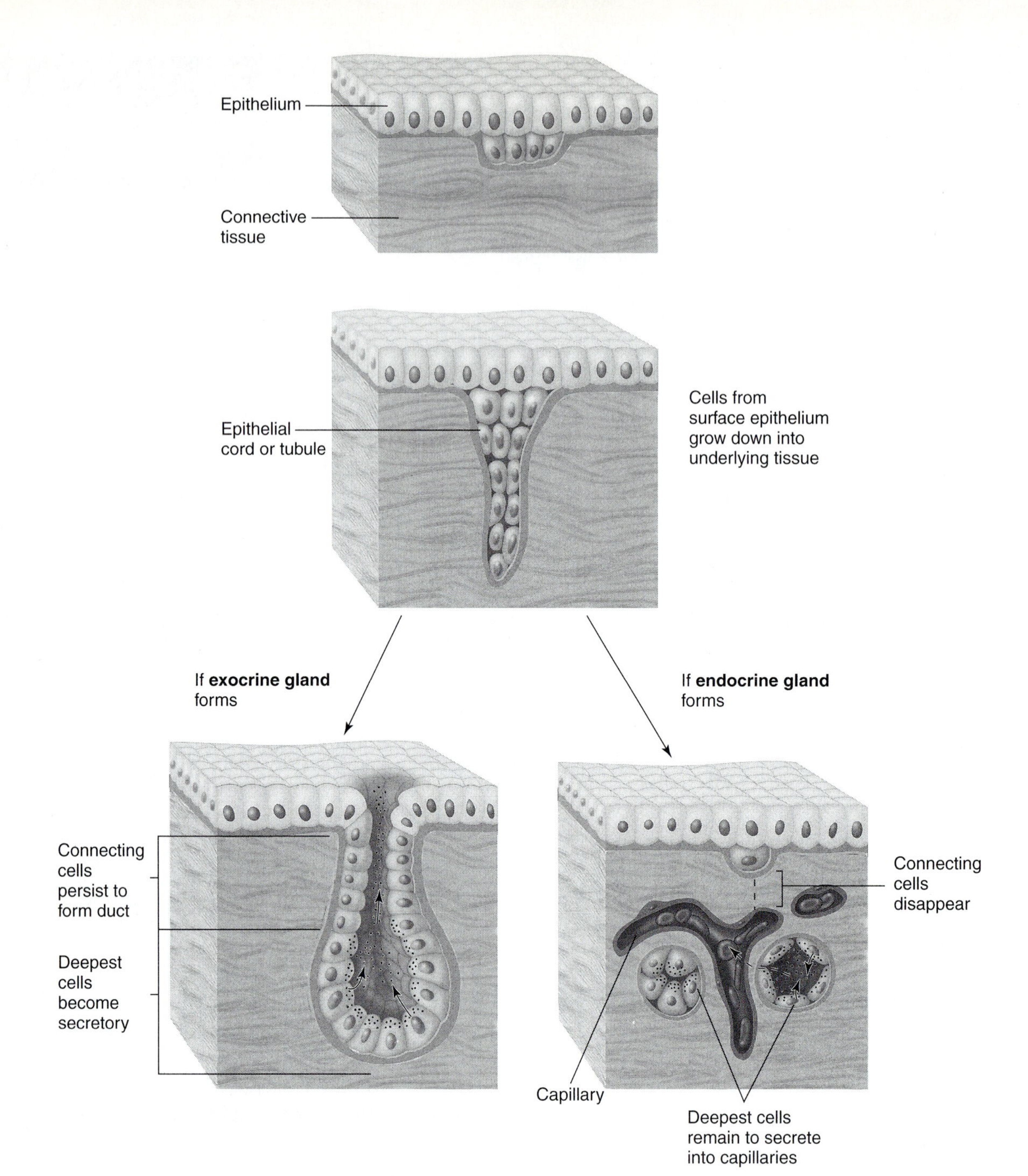


**Figure 4.1 The formation of exocrine and endocrine glands from epithelial membranes.** Note that exocrine glands retain a duct that can carry their secretion to the surface of the epithelial membrane, whereas endocrine glands are ductless.

**(For a full-color version of this figure, see fig. 1.14 in *Human Physiology,* eighth edition, by Stuart I. Fox.)**

## Table 4.1 A Partial Listing of the Endocrine Glands

| Endocrine Gland | Major Hormones | Primary Target Organs | Primary Effects |
|---|---|---|---|
| Adipose tissue | Leptin | Hypothalamus | Suppresses appetite |
| Adrenal cortex | Glucocorticoids<br>Aldosterone | Liver and muscles<br>Kidneys | Glucocorticoids influence glucose metabolism; aldosterone promotes $Na^+$ retention, $K^+$ excretion |
| Adrenal medulla | Epinephrine | Heart bronchioles and blood vessels | Causes adrenergic stimulation |
| Heart | Atrial natriuretic hormone | Kidneys | Promotes excretion of $Na^+$ in the urine |
| Hypothalamus | Releasing and inhibiting hormones | Anterior pituitary | Regulates secretion of anterior pituitary hormones |
| Small intestine | Secretin and cholecystokinin | Stomach, liver, and pancreas | Inhibits gastric motility and stimulates bile and pancreatic juice secretion |
| Islets of Langerhans (pancreas) | Insulin<br>Glucagon | Many organs<br>Liver and adipose tissue | Insulin promotes cellular uptake of glucose and formation of glycogen and fat; glucagon stimulates hydrolysis of glycogen and fat |
| Kidneys | Erythropoietin | Bone marrow | Stimulates red blood cell production |
| Liver | Somatomedins | Cartilage | Stimulates cell division and growth |
| Ovaries | Estradiol-17$\beta$ and progesterone | Female reproductive tract and mammary glands | Maintains structure of reproductive tract and promotes secondary sex characteristics |
| Parathyroid glands | Parathyroid hormone | Bone, small intestine, and kidneys | Increases $Ca^{2+}$ concentration in blood |
| Pineal gland | Melatonin | Hypothalamus and anterior pituitary | Affects secretion of gonadotrophic hormones |
| Pituitary anterior | Trophic hormones | Endocrine glands and other organs | Stimulates growth and development of target organs; stimulates secretion of other hormones |
| Pituitary posterior | Antidiuretic hormone<br>Oxytocin | Kidneys and blood vessels<br>Uterus and mammary glands | Antidiuretic hormone promotes water retention and vasoconstriction; oxytocin stimulates contraction of uterus and mammary secretory units |
| Skin | 1,25-Dihydroxyvitamin $D_1$ | Small intestine | Stimulates absorption of $Ca^{2+}$ |
| Stomach | Gastrin | Stomach | Stimulates acid secretion |
| Testes | Testosterone | Prostate seminal vesicles, and other organs | Stimulates secondary sexual development |
| Thymus | Thymopoietin | Lymph nodes | Stimulates white blood cell production |
| Thyroid gland | Thyroxine ($T_4$) and ($T_3$) calcitonin | Most organs | Thyroxine and triiodothyronine promote growth and development and stimulate basal rate of cell respiration (basal metabolic rate or BMB); calcitonin may participate in the regulation of blood $Ca^{2+}$ levels |

# Histology of the Endocrine Glands

EXERCISE 

## Materials

1. Microscopes
2. Prepared slides

Endocrine glands vary greatly in structure but have some features in common because they all secrete hormones into the blood. The histological structure of the endocrine glands shows how these glands function and how they are related to surrounding tissues.

### OBJECTIVES

1. Describe the structure of the ovaries and testes and the functions performed by their component parts.
2. Describe the histological structure of the pancreas and identify both its endocrine and exocrine structures.
3. Describe the histological structure of the adrenal and thyroid glands and the functions of their component parts.
4. Describe the embryological origin and structure of the anterior pituitary and posterior pituitary and their relationships to the hypothalamus.
5. List the hormones secreted by the anterior pituitary and posterior pituitary and explain how the secretion of these hormones is regulated.

### Textbook Correlations*

Before performing this exercise, you should study the introductory material presented here. Further information relating to this exercise can be found in these pages of *Human Physiology,* eighth edition, by Stuart I. Fox:

- *Pituitary Gland,* Chapter 11, pp. 299–305.
- *Adrenal Glands,* Chapter 11, pp. 305–308.
- *Thyroid and Parathyroid Glands,* Chapter 11, pp. 308–312.
- *Pancreas and Other Endocrine Glands,* Chapter 11, pp. 312–316.

Knowledge of the normal histology of the endocrine glands helps in understanding their normal physiology and in diagnosing various pathological states. Endocrine glands may *atrophy* (lose structure) or develop hormone-secreting tumors known as *adenomas.*

In the testes, various diseases, such as those resulting from *Klinefelter's syndrome* (XXY genotype) or from the *mumps,* are associated with atrophy of the seminiferous (sperm-producing) epithelium. In the ovaries, granulosa cell tumors may secrete excessive estrogen. In *diabetes mellitus,* the beta cells within the islets of Langerhans in the pancreas may decrease in number and have decreased numbers of insulin-containing granules per cell. *Pheochromocytoma,* a tumor of the adrenal medulla, secretes excess epinephrine. Tumors of the anterior pituitary may result in *gigantism* and *acromegaly* (from elevated levels of growth hormone), hyperpigmentation (from excessive adrenocorticotropic hormone), or persistent lactation (from elevated prolactin secretion). These examples are only a few of the disorders associated with an abnormal histology of the various endocrine glands.

Endocrine glands may be independent organs, or they may be part of an organ that also performs nonendocrine functions (table 4.1). Organs that perform both endocrine and nonendocrine functions include the adipose tissue, brain, stomach, pancreas, liver, and skin.

Hormones are carried by the blood to all organs of the body; only certain organs, however, can respond to a given hormone. These are called the **target organs** for the hormone. Hormones affect the metabolism of their target organs and in so doing help to regulate growth and development, total body metabolism, and reproduction.

*Multimedia Correlations (also see Appendix 3)

- *MediaPhys 2.0:* Topics 12.17–12.51

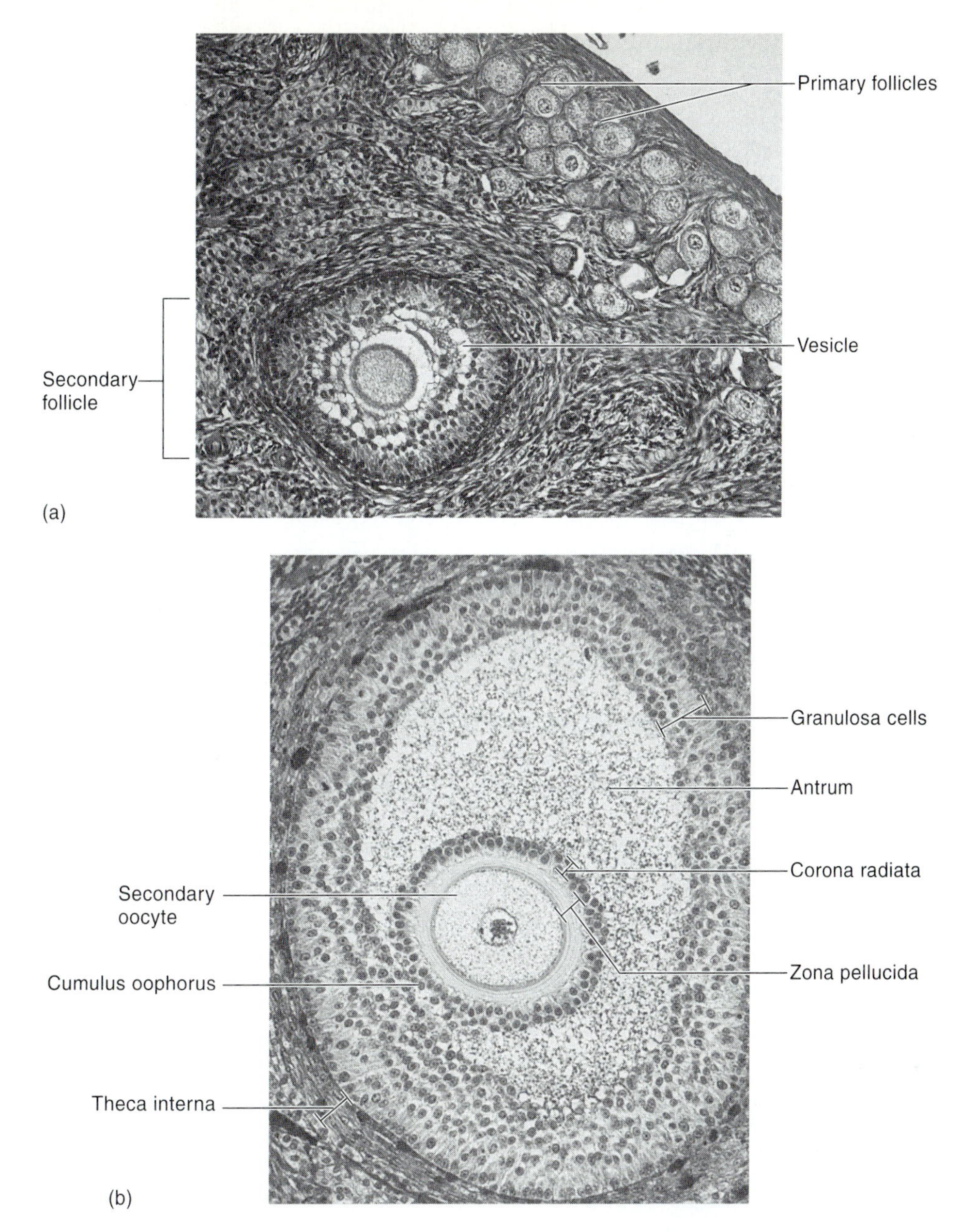


**Figure 4.2** **Ovarian follicles.** Photomicrographs of (a) a secondary follicle and (b) a graafian follicle within an ovary (450×).
**(For a full-color version of this figure, see fig. 20.28 in *Human Physiology,* eighth edition, by Stuart I. Fox.)**

## A. Ovary

The ovary is an endocrine gland as well as the producer of female gametes *(ova)*. The ovum (egg cell) can be thought of as an exocrine secretion because it enters a duct—the uterine tube—after leaving the ovary. The primary hormones of the ovary, *estrogen* and *progesterone*, are secreted directly into the blood of the circulatory system.

The **ovarian follicles** are brought to maturity under the influence of the *gonadotropic hormones* (FSH and LH, mentioned later) secreted by the anterior pituitary. In every cycle, one of the mature follicles, or **graafian follicle,** eventually ruptures through the surface of the ovary to release its ovum (a process called *ovulation*). The empty follicle is then converted into a new endocrine structure called the **corpus luteum.** If fertilization does not occur, the corpus luteum regresses and the cycle is ready to begin again.

The microscopic appearance of the ovary is thus continuously changing as the cycle progresses. A single slide of the ovary will reveal many follicles at different stages of maturation, including primary, secondary, and graafian follicles (fig. 4.2).

### PROCEDURE

Using the low-power objective, scan the slide of the ovary and try to locate a circular-to-elliptical structure, the follicle, that encloses a space filled with fluid and scattered cells. Identify the following parts of the follicle (fig. 4.2):

1. **Ovum.** The ovum (egg cell) is the largest cell in the follicle and, at this stage of development, is called a *secondary oocyte*.
2. **Granulosa cells.** Granulosa cells are the numerous small cells found within the follicle.
3. **Antrum.** The antrum is the central fluid-filled cavity of the follicle.
4. **Cumulus oophorus.** Cumulus oophorus means "egg-bearing hill." This is the mound of granulosa cells that supports the ovum.
5. **Corona radiata.** The corona radiata is the layer of granulosa cells that surrounds the ovum. The ovum continues to be surrounded by its corona radiata after ovulation, and this layer of cells presents the first barrier to sperm penetration during fertilization.
6. **Zona pellucida.** The zona pellucida is a clear region containing glycoproteins between the plasma membrane of the ovum and the corona radiata.

## B. Testis

The testis produces both the male gametes *(sperm)* and the male sex hormone *(testosterone)*. Sperm are produced within the **seminiferous tubules** and travel through these tubules to the **epididymis,** where the sperm are passed into a single tubule that becomes the **ductus (vas) deferens** (fig. 4.3). The ductus deferens picks up fluid from the *seminal vesicles* and the *prostate* and passes its contents, now called **semen,** to the *ejaculatory duct*.

The seminiferous tubules are highly convoluted and tightly packed within the testis. The small spaces, or interstices, between adjacent convolutions of the tubules are filled with connective tissue known as interstitial tissue. Within this connective tissue are the interstitial **cells of Leydig,** endocrine cells that produce the *androgens* (male sex steroid hormones). The major androgen secreted by the Leydig cells of sexually mature males is *testosterone*.

Because the seminiferous tubules are highly convoluted, the chances of seeing a longitudinal section of a tubule are remote. Most of the tubules will be cut more or less in cross section, giving a circular or oblong appearance (fig. 4.3*b*).

### PROCEDURE

Using the *low-power* objective, scan the slide of the testis and locate one seminiferous tubule within a section, or lobule. Switch to a high-power objective and observe these structures (fig. 4.3):

1. **Spermatogenic cells.** Spermatogenic cells form the *germinal epithelium* found along the outer wall of the tubules. These cells divide by **meiosis** to produce the sperm. The outermost cells are *diploid* (forty-six

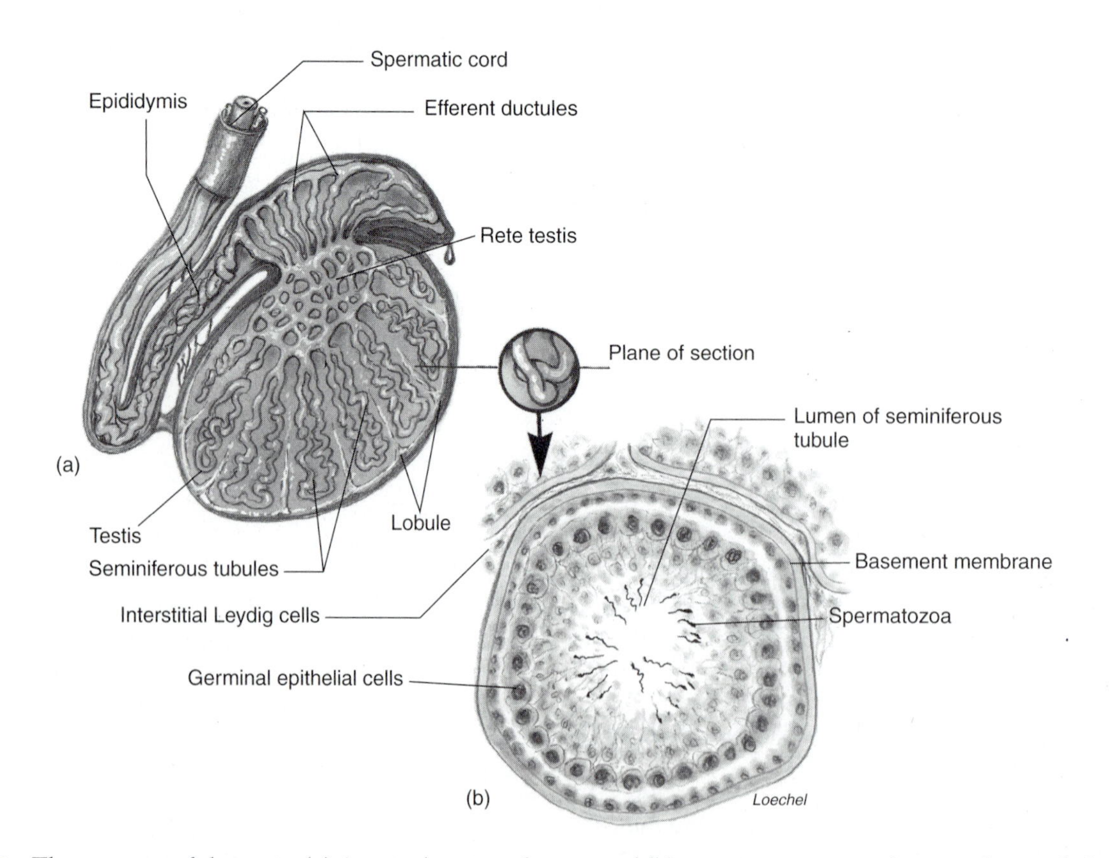


**Figure 4.3 The structure of the testis.** (a) A sagittal section of a testis and (b) a transverse section of a seminiferous tubule.

**(For a full-color version of this figure, see fig. 20.12 in *Human Physiology,* eighth edition, by Stuart I. Fox.)**

chromosomes), whereas the cells towards the lumen have completed meiotic division and are *haploid* (twenty-three chromosomes). Within the tubular epithelium, chromosomes at various stages of meiosis can be seen as darkened structures. Haploid **spermatozoa** can often be seen within the tubular lumen.
2. **Leydig cells.** These endocrine cells can be seen in the interstitial connective tissue between adjacent convolutions of the tubules.

## C. Pancreatic Islets (of Langerhans)

The pancreas has both an exocrine and an endocrine function. The exocrine secretion *(pancreatic juice)* is produced by pancreatic cells called **acini** that are arranged in clusters around a central duct. Pancreatic juice, containing digestive enzymes and bicarbonate, drains into *interlobular ducts* located in bands of connective tissue. From here the secretion flows into the pancreatic duct and empties into the duodenum.

The endocrine secretions of the pancreas, **insulin** and **glucagon,** are produced by scattered groups of cells called the **islets of Langerhans.** These hormones do not enter the interlobular ducts, but rather leave the pancreas by way of the circulatory system.

### PROCEDURE

Using the *low-power* objective, scan the slide of the pancreas and attempt to identify these structures (fig. 4.4):

1. **Acini.** The pancreatic acini are dark-staining clusters of cells that form most of the body of the pancreas.
2. **Interlobular ducts.** The interlobular ducts may be mistaken for veins because of their large size, thin walls, and flattened, irregular shape. Unlike veins, however, their walls are composed of only a single layer of columnar epithelial cells, and no red blood cells will be seen in the lumina.
3. **Pancreatic islets (of Langerhans).** Under low power, the pancreatic islets will appear as light patches, circular in shape, against the dark background of the acini. Under high power, the *alpha cells* (which secrete glucagon) can easily be distinguished from the *beta cells* (which secrete insulin). The alpha cells are smaller and contain pink-staining granules, whereas the beta cells are larger and stain blue.

## D. Adrenal Gland

The **adrenal gland** is actually two different glands located in the same organ. In lower organisms, these glands are separated; in higher organisms (including humans), they are closely associated as the *adrenal cortex* (outer part) and the *adrenal medulla* (inner part) (fig. 4.5).

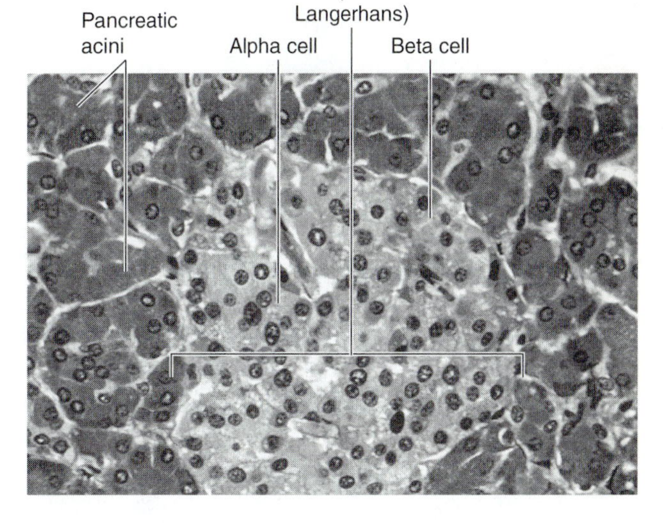


**Figure 4.4 The histology of the pancreas.** The exocrine pancreatic acinar cells (that form secretory structures called acini) and endocrine pancreatic islet (of Langerhans) are seen.

The **adrenal cortex** secretes **corticosteriod hormones.** These include hormones that regulate salt balance *(mineralocorticoids)* and hormones that regulate glucose homeostasis *(glucocorticoids)*. The **adrenal medulla** secretes two hormones, **epinephrine** and **norepinephrine,** that act together with sympathetic nerve stimulation to enhance the response of the cardiovascular system to increased physical demand. The cells of the adrenal medulla are derived from the same embryonic tissue (neural crest ectoderm) as postganglionic sympathetic neurons, whereas the adrenal cortex is derived from a different embryonic tissue (mesoderm). Therefore, these two regions of the adrenal gland are different both physiologically and histologically.

### PROCEDURE

Before observing the slide of the adrenal gland under the microscope, hold it up to the light and note the clear distinction between the adrenal cortex and adrenal medulla. Using the *low-power* objective, focus on the outer edge of the gland. Scan from this point inward and identify these structures (fig. 4.5):

1. **Capsule.** The capsule is a thin, tough layer of connective tissue that surrounds the gland.
2. **Zona glomerulosa.** The zona glomerulosa is the outer layer of the adrenal cortex; its cells are tightly packed in an irregular arrangement. The z. glomerulosa secretes the mineralocorticoids (mainly *aldosterone* and *deoxycorticosterone*). The secretion of these hormones is largely under the control of angiotensin II.

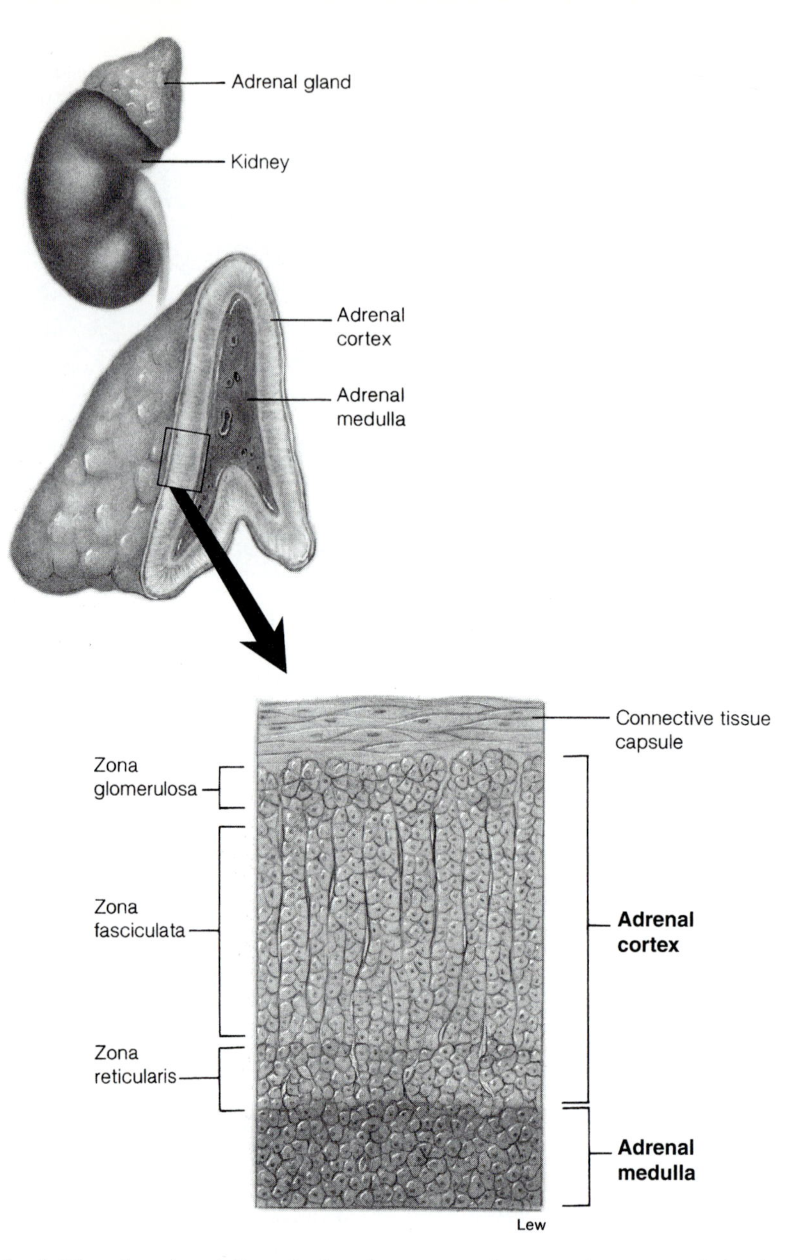


**Figure 4.5 The adrenal gland.** The adrenal medulla and adrenal cortex are shown in the sectioned gland, and the histology of the adrenal cortex, showing its three zones, is illustrated as it would appear at a magnification of 450×.

**(For a full-color version of this figure, see fig. 11.18 in *Human Physiology,* eighth edition, by Stuart I. Fox.)**

3. **Zona fasciculata.** The zona fasciculata is located inside the z. glomerulosa. This layer is the thickest part of the adrenal cortex, with its cells arranged in columns. They secrete the glucocorticoids when stimulated by adrenocorticotropic hormone (ACTH) secreted by the anterior pituitary. The most important glucocorticoids are *hydrocortisone (cortisol)* and *corticosterone*.
4. **Zona reticularis.** The z. reticularis is the innermost layer of the adrenal cortex, and is also involved in the secretion of the glucocorticoids. The epithelial cells in this layer form interconnections (anastomoses) with one another and stain a darker color than those of the z. fasciculata.
5. **Adrenal medulla.** The adrenal medulla forms the distinctive central region of the gland that stains a lighter color than the surrounding z. reticularis. It is composed of tightly packed clusters of *chromaffin cells*.

## E. Thyroid

Like the ovary, the functional units of the thyroid are called **follicles** (fig. 4.6). Each thyroid follicle is composed of a single layer of epithelial cells surrounding a homogenous protein-rich fluid, the *colloid*.

The hormones secreted by the thyroid follicles are *triiodothyronine* ($T_3$) and *tetraiodothyronine* ($T_4$, or thyrox-

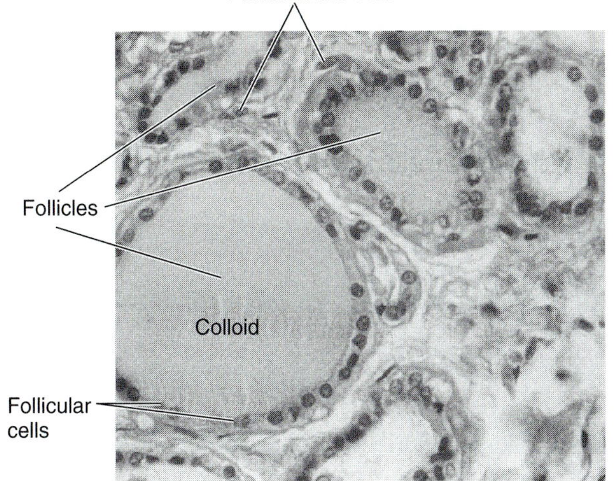


**Figure 4.6** The histology of the thyroid gland.

(For a full-color version of this figure, see fig. 11.22 in *Human Physiology,* eighth edition, by Stuart I. Fox.)

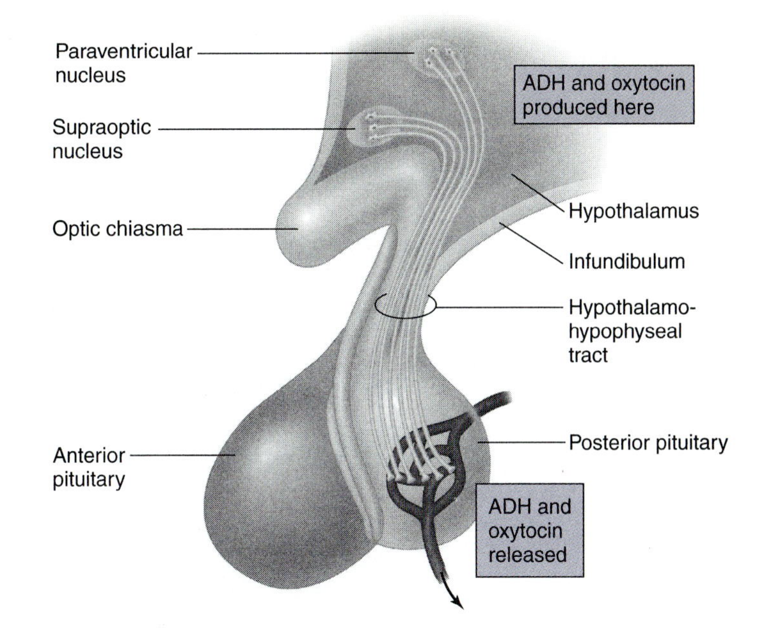


**Figure 4.7** Hypothalamic control of the posterior pituitary. The posterior pituitary, or neurohypophysis, stores and releases hormones—vasopressin and oxytocin—that are actually produced in neurons within the supraoptic and paraventricular nuclei of the hypothalamus. These hormones are transported to the posterior pituitary by axons in the hypothalamo-hypophyseal tract.

(For a full-color version of this figure, see fig. 11.13 in *Human Physiology,* eighth edition, by Stuart I. Fox.)

ine). These hormones are released from the epithelial cells of the follicle into the adjacent capillaries and are important regulators of growth and metabolism.

The thyroid gland also contains **parafollicular cells** that secrete the hormone *calcitonin* (also called *thyrocalcitonin*). This hormone is believed to play a relatively minor role in the regulation of blood calcium concentrations.

### PROCEDURE

Scan the slide under *low* power and observe the follicles (fig. 4.6). Note the clear space between the lighter, inner colloid and the darker, surrounding epithelial cells. This space is an artifact (produced by manipulation of the tissue when the slide was prepared) and is not present *in vivo.*

## F. Pituitary Gland

The **pituitary,** or *hypophysis,* like the adrenal gland, is derived from two distinct embryonic origins. The **anterior pituitary,** also known as the **adenohypophysis** (*adeno* means "glandular"), originates in the embryo from a dorsal outpouching *(Rathke's pouch)* of oral epithelium.

Sometimes referred to as a master gland, the anterior pituitary secretes hormones that control other glands. These hormones include *adrenocorticotropic hormone (ACTH),* which stimulates the adrenal cortex, *thyroid-stimulating hormone (TSH),* which stimulates the thyroid, *gonadotropic hormones* (*FSH* and *LH—follicle-stimulating hormone* and *luteinizing hormone*), which stimulate the gonads), and *prolactin,* which stimulates the mammary glands. In addition, the anterior pituitary secretes *somatotropic hormone,* or *growth hormone (GH),* which stimulates growth in children.

In contrast, the **posterior pituitary,** also known as the **neurohypophysis,** is derived in the embryo from a ventral outpouching of the floor of the brain and secretes only two hormones: *vasopressin* (also called *antidiuretic hormone, ADH*) and *oxytocin.*

The secretion of hormones from both the anterior and the posterior pituitary is controlled by a part of the brain known as the **hypothalamus.** The posterior pituitary is derived as a downgrowth of the hypothalamus, providing the direct neural connection between them (fig. 4.7). Vasopressin (ADH) and oxytocin are manufactured in cell bodies of hypothalamic neurons and packaged into vesicles that travel down the axons of these neurons to the posterior pituitary. Here, these hormones are stored until stimulated for release by axons of the neurons in the hypothalamus. The posterior pituitary is therefore simply a storage organ.

Since the anterior pituitary is derived from oral epithelium and not from brain tissue, there is no direct neural connection between the hypothalamus and the anterior pituitary. There is, however, a special vascular connection between these two organs. A capillary bed in the hypothalamus is connected to a capillary bed in the anterior pituitary by means of venules that run between them. This vascular connection is known as the **hypothalamo-hypophyseal portal system** (fig. 4.8).

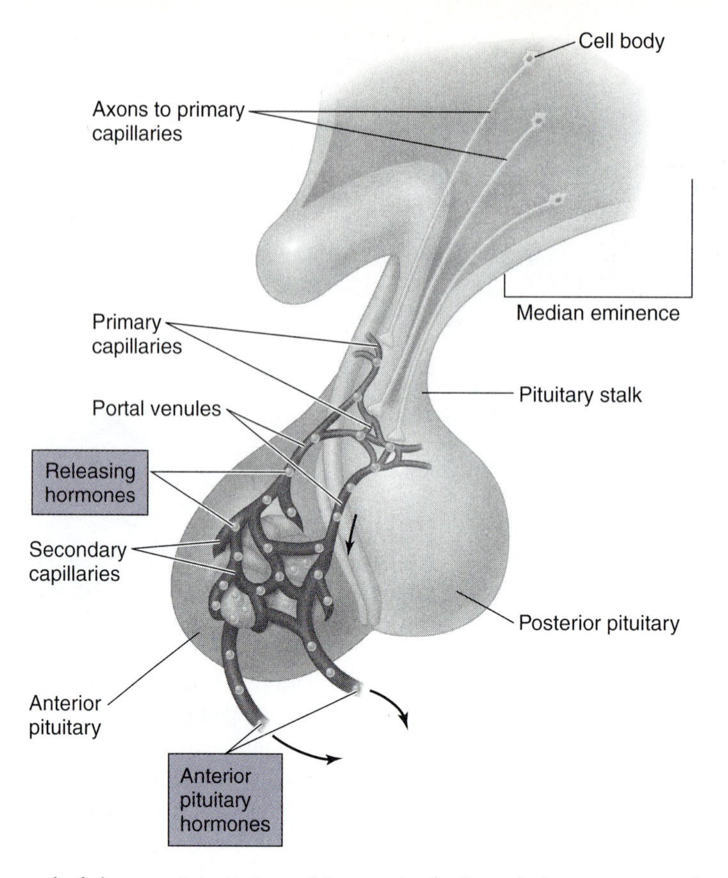


**Figure 4.8 Hypothalamic control of the anterior pituitary.** Neurons in the hypothalamus secrete releasing hormones (shown as dots) into the blood vessels of the hypothalamo-hypophyseal portal system. These releasing hormones stimulate the anterior pituitary to secrete its hormones into the general circulation.

**(For a full-color version of this figure, see fig. 11.15 in *Human Physiology,* eighth edition, by Stuart I. Fox.)**

Unlike the posterior pituitary, the anterior pituitary manufactures its own hormones. These hormones are released upon the arrival of specific chemical messengers, called *releasing hormones*, and secreted into the hypothalamo-hypophyseal portal system by the hypothalamus, as illustrated in figure 4.8. The anterior pituitary, then, is not actually the "master gland," since the secretion of its hormones is in turn controlled by releasing hormones secreted by the hypothalamus. Actually, both the hypothalamus and anterior pituitary are controlled by negative feedback inhibition from their target glands, so the entire concept of a "master gland" is erroneous.

## PROCEDURE

Scan the slide of the pituitary gland under *low* power (fig. 4.9). Distinguish the anterior pituitary (observe capillaries and darkly stained blood cells) from the posterior pituitary (characteristic nerve tissue, lightly stained). Next, switch to the *high*-power objective and identify these structures:

1. **Anterior pituitary**
   (a) **Sinusoids.** Sinusoids are modified capillaries that lack an endothelial wall. They can easily be identified by the presence of red blood cells.

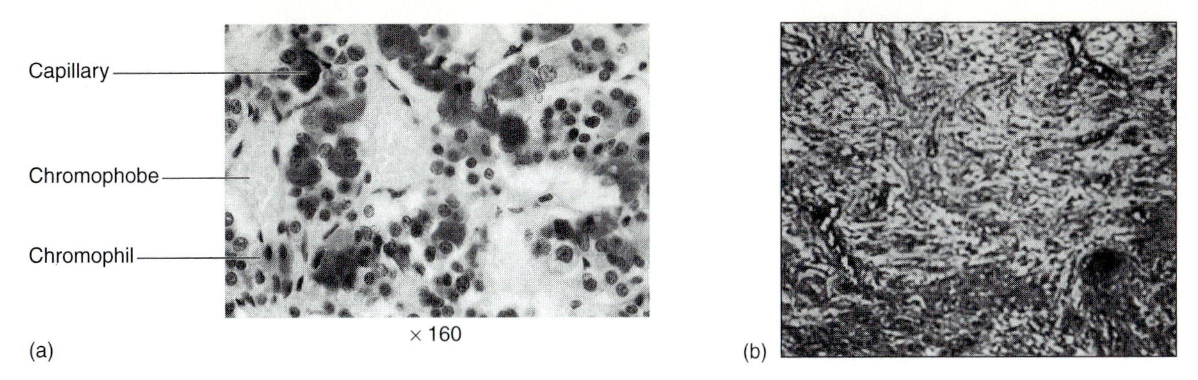


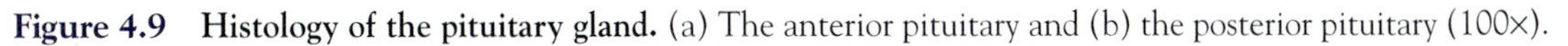
**Figure 4.9 Histology of the pituitary gland.** (a) The anterior pituitary and (b) the posterior pituitary (100×).

(b) **Chromophils.** "Color-loving" chromophils are pituitary cells that readily take up stain into granules present in the cytoplasm. They are divided into two general categories on the basis of their staining properties: *acidophils* contain red-staining granules; *basophils* contain blue-staining granules. These two categories of cells produce different hormones.

(c) **Chromophobes.** The cytoplasm of chromophobes ("color-fearing" cells) does not pick up stain, and hence these cells appear quite dull next to the chromophils. It is believed that the chromophobes are not involved in hormone production.

2. **Posterior pituitary**

(a) **Nerve fibers.** Nerve fibers are the axons of neurons extending from the hypothalamus and compose most of the mass of the gland.

(b) **Pituicytes.** Pituicytes are randomly distributed among the nerve fibers and lack the bright color of the anterior pituitary cells.

# Laboratory Report 4.1

Name ______________________________

Date ______________________________

Section ______________________________

## REVIEW ACTIVITIES FOR EXERCISE 4.1

### *Test Your Knowledge of Terms and Facts*

Match the gland with the hormone it secretes:

| | |
|---|---|
| _____1. ovarian follicle | (a) growth hormone |
| _____2. interstitial cells of Leydig | (b) glucagon |
| _____3. alpha cells of islets of Langerhans | (c) hydrocortisone |
| _____4. beta cells of islets of Langerhans | (d) insulin |
| _____5. zona glomerulosa of adrenal cortex | (e) testosterone |
| _____6. zona fasciculata of the adrenal cortex | (f) aldosterone |
| _____7. adrenal medulla | (g) estrogen |
| _____8. posterior pituitary | (h) oxytocin |
| _____9. anterior pituitary | (i) epinephrine |
| | (j) thyroxine |

10. The fluid-filled central cavity of an ovarian follicle is called the ______________________.
11. Most of the mass of the testis is composed of the ______________________.
12. Another name for tetraiodothyronine is ______________; it is secreted by the ______________ gland.
13. Two glands derived from neural tissue are the ______________ and the ______________.
14. Another name for vasopressin is ______________; it is secreted by the ______________.

### *Test Your Understanding of Concepts*

15. Distinguish between exocrine and endocrine glands.

16. Describe the structures involved in the production, transport, and secretion of oxytocin and vasopressin.

17. Where is ACTH produced? What does it do? Explain how its secretion is regulated.

18. The anterior pituitary has sometimes been called a "master gland." Why was this term used? Why is this erroneous?

## Test Your Ability to Analyze and Apply Your Knowledge

19. Do you think that ligation (tying) of the vas deferens will affect the blood concentration of testosterone? Explain.

20. In *Cushing's syndrome*, the adrenal cortex secretes excessive amounts of glucocorticoids. Which hormones are these? Propose three different possible causes of excessive secretion of glucocorticoids.

# Thin-Layer Chromatography of Steroid Hormones

EXERCISE 4.2

## Materials

1. Thin-layer plates (silica gel, F-254 [F = fluorescent]), chromatography developing chambers, capillary tubes
2. Driers (chromatography or hair driers), ultraviolet viewing box (short wavelength), rulers or spotting template (optional)
3. Steroid solutions: 1.0 mg/mL in absolute methanol of testosterone, hydrocortisone, cortisone, corticosterone, and deoxycorticosterone; 5 mg/mL of estradiol
4. Unknown steroid solution containing any two of the steroids previously described
5. Developing solvent: 60 mL toluene plus 10 mL ethyl acetate plus 10 mL acetone, or a volume containing a comparable 6:1:1 ratio of solvents

## Textbook Correlations

Before performing this exercise, you should study the introductory material presented here. Further information relating to this exercise can be found in these pages of *Human Physiology,* eighth edition, by Stuart I. Fox:

- *Chemical Classification of Hormones.* Chapter 11, pp. 287–289.
- *Mechanism of Steroid Hormone Action.* Chapter 11, pp. 293.
- *Functions of the Adrenal Cortex.* Chapter 11, pp. 306–307.
- *Gonads and Placenta.* Chapter 11, p. 315–316.

Slight differences in steroid structure are responsible for significant differences in biological effects. Differences in structure and solubility can be used to separate a mixture of steroids and to identify unknown molecules.

### OBJECTIVES

1. Identify the major classes of steroid hormones and the glands that secrete them.
2. Describe the primary differences between different functional classes of steroid hormones.
3. Demonstrate the technique of thin-layer chromatography and explain how this procedure works.

## A. Steroid Hormones

The steroid hormones, secreted by the adrenal cortex and the gonads (fig. 4.10), are characterized by a common four-ring structure. The carbon atoms in this structure are numbered as shown here:

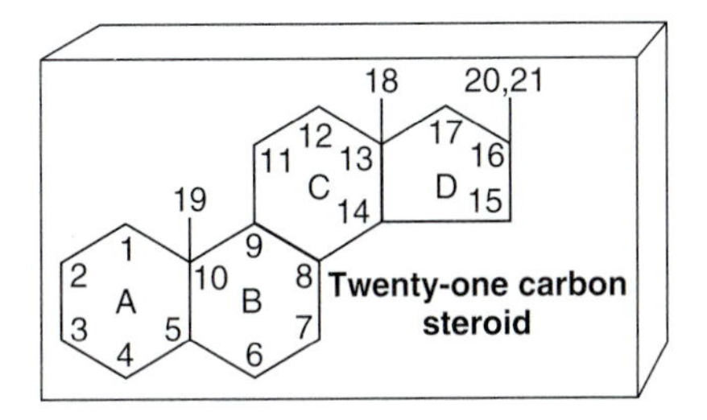


Seemingly slight modifications in chemical structure result in very great differences in biological activity. On the basis of their activity and their structure, the steroid hormones can be grouped into the following functional categories: (1) androgenic hormones, (2) estrogenic hormones, (3) progestational hormones, and (4) corticosteroid hormones. These can be divided further into the subcategories of glucocorticoids and mineralocorticoids.

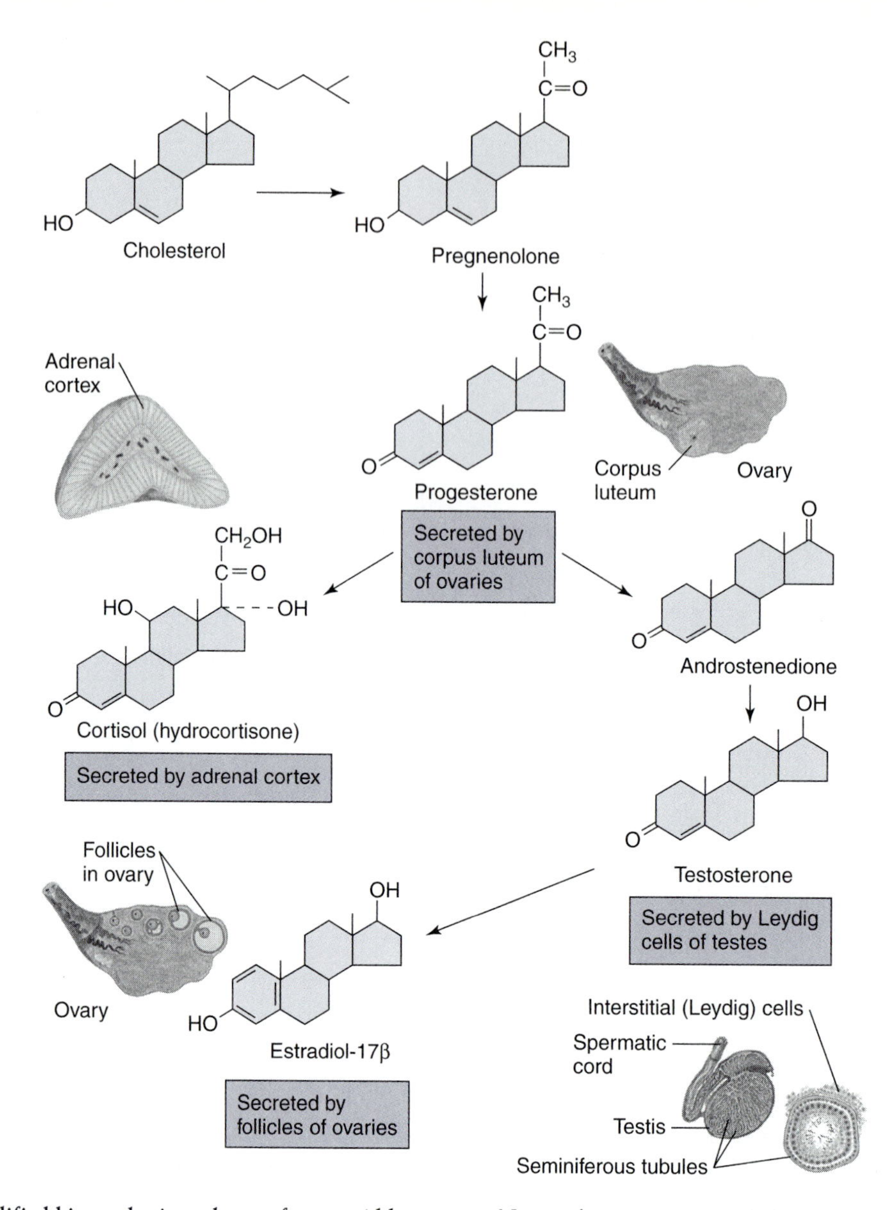


**Figure 4.10** **Simplified biosynthetic pathways for steroid hormones.** Notice that progesterone (a hormone secreted by the ovaries) is a common precursor of all other steroid hormones and that testosterone (the major androgen secreted by the testes) is a precursor of estradiol-17β, the major estrogen secreted by the ovaries.

**(For a full-color version of this figure, see fig. 11.2 in *Human Physiology,* eighth edition, by Stuart I. Fox.)**

The **androgenic hormones** have a characteristic nineteen-carbon steroid structure and function in the development of male secondary sex characteristics. The most potent androgenic hormone is **testosterone,** secreted by the testes.

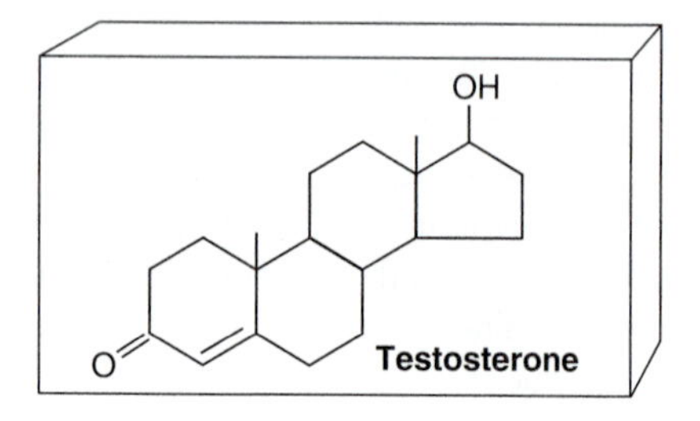


Although the primary source of androgens is the testes, the adrenal cortex also secretes small amounts (adrenal androgens are *dehydroepiandrosterone*, or *DHEA*, and *androstenedione*). Adrenal hyperplasia (Cushing's syndrome) and tumors of the adrenal cortex can also cause excessive levels of plasma androgen, which can have a masculinizing effect in females.

Testosterone and the other androgens are secreted by the interstitial Leydig cells of the testes. This secretion is stimulated by a gonadotropic hormone of the anterior pituitary, *interstitial cell-stimulating hormone (ICSH)*, which is identical to *luteinizing hormone (LH)*.

The structural difference between the androgens and **estrogens** is seemingly slight. The estrogens are eighteen-carbon steroids with three points of unsaturation (double bonds, see appendix 1) in the A ring. These two categories of steroids, however, stimulate the development of markedly different (male and female) secondary sex characteristics. The chief estrogenic hormone is **estradiol.**

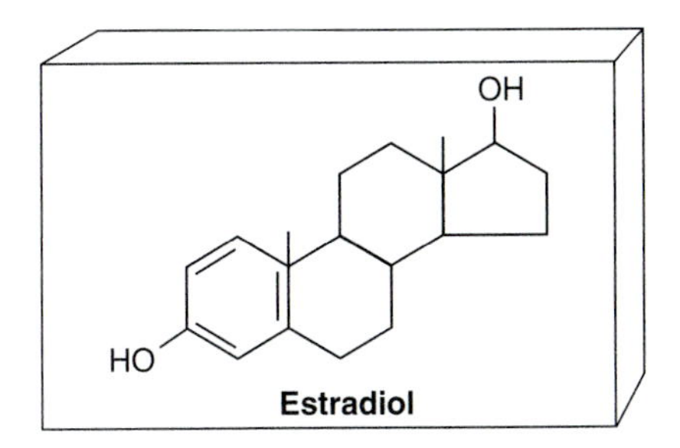


The estrogens are normally secreted in cyclically increasing and decreasing amounts by the ovaries, reaching a peak about the time of ovulation. The cyclical secretion of estrogens is stimulated, in turn, by the cyclical secretion of follicle-stimulating hormone (FSH), a gonadotropic hormone of the anterior pituitary. Abnormally high concentrations of circulating estrogenic hormones may be due to tumors of the adrenal cortex or the ovary. Excessive levels of these hormones can have a feminizing effect in males.

In the normal female cycle, estrogens stimulate growth and development of the inner lining (endometrium) of the uterus. The final maturation of the endometrium is under the control of the hormone **progesterone,** secreted primarily in the phase of the cycle after ovulation (luteal phase) by the corpus luteum of the ovaries. The cyclical secretion of progesterone is, in turn, stimulated by the cyclical secretion of luteinizing hormone (LH) from the anterior pituitary.

During pregnancy, the placenta secretes increasing amounts of progesterone in accordance with the development of the fetus. Progesterone is a twenty-one-carbon steroid.

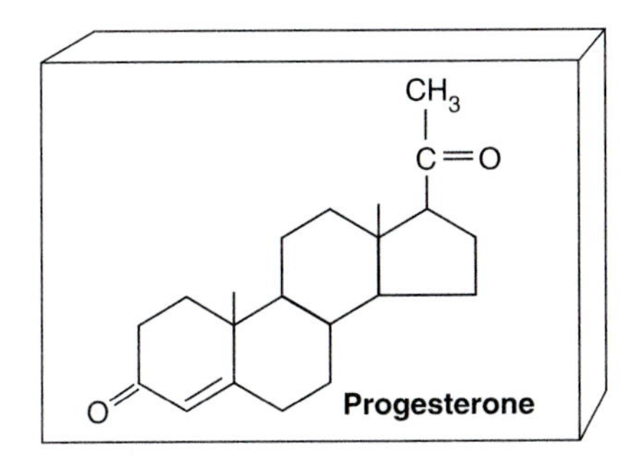


The **corticosteroid hormones** are steroid hormones of the adrenal cortex. Also composed of twenty-one carbons, corticosteroids differ from progesterone by the presence of three or more oxygen groups. These hormones are divided into two functional classes, mineralocorticoids and glucocorticoids, that are secreted by two distinct regions (zona) of the cortex.

The **mineralocorticoids,** secreted by the zona glomerulosa of the adrenal cortex, are involved in the regulation of sodium and potassium balance. Secretion of the mineralocorticoid aldosterone, for example, is stimulated by angiotensin II, which, in turn, is regulated by the secretion of renin from the kidneys. The most potent mineralocorticoids are **aldosterone** and, to a lesser degree, *deoxycorticosterone (DOC)*.

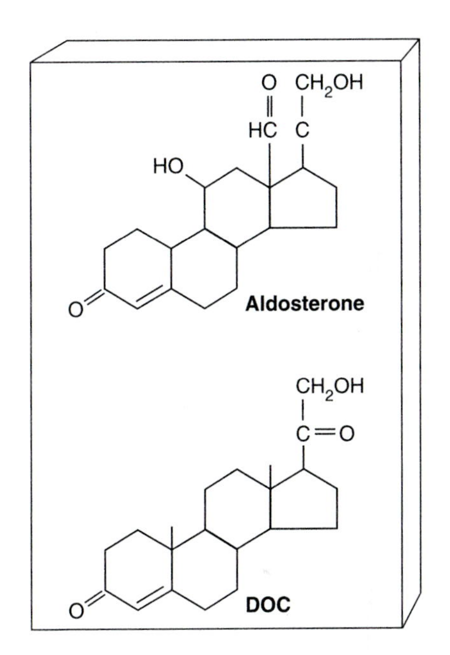


An abnormal secretion of the mineralocorticoids is usually associated with hypertension and may be produced by primary aldosteronism or by secondary aldosteronism due to low blood sodium, high blood potassium, hypovolemia, cardiac failure, kidney failure, or cirrhosis of the liver.

The **glucocorticoids,** secreted by the zona fasciculata and the zona reticularis of the adrenal cortex, stimulate the breakdown of muscle proteins and the conversion of amino acids into glucose (gluconeogenesis). The secretions of the z. fasciculata and the z. reticularis are stimulated by adrenocorticotropin (ACTH) from the anterior pituitary. The most potent glucocorticoid in humans is **cortisol** (or **hydrocortisone**), whereas in many other mammals it is **corticosterone.** *Cortisone* is a glucocorticoid used medically to inhibit inflammation and suppress the immune system.

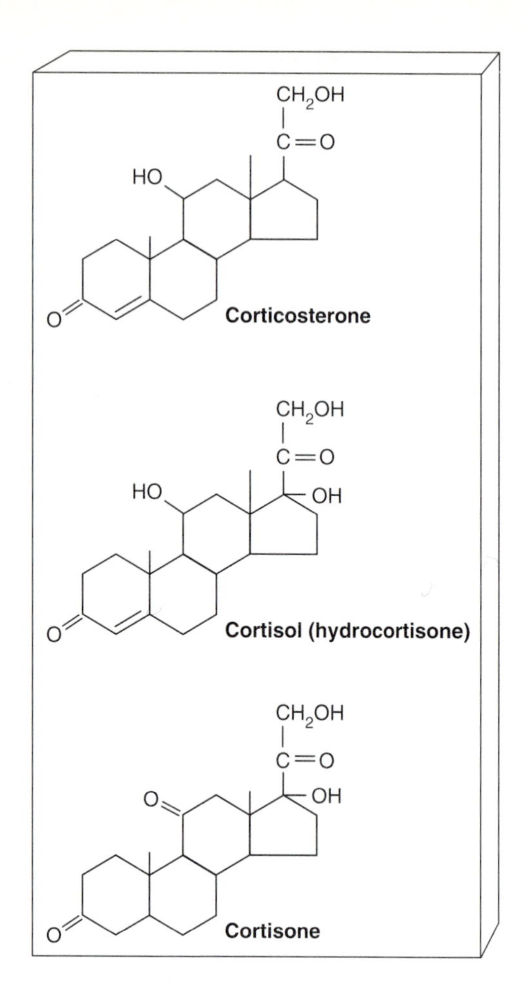


An increased secretion of the glucocorticoids is associated with Cushing's syndrome (adrenal hyperplasia), pregnancy, and stress due to disease, surgery, and burns.

## B. Thin-Layer Chromatography

In this exercise, an attempt will be made to identify two unknown steroids that are present in the same solution. To accomplish this task, you must (1) separate the two steroids and (2) identify these steroids by comparing their behavior with that of known steroids.

Since each steroid has a different structure, each will dissolve in a given solvent to a different degree. These differences in solubility will be used to separate and identify the steroids on a *thin-layer chromatography plate*.

The thin-layer plate consists of a thin layer of porous material (in this procedure, silica gel) that is coated on one side of a plastic, glass, or aluminum plate. The solutions of steroids are applied on different spots along a horizontal line near the bottom of the plate (a procedure called "spotting"), and the plate is placed on edge in a solvent bath *with the spots above the solvent*.

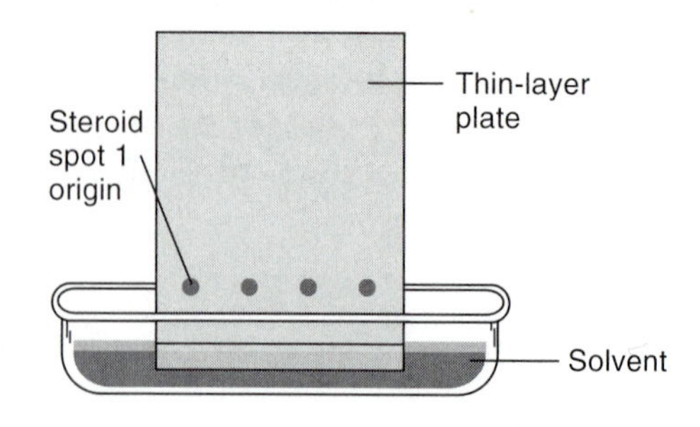


As the solvent creeps up the plate by capillary action, the steroids dissolve off their original spots (or *origins*) and are carried upward with the solvent toward the top end of the plate. Because the solubility of each steroid is different, the time required for the solvent to dissolve and transport the steroids will vary. The relative distance traveled by each steroid will also differ. If the process is halted before all the steroids have reached the top of the plate, some will have migrated farther from the origin than others.

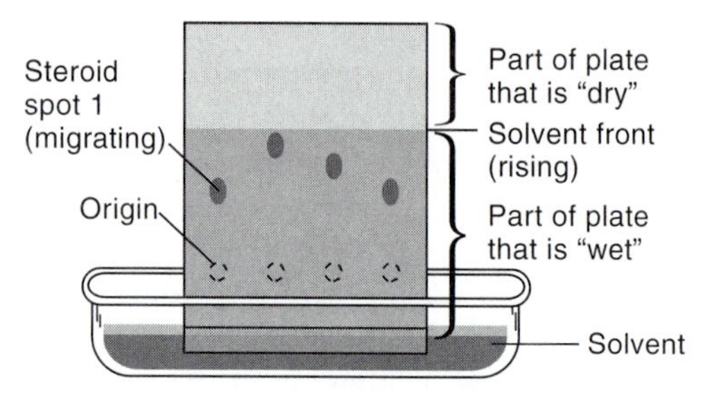


The final pattern (**chromatogram**) obtained using the same steroids, solvent, and conditions is predictable and reproducible. In other words, the distance that a given steroid migrates in a given solvent relative to the travel of the solvent front can be used as an identifying characteristic of that steroid. Each steroid can then be identified by a characteristic numerical value (**$R_f$** value) determined by calculating the distance the steroid traveled (**$D_s$**) relative to the distance traveled by the solvent front (**$D_f$**), as shown here:

$$R_f = \frac{\text{distance from origin to steroid spot}\left(D_s\right)}{\text{distance from origin to solvent front}\left(D_f\right)}$$

Using this method, an unknown steroid can be identified by comparing its $R_f$ value in a given solvent with the $R_f$ values of known steroids in the same solvent.

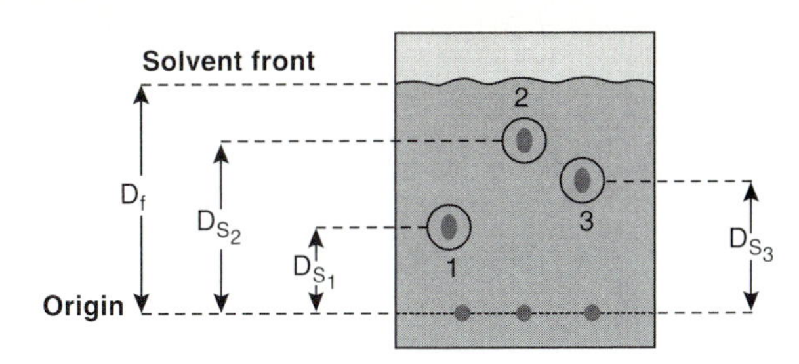

The chromatographic separation and identification of steroid hormones has revealed much about endocrine physiology. The placenta, for example, secretes estrogens that are more polar (more water-soluble) than estradiol, the predominant ovarian estrogen. These polar placental estrogens, *estriol* and *estetrol,* are now measured clinically during pregnancy to assess the health of the placenta.

Chromatograms of androgens recovered from their target tissues (such as the prostate) has revealed that these tissues convert testosterone into other active products. Further, these compounds appear to be more biologically active (more androgenic) than testosterone itself. Testosterone secreted by the testes, therefore, is a *prehormone* that is enzymatically converted in the target tissue into more active products, such as **dihydrotestosterone (DHT)** in many tissues. Males who have a congenital deficiency in *5α-reductase,* the enzyme responsible for this conversion, exhibit many symptoms of androgen deficiency even though their testes secrete large amounts of testosterone.

## PROCEDURE

1. Using a pencil, make a tiny notch on the left margin of the thin-layer plate, approximately 1 1/2 inches from the bottom. The origin of all the spots will lie on an imaginary line extending across the plate from this notch.
2. Using a capillary pipette, carefully spot steroid solution 1 (estradiol) about 1/2 inch in from the left-hand margin of the plate, along the imaginary line. Repeat this procedure, using the same steroid at the *same* spot, two more times. Allow the spot to dry between applications.
3. Repeat step 2 with each of the remaining steroid solutions (2, testosterone; 3, hydrocortisone; 4, cortisone; 5, corticosterone; 6, deoxycorticosterone; 7, unknown), spotting each steroid approximately 1/2 inch to the right of the previous steroid, along the imaginary line.

4. Observe the steroid spots at the origin under an ultraviolet lamp. (**Caution:** Do not look directly at the UV light.)
5. Place the thin-layer plates in a developing chamber filled with solvent (toluene/ethyl acetate/acetone, 6:1:1), and allow the chromatogram to develop for 1 hour.
6. Remove the thin-layer plate, dry it, and observe it under the UV light. Using a pencil, outline the spots observed under the UV light.
7. In the laboratory report, record the $R_f$ values of the known steroids and determine the steroids present in the unknown solution.

# Laboratory Report 4.2

Name ______________________

Date ______________________

Section ______________________

## DATA FROM EXERCISE 4.2

1. Record your data in this table and calculate the $R_f$ value for each spot.

| Steroid | Distance to Front | Distance to Spot | $R_f$ Value |
|---|---|---|---|
| 1. Estradiol | | | |
| 2. Testosterone | same | | |
| 3. Hydrocortisone | same | | |
| 4. Cortisone | same | | |
| 5. Corticosterone | same | | |
| 6. Deoxycorticosterone | same | | |
| 7. **Unknown 1** | same | | |
| 8. **Unknown 2** | same | | |

2. The unknown solutions contained these steroid hormones:
   Unknown 1: ______________________
   Unknown 2: ______________________

## REVIEW ACTIVITIES FOR EXERCISE 4.2

### *Test Your Knowledge of Terms and Facts*

1. The chief estrogenic hormone is ______________, secreted by the ______________.
   ______________________
2. The chief androgenic hormone is ______________, secreted by the ______________.
3. Dehydroepiandrosterone is what kind of a hormone? ______________? Which gland secretes it?
   ______________.
4. Name two different endocrine glands that secrete progesterone: the ______________ and the ______________.
5. The major mineralocorticoid is ______________. It is secreted by the ______________ of the ______________.
6. The major glucocorticoid in humans is ______________. Another name for this hormone is ______________. It is secreted by the ______________ of the ______________.
7. Name these steroids:
   (a) eighteen-carbon sex steroid ______________
   (b) nineteen-carbon sex steroid ______________
   (c) twenty-one-carbon sex steroid ______________

## *Test Your Understanding of Concepts*

8. Create an outline or flow chart of the categories and subcategories of the steroid hormones. Indicate the gland that secretes each hormone.

9. Which of the steroid hormones used in this exercise was most soluble in the solvent? Which was least soluble? Explain.

## *Test Your Ability to Analyze and Apply Your Knowledge*

10. Suppose a man took a drug that acted as a 5α-reductase inhibitor. What effects might this drug have on the prostate? Explain.

11. Suppose the 5α-reductase inhibitor drug described in question 10 also caused hair to grow in a man with male-pattern baldness. (*Note:* There is such a drug—*Propecia.*) What could you conclude about the cause of male-pattern baldness? Explain.

12. The concentration of estriol and estetrol increases in the blood of pregnant women as the pregnancy progresses. How would the migration and $R_f$ values of these two hormones compare with the migration and $R_f$ value of estradiol in this chromatography exercise? Explain.

# Insulin Shock

EXERCISE

## Materials

1. Large beaker filled with water
2. Guppy, goldfish, or another fish of comparable size
3. Insulin (insulin, zinc—100 IU), glucose

Insulin stimulates the tissue uptake of blood glucose and thus acts to lower the blood glucose concentration. Excessive insulin secretion can cause hypoglycemia, which, because of the brain's reliance on blood glucose, can affect brain function and even produce coma and death.

### OBJECTIVES

1. Describe the mechanism by which insulin regulates the blood glucose concentration.
2. Demonstrate the effects of excessive insulin on a small fish and explain the clinical significance of these effects.

### Textbook Correlations

Before performing this exercise, you should study the introductory material presented here. Further information relating to this exercise can be found in these pages of *Human Physiology,* eighth edition, by Stuart I. Fox:

- *Pancreatic Islets (Islets of Langerhans).* Chapter 11, pp. 312–313.
- *Energy Regulation by the Islets of Langerhans.* Chapter 19, pp. 611–615.
- *Diabetes Mellitus and Hypoglycemia.* Chapter 19, pp. 615–618.

**Insulin** is a polypeptide hormone secreted by the *beta cells of the islets of Langerhans.* Insulin stimulates the transport of glucose from the blood into the muscles, liver, and adipose tissue, thus lowering the blood glucose concentration. The effects of insulin injection on the blood glucose concentration of humans is depicted in figure 4.11.

When the islets are incapable of secreting an adequate amount of insulin, a condition known as *diabetes mellitus* develops, in which the transport of glucose from the blood into the body tissues is impaired. This results in an increase in the blood sugar level *(hyperglycemia)* and the appearance of glucose in the urine *(glucosuria).*

Under these conditions, the body tissues cannot obtain sufficient glucose for cellular respiration and increasingly rely on the metabolism of fat for energy. The intermediate products of fat metabolism (ketone bodies) accumulate in the blood, resulting in *ketoacidosis.* Because one of these products is a volatile compound called *acetone* (which has a fruity odor), a person with this condition has fruity-smelling breath. The excretion of ketone bodies and glucose in the urine results is accompanied by the excretion of large amounts of water, causing dehydration. The combination of acidosis and dehydration that results from insufficient insulin may produce a diabetic coma.

People with **type 1 diabetes mellitus** must be given insulin injections to maintain blood glucose homeostasis. If a person with this type of diabetes is given too much insulin, however, the blood sugar level will fall below normal *(hypoglycemia).* Because the central nervous system can only use plasma glucose for energy, the lowering of blood sugar essentially starves the brain. The ensuing condition is called *insulin shock.* The symptoms of insulin shock can be illustrated by the reaction of fish to insulin in this exercise.

Hypoglycemia can, however, have other causes and can result in symptoms less severe than those accompanying insulin shock. *Reactive hypoglycemia* may occur after a meal if the beta cells secrete excessive insulin in response to carbohydrates in the digested food. This type of hypoglycemia sometimes occurs in the beginning stages of diabetes mellitus and can often be controlled by eating smaller, more frequent meals lower in carbohydrates. Hypoglycemia can also occur as a result of alcohol ingestion for reasons that are not well understood. Other causes of

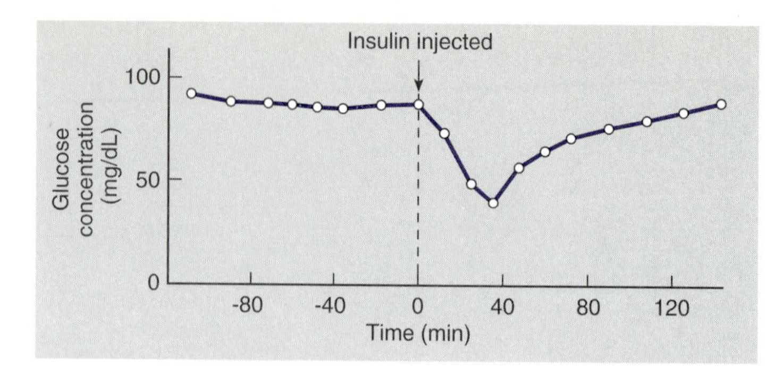


**Figure 4.11 Homeostasis of the blood glucose concentration.** Average blood glucose concentrations of five healthy individuals are graphed before and after a rapid intravenous injection of insulin. The "0" indicates the time of the injection. Notice that, following injection of insulin, the blood glucose is brought back up to the normal range. This occurs as a result of the action of hormones antagonistic to insulin, which cause the liver to secrete glucose into the blood. In this way, homeostasis is maintained.

**(For a full-color version of this figure, see fig. 1.5 in *Human Physiology,* eighth edition, by Stuart I. Fox.)**

hypoglycemia include tumors of the beta cells (insulinomas), which secrete excessive insulin, and liver diseases in which the ability to produce glucose from glycogen and noncarbohydrate molecules is impaired.

The symptoms of hypoglycemia appear when the blood glucose concentration is about 45 mg/dL (normal serum glucose concentration ranges from 70 to 100 mg/dL). Symptoms can appear, however, at higher glucose concentrations if the cerebral circulation is impaired, such as observed in elderly people with atherosclerosis. The symptoms of glucose deficiency—faintness, weakness, nervousness, hunger, muscular trembling, and tachycardia—are symptoms similar to those seen when the brain lacks sufficient oxygen. More prolonged hypoglycemia may damage other parts of the brain, resulting in behavior resembling neuroses and psychoses. Indeed, severe brain damage may result in coma and death.

Type 1 diabetes mellitus is caused by an insulin deficiency, whereas **type 2 diabetes mellitus** is associated with a decreased tissue sensitivity to insulin, called an *insulin resistance*. Indeed, most people with type 2 diabetes mellitus have normal or even elevated levels of insulin. Type 2 diabetes mellitus is more common than type 1 (table 4.2). It is generally associated with obesity, which increases the insulin resistance, and can usually be controlled by moderate weight loss and exercise.

## Table 4.2 Comparison of Type 1 and Type 2 Diabetes Mellitus

| Feature | Type 1 | Type 2 |
|---|---|---|
| Usual age at onset | Under 20 years | Over 40 years |
| Development of symptoms | Rapid | Slow |
| Percentage of diabetic population | About 10% | About 90% |
| Development of ketoacidosis | Common | Rare |
| Association with obesity | Rare | Common |
| Beta cells of islets (at onset of disease) | Destroyed | Not destroyed |
| Insulin secretion | Decreased | Normal or increased |
| Autoantibodies to islet cells | Present | Absent |
| Treatment | Insulin injections | Diet and exercise; oral stimulators of insulin sensitivity |

## PROCEDURE

1. Place a small fish (guppy or goldfish) in a large beaker of water to which a few hundred units of insulin have been added.
2. Observe the effects of insulin overdose. (If no effects are seen in 30 minutes, repeat this step with another fish.)
3. Remove the fish to a second beaker of water containing 5% glucose. Observe the recovery.

# Laboratory Report 4.3

Name ____________________

Date ____________________

Section ____________________

## REVIEW ACTIVITIES FOR EXERCISE 4.3

### *Test Your Knowledge of Terms and Facts*

1. Insulin is secreted by the ____________________ cells of the ____________________.
2. Insulin stimulates the ____________________.
3. Insulin action ____________________ the blood glucose concentration.
4. A deficiency of insulin causes ____________________ (term for blood glucose concentration); by contrast, an excess of insulin causes ____________________.
5. The disease in which there is inadequate action of insulin is called ____________________.
6. A person with the disease named in question 5 would, if untreated, have ____________________ (a term describing urine).

### *Test Your Understanding of Concepts*

7. Explain the meaning of the term *insulin shock*, and why it is it dangerous.

8. Explain why a person with type 1 diabetes mellitus may have a "fruity" breath. Why is this person in danger of dehydration?

## *Test Your Ability to Analyze and Apply Your Knowledge*

9. Explain why a person with type 1 diabetes mellitus can lose weight rapidly.

10. The most common type of diabetes mellitus is type 2. In this condition, the person has beta cells that secrete insulin, sometime even in high amounts. Despite this, the person can still have hyperglycemia. Propose an explanation as to how this might be possible. Most people with type 2 diabetes mellitus are overweight. What does that suggest about large adipocytes compared to small adipocytes?

# Skeletal Muscles

# Section 5

**T**he basic mechanism of contraction for **striated muscles** (*skeletal* and *cardiac* muscles) can be divided into three phases: (1) electrical excitation of the muscle cell, (2) excitation-contraction coupling, and (3) sliding of the muscle filaments, or contraction.

At rest, there is an electrical *potential difference* across the *muscle fiber* (cell) membrane equal to approximately –80 mV (millivolts). The negative sign indicates that the inside of the membrane is negatively charged in comparison to the outside of the cell. When the muscle fiber is appropriately stimulated, either by a direct electric shock or by the motor nerve that innervates the muscle, the permeability of the membrane to cations changes. The diffusion of $Na^+$ into the cell *depolarizes* the membrane and its polarity momentarily reverses. This is immediately followed by the outward diffusion of $K^+$, which *repolarizes* the membrane and reestablishes the more negative resting membrane potential. This rapid depolarization and repolarization of the membrane at the stimulated point is called an **action potential.**

As action potentials are conducted along the muscle fiber membrane, they stimulate a rise in the cytoplasmic concentration of $Ca^{2+}$. In skeletal muscles, this $Ca^{2+}$ comes from a system of intracellular tubules called the *sarcoplasmic reticulum.* In the resting muscle, the absence of this calcium allows two proteins, *troponin* and *tropomyosin,* which are part of the thin actin filaments within the sarcomeres (described below), to inhibit contraction. As a result of electrical stimulation, $Ca^{2+}$ is released into the cell and becomes attached to troponin so that the troponin-tropomyosin complex no longer has an inhibitory effect. Thus, the influx of calcium ions is said to couple electrical excitation to muscle contraction.

Within the muscle fiber, there are numerous subunits *(fibrils)* that are oriented parallel to the long axis of the fiber (fig. 5.1*a*). Each fibril, in turn, is composed of numerous repeating subunits called **sarcomeres.** The sarcomere is the functional unit of contraction. When contraction is stimulated by $Ca^{2+}$, the **thick** and **thin filaments** (composed of **myosin** and **actin,** respectively), within the sarcomeres slide over one another. This sliding of the filaments allows each sarcomere to shorten while its filaments remain the same length. As the sarcomeres become shorter, the fibrils and thus the entire muscle fiber shorten, resulting in muscle contraction (fig. 5.1*b*).

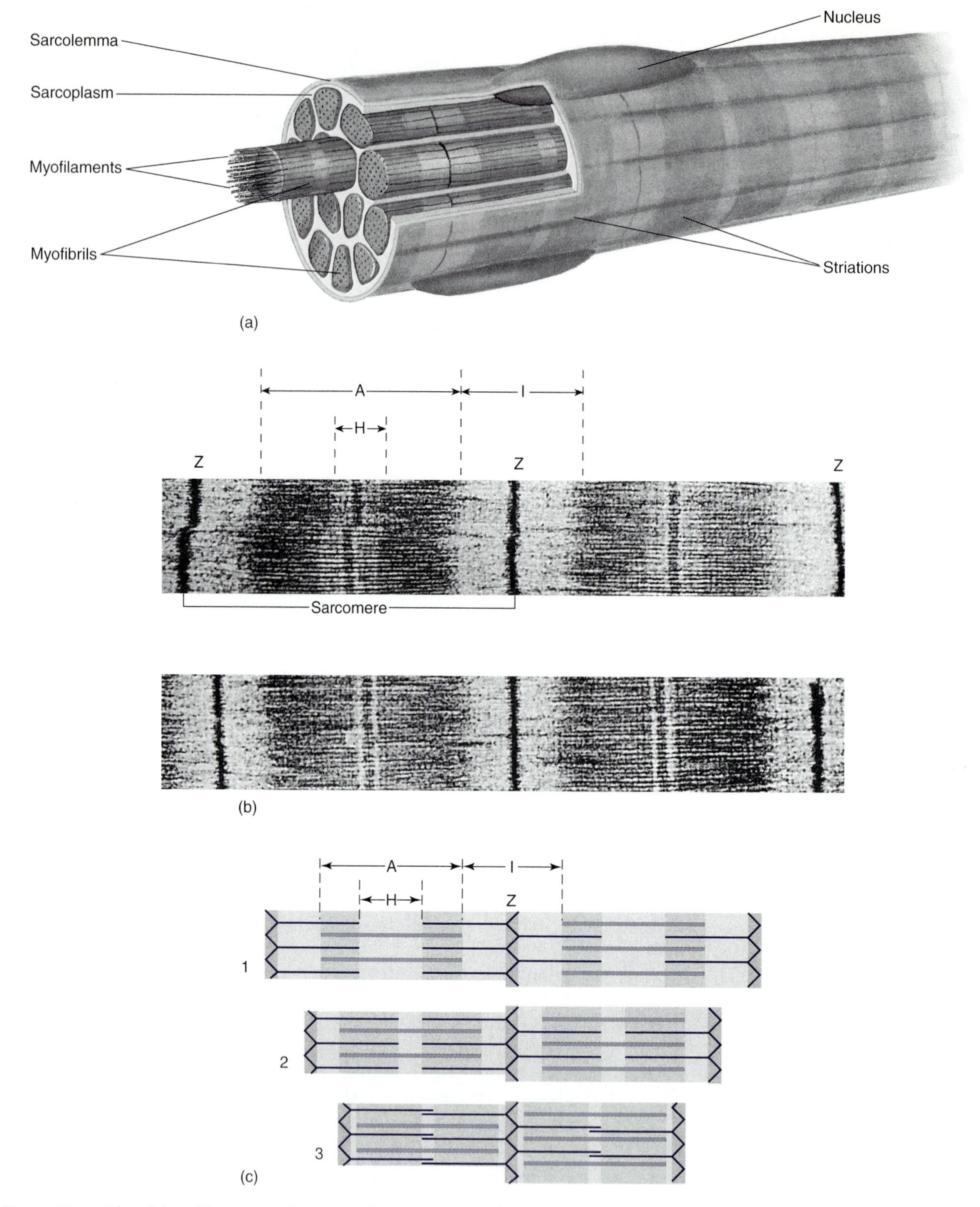


**Figure 5.1 The sliding filament model of muscle contraction.** (a) The structure of a muscle fiber. The changes in band patterns during contraction are shown in an electron micrograph (b) and a diagram (c). As the filaments slide, the Z lines are brought closer together. The A bands remain the same length during contraction, but the I and H bands become progressively narrower and may eventually become obliterated.

**(For full-color versions of (a) and (c) in this figure, see fig. 12.5 and 12.9b in *Human Physiology,* eighth edition, by Stuart I. Fox.)**

# Neural Control of Muscle Contraction

EXERCISE **5.1**

## Materials

1. Frogs
2. Surgical scissors, forceps, sharp probes, dissecting trays, glass probes
3. Recording equipment (either kymograph or electrical recorder, such as physiograph) and electrical stimulators
4. As an alternative to pen-and-paper recording equipment, a computerized data acquisition and analysis system, such as those provided by Biopac, iWorx, and others, may be used to perform the exercises in this section.
5. Straight pins (bent into a "Z" shape), or thin-wired fish hooks (barbless or debarbed) may be used instead; thread
6. Bone clamp (if kymograph is used) or myograph transducer (if physiograph is used)
7. Frog Ringer's solution. (Dissolve 6 g of NaCl, 0.075 g of KCl, 0.10 g of $CaCl_2$, and 0.10 g of $NaHCO_3$ in a liter of water.)

Isolated muscles from a pithed frog can be used to study the physiology of muscle contraction. Isolated frog muscles can be stimulated directly by an electric shock and indirectly through the activation of the appropriate motor nerve.

### OBJECTIVES

1. Prepare a pithed frog for the study of muscle physiology.
2. Describe how muscle contraction can be stimulated by a direct electric shock.
3. Explain how motor nerves stimulate the contraction of skeletal muscles.

## Textbook Correlations*

Before performing this exercise, you should study the introductory material presented here. Further information relating to this exercise can be found in these pages of *Human Physiology,* eighth edition, by Stuart I. Fox:

- *Skeletal Muscles.* Chapter 12, pp. 326–329.
- *Regulation of Contraction.* Chapter 12, pp. 335–339.
- *Neural Control of Skeletal Muscles.* Chapter 12, pp. 347–353.

## Recording Procedures

The **physiograph** is a device for recording the mechanical aspects of muscular contraction (fig. 5.2). It is very sensitive because the mechanical movements of the muscle are first transduced (converted) into electrical current and then greatly amplified prior to recording. Mechanical events with different energies (from muscle contraction to sound waves) can also be recorded, as can nerve impulses and other primarily electrical activity, such as the *electrocardiograph* (ECG), *electromyograph* (EMG), or the *electroencephalograph* (EEG) recordings. Because a number of physiological parameters can be simultaneously recorded on different channels of the physiograph, the temporal (time) relationship between events can be studied.

The physiograph consists of four basic parts: (1) the **transducer** changes the original energy of the physiological event into electrical energy; (2) the **coupler** makes the input energy from the transducer compatible with the built-in amplifier (fig. 5.3); (3) the **amplifier** then increases the strength of the electrical current and forwards the signal to a galvanometer; (4) the **galvanometer** responds to the current generated by directing movement of a pen. The movement of the pen is proportional to the strength of the electrical current generated by the physiological event being measured. Because recording paper moves continuously at a known speed under the pen, both the *frequency* (number per unit time) and the *strength* (amplitude of the pen deflection from a baseline) of the physiological event can be continuously recorded (fig. 5.2).

*Multimedia Correlations (also see Appendix 3)
- A.D.A.M. *InterActive Physiology* (Muscular System): The Neuromuscular Junction
- *MediaPhys 2.0:* Topic 5.10

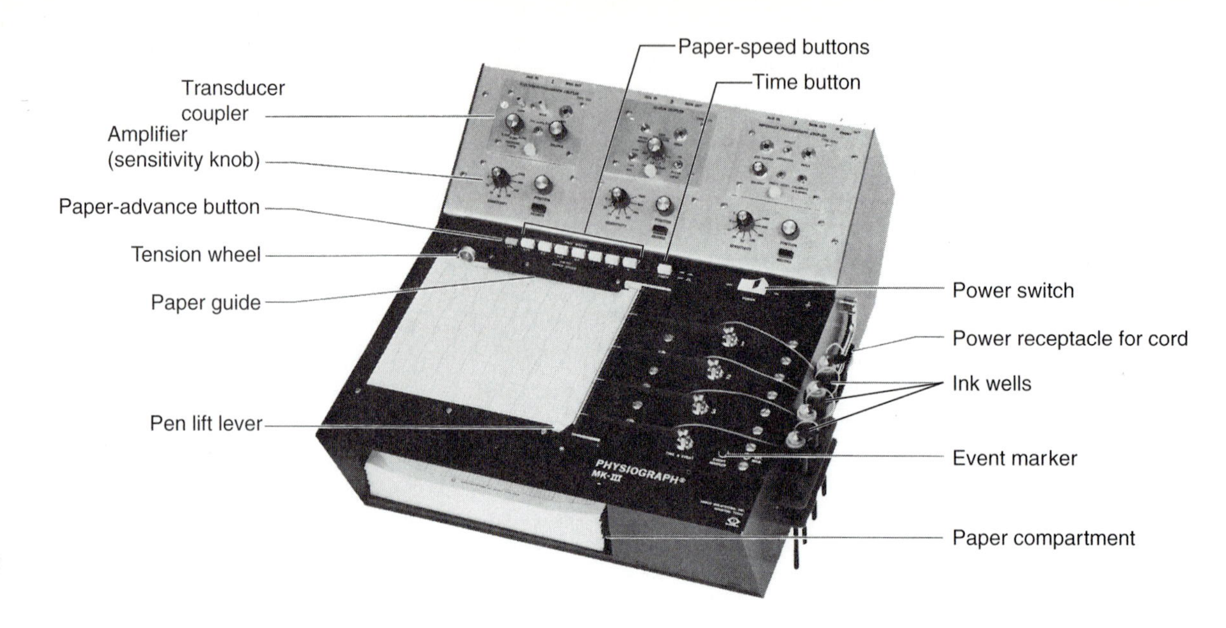


**Figure 5.2** The Physiograph Mark III recorder.

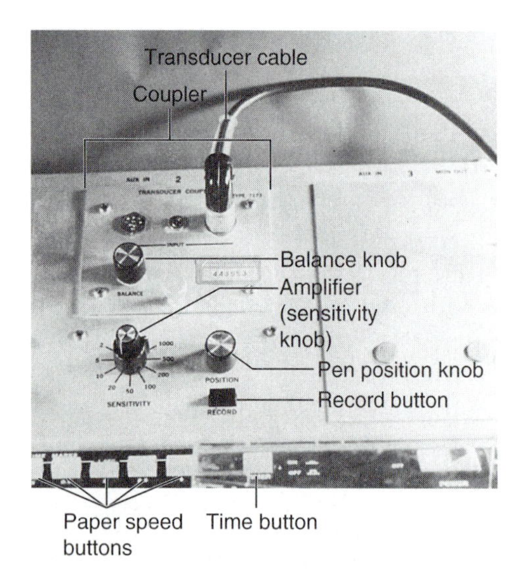


**Figure 5.3 The Narco Mark III Physiograph with an inserted transducer coupler.** The transducer coupler is connected by means of a cable to a myograph transducer.

Similar principles are involved when the physiological event (such as the contraction of a frog skeletal muscle or heart) is transduced into an electric current, and this is now fed into a computer rather than into a pen-and-paper recorder. Unlike the pen-and-paper recorder, the computer can not only display the data but can also analyze it. **Computerized data acquisition and analysis** is usually performed with particular commercial products, such as those produced by Biopac and iWorx. These companies provide exercises comparable to some in this laboratory guide, and these can be used in conjunction with this guide as determined by the instructor.

### *For Physiograph Recording*

1. Insert the *transducer coupler* into the physiograph and plug the *myograph transducer* into the coupler (fig. 5.3).*
2. Raise the inkwells and lower the pens onto the paper by lowering the pen lifter (fig. 5.2). With an index finger covering the hole on the rubber bulb, squeeze to force ink into each pen.
3. Turn the physiograph on with the rocking power switch. Set the paper speed by depressing the *paper-speed* button marked 0.5 cm per second. Turn on the paper drive by depressing the *paper-advance* button and releasing it, allowing it to rise (the *up* position is on).
4. Move the *time switch* to "on." The bottom pen, labeled *time & event*, will make upward deflections every second.

**Note:** *At a paper speed of 0.5 cm/sec, these deflections will be separated by a distance equal to the width of one small box on physiograph recording paper. If the paper speed is increased to 1.0 cm/sec, the deflections of the time-and-event pen will be two small boxes apart.*

5. Turn the outer knob of the amplifier sensitivity control to its lowest number (this will be its greatest sensitivity) (fig. 5.3). With the *record* button off (in the up position), adjust the position of the recording pen for the appropriate channel with the *position* knob, so that the pen writes on the heavy horizontal line closest to the center of the channel being recorded.

---

*For Physiograph Mark III, Narco Bio-Systems

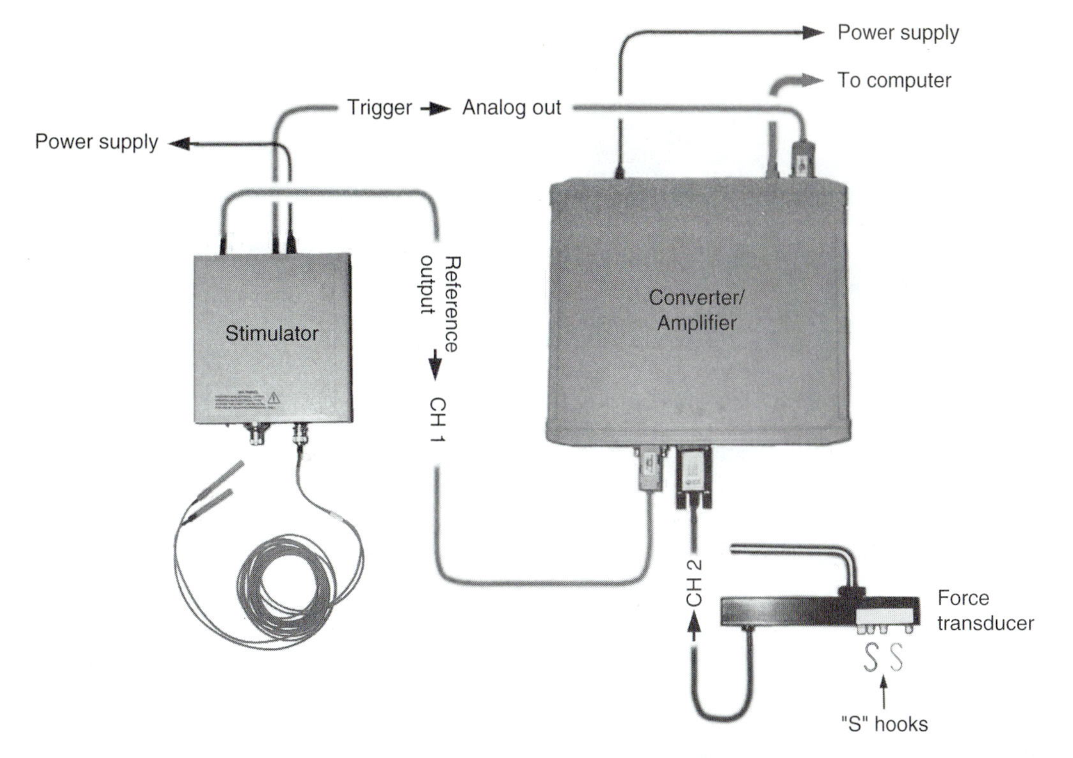


**Figure 5.4 Connection of hardware for use of Biopac system with frog muscle exercises.** The details of the setup are described in the Biopac *PRO* Lesson A02. Once the equipment is set up, it can be used to perform the exercises in this section.

6. Depress the *record* button (the *down* position is on), causing the pen to move away from the heavy horizontal line. Bring the pen back to the line by rotating the *balance* knob. The pen should now remain on the heavy line whether or not the record button is depressed and regardless of the setting of the sensitivity control.

### For Computerized Recording

1. If the Biopac system is used, their force transducer (SS12LA), stimulator (BSLSTM), and amplifier (MP30) can be connected to each other and to the computer as illustrated in figure 5.4.
   a. Follow the Biopac Setup procedure as outlined in their BSL *PRO* Lesson A02 (available on the Internet).
   b. Once the equipment is set up, the exercises in this section can be performed separately or in conjunction with the Biopac *PRO* lesson.
2. If the iWorx system is used (fig. 5.5), the iWorx data acquisition unit, cable, and displacement transducer should be connected to each other and to the computer as shown in iWorx Experiments 7 and 8 (available on the Internet). The exercises in this section can then be performed separately or in conjunction with iWorx Experiments 7 and 8.

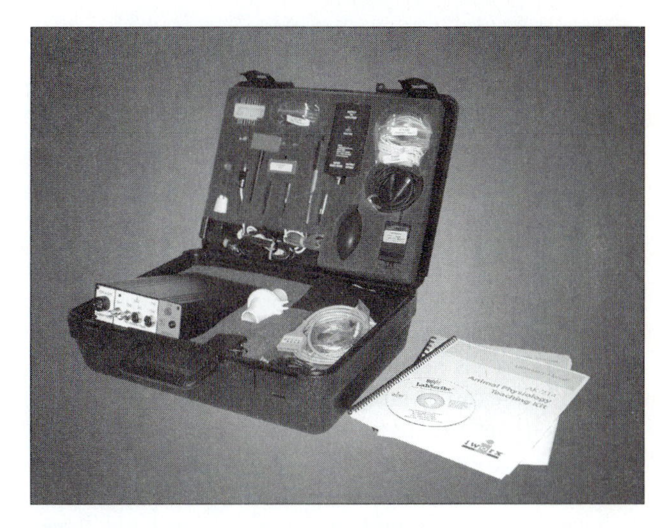

**Figure 5.5 The iWorx system components.** Together with their force transducer and stimulator, this system can be set up as described in iWorx Experiments 7 and 8, and used to perform the exercises in this section.

## A. Frog Muscle Preparation and Stimulation

To study the physiology of frog muscle and nerve, the frog must be killed but its tissues kept alive. This can be accomplished by destroying or **pithing** the frog's central nervous system. The frog is clinically dead (clinical death is defined

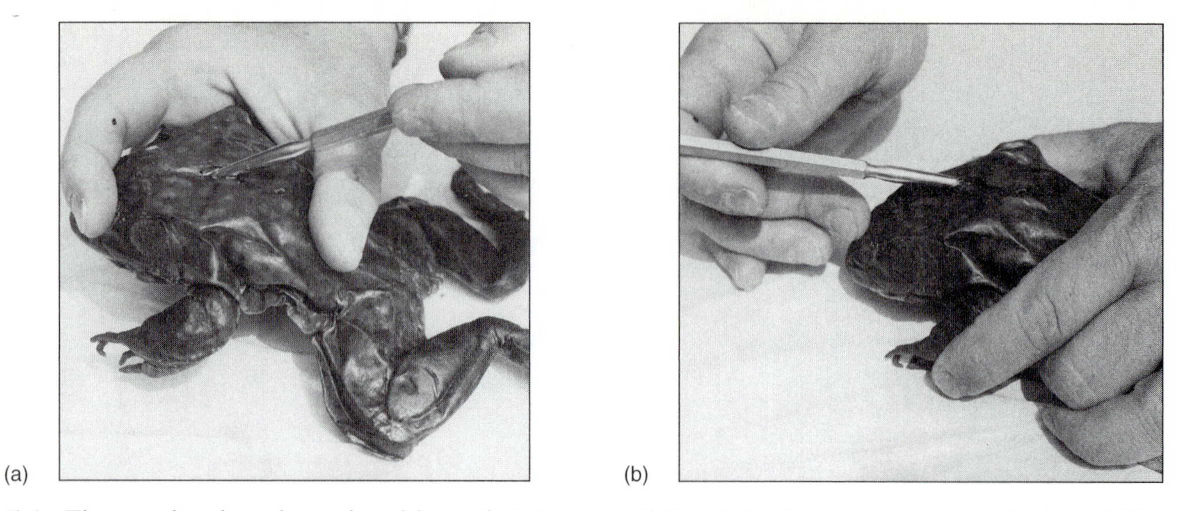


**Figure 5.6 The procedure for pithing a frog.** (a) A probe is first inserted through the foramen magnum into the skull. (b) The probe is then inserted through the spinal cord. (Procedure was simulated with a preserved frog.)

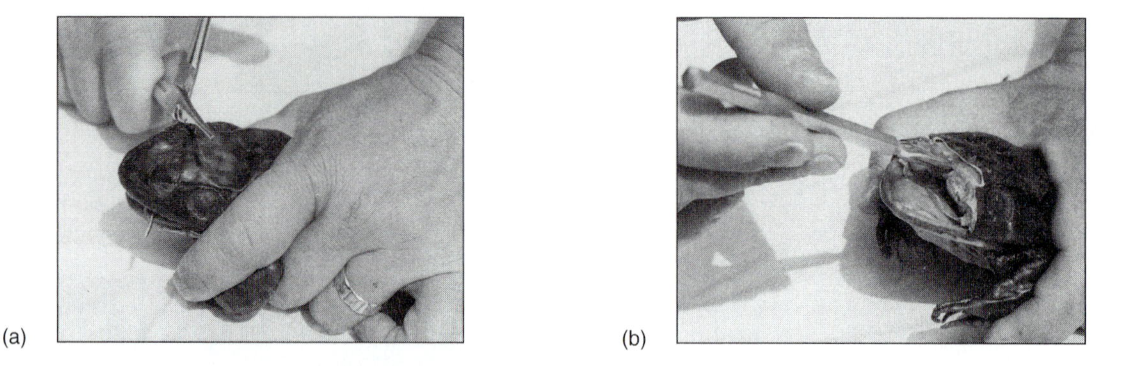


**Figure 5.7 An alternate pithing procedure.** (a) The frog is first decapitated. (b) A probe is then inserted into the spinal cord. (Procedure was simulated with a preserved frog.)

as the irreversible loss of higher brain function), but its muscles and peripheral nerves will continue to function as long as their cells remain alive. Under the proper conditions, this state can be prolonged for several hours.

There are two techniques for pithing a frog. One technique involves grasping the frog securely with one hand and flexing its head forward so that the base of the skull can be felt with the fingers of the other hand. Then perform these steps:

1. Quickly insert a sturdy metal probe into the skull through the foramen magnum (the opening in the skull where the spinal cord joins the brain stem) as in figure 5.6*a*.
2. Move the probe around in the skull, destroying the brain and preventing the frog from feeling any pain (it is now clinically dead).
3. Keeping the head flexed, partially withdraw the probe and turn it so that it points toward the hind end of the frog. Insert the probe downward into the spinal canal (fig. 5.6*b*), destroying the spinal cord and its reflexes. The frog's legs will straighten out as the inserted probe causes reflex stimulation of the spinal nerves. When the spinal cord is destroyed, the frog will become limp.

Alternatively, the following procedure may be employed. Force one blade of a pair of sharp scissors into the frog's mouth as shown in figure 5.7*a*. Quickly decapitate the frog by cutting behind its eyes. It should be understood that the frog is dead as soon as its brain has been severed from its spinal cord. Insert a probe down into the exposed spinal cord as described above to destroy the frog's spinal reflexes (fig. 5.7*b*).

After the frog has been pithed, completely skin one of its legs to expose the underlying muscle. Discard the skin. Then run one blade of a pair of scissors under the Achilles tendon and cut it, leaving part of the tendon still attached to the gastrocnemius muscle (fig. 5.8).

(a)

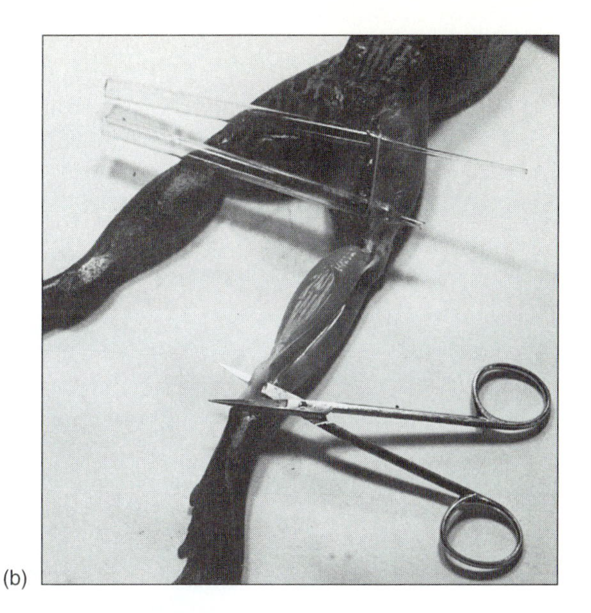

(b)

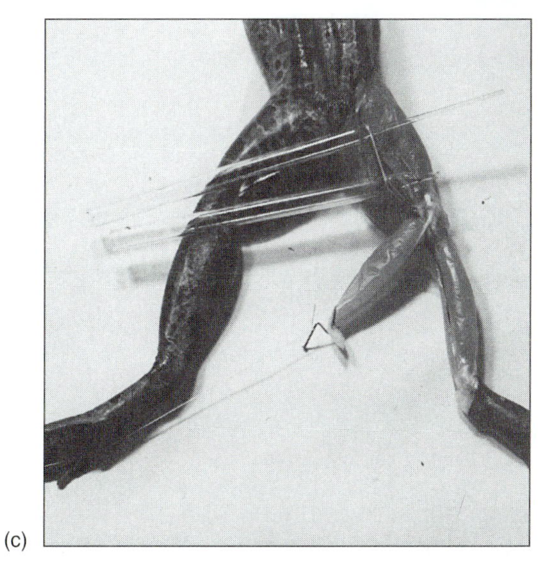

(c)

**Figure 5.8 Preparation of gastrocnemius.** After the frog's leg is skinned (a), the Achilles tendon is cut (b) and a bent pin is inserted into it (c). A length of thread attaches this pin to the hook of the myograph transducer (not shown). In (b), the sciatic nerve is shown between two glass probes in preparation for exercise 5.1.

## SETUP FOR DIRECT STIMULATION OF THE MUSCLE

1. Secure the frog to a dissecting tray by inserting sharp pins through the arms and legs.
2. Push a Z-bent pin through the Achilles tendon. Tie one end of a cotton thread to the bent pin and the other end to the hook of a myograph transducer. Position the myograph so that it is directly above the muscle and adjust the height of the myograph so that the muscle is under tension. Insert two stimulating electrodes directly into the gastrocnemius muscle (fig. 5.8).
3. Establish the **threshold stimulus** (the minimum stimulus that will evoke a particular response). To do this, set the stimulus intensity on 1.0 V and deliver a single shock. Increase the strength of the stimulus in 0.5-V increments until the muscle responds with a contraction *(twitch)* that is recorded clearly on the kymograph or physiograph. Record this voltage and enter it in your laboratory report. **Threshold:** _____ V

**Note:** *Rinse the muscle periodically with Ringer's solution (a salt solution balanced to the extracellular fluid of the frog). Do not allow the muscle to dry out.*

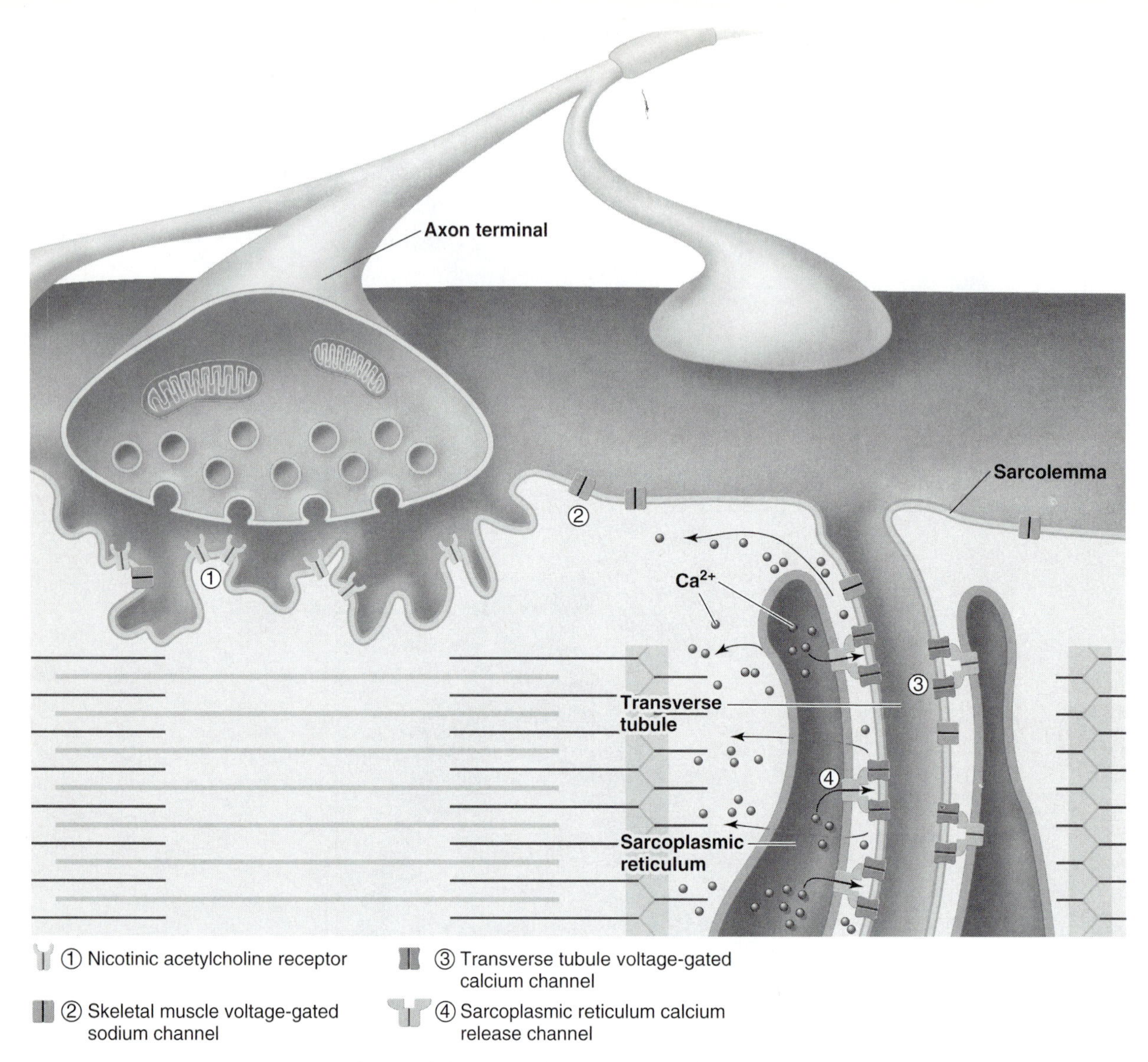


**Figure 5.9 Excitation-contraction coupling in skeletal muscles.** The numbered structures depict the molecules involved in *excitation-contraction coupling,* which refers to the release of $Ca^{2+}$ from the sarcoplasmic reticulum in response to electrical excitation (action potentials), and the stimulation of muscle contraction in response to the released $Ca^{2+}$. Voltage-gated $Ca^{2+}$ channels in the transverse tubules (activated by action potentials) interact with $Ca^{2+}$ release channels in the sarcoplasmic reticulum, leading to the diffusion of $Ca^{2+}$ into the sarcoplasm. The $Ca^{2+}$ can then bind to troponin to stimulate contraction.

**(For a full-color version of this figure, see fig. 12.15 in *Human Physiology,* eighth edition, by Stuart I. Fox.)**

## B. Stimulation of a Motor Nerve

In the body *(in vivo)*, skeletal muscles are stimulated to contract by somatic motor nerves. Action potentials in the motor nerve fibers cause the release of a chemical neurotransmitter called **acetylcholine (ACh)** from the axon endings. This transmitter combines with a receptor protein in the muscle cell membrane and stimulates the production of new action potentials in the muscle fiber. Electrical stimulation of the muscle fibers causes $Ca^{2+}$ to be released from the sarcoplasmic reticulum (fig. 5.9).

The $Ca^{2+}$ released into the sarcoplasm binds to a regulatory protein called **troponin,** which is also bound to a protein known as **tropomyosin** attached to the thin actin filaments. When troponin binds to $Ca^{2+}$, the troponin-tropomyosin complex shifts position on the thin filaments. This exposes binding sites on the actin for the cross-bridges of the myosin thick filaments. Binding of the cross-bridges so actin causes power strokes, producing sliding of the filaments and contraction (fig. 5.10).

The electrical activity of somatic motor neurons is normally stimulated in the spinal cord by synapses with other neurons. These other neurons may be association neurons located in the brain or spinal cord, or they may be sensory, or afferent, neurons. Alternatively, action potentials in motor, or efferent, nerve fibers may be stimulated by damage to the fibers peripherally. This damage produces an *injury current* that stimulates action potentials and subsequent muscle contractions when, for example, a somatic nerve is pinched.

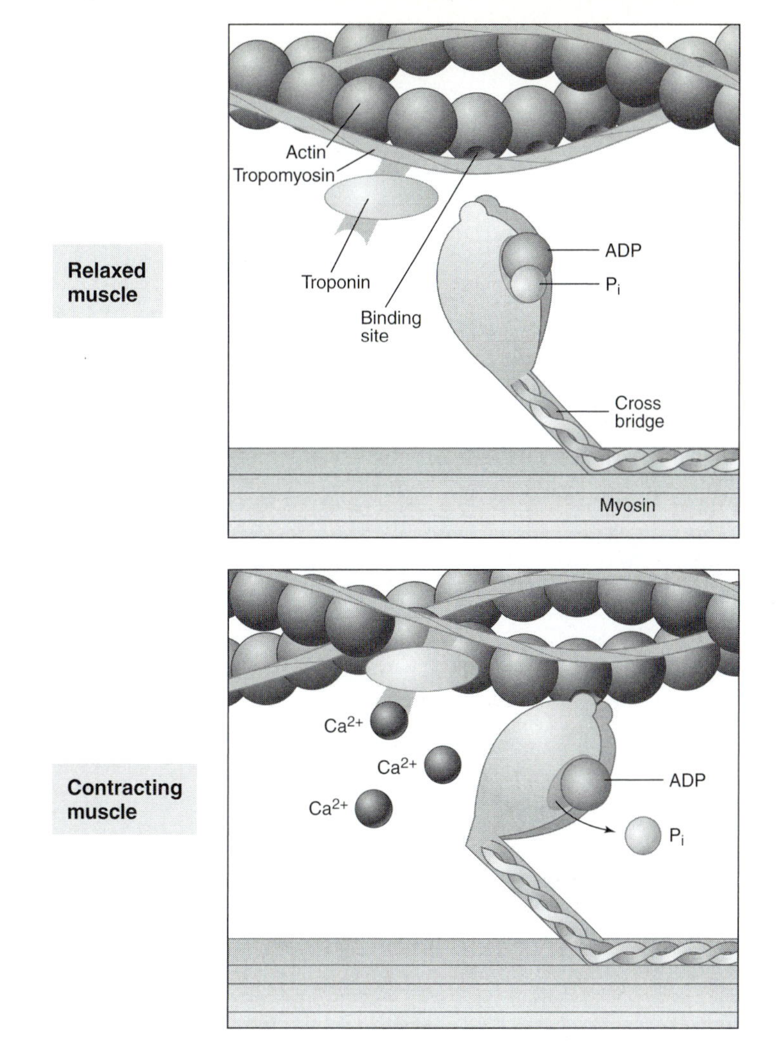


**Figure 5.10 The role of $Ca^{2+}$ in muscle contraction.** The attachment of $Ca^{2+}$ to troponin causes movement of the troponin-tropomyosin complex, which exposes binding sites on the actin. The myosin cross bridges can then attach to actin and undergo a power stroke.

**(For a full-color version of this figure, see fig. 12.14 in *Human Physiology,* eighth edition, by Stuart I. Fox.)**

Some types of muscle degeneration are secondary to nerve damage or to dysfunctions at the neuromuscular junctions. Muscle degeneration follows damage to the motor nerve pathway because proper neuromuscular activity and resulting muscle tone are required for the health of the muscle. In the disease **myasthenia gravis** (*myasthenia* means abnormal muscle weakness or fatigue), antibodies secreted by the immune system block the muscle membrane receptors for acetylcholine, the neurotransmitter of somatic motor neurons. This autoimmune disease prevents the muscle from being properly stimulated by somatic motor neurons.

## PROCEDURE

1. With the frog on its belly, part the posterior muscles of the thigh around the femur to reveal the white sciatic nerve (fig. 5.8*a*).
2. Using glass probes, gently free the nerve from its attached connective tissue and raise it between two glass probes (fig. 5.8*b*).
3. Place both stimulating electrodes under the sciatic nerve. Starting with the stimulator set at 0 V, gradually increase the stimulus voltage in small increments until the minimum stimulus that will produce a muscle twitch is attained. Record this threshold voltage in the space provided in your laboratory report.
   **Threshold:** _____ V
4. Turn off the stimulator but leave the recorder running. Using a length of cotton thread, tie a knot in the nerve. Record and observe the response of the gastrocnemius muscle.

# Laboratory Report 5.1

Name ______________________________

Date ______________________________

Section ______________________________

## DATA FOR EXERCISE 5.1

1. Your threshold stimulus voltage when the electrodes were placed in the muscle: __________ V.
2. Your threshold stimulus voltage when the electrodes were placed on the nerve:__________ V.

## REVIEW ACTIVITIES FOR EXERCISE 5.1

### *Test Your Knowledge of Terms and Facts*

1. Arrange these structures in decreasing order of size: sarcomere, fibril, filaments, fiber:
   (a) ______________________
   (b) ______________________
   (c) ______________________
   (d) ______________________
2. The electrical events conducted along the cell (plasma) membrane that stimulate contraction are called ______________________________.
3. Actin and myosin comprise the ______________________ and ______________________ filaments, respectively.
4. What substance couples electrical excitation to muscle contraction? ______________________
5. The substance named in question 4 is stored in which intracellular organelle? ______________________
6. The substance named in question 4 binds to a regulatory protein known as ______________________, which in turn is bound to an inhibitory protein called ______________________.
7. The neurotransmitter chemical that stimulates contraction of skeletal muscles: ______________________.

### *Test Your Understanding of Concepts*

8. Draw a sarcomere and label the parts and the bands. Then, describe and illustrate how the structure of a sarcomere changes during muscle contraction.

9. Trace the course of events from the release of ACh by a motor neuron to the binding of myosin cross-bridges to actin.

## *Test Your Ability to Analyze and Apply Your Knowledge*

10. Which had the lower threshold for stimulation of muscle contraction—stimulation of the muscle directly, or stimulation of the nerve that innervates the muscle? Propose an explanation for these results.

11. Using your knowledge of the regulation of muscle contraction, predict what might happen to the beating of a heart if the blood concentration of $Ca^{2+}$ were abnormally increased.

12. Predict the effects on muscles of a drug that blocks the action of acetylcholinesterase, an enzyme that breaks down acetylcholine. Compare that to the effects on muscles of a drug that blocks acetylcholine receptors.

# Summation, Tetanus, and Fatigue

EXERCISE **5.2**

## Materials

1. Frogs
2. Equipment and setup used in exercise 5.1
3. Electrocardiograph plates and electrolyte gel
4. Alternative equipment: Physiogrip, (Intellitool, Inc.); or Biopac system with hand dynamometer

Twitch, summation, and tetanus can be produced by direct electrical stimulation of frog muscles *in vitro* (in the laboratory) and by stimulation of human muscles *in vivo* (within the living body). These procedures demonstrate how normal muscular movements are produced.

### OBJECTIVES

1. Define the terms *twitch, summation, tetanus,* and *fatigue.*
2. Demonstrate twitch, summation, and tetanus in both frog and human muscles; and demonstrate fatigue in the frog muscle preparation.
3. Explain how a smooth, sustained contraction is normally produced.

## Textbook Correlations*

Before performing this exercise, you should study the introductory material presented here. Further information relating to this exercise can also be found in these pages of *Human Physiology,* eighth edition, by Stuart I. Fox:

- *Motor Units.* Chapter 12, pp. 328–329.
- *Contractions of Skeletal Muscles.* Chapter 12, pp. 340–342.
- *Muscle Fatigue.* Chapter 12, pp. 346.

Individual skeletal muscle fibers cannot sustain a contraction; they can only twitch. Muscle fibers also do not usually produce a graded contraction; instead, they usually contract maximally to stimulation. Consequently, smooth, graded skeletal muscle contractions are produced by the **summation** of fiber twitches. Summation occurs when some fibers within a muscle are in the process of contraction before other fibers in the muscle have had time to relax completely from a previous twitch. This results in a second, stronger twitch that may partially "ride piggyback" on the first (fig. 5.11*a*). The strength of skeletal muscle contraction, therefore, depends on the number of fibers stimulated rather than on the strength of the individual muscle fiber contractions. Maintenance of a sustained muscle contraction is called **tetanus.** (The term *tetanus* should not be confused with the disease tetanus, which is accompanied by a painful state of muscle contracture, or *tetany*).

Tetanus can be demonstrated in the laboratory by setting the stimulator to deliver shocks automatically to the muscle at an ever-increasing frequency until the twitches fuse into a smooth contraction (fig. 5.11*b*). This is similar to what occurs in the body when different motor neurons in the spinal cord are activated to stimulate a muscle at slightly different times.

If the stimulator is left on so that the muscle remains in tetanus, a gradual decrease in contraction strength will be observed. This is due to **muscle fatigue.** Fatigue during a sustained maximal contraction, as when lifting a very heavy weight, appears to be due to an accumulation of extracellular $K^+$. This depolarizes the membrane potentials and interferes with the ability of the muscle fiber to produce action potentials. Fatigue (and muscle pain) during more moderate exercise is related to the production of lactic acid during anaerobic respiration, which decreases muscle fiber pH. The fall in pH is believed to interfere with the storage or release of $Ca^{2+}$ from the sarcoplasmic reticulum, and thus to interfere with excitation-contraction coupling.

*Multimedia Correlations (also see Appendix 3)
- Intelitool: Physiogrip
- A.D.A.M. *InterActive Physiology* (Muscular System): Contraction of Motor Units; Contraction of Whole Muscle
- *MediaPhys 2.0:* Topics 5.16–5.18

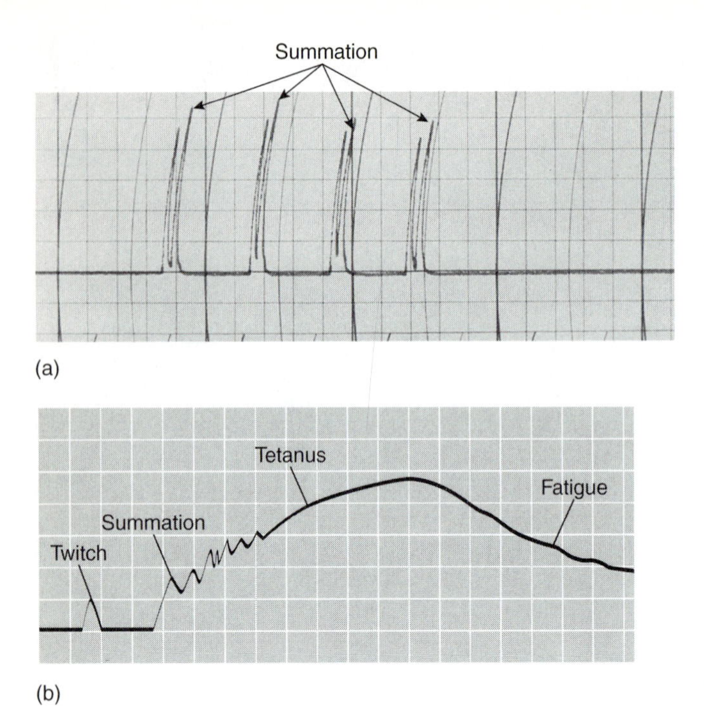


**Figure 5.11 Recording of muscle contractions.** (a) A recording of the summation of two muscle twitches on a physiograph recorder. Notice that contraction to the second stimulus is greater than contraction to the first stimulus (the intensity of the first and second stimuli is the same). (b) Diagram illustrating twitch, summation, tetanus, and fatigue.

## A. Twitch, Summation, Tetanus, and Fatigue in the Gastrocnemius Muscle of the Frog

Summation, tetanus, and fatigue can be demonstrated with the frog gastrocnemius muscle preparation used in exercise 5.1.

### PROCEDURE

1. Set the stimulus voltage above threshold and press down on the switch that delivers a single pulse to the muscle two or three times in rapid succession. If this is done with enough speed, successive twitches can be made to "ride piggyback" on preceding twitches, demonstrating *summation* (fig. 5.11*a*).
2. Set the stimulus switch to deliver shocks to the muscle automatically at a frequency of about one per second. Gradually increase the frequency of stimulation until the twitches fuse into a smooth, sustained contraction, demonstrating *tetanus* (fig. 5.11*b*).
3. Maintain stimulation until the strength of contraction gradually diminishes as a result of muscle *fatigue* (fig. 5.11*b*).
4. Enter your recordings in the laboratory report.

## B. Twitch, Summation, and Tetanus in Human Muscle

The properties of frog muscle contractions observed *in vitro* reflect the behavior of human muscle *in vivo* in many ways. A single pulse of electrical stimulation produces a single short contraction (twitch), and many pulses of stimulation delivered in rapid succession produce a summation of twitches resulting in a smooth, graded muscular contraction, and eventually in tetanus.

Sustained muscular spasm *(tetany)* may be produced by hypocalcemia and by alkalosis. (The most common cause of tetany is alkalosis produced by hyperventilation.) Cramps may be due to a variety of conditions, including salt depletion. General muscle weakness may be caused by alterations in plasma potassium levels (due, for example, to excessive diarrhea or vomiting).

**Muscular dystrophy** refers to any of a variety of diseases in which there is a progressive lack of support (dystrophy) and weakness of skeletal muscles (although heart muscle may also be involved) that does not seem to be caused by inflammation or neural disease. In severe forms of these diseases, large numbers of filaments and sarcomeres are replaced with fibrous connective tissue and fatty deposits.

### PROCEDURE

1. Rub a small amount of electrolyte gel on the skin near the wrist and attach an ECG electrode plate to this area with an elastic band. Rub electrolyte gel on a second ECG electrode plate and place it on the anterior, medial area of the forearm, just below (distal to) the elbow. Do not attach this electrode to the arm, as this will be the *exploring electrode* (fig. 5.12).
2. Connect the electrode plates to a stimulator, making sure that the stimulator is *off* at this time.
3. Set the stimulus intensity at 15 V and deliver a single pulse of stimulation. If no twitch is observed or felt in the fingers, move the exploring electrode around the medial area of the forearm until an effect is seen or felt. (See fig. 5.12 for the approximate position of the electrode.)

Laboratory Report 5

A. Twitch, Summation, Tetanus, anc
1. In this space, tape your recording c

2. Label twitch, summation, tetanus,
figure 5.11.

B. Twitch, Summation, and Tetanus
1. Describe the results of your proced
have your data in a computer file, |
fatigue (if obtained).

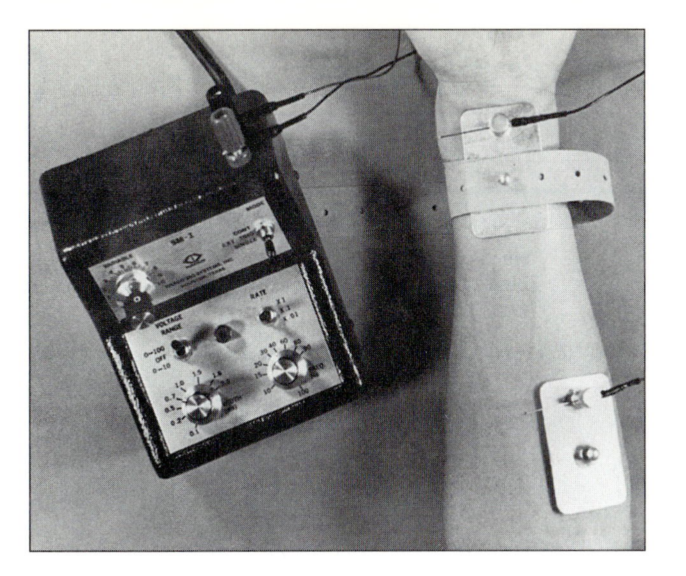

**Figure 5.12** Placement of electrodes for eliciting finger twitches in response to electrical stimulation.

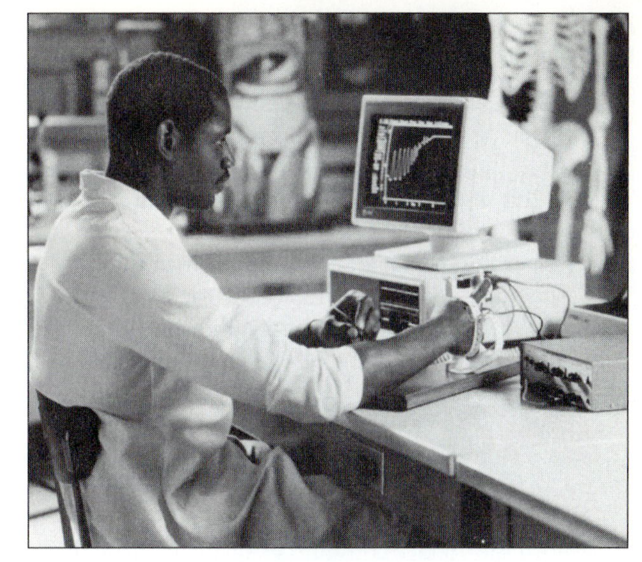

**Figure 5.13** Student using the Physiogrip to demonstrate muscle twitches, summation, and tetanus.

**Caution:** *The stimulus intensity may have to be increased for some people, but do not exceed 30 V! An effect can generally be obtained at a lower voltage by moving the exploring electrode to a slightly different position or by adding electrolyte gel. A tingling sensation means that the stimulus intensity is adequate, although the position may have to be changed.*

4. Once a muscle twitch has been observed, set the stimulator so that it automatically delivers one pulse of stimulation per second. Adjust the exploratory plate so that only one finger twitches.
5. Keeping the stimulus intensity constant, gradually increase the frequency of stimulation until a maximum contraction is reached. Gradually decrease the stimulus frequency until the individual twitches are reproduced.

## ALTERNATIVE PROCEDURE FOR INTELITOOL PHYSIOGRIP

1. As in the previous procedure, determine the correct points for placement of the electrodes on the forearm so that the flexor digitorum superficialis muscle is stimulated. This will result in flexion of the finger that grips the trigger of the Physiogrip.
2. Hold the Physiogrip (fig. 5.13) with light pressure while another student delivers electrical shocks using the stimulator. Start with about 15 V at a duration of 1 millisecond and gradually increase the voltage until threshold is observed.
3. Continue to increase the voltage in small increments, demonstrating the graded increase in contraction strength in response to stronger electrical stimuli.
4. With the voltage constant, gradually increase the frequency of stimulation to demonstrate tetany.

## ALTERNATIVE PRO
## FOR BIOPAC SYST

1. The hand dynamometer (fig. 5.1 Biopac Lesson 2 (EMG II) can b exercise, once the stimulating el on the forearm to stimulate the superficialis muscle.
2. As per the previous instructions, electrical shocks using the stimu about 14 V at a duration of 1 mi gradually increase the voltage u observed.
3. Continue to increase the voltag increments, demonstrating the g contraction strength in response electrical stimuli.
4. With the voltage constant, grad frequency of stimulation to dem and tetany.



# REVIEW ACTIVITIES FOR EXERCISE 5.2

## *Test Your Knowledge of Terms and Facts*

1. Define these terms:
   (a) twitch ______________________
   (b) summation ______________________
   (c) tetanus ______________________
2. Sustained muscular spasm is called ______________________. The two most common causes of this are ______________________ and ______________________.

## *Test Your Understanding of Concepts*

3. Describe how the summation of twitches is produced. Using this information, explain how variations in the strength of muscle contractions are produced.

4. Explain how tetanus was produced *in vitro* (in this laboratory exercise). Contrast this with the way that sustained contractions are normally produced *in vivo* (in the body).

## *Test Your Ability to Analyze and Apply Your Knowledge*

5. Suppose you hold a 10-pound weight steadily in your right hand with your elbow slightly bent, so that you maintain a contraction of your biceps brachii muscle. After a period of time, you experience pain and it becomes increasingly difficult to keep the weight at the same level. Explain why this is so.

6. Continuing with the situation described in question 5, suppose your hand starts to shake. Describe the changes in the state of your skeletal muscle fibers as your hand goes from being held steady to shaking. Relate these changes to the recordings obtained in this exercise.

# Electromyogram (EMG)

EXERCISE 5.3

## MATERIALS

1. Physiograph or another electrical recorder and high-gain coupler
2. EMG plates and disposable adhesive paper washers for EMG plates
3. Electrolyte (ECG) gel or paste and alcohol swabs
4. Alternative: Biopac system equipment for EMG I and II (Lessons 1 and 2)

The electrical activity produced by muscles can be recorded using surface electrodes. This recording may demonstrate the action of antagonistic muscles and may be useful in biofeedback training of muscles.

## OBJECTIVES

1. Demonstrate the antagonism between the action of the biceps and triceps muscles using the electromyogram (EMG).
2. Distinguish between isotonic and isometric muscle contractions, with examples of each.
3. Describe the EMG of the biceps and triceps during flexion and extension of the arm.
4. Explain the importance of antagonist inhibition in skeletal movements.
5. Explain how the EMG can be used in biofeedback techniques.

## Textbook Correlations*

Before performing this exercise, you should study the introductory material presented here. Further information relating to this exercise can be found in these pages of *Human Physiology,* eighth edition, by Stuart I. Fox:

- *Motor Units.* Chapter 12, pp. 328–329.
- *Contractions of Skeletal Muscles.* Chapter 12, pp. 340–342.
- *Alpha and Gamma Motoneurons.* Chapter 12, pp. 349–350.

Skeletal muscle contraction occurs in response to the electrical stimulation of the muscle fibers. In the previous exercises, following stimulation the mechanical response of the muscles was observed and recorded on a myograph transducer as tension exerted. Although the muscles were stimulated by electric shocks, the electrical activity of the individual muscle cells was not recorded. Notice that electric shocks delivered to the muscle fibers induce the formation of membrane action potentials, and that it is these action potentials (acting via the release of $Ca^{2+}$ from the sarcoplasmic reticulum) that lead to contraction of the muscles.

In the body *(in vivo)*, muscle fibers are stimulated to contract by motor neurons. The axon of a motor neuron branches to innervate a number of muscle fibers, all of which contract when the axon is stimulated. The axon and the muscle fiber it stimulates is known as a **motor unit** (fig. 5.15). When more strength is required for a muscle contraction, more motor units are enlisted, or *recruited,* into the contraction.

The contraction that results in muscles shortening is called **isotonic** ("equal tension") because the force of contractions remains relatively constant throughout the movement. Isotonic contractions are easily observed since the muscle becomes shorter and the corresponding limb or object attached is moved, such as movements when walking, lifting a chair, or doing push-ups (fig 5.16*b*). In **isometric** ("equal measure") contractions, the length of a muscle remains constant and no movement is seen because the force in the muscle being contracted is opposed by an equal opposing force, such as gravity. For example, an isometric contraction occurs when a person supports an object in a fixed position, such as in postural muscles when standing or sitting motionless (fig. 5.16*a*). An isometric contraction can be converted to an isotonic contraction when an increased force generated within the muscle overcomes the opposing resistance and results in muscle movement. This occurs, for example, when a straining body successfully clears the floor during the first push-up exercise.

*Multimedia Correlations (also see Appendix 3)
- Biopac: Student Lab Lessons 1 and 2
- Intelitool: Flexicomp
- A.D.A.M. *InterActive Physiology* (Muscular System): The Neuromuscular Junction; Contraction of Motor Units

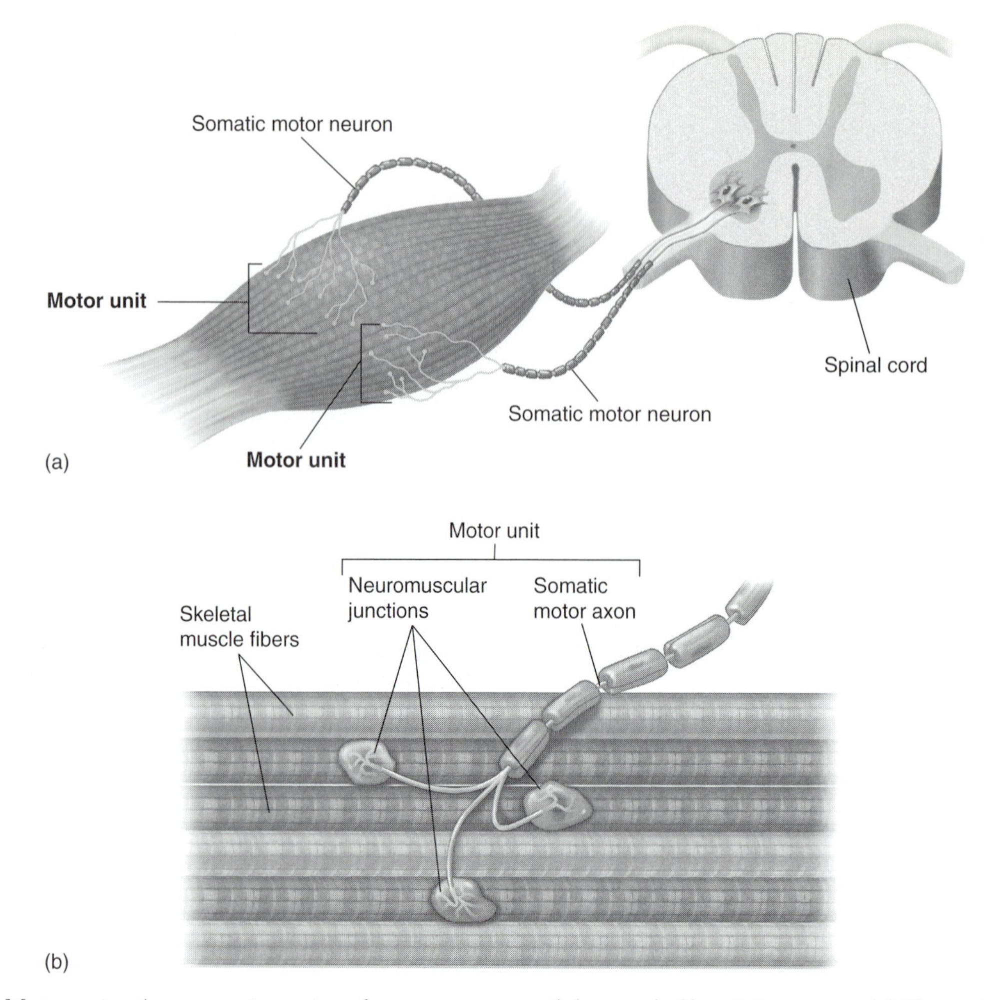


**Figure 5.15** **Motor units.** A motor unit consists of a motor neuron and the muscle fibers it innervates. (a) Illustration of a muscle containing two motor units. In reality, a muscle would contain many hundreds of motor units, and each motor unit would contain many more muscle fibers than are shown here. (b) A single motor unit consisting of a branched motor axon and the three muscle fibers it innervates (the fibers that are highlighted) is depicted. The other muscle fibers would be part of different motor units and would be innervated by different neurons (not shown).

**(For a full-color version of this figure, see fig. 12.4 in *Human Physiology,* eighth edition, by Stuart I. Fox.)**

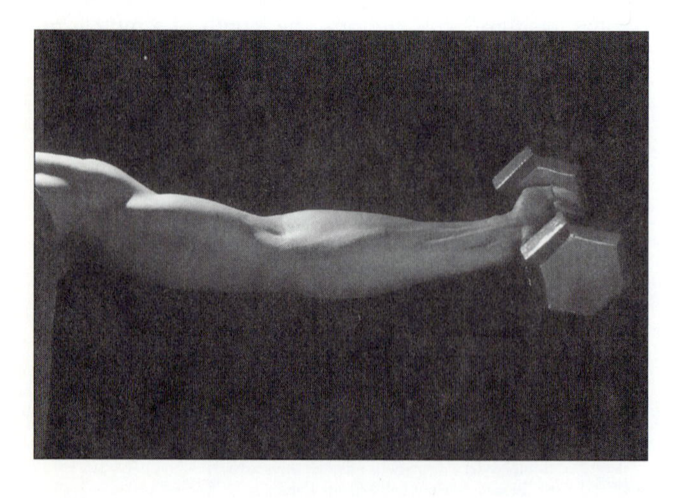

(a)

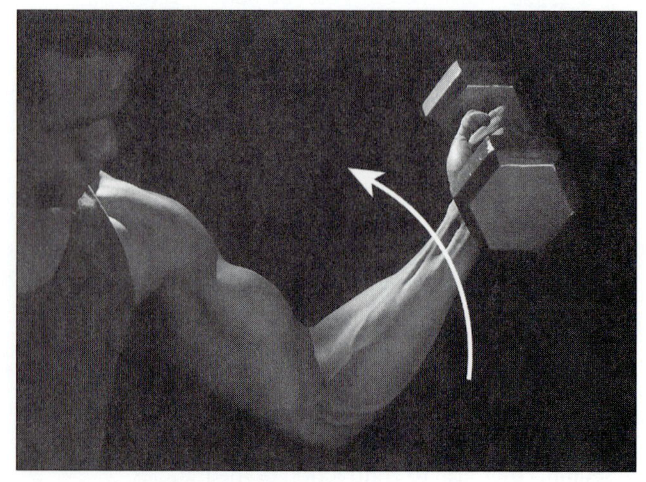

(b)

**Figure 5.16** Isometric (a) and isotonic (b) muscle contractions.

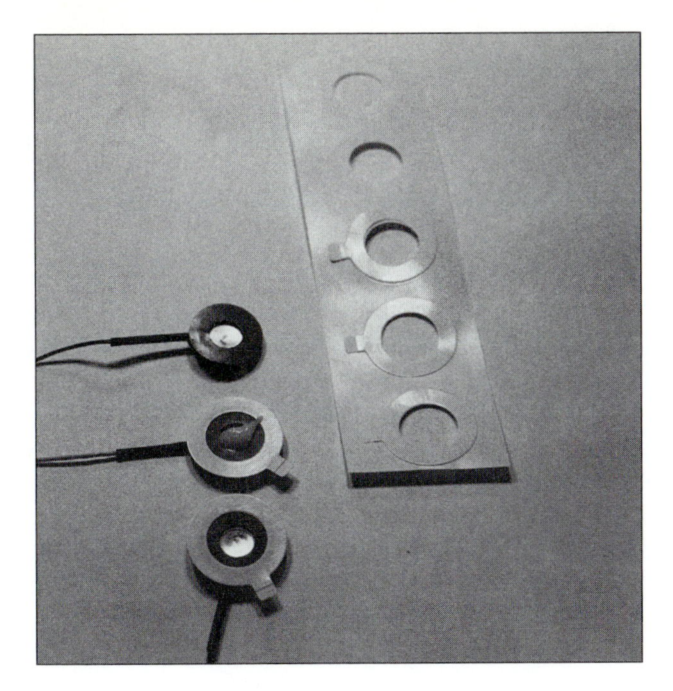

**Figure 5.17** Electrode plates and adhesive washers needed for electromyograph (EMG) procedure.

## A. Electromyogram Recording

When somatic motor nerves stimulate skeletal muscles to contract, the action potentials produced by the muscles transmit potential differences to the overlying skin that can be recorded by a pair of surface electrodes on the skin. The recording obtained is called an **electromyogram (EMG).** When the pair of electrodes are placed on the anterior surface of the upper arm, they record potentials generated by the *biceps brachii* muscle; and when they are placed over the posterior surface of the upper arm, they record the electrical activity of the *triceps brachii* muscle. Contraction of the biceps flexes the arm, whereas contraction of the triceps extends the arm. These two groups of muscles, therefore, are **antagonistic.** Notice that during arm flexion the activity of the triceps muscle is inhibited, and during arm extension the activity of the biceps muscle is inhibited. In this way, flexion stretches the extensor muscles, whereas extension stretches the flexor muscles.

### Procedure: Recruitment of Motor Units

1. Using cotton or a paper towel soaked in alcohol, cleanse the skin over the biceps and triceps muscles.
2. Apply the self-sticking paper washers to the raised plastic area surrounding the electrode plates (fig. 5.17). Squeeze *electrolyte gel* onto the metal electrode plates. Use a paper towel to smooth the gel so that it completely fills the well between the electrode and the surrounding plastic.
3. Remove the paper coverings over the adhesive area of the washers and apply the electrodes to the skin over the biceps muscle. Apply one electrode to the skin over the proximal portion of the biceps and the other over the distal portion, aligned with the first (fig. 5.18). Apply the ground electrode over the triceps muscle.
4. Plug the electrodes into the *high-gain coupler* module of the physiograph. Set this module to a *gain* of ×100, a *time constant* of 0.03, and a *sensitivity* between 20 and 100.
5. Set the *chart speed* at 0.5 cm/sec. With the arm relaxed and hanging down, establish a baseline, or control, in the recording. Then flex the arm (bringing the hand upward), and observe the recording of an isotonic contraction. Extend the arm back to its previous position, and then flex it again so that the difference between flexion and extension can be seen in the recording.

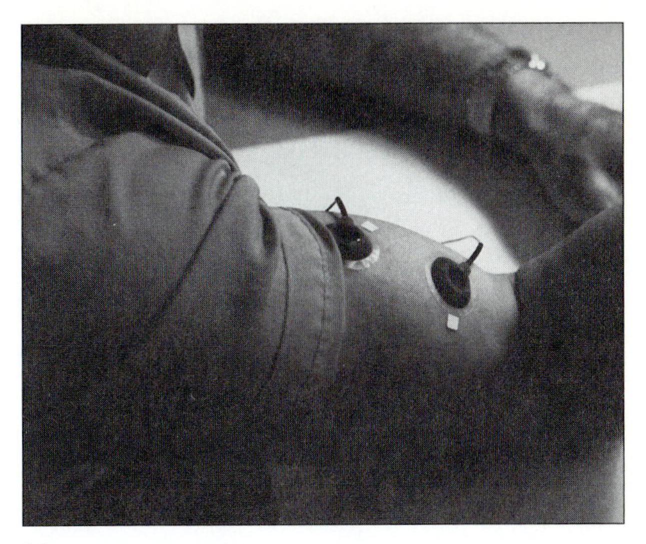

(a)

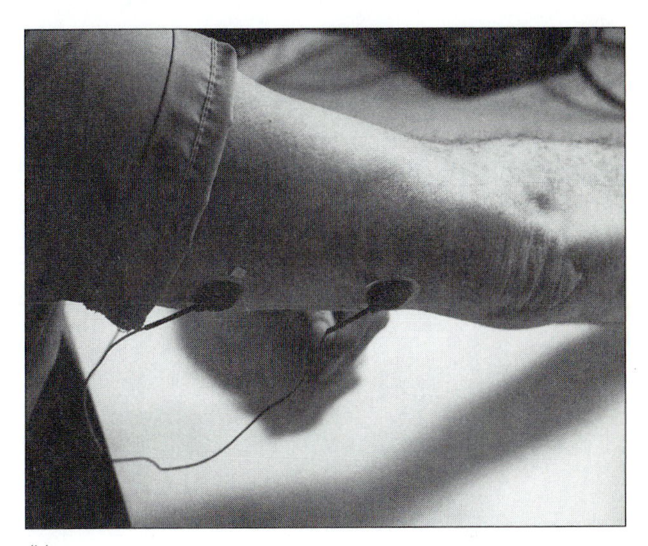

(b)

**Figure 5.18** Placement of the EMG electrodes.
The positions are shown for recording from (a) the biceps and (b) the triceps muscles.

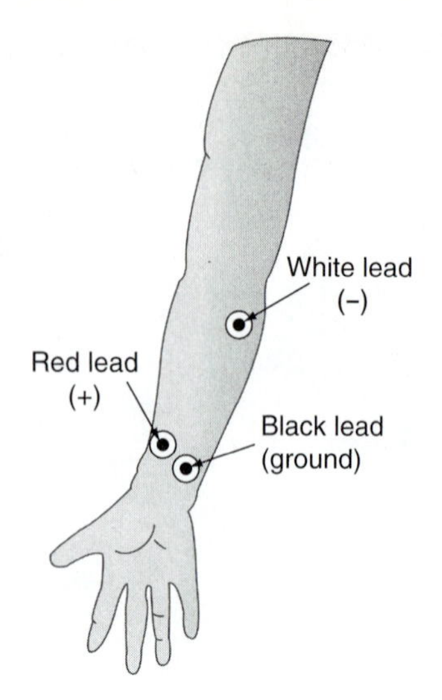


**Figure 5.19** Biopac electrode positions for the EMG.

6. Flex the arm again, this time lifting a chair or other weight. Observe the effect of this activity on the height (amplitude) of the recording. Now, "make a muscle" of increasing strength up to a maximum, demonstrating the recruitment of motor units into the isometric contraction.

## ALTERNATIVE PROCEDURE USING THE BIOPAC SYSTEM

1. Stick three electrodes on the forearm in the positions indicated in figure 5.19, and attach the correctly colored electrodes as shown in that figure.
2. Proceed with the set-up instructions as per Lesson 1 (EMG I) for the Biopac student lab.
3. Demonstrate recruitment of motor units by clenching your fist tighter and tighter, up to your maximum strength.
4. Repeat this procedure after moving the electrodes to your other forearm.

## PROCEDURE: EMG DURING ARM FLEXION AND EXTENSION

1. With the electrodes placed over the biceps brachii muscle as previously described, flex your arm as if you were trying to lift the table.
2. Change the position of the electrodes so that the two recording electrodes are lined up over the triceps muscle (one proximal and one distal) and the ground electrode is over the biceps muscle. Flex and extend the arm as before and observe the recording.
3. Place the hand on a table with the elbow bent and slowly extend the arm, as if doing a pushup. Observe the effect of this action on the EMG. Note the change in the recording as an isometric contraction becomes isotonic. Enter your recordings in the laboratory report.

The activity of antagonistic muscle groups is controlled in the central nervous system, so that when one group of muscles (the *agonist*) is stimulated to contract, the *antagonist* muscle group is inhibited, and will be stretched. This inhibition of antagonistic muscle groups occurs largely through the action of descending motor tracts that originate in the brain. When a person has spinal cord damage that blocks these descending inhibitory influences, the antagonistic muscles may contract when they are stretched by the movement of a limb. Without inhibitory influences, stretch reflexes in a person with spinal cord damage may cause antagonistic muscles to alternately stretch and contract, producing a **flapping tremor,** or **clonus.**

## B. Biofeedback and the Electromyograph

Our behavior changes as a result of the pleasant or unpleasant consequences of our actions. That is, positive and negative reinforcements modify behavior; this represents a type of learning that experimental psychologists call *operant conditioning*. The pairing of a particular behavior, such as cigarette smoking, with unpleasant sensations has been used successfully to shape human behavior through *aversion conditioning*.

**Biofeedback techniques** similarly effect learning, usually through feedback provided by electronic monitors of specific physiological states. The electromyogram, for example, provides a visual display of muscle stimulation that can be used as a psychological reward to reinforce effort spent attempting to contract specific muscle groups. In this exercise, the EMG will be used to demonstrate biofeedback techniques involved in learning how to increase the strength of contraction of the triceps muscle.

**Biofeedback techniques** serve a variety of clinical functions. The EMG is sometimes used to train people with neuromuscular disorders to regain use of affected limbs. Physiological monitoring of the heart rate and blood pressure has enabled patients with high blood pressure to lower their pulse; and the production of alpha rhythms on *electroencaphalogram (EEG)* recordings has been used to teach people with stress techniques for relaxation.

## PROCEDURE

1. Prepare the EMG electrodes as described in the previous procedure. Cleanse the skin with alcohol and place the two recording electrodes over the triceps muscle and the ground electrode over the biceps muscle.
2. Set the *high-gain coupler* to a *gain* of ×100, a *time constant* of 0.03, and a *sensitivity* between 20 and 100. Set the *chart speed* of the recorder to 0.5 cm/sec.
3. Extend the arm and observe the highest amplitude of the recording. Attempt a forced extension and observe the amplitude (height) of the recording. Attempt to increase the amplitude of the EMG by various procedures. (*Hint:* Try to extend the arm with the back of the hand against a table.)
4. Alternatively, the Biopac system can be used with the same setup previously described. Demonstrate biofeedback by attempting to reproduce a particular muscle tension while watching the recording on the computer screen.

# Laboratory Report 5.3

Name ______________________

Date ______________________

Section ______________________

## DATA FROM EXERCISE 5.3

A. **Electromyogram Recording**

1. **Recruitment of Motor Units**

   In this space, tape your recordings or draw facsimiles. Label the parts of your recording.

2. **Electromyogram During Arm Flexion and Extension**

   In this space, tape your recordings or draw facsimiles.

   (a) Label flexion and extension for both the biceps and triceps muscles in your recording.

   (b) Compare your reading of the *isotonic* contraction to that of the *isometric* contraction.

B. **Biofeedback and the Electromyograph**

In this space, tape your recordings or draw a facsimiles. Label the region of your recording that demonstrates biofeedback.

# Blood: Gas Transport, Immunity, and Clotting Functions

# Section 6

**T**he various components of the blood serve different physiological functions. The **plasma,** or fluid portion of the blood (fig. 6.1), provides the major means for distributing chemicals between organs. For example, it transports food molecules absorbed through the small intestine, hormones secreted by the endocrine glands, and antibodies produced by certain white blood cells. The plasma also helps to eliminate metabolic wastes by carrying these unwanted molecules to the liver (for excretion in the bile), to the kidneys (for excretion in the urine), and, in the case of $CO_2$ gas, to the lungs (for excretion in the exhaled air).

In addition to plasma, the blood also contains two major types of cells: **red blood cells** *(erythrocytes)* and **white blood cells** *(leukocytes)* (fig. 6.1*a*). Red blood cells contribute to the respiratory function of the blood by providing transport for oxygen and, to a lesser degree, for carbon dioxide. Blood is "typed" based upon the presence or absence of specific molecules displayed by red blood cells (the ABO system and the Rh factor are examples). White blood cells and their products help to provide immunity from infection by recognizing and attacking foreign molecules and cells. The blood also contains **platelets,** which are membrane-bound fragments derived from a bone marrow cell called a *megakaryocyte.* Platelets, together with proteins in the plasma, help to maintain the integrity of blood vessels by forming *blood clots.*

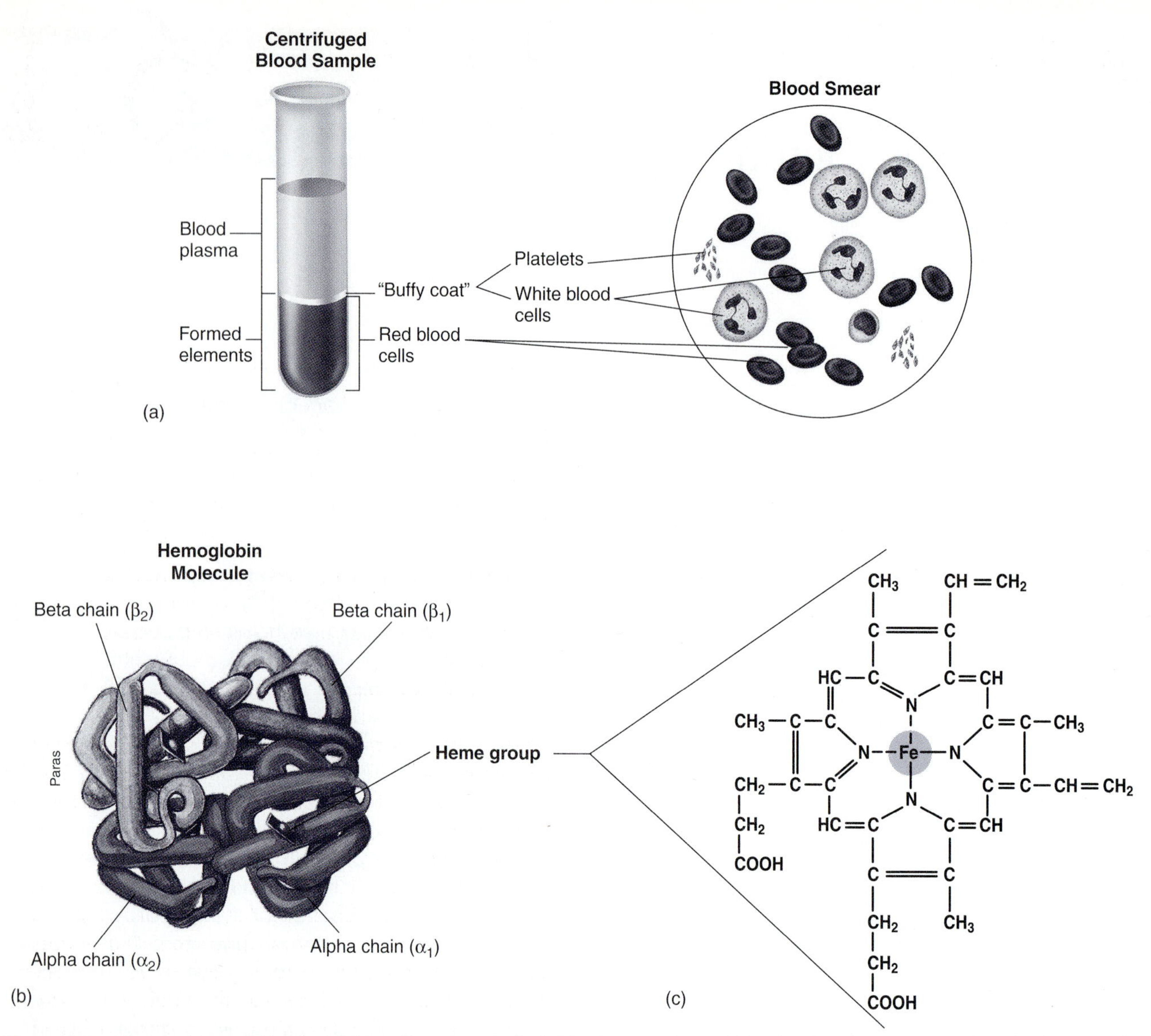


**Figure 6.1 Composition of blood.** (a) Centrifugation of whole blood causes the red blood cells to become packed at the bottom of the tube; this layer is covered by a thin, buffy coat of leukocytes and platelets, and all of these formed elements of blood are separated from the plasma. (b) The structure of hemoglobin within red blood cells, and (c) the structure of each heme group. There are four heme groups per hemoglobin molecule.

**(For a full-color version of this figure, see fig. 13.1 for part (a) and 16.33 for part (b) in *Human Physiology,* eighth edition, by Stuart I. Fox.)**

# Red Blood Cell Count, Hemoglobin, and Oxygen Transport

EXERCISE 6.1

## Materials

1. Hemocytometer
2. Unopettes (Becton-Dickinson) for manual red blood cell count and hemoglobin measurements
3. Heparinized capillary tubes, clay capillary tube sealant (Seal-ease), microcapillary centrifuge, hematocrit reader
4. Microscope
5. Sterile lancets and 70% alcohol for preparing fingertip blood. Alternatively, dog or cat blood (obtained from a veterinarian) may be used.
6. Colorimeter and cuvettes
7. Container for disposal of blood-containing items

Almost all of the oxygen transported by the blood is carried within the red blood cells attached to hemoglobin. Measurements of the oxygen-carrying capacity of blood include the red blood cell count, hemoglobin concentration, and hematocrit. Anemia results when one or more of these measurements is abnormally low.

## OBJECTIVES

1. Describe the composition of blood.
2. Describe the composition of hemoglobin and explain how hemoglobin participates in oxygen transport.
3. Demonstrate the procedures for taking the red blood cell count and hemoglobin and hematocrit measurements, and list the normal values for these measurements.
4. Explain how measurements of the oxygen-carrying capacity of blood can be used to diagnose anemia and polycythemia.

## Textbook Correlations*

Before performing this exercise, you should study the introductory material presented here. Further information relating to this exercise can be found in these pages of *Human Physiology*, eighth edition, by Stuart I. Fox:

- *The Formed Elements of Blood.* Chapter 13, pp. 368–370.
- *Hemoglobin and Oxygen Transport.* Chapter 16, pp. 504–510.

Each ventilation cycle delivers a fresh supply of oxygen to the alveoli of the lungs. The amount of oxygen that leaves the lungs dissolved in plasma is equal to 0.3 mL of $O_2$ per 100 mL of blood. The amount of oxygen leaving the lungs in whole blood, however, is equal to 20 mL of $O_2$ per 100 mL of blood. Most of the oxygen (19.7 mL $O_2$ 100 mL blood), therefore, must be carried within the cellular elements of the blood. This oxygen is carried by **hemoglobin** molecules within the red blood cells (fig. 6.1*b*). In this way the oxygen is transported to the body cells and used for aerobic respiration and the production of cellular energy.

Each hemoglobin molecule consists of two pairs of polypeptide chains (one pair called the *alpha chains* and one pair called the *beta chains*) and four disc-shaped organic groups called *heme groups*. Each heme group contains one central ferrous ion ($Fe^{2+}$) capable of bonding with one molecule of oxygen (fig. 6.1*b*). Thus, one molecule of hemoglobin can combine with four molecules of oxygen.

The hemoglobin within the red blood cells load up with oxygen in the capillaries of the lungs and unload oxygen in the tissue capillaries. In both cases, oxygen moves according to its diffusion gradient. Since red blood cells always respire anaerobically (so they do not consume the oxygen they carry), a maximum diffusion gradient for oxygen is maintained between the red blood cells and the tissues.

The **oxygen-carrying capacity** of the blood is dependent on the total number of red blood cells and, consequently, on the total amount of hemoglobin. The total number of red blood cells is dependent on a balance between the rates of red blood cell production and destruction. The rate of red blood cell production by the bone marrow is regulated by the hormone **erythropoietin,** secreted by the kidneys. Erythropoietin is secreted when blood oxygen levels fall, such as when traveling in

*Multimedia Correlations (also see Appendix 3)
- *MediaPhys 2.0*: Topics 10.37–10.44

high-altitude environments. The rate of renal erythropoietin secretion is, therefore, regulated by the oxygen requirements of the body.

Older red blood cells (those that are approximately 120 days old) are routinely destroyed by the action of phagocytic cells fixed to the sides of blood channels (sinusoids) by a meshwork (reticulum) of fibers. Located in the spleen, liver, and bone marrow, these fixed phagocytes compose the **reticuloendothelial system.** These reticuloendothelial cells digest the hemoglobin within the old red blood cells into the component parts of protein, iron, and the *heme* pigment. The protein is hydrolyzed and returned to the general amino acid pool of the body, the iron is recycled to the bone marrow, and the heme is changed into a new pigment called **bilirubin.**

Bilirubin is released into the blood by the reticuloendothelial cells, then picked up by the liver and secreted into the bile as bile pigment. An abnormal increase in the amount of bilirubin in the blood, due to an increased rate of red blood cell destruction, liver dysfunction, or bile duct obstruction, results in the condition known as *jaundice* (yellowing of the skin and sclera of the eyes).

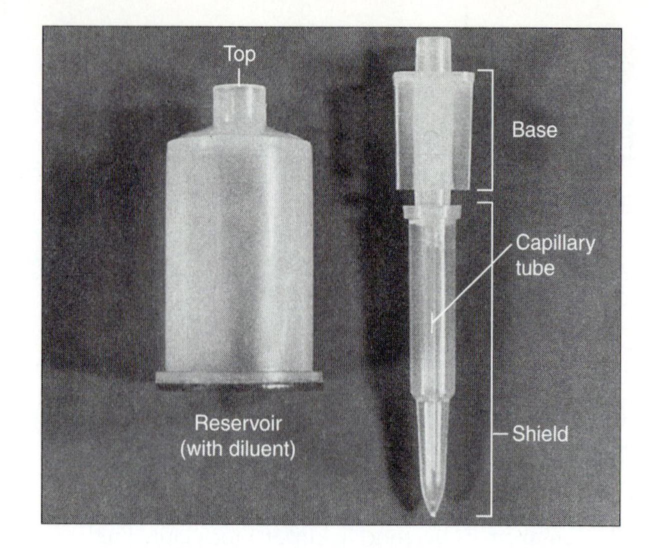


**Figure 6.2 The Unopette system (Becton-Dickinson).** This consists of a reservoir containing the premeasured amount of diluent (left) and a plastic capillary tube within a shield (right) for puncturing the reservoir top and delivering a measured amount of whole blood to the reservoir.

## A. Red Blood Cell Count

The object of this exercise is to determine the number of red blood cells in a cubic millimeter of blood. Because this number is very large, it is practical to dilute a sample of blood with an isotonic solution, count the number of red blood cells in a fraction of this diluted blood, and then multiply by a correction factor. This procedure is accurate only when (1) the blood diluted is a representative fraction of all the blood in the body, (2) the dilution volumes are accurate, and (3) the sample counted is representative of the total volume of diluted blood.

### PROCEDURE

#### Obtaining and Diluting Blood Samples

1. The Unopette reservoir (fig. 6.2) contains a premeasured amount of diluting solution (Hayem's solution). Use the pointed end of the shielded capillary tube (fig. 6.2) to puncture the plastic top of this reservoir. Turn the shielded capillary tube and attach it to the reservoir opening until needed.
2. Swing your hand around until your fingers become engorged with blood *(hyperemia)*. Cleanse the tip of your index or third finger with 70% alcohol and prick it with a sterile lancet.

**Caution:** *Because of the danger of exposure to the AIDS virus and other harmful agents when handling blood, each student should perform this and other blood exercises with his or her own blood only. All objects that have been in contact with blood must be discarded in a container indicated by the instructor.*

3. Discard the first drop of blood and point your finger downward to collect the next large drop of blood. Remove and discard the shield over the capillary pipette of the Unopette, and simply touch the tip of the pipette to the drop of blood. Allow the pipette to fill by capillary action (fig. 6.3*a*).
4. Squeeze the previously punctured reservoir with the fingers of one hand and, while squeezing, insert the pipette of blood into the punctured top of the reservoir. Releasing the pressure on the reservoir will draw the blood into the premeasured Hayem's solution within the reservoir (fig. 6.3*b*).
5. Gently mix the blood with the Hayem's solution for approximately 1 minute.

### PROCEDURE

#### Filling the Hemocytometer and Determining Red Blood Cell Count

1. Place a coverslip on the hemocytometer so that it covers one of the silvered areas.
2. Remove the capillary pipette from the reservoir, turn it around, and reinsert it backwards into the reservoir so that the capillary is pointing out of the reservoir (like the needle of a syringe—see fig. 6.4). Discard the first 3 drops of blood from the Unopette (properly onto a piece of cotton, and disposed into a designated container). Place the next drop of diluted blood in the "V" region of the hemocytometer, at the edge of the coverslip. The diluted blood will be drawn underneath the coverslip by capillary action (fig. 6.4).

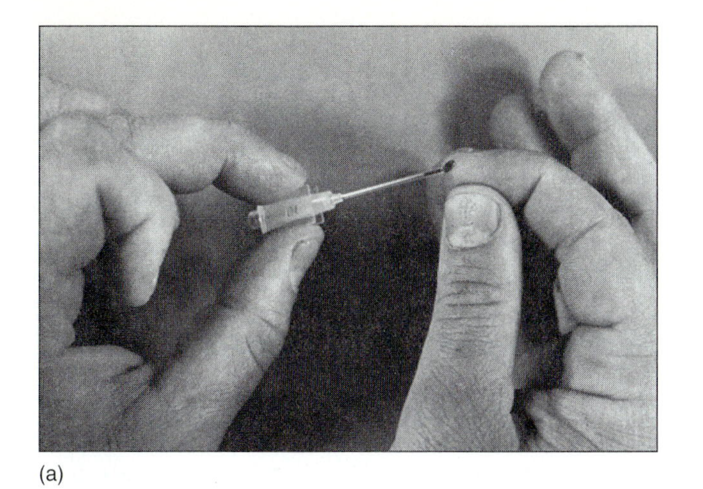
(a)

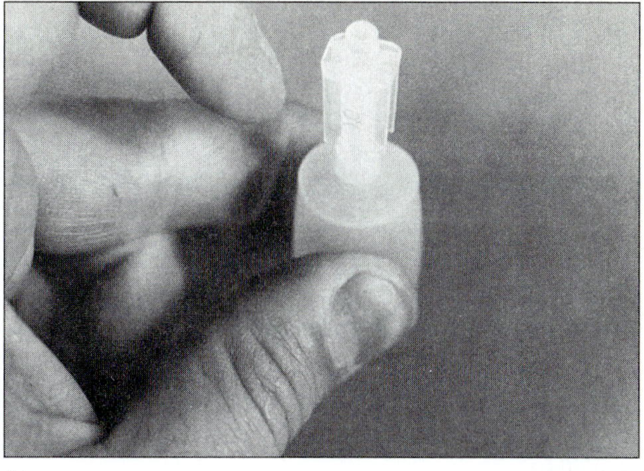
(b)

**Figure 6.3 The Unopette method for measuring a red blood cell count or hemoglobin concentration.** (a) Fill the plastic capillary pipette with fingertip blood. Then (b), squeeze the reservoir to draw blood out of the pipette into diluent within the reservoir.

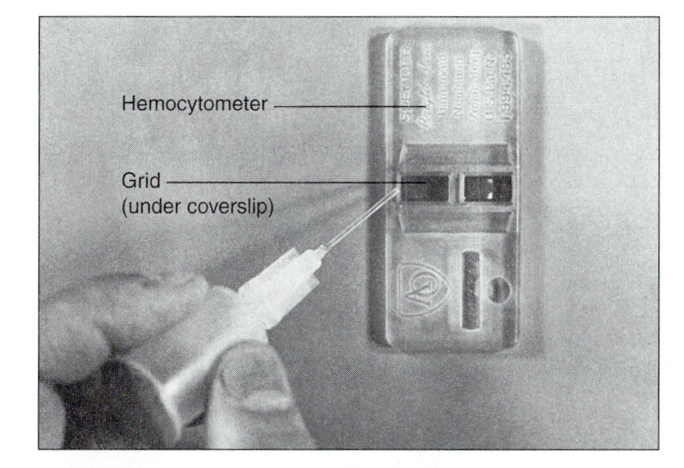


**Figure 6.4 Procedure for filling a hemocytometer.** A Unopette reservoir is used to fill the hemocytometer with diluted blood. The squeezing of the reservoir places a drop of diluted blood at the edge of the coverslip, whereupon the drop of blood moves under the hemocytometer grid by capillary action.

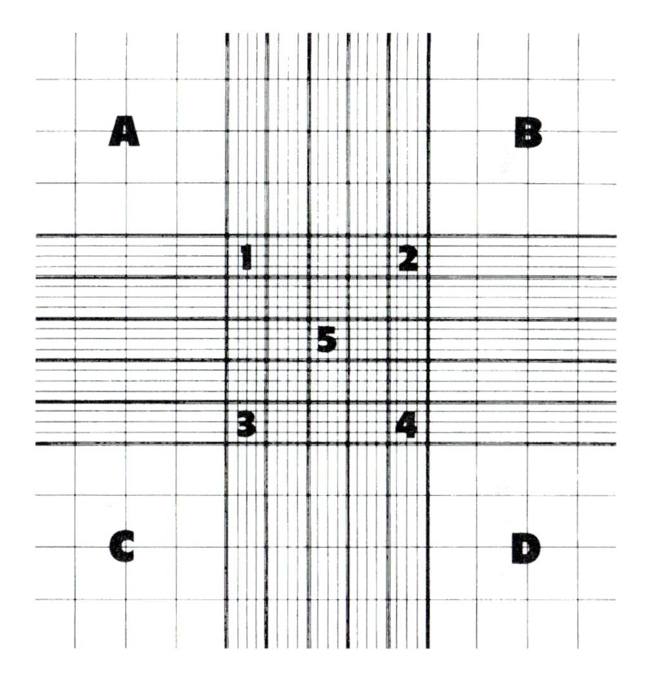


**Figure 6.5 The hemocytometer grid.** Squares 1–5 are used for red blood cell counts; squares A–D are used for white blood cell counts.

3. Locate the grid on the hemocytometer using the 10× objective. Focus first, then change to 45× and count the total number of red blood cells in the squares numbered 1 through 5 *only* (fig. 6.5).

**Note:** *If a red blood cell lies on the upper or the left-hand lines of the square, include it in your count. Do not count those that lie on the lower or the right-hand lines.*

4. The central grid of twenty-five squares is 1 square mm ($mm^2$) in area and 0.10 mm deep. The dilution factor is 1:200. To convert the number of red blood cells that you counted in 5 squares to the number of red blood cells per cubic millimeter, multiply your count by 10,000 (the product of 5 × 10 × 200).

5. Record your total count of red blood cells in 5 indicated squares.
   **Red blood cells/5 squares:** ____________________
   Calculate the number of red blood cells in a cubic millimeter of your blood; and enter this number in the laboratory report.

Normal red blood cell counts: For an adult **male** is 4.5–6.0 million per cubic mm ($mm^3$); and for an adult **female** is 4.0–5.5 million/$mm^3$.

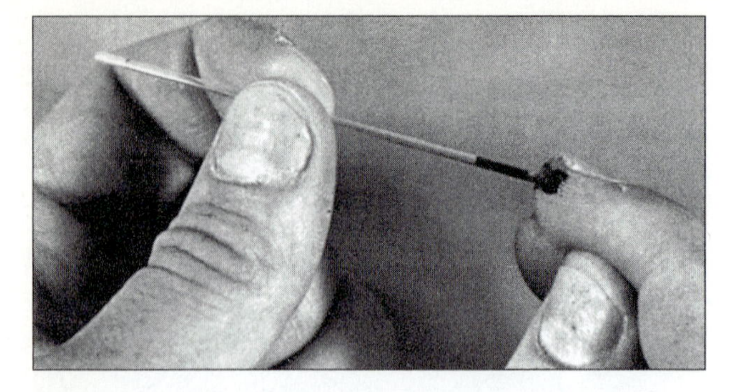

**Figure 6.6** The method for filling a capillary tube with fingertip blood.

## B. Hematocrit

When whole blood is centrifuged, the red blood cells become packed at the bottom of the tube, leaving the plasma at the top. The ratio of the volume of packed red blood cells to the total blood volume is called the **hematocrit.**

### PROCEDURE

1. Prick your finger with a sterile lancet to obtain a drop of blood as described in the previous procedure. Again, discard the first drop onto an alcohol swab, and dispose of this properly in a designated container.
2. Obtain a heparinized capillary tube (**heparin** is an *anticoagulant*). Notice that one end of the tube is marked with a red band. Touch the end of the capillary tube opposite the marked end to the drop of blood, allowing blood to enter the tube by capillary action and gravity (fig. 6.6). The tube does not have to be completely full (half full or more is adequate), and air bubbles are not important (they will disappear during centrifugation).
3. Seal the red-banded (fire-polished) end of the capillary tube by gently pushing it upright into clay capillary sealant. Carefully, rotate and remove the tube.
4. Place the sealed capillary tube in a numbered slot of the microcapillary centrifuge, with the plugged end of the capillary tube facing outward against the rubber gasket. Screw the top plate onto the centrifuge head and centrifuge for 3 minutes. At the end of the centrifugation, determine the hematocrit with the hematocrit reader provided, and enter this value in your laboratory report.

Normal hematocrit values: For an adult **male**, is 47 ± 7%; and for an adult **female** is 42 ± 5% (of the total blood volume).

## C. Hemoglobin Concentration

**Hemoglobin** absorbs light in the visible spectrum and hence is a *pigment* (a colored compound). It should therefore be possible to measure the concentration of hemoglobin in a hemolyzed sample of blood by measuring the intensity of its color. This procedure, however, is complicated by the fact that red blood cells contain different types of hemoglobin, and each type absorbs light in a slightly different region of the visible spectrum (i.e., has a slightly different color).

When the oxygen concentration of the blood is high, such as in the capillaries of the lungs, normal **deoxyhemoglobin** combines with oxygen to form the compound **oxyhemoglobin.** When the concentration of oxygen in the blood is low, such as in the capillaries of the tissues, the oxyhemoglobin dissociates (comes apart) to form reduced hemoglobin and oxygen.

$$\text{deoxyhemoglobin} + \text{oxygen} \underset{\text{tissues}}{\overset{\text{lungs}}{\rightleftarrows}} \text{oxyhemoglobin}$$

Arterial blood is bright red due to the predominance of the oxyhemoglobin pigment, whereas venous blood has the darker hue characteristic of deoxyhemoglobin. It should be emphasized, however, that venous blood, although darker in color, still contains a large amount of oxyhemoglobin; this functions as an oxygen reserve.

A less common though clinically important form of hemoglobin is **carboxyhemoglobin,** which is a complex of hemoglobin and *carbon monoxide*. This complex, unlike oxyhemoglobin, does not readily dissociate; thus, the hemoglobin that is bonded to carbon monoxide cannot participate in oxygen transport. The carboxyhemoglobin complex has a bright, cranberry red color.

A small percentage of hemoglobin contains iron that is oxidized to the *ferric* state ($Fe^{3+}$) instead of being in the normal *ferrous* state ($Fe^{2+}$). Hemoglobin in this oxidized state is called **methemoglobin** and is incapable of bonding with either oxygen or carbon monoxide. An increase in the amount of methemoglobin is associated with some genetic diseases, or it may result from the action of certain drugs, such as nitroglycerin.

The following exercise describes how the concentration of hemoglobin in a solution of hemolyzed blood can be assessed by measuring the intensity of its color with a spectrophotometer. To do this accurately, all the hemoglobin must be converted into one form (methemoglobin) and then combined with cyanide to make it more stable. The unknown hemoglobin concentration can then be determined by comparing its absorbance with the absorbance of a standard hemoglobin solution of known concentration.

## PROCEDURE

### *Measurement of Blood Hemoglobin Concentration*

**Note:** *Alternative procedures for estimating hemoglobin concentration in blood include the Tallquist paper blot method and the use of the hemoglobinometer.*

1. A different Unopette will be used for this procedure. This Unopette has a solution of cyanmethemoglobin reagent (yellow color); and has a capillary pipette that delivers twice as much blood as the one used for the red blood cell count procedure. The correct capillary pipette can be identified by the yellow number 20 (for 20 μL) on its side. As in the previous procedure, puncture the top of the reservoir with the shielded capillary pipette.

2. Wipe the tip of your third finger with 70% alcohol and puncture it with a sterile lancet. Discard the first drop of blood onto an alcohol swab, and properly dispose in a designated container. Fill the plastic capillary pipette in the Unopette with 0.02 mL of blood. Fill the capillary tube by simply touching the tip of the pipette to the drop of blood and allowing the pipette to completely fill by capillary action (see fig. 6.3*a*).
3. Squeeze the reservoir first, then insert the pipette and release the reservoir, aspirating the blood into the reservoir (see fig. 6.3*b*). Squeeze and release the reservoir a few more times to completely aspirate blood from the pipette. Mix and allow it to stand at room temperature for 10 minutes.
4. Some standard hemoglobin solutions come full strength and must be diluted with cyanmethemoglobin reagent to be at the same dilution as the unknown. Use the same procedure as in steps 2 and 3 to make this dilution (using a new Unopette).

   Some purchased hemoglobin standards arrive already diluted. Using these may require calculating the hemoglobin concentration of blood that would be equivalent to that of the diluted standard solution.
5. Set the colorimeter at a wavelength of 540 nm and standardize the instrument using plain cyanmethemoglobin solution as the blank. Record the absorbance values of the unknown and standard.

   Absorbance of **unknown:** ____________________

   Absorbance of **standard:** ____________________
6. Calculate the hemoglobin concentration of the unknown using the *formula*

$$\text{Concentration}_{\text{unknown}} = \frac{\text{Concentration}_{\text{standard}} \times A_{\text{unknown}}}{A_{\text{standard}}}$$

Enter your hemoglobin concentration in the laboratory report.

Normal hemoglobin concentration for an adult **male** is 13–16 g/dL, and for an adult **female** is 12–15 g/dL.

## D. Calculation of Mean Corpuscular Volume (MCV) and Mean Corpuscular Hemoglobin Concentration (MCHC)

An abnormally low hemoglobin, hematocrit, or red blood cell count may indicate a condition known as anemia. Anemia may be caused by iron deficiency, vitamin $B_{12}$ and folic acid deficiencies, bone marrow disease, hemolytic disease (e.g., sickle-cell anemia), loss of blood through hemorrhage, or infections. Diagnosis of a specific type of anemia is aided by relating the measurements of hemoglobin, hematocrit, and red blood cell count to derive the **mean corpuscular volume (MCV)** and the **mean corpuscular hemoglobin concentration (MCHC).**

**Anemia** is subdivided into a number of categories on the basis of the **MCV** and **MCHC.** *Macrocytic anemia* (MCV greater than 94, MCHC within normal range) may be caused by folic acid deficiency or by vitamin $B_{12}$ deficiency associated with the disease *pernicious anemia.* In this condition, a polypeptide "intrinsic factor" that is necessary for vitamin $B_{12}$ absorption is not secreted as it normally is by the stomach. *Normocytic normochromic anemia* (normal MCV and MCHC) may be due to acute blood loss, hemolysis, aplastic anemia (damage to the bone marrow), or a variety of chronic diseases. *Microcytic hypochromic anemia* (abnormally low MCV and low MCHC), the most common type, is caused by inadequate amounts of iron.

## PROCEDURE

1. Calculate the **mean corpuscular volume** (MCV) according to the following formula:

$$\text{MCV} = \frac{\text{hematocrit} \times 10}{\text{RBC count (millions per mm}^3\text{ blood)}}$$

### *Example*

Hematocrit = 46
RBC count = 5.5 million

$$\text{MCV} = \frac{46 \times 10}{5.5} = 84$$

Calculate your mean corpuscular volume (MCV) and enter it in the laboratory report.

> The normal adult **male** and **female** mean corpuscular volume (MCV) ranges from **82–92** cubic micrometers

2. Calculate your **mean corpuscular hemoglobin concentration** (MCHC) according to the following formula:

$$\text{MCHC} = \frac{\text{Hemoglobin}\left(\text{g/dl}\right) \times 100}{\text{Hematocrit}}$$

***Example***

Hematocrit = 46

Hemoglobin = 16 g/dL

$$\text{MCHC} = \frac{16 \times 100}{46} = 35$$

Calculate your mean corpuscular hemoglobin concentration (MCHC) and enter it in the laboratory report.

> The average normal adult **male** and **female** mean corpuscular hemoglobin concentration (MCHC) is **32–36** (in percent).

# Laboratory Report 6.1

Name ______________________________

Date ______________________________

Section ______________________________

## DATA FROM EXERCISE 6.1

**A. Red Blood Cell Count**

1. Enter your red blood cell count per cubic millimeter ($mm^3$) of blood: ______________ per $mm^3$ of blood.

**B. Hematocrit**

1. Enter your hematocrit (in percent): ______________.

**C. Hemoglobin Concentration**

1. Enter your hemoglobin concentration in g per 100 mL (or deciliter, dL) of blood: ______________ g/dL of blood.

**D. MCV and MCHC**

1. Calculate your mean corpuscular volume (MCV) and mean corpuscular hemoglobin concentration (MCHC) and enter these values here.

   MCV: ______________ cubic micrometers ($\mu m^3$)

   MCHC: ______________ percent (%)

2. Compare your values to the normal values and write your conclusions here.

# REVIEW ACTIVITIES FOR EXERCISE 6.1

## *Test Your Knowledge of Terms and Facts*

1. One hemoglobin molecule contains _____________________ heme groups; each heme group normally combines with one molecule of _____________________.
2. The hormone _____________________ stimulates the bone marrow to produce red blood cells; this hormone is secreted by the _____________________.
3. Old red blood cells are destroyed by the _____________________ system, which includes these three organs:
   (a) ___________________________________________
   (b) ___________________________________________
   (c) ___________________________________________
4. Heme derived from hemoglobin, minus the iron, is converted into a different pigment, known as ______________________________; an accumulation of this pigment can cause a yellowing known as ______________________________.
5. Define the term *hematocrit*. ________________________________________________________________
6. The molecule formed by the binding of oxygen to deoxyhemoglobin: ______________________________.
7. A hemoglobin molecule containing oxidized iron ($Fe^{3+}$) is called ______________________________.
8. A molecule formed from the combination of hemoglobin and carbon monoxide is ______________________________________________________________.
9. A general term for an abnormally low red blood cell count or hemoglobin concentration: ____________________________________________________.
10. The most common cause of the condition described in question 9 is ______________________________.

## *Test Your Understanding of Concepts*

11. Describe some of the causes of anemia. Why is anemia dangerous?

12. Newborn babies, particularly premature ones, often have a rapid rate of red blood cell destruction and have jaundice. What is the relationship between these two conditions?

13. Could a person have a low hematocrit yet have a normal red blood cell count? Explain what might cause this condition.

## *Test Your Ability to Analyze and Apply Your Knowledge*

14. Results of blood tests performed in this exercise would be different for anemia and for carbon monoxide poisoning, yet in one respect these two conditions are similar. Explain why this statement is true.

15. People who live at high altitudes often have a high red blood cell count, a condition called *polycythemia*. Explain the cause of the polycythemia, and its possible benefit. Do you think it could have any adverse effects? Explain.

Neutrophils

Eosinophils

Basophils

Lymphocytes

Monocytes

Platelets (thrombocytes)

Erythrocytes

The average differential count in the normal adult is as follows:

| | |
|---|---|
| Neutrophils | 55%–75% |
| Eosinophils | 2%–4% |
| Basophils | 0.5%–1% |
| Lymphocytes | 20%–40% |
| Monocytes | 3%–8% |

| Leukocyte | Cells Counted | Total | Percent |
|---|---|---|---|
| Neutrophils | | | |
| Eosinophils | | | |
| Basophils | | | |
| Lymphocytes | | | |
| Monocytes | | | |

**Plate 1** Formed elements of blood.

# White Blood Cell Count, Differential, and Immunity

EXERCISE **6.2**

### Materials

1. Microscopes, hemocytometer slides
2. Thoma diluting pipettes
3. Lancets and alcohol swabs, for preparing fingertip blood. Alternatively, dog or cat blood (obtained from a veterinarian) may be used instead.
4. For total white blood cell count: Methylene blue in 1% acetic acid; and for differential count: Wright's stain (or Harleco Diff-Quik or VWR Statstain)
5. Heparinized capillary tubes and glass slides

White blood cells—lymphocytes, monocytes, neutrophils, eosinophils, and basophils—are agents of the immune system. Lymphocytes provide immunity against specific antigens, whereas the other leukocytes are phagocytic. The total white blood cell count and the relative proportion of each type of white blood cell (differential count) change in a characteristic way in different disease states.

## OBJECTIVES

1. Distinguish the different types of leukocytes by the appearance of their nuclei and their cytoplasm.
2. Describe the origin and function of B and T lymphocytes.
3. List the phagocytic white blood cells and explain their functions during local inflammation.
4. Perform a total and a differential white blood cell count and explain the importance of this information in the diagnosis of diseases.

### Textbook Correlations

Before performing this exercise, you should study the introductory material presented here. Further information relating to this exercise can be found in these pages of *Human Physiology,* eighth edition, by Stuart I. Fox:

- *The Formed Elements of Blood.* Chapter 13, pp. 368–370.
- *Defense Mechanisms.* Chapter 15, pp. 446–453.
- *Functions of B Lymphocytes.* Chapter 15, pp. 453–457.
- *Functions of T Lymphocytes.* Chapter 15, pp. 457–463.

The **white blood cells (leukocytes)** are divided into two general categories on the basis of their histological appearance: granular (or polymorphonuclear) and agranular. Leukocytes in the granular category have granules in the cytoplasm and lobed or segmented nuclei, whereas those in the agranular category lack visible cytoplasmic granules and have unlobed nuclei (see **plate 1**, following p. 212).

The *granular leukocytes* are distinguished by their affinity for specific stains. The cytoplasmic granules of **eosinophils** stain bright red (the color of eosin stain), and the granules of **basophils** stain dark blue (the color of basic stain). The granules of **neutrophils** have a low affinity for stain; therefore, the cytoplasm of these cells appears relatively clear.

The *agranular leukocytes* include **lymphocytes** and **monocytes.** Lymphocytes are the smaller of these two cell types and are easily identified by their round nuclei and scant cytoplasm. Larger monocytes have kidney-bean-shaped nuclei, often with brainlike convolutions, and their cytoplasm has a ground glass appearance. Monocytes may also sometimes be identified by the appearance of short, blunt, cytoplasmic extensions (pseudopods).

Leukocytes can leave the vascular system and enter the connective tissues of the body by squeezing through capillaries (a process known as *diapedesis* or *extravasation*). During an inflammation response, the release of *histamine* from tissue mast cells and basophils increases the permeability of the capillaries, and consequently promotes the process of diapedesis. This sequence of events also produces the local edema, redness, and pain associated with inflammation.

Neutrophils and, to a lesser degree, eosinophils, destroy the invading pathogens by phagocytosis. The battle is then joined by monocytes, which also enter the connective tissues and are transformed into voracious phagocytic cells known as *tissue macrophages*. The engorged white blood cells form pus in the inflamed area.

If these nonspecific immunological defenses are not sufficient to destroy the pathogens, lymphocytes may be recruited, and their specific actions used to reinforce the nonspecific immune responses. Lymphocytes are first produced in the embryonic bone marrow, which then seeds the other lymphopoietic sites: the *thymus, lymph nodes,* and *spleen.* The thymus, in turn, sends cells to other locations and apparently regulates the general rate of lymphocyte production at all these sites through the release of a hormone. Therefore, all lymphocytes may be categorized in terms of their ancestry as either bone marrow-derived **B cells** or thymus-derived **T cells.**

**Antigens** are molecules that activate the immune system. A specific *receptor protein* displayed on the outer membrane of each lymphocyte is capable of recognizing and binding to a specific antigen. By means of this bonding, the antigen selects the lymphocyte that is capable of attacking it. Bonding of the antigen to its membrane receptor protein stimulates that lymphocyte to divide numerous times, until a large population of genetically identical cells (a *clone*) is produced. This **clonal selection theory** accounts for the fact that the immune response to a second and subsequent exposures to an antigen is greater than the immune response to the initial exposure to the antigen.

When stimulated by antigens (generally bacterial), B lymphocytes (or B cells) develop into **plasma cells** that secrete large numbers of **antibody** molecules into the plasma, thus providing **humoral immunity.** These antibodies may destroy bacteria in one of two ways: (1) the antibodies coat the bacterial cell, making it more easily attacked by the phagocytic neutrophils and tissue macrophages; (2) the attachment of antibody to antigen on the bacterial surface activates a system of plasma proteins—*complement*—that lyses the bacterial cell. These two systems work together, since a chemical released from complement attracts the phagocytic white blood cells and increases capillary permeability. Inflammation can later be suppressed by eosinophils, which engulf free antigen-antibody complexes, thus preventing the complement reaction.

T lymphocytes (or T cells) do not secrete antibodies. Instead, they must move into close proximity with their victim cells in order to destroy them. Consequently, T lymphocytes are said to provide **cell-mediated immunity,** often involving the secretion of chemicals, *lymphokines,* released by some T cells. This cell-mediated immunity against cells infected with viruses, cancer cells, and cells of tissue transplants is again directed against specific antigens on the victim cell surface. Therefore, both T and B lymphocytes are specific in their immune attack, and indeed cooperate with each other in the immune defense against disease.

## A. Total White Blood Cell Count

In this procedure, a small amount of blood is diluted with a solution that disintegrates the red blood cells (causes hemolysis) and lightly stains the white blood cells (WBC or leukocytes). The stained white blood cells are counted in the four large corner squares of a hemocytometer (see exercise 6.1, red blood cell count).

Because the dilution factor is 20 and each of the four squares counted has a volume of 0.1 cubic millimeters ($mm^3$), the number of white blood cells per cubic millimeter of blood can be calculated as:

$$\text{WBC per mm}^3 = \frac{\text{white cells} \times 20}{4 \times 0.1\ \text{mm}^3}$$

or,

$$\text{WBC per mm}^3 = \text{white cells} \times 50$$

### PROCEDURE

**Caution:** *Because of the danger of exposure to the AIDS virus and other harmful agents when handling blood, each student should perform this and other blood exercises with his or her own blood only. All objects that have been in contact with blood must be discarded in a container indicated by the instructor.*

1. As described in exercise 6.1A, obtain a drop of blood after discarding the first drop, and fill the diluting pipette (the one with the white bead) to the *0.5 mark.* Avoid air bubbles; if too much blood is drawn into the pipette, remove it by touching the tip of the pipette to a filter paper.
2. Draw the diluting fluid to the *11 mark* on the pipette.
3. Shake or roll the pipette for 3 minutes.
4. Discard the first 4 drops, and fill the hemocytometer as described in exercise 6.1.
5. Allow the cells to settle for 1 minute; then using the low-power objective, count the number of white blood cells in the four large corner squares (labeled A, B, C, and D in fig. 6.5). If a cell is lying on the upper or left-hand line, include it in your count, but do not include cells that are touching the lower or right-hand line.
6. Calculate the number of white blood cells per cubic millimeter of blood and enter this value in the laboratory report.

**The normal white blood cell count is 5,000–10,000 cells per cubic millimeter ($mm^3$) of blood.**

## B. Differential White Blood Cell Count

Clinically, it is also important to determine the relative quantity (percentage) of each leukocyte type within a population of white blood cells. This percentage is obtained by microscopic identification of each leukocyte type out of a total count of 100 white blood cells (see plate 1).

An increase in the white blood cell count **(leukocytosis)** may be produced by an increase in any one of the leukocyte types. These include: (1) *neutrophil leukocytosis,* due to appendicitis, rheumatic fever, smallpox, diabetic acidosis, or hemorrhage; (2) *lymphocyte leukocytosis,* due to infectious mononucleosis or chronic infections (such as syphilis); (3) *eosinophil leukocytosis,* due to parasitic diseases (such as trichinosis), psoriasis, bronchial asthma, or hay fever; (4) *basophil leukocytosis,* due to hemolytic anemia, chicken pox, or smallpox; and (5) *monocyte leukocytosis,* due to malaria, Rocky Mountain spotted fever, bacterial endocarditis, or typhoid fever. In certain cases, an increase in the relative abundance of one type of leukocyte may occur in the absence of an increase in the total white blood cell count—for example, lymphocytosis due to pernicious anemia, influenza, infectious hepatitis, rubella ("German measles"), or mumps.

A decrease in the white blood cell count **(leukopenia)** is usually due to either a decrease in the number of neutrophils or a decrease in the number of eosinophils. A decrease in the number of neutrophils occurs in typhoid fever, measles, infectious hepatitis, rubella, and aplastic anemia. *Eosinopenia* is produced by an elevated secretion of the corticosteroids, which occurs under various conditions of stress, such as severe infections and shock, and in adrenal hyperfunction (Cushing's syndrome).

### PROCEDURE

#### *Making a Blood Smear*

1. Fill a heparinized capillary tube with blood. This can serve as a reservoir of blood for making a number of slides.
2. Using the capillary tube, apply a small drop of blood on one end of a glass slide that is *absolutely clean* and free of grease (fig. 6.7*a*). Place this slide flat on a laboratory bench.
3. Lower a second glass slide at an angle of 30° to the first slide, so that it is lightly touching the first slide *in front of* the drop of blood (fig. 6.7*b*).
4. Gently pull the second slide backwards into the drop of blood, maintaining the pressure and angle that allows the blood to spread out along the edge of the second slide (fig. 6.7*b*).
5. Keeping the same angle and pressure, push the second slide across the first in a rapid, smooth motion. The blood should now be spread in a thin film across the first slide. Done correctly, the concentration of blood in the smear should diminish toward the distal end, producing a feathered appearance (fig. 6.7*c*,*d*).

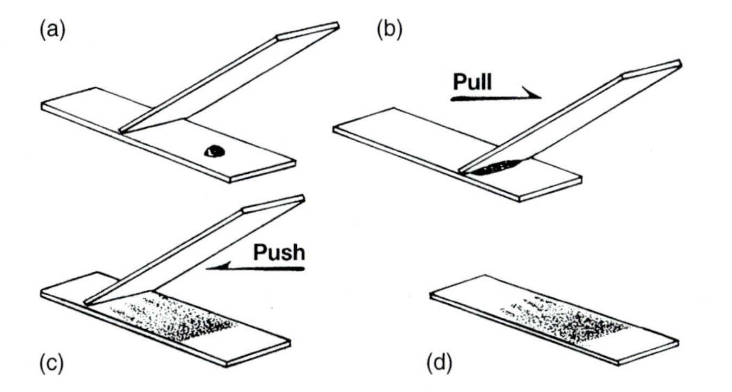


**Figure 6.7** A procedure (a–d) for making a blood smear for a differential white blood cell count.

### PROCEDURE

#### *Staining a Slide Using Wright's Stain*

1. Place the slide on a slide rack and flood the surface of the slide with Wright's stain. Rock the slide back and forth gently for 1 to 3 minutes.

**Note:** *The stain is dissolved in methyl alcohol, which evaporates easily. If any part of the slide should dry during this procedure, the stain will precipitate, ruining the slide.*

2. Drip buffer or distilled water on top of the Wright's stain, being careful not to wash the stain off the slide. Mixing Wright's stain with water is crucial for proper staining; this mixing can be aided by gently blowing on the surface of the stain. Proper staining is indicated by the presence of a metallic sheen on the surface of the stain. The diluted stain should be left on the slide for a full 5 minutes.

3. Wash the stain off the slide with a jet of distilled water from a water bottle and allow the slide to drain at an angle for a few minutes.
4. Using the oil-immersion objective, count the different types of white blood cells. Start at one point in the feathered-tip area of the blood smear and systematically scan the slide until you have counted a total of 100 leukocytes.
5. Keep a running count of the different leukocytes in the table provided on plate 1 and indicate the total number of each. Calculate the percentage of the total count contributed by each type of leukocyte, and enter these values in your laboratory report.

## PROCEDURE

### *Alternative: Staining a Slide Using Diff-Quik (Harleco)*

1. Dip the slide in fixative solution (light blue) five times, allowing 1 second per dip.
2. Dip the slide in solution 1 (orange) five times, allowing 1 second per dip.
3. Dip the slide in solution 2 (dark blue) five times, allowing 1 second per dip.
4. Rinse the slide with distilled water and count the white blood cells using the oil-immersion objective as described in steps 4 and 5 of the previous procedure. Enter these values in your laboratory report.

# Laboratory Report 6.2

Name ______________________

Date ______________________

Section ______________________

## DATA FROM EXERCISE 6.2

**A. Total White Blood Cell Count**

1. Enter your white blood cell count: __________ WBC per $mm^3$.

2. Compare your measured values to the normal range and write your conclusions in this space.

**B. Differential White Blood Cell Count**

1. Count each type of white blood cell (leukocyte) and record the number in the table below. Next, determine your grand total of leukocytes counted (roughly 100). Now, calculate the percentage of each type of white blood cell and enter these values in this table (see plate 1 for normal values).

| Leukocytes | Cells Counted | Total WBCs Counted | Percentage |
|---|---|---|---|
| Neutrophils | | | |
| Eosinophils | | | |
| Basophils | | | |
| Lymphocytes | | | |
| Monocytes | | | |

2. Compare your values to the normal range and write your conclusions in this space.

## REVIEW ACTIVITIES FOR EXERCISE 6.2

### *Test Your Knowledge of Terms and Facts*

Identify the leukocyte by these descriptions:

| | |
|---|---|
| ______1. polymorphonuclear with poorly staining granules | (a) eosinophil |
| ______2. agranular with round nucleus, little cytoplasm | (b) neutrophil |
| ______3. granules with affinity for red stain | (c) monocyte |
| ______4. rarest white blood cell | (d) lymphocyte |
| ______5. agranular and phagocytic | (e) basophil |

6. Antibodies are produced by ______________________________ lymphocytes; cell-mediated immunity is provided by ______________________________ lymphocytes.
7. While blood cells leave capillaries by a process called ______________________________.
8. The major phagocytic white blood cells are the ______________________________.
9. Molecules that activate the immune system are called ______________________________.

### *Test Your Understanding of Concepts*

10. Distinguish between humoral and cell-mediated immunity, identifying the cells involved, their origin, and their functions.

11. Describe the clonal selection theory and explain how it accounts for the ability to defend against subsequent exposure to a particular antigen.

## *Test Your Ability to Analyze and Apply Your Knowledge*

12. *Active immunizations* involve the exposure of a person to a pathogen whose *virulence* (ability to cause disease) has been reduced without altering its *antigenicity* (nature of its antigens). How do you think this might be accomplished? What are the benefits and dangers of this procedure?

13. *Passive immunizations* involve injecting a person who has been exposed to a pathogen with serum containing antibodies, called *antiserum* or *antitoxin*. Antiserum is developed by injecting an animal with a pathogen. What happens in that animal? What are the benefits and shortcomings of passive immunization compared to active immunization?

# Blood Types

EXERCISE 6.3

## Materials

1. Sterile lancets, 70% alcohol
2. Anti-A, anti-B, and anti-Rh sera (Hardy Diagnostics)
3. Slide warmer, glass slides, and toothpicks
4. Container for the disposal of blood-containing objects

Red blood cells (RBCs) have characteristic molecules on the surface of their membranes that can be different in different people. These genetically determined membrane molecules can function as antigens—capable of bonding to specific antibodies when exposed to plasma from a person with a different blood type. The major blood group antigens are the Rh antigen and the antigens of the ABO system.

### OBJECTIVES

1. Explain what is meant by the term *blood type,* and identify the major blood types.
2. Explain how agglutination occurs, and how agglutination tests can be used to determine a person's blood type.
3. Identify the different genotypes that can produce the different blood group phenotypes, and explain how different blood types can be inherited.
4. Explain how erythroblastosis fetalis is produced.
5. Explain the dangers of mismatched blood types in blood transfusions.

### Textbook Correlations

Before performing this exercise, you should study the introductory material presented here. Further information relating to this exercise can be found in these pages of *Human Physiology,* eighth edition, by Stuart I. Fox:

- *Red Blood Cell Antigens and Blood Typing,* Chapter 13, pp. 372–374.

When blood from one person is mixed with plasma from another person, the red blood cells will sometimes **agglutinate,** or clump together (fig. 6.8). This agglutination reaction, which is very important in determining the safety of transfusions (agglutinated cells can block small blood vessels), is due to a mismatch of genetically determined blood types.

On the surface of each red blood cell are a number of molecules that have antigenic properties, and in the plasma each antibody molecule has two combining sites for antigens. In a positive agglutination test, the red blood cells clump together because they are combined through antibody bridges.

## A. The Rh Factor

One of the antigens on the surface of red blood cells is the **Rh factor** (named because it was first discovered in rhesus monkeys). The Rh factor is found on the red blood cell membranes of approximately 85% of the people in the United States. The presence of this antigen on the red blood cells (an **Rh positive** phenotype) is inherited as a dominant trait and is produced by both the *homozygous (RR)* genotype and the *heterozygous (Rr)* genotype. Individuals who have the *homozygous recessive* genotype *(rr)* do not have this antigen on their red blood cells and are said to have the **Rh negative** phenotype.

Suppose an Rh positive man who is heterozygous *(Rr)* mates with an Rh positive woman who is also heterozygous.

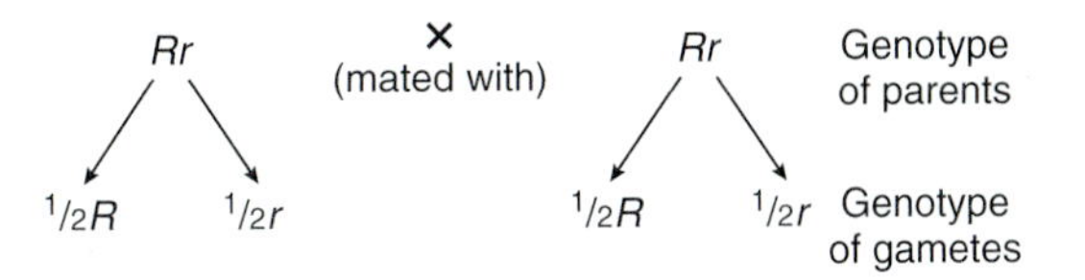


Since the mother is Rh positive, her immune system cannot be stimulated to produce antibodies by the presence of an Rh positive fetus. The development of **immunological competence** does not occur until shortly after birth, so that an Rh negative fetus in an Rh positive mother would not yet have an immune response during normal gestation (pregnancy).

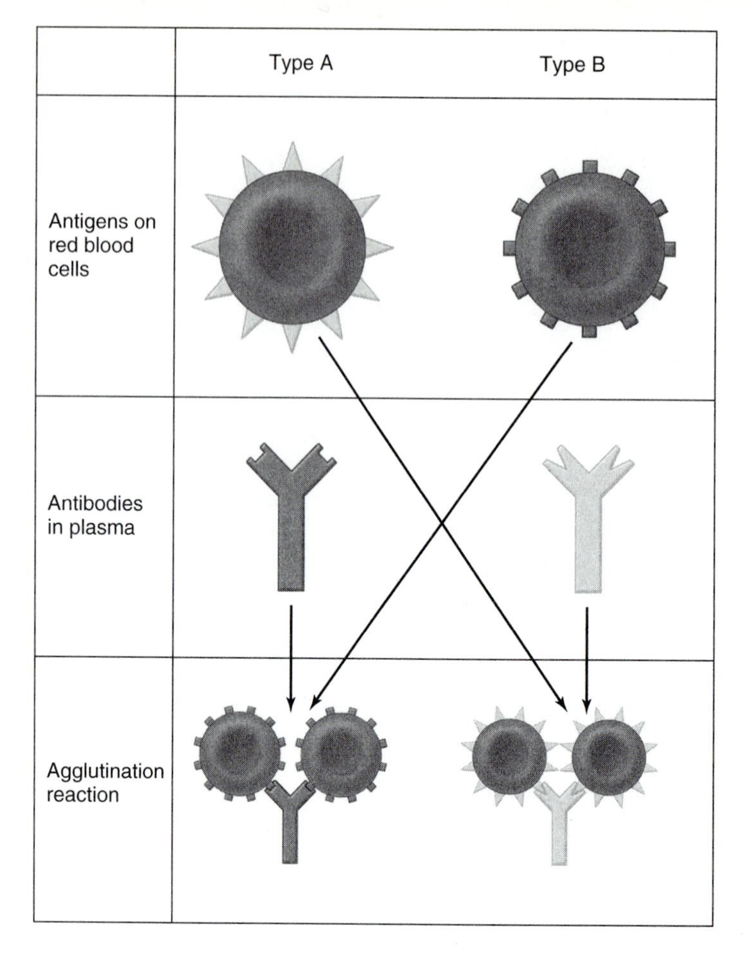


**Figure 6.8 Agglutination reaction.** A person with type A blood has type A antigens on their red blood cells and antibodies in their plasma against the type B antigen. A person with type B blood has type B antigens on their red blood cells and antibodies in their plasma against the type A antigen. Therefore, if red blood cells from one blood type are mixed with antibodies from the plasma of the other blood type, an agglutination reaction occurs. In this reaction, red blood cells stick together because of antigen-antibody binding.

**(For a full-color version of this figure, see fig. 13.5 in *Human Physiology,* eighth edition, by Stuart I. Fox.)**

However, when an Rh negative mother is carrying an Rh positive fetus, some of the Rh antigens may enter her circulation when the placenta tears at birth (red blood cells do not normally cross the placenta during pregnancy). Since these red blood cells express an antigen (the Rh factor) that is foreign to the mother, her immune system will eventually be stimulated to produce antibodies that are capable of destroying the red blood cells of subsequent Rh positive fetuses, a condition known as *hemolytic disease of the newborn*, or **erythroblastosis fetalis.** However, erythroblastosis fetalis can be prevented by the administration of exogenous Rh antibodies known as **Rho(D) immune globulin** (e.g., *Rho*GAM) to the mother within 72 hours after delivery. These antibodies destroy the fetal Rh positive red blood cells that have entered the maternal circulation before they can stimulate an immune response in the mother.

## PROCEDURE

1. Place one drop of anti-Rh serum on a clean glass slide.
2. Add an equal amount of fingertip blood and mix it with the antiserum (use an applicator stick or a toothpick).
3. Place the slide on a slide warmer (45° C to 50° C) and rock it back and forth.
4. Examine the slide for agglutination. If no agglutination is observed after a 2-minute period, examine the slide under the low-power objective of the microscope. The presence of grains of agglutinated red blood cells indicates Rh positive blood.

**Caution:** *Handle only your own blood and be sure to discard the slide, toothpicks, and lancet in the container provided by the instructor.*

5. Enter your Rh factor type (positive or negative) in the laboratory report.

**Table 6.1** Incidence of Blood Types—Approximate Incidence in the U.S. (%)

| Blood Types | Caucasian | Black | Asian |
|---|---|---|---|
| O | 45 | 48 | 36 |
| A | 41 | 27 | 28 |
| B | 10 | 21 | 23 |
| AB | 4 | 4 | 13 |

## B. The ABO Antigen System

Each individual inherits two genes, one from each parent, that control the synthesis of red blood cell antigens of the ABO classification. Each gene contains the information for one of three possible phenotypes: antigen A, antigen B, or no antigen (written O). Thus, an individual may have one of six possible genotypes: **AA, AO, BB, BO, AB,** or **OO.**

An individual who has the genotype AO will produce type A antigens just like an individual who has the genotype AA; and therefore both are said to have **type A** blood. Likewise, an individual with the genotype BO and one with the genotype BB will both have **type B** blood. Since lack of antigen is a recessive trait, an individual with **type O** blood must have the genotype OO.

Unlike many other traits, the heterozygous genotype AB has a phenotype that is different from either of the homozygous genotypes (AA or BB). Since there is no dominance between A and B, individuals with the genotype AB produce red blood cells with *both* the A and B antigens (a condition known as *codominance*) and have **type AB** blood. The most common blood types are type O and type A; the rarest is type AB (table 6.1).

Also, unlike the other immune responses considered, antibodies against the A and B antigens are not induced by prior exposure to these blood types. A person with type A blood, for example, has antibodies in the plasma against type B blood even though that person may never have been exposed to this antigen. A transfusion with type B blood into the type A person would be extremely dangerous because the anti-B antibodies in the recipient's plasma would agglutinate the red blood cells in the donor's blood. The outcome would be the same if the donor were type A and the recipient type B (see **plate 2,** following p. 232).

| Antigen on RBC Surface | Antibody in Plasma |
|---|---|
| A (type A) | Anti-B |
| B (type B) | Anti-A |
| O (type O) | Anti-A and anti-B |
| AB (type AB) | No antibody |

### PROCEDURE

1. Draw a line down the center of a clean glass slide with a marking pencil and label one side **A** and the other side **B.**
2. Place a drop of anti-A serum on the side marked **A** and a drop of anti-B serum on the side marked **B.**
3. Add a drop of blood to each antiserum and mix each with a separate applicator stick.
4. Tilt the slide back and forth and examine for agglutination over a 2-minute period. *Do not heat the slide on the slide warmer.*
5. Enter your ABO blood type in the laboratory report.

# Laboratory Report 6.3

Name ______________________

Date ______________________

Section ______________________

## DATA FROM EXERCISE 6.3

1. Did your blood agglutinate with the anti-Rh serum? ______________________
   Are you Rh positive or negative? ______________________
2. Indicate (with a yes or no) whether your blood agglutinated with the anti-A and anti-B sera.
   Anti-A: ______________________
   Anti-B: ______________________
   What is your blood type? ______________________

## REVIEW ACTIVITIES FOR EXERCISE 6.3

### *Test Your Knowledge of Terms and Facts*

1. Name the antigens present and absent on the surface of a red blood cell if the person is:
   (a) type A negative ______________________
   (b) type O positive ______________________
   (c) type AB negative ______________________
2. If a person has blood type A, the possible genotypes that the person may have are __________ and __________.
3. If a person who is blood type O marries a person who is blood type A, what are the possible blood types their children could have? ______________ or ______________
4. The universal blood donor is blood type ______________.
5. The rarest blood type is blood type ______________.
6. The most common Rh type is ______________.
7. The person most in danger of having a child who develops erythroblastosis fetalis is a woman who has the blood type ______________ when her husband has the blood type ______________.

### *Test Your Understanding of Concepts*

8. What are the dangers of giving a person a transfusion when the blood types don't match?

9. Explain how hemolytic disease of the newborn is produced. How may the disease be prevented?

## *Test Your Ability to Analyze and Apply Your Knowledge*

10. Can blood types be used in paternity cases to prove or disprove possible fatherhood? Give examples to support your answer.

11. Suppose a person who has type A blood receives large amounts of whole blood from a person who has the universal donor blood type. Will that be safe? Explain.

# Blood Clotting System

EXERCISE 6.4

MATERIALS

1. Pipettes (0.10–0.20 mL), and small test tubes
2. Constant-temperature water bath set at 37° C
3. 0.02 M calcium chloride, activated thromboplastin, activated cephaloplastin (Dade), fresh plasma

Two interrelated clotting pathways—the intrinsic system and the extrinsic system—require the successive activation of specific plasma clotting factors. Defects in these factors can be detected by means of two clotting-time tests.

## OBJECTIVES

1. Describe the intrinsic and extrinsic clotting systems.
2. Describe why bleeding time is prolonged in cases of hemophilia and vitamin K deficiency.
3. Demonstrate the tests for prothrombin time and for activated partial thromboplastin time (APTT). Identify the normal values for each, and explain how these tests are used to diagnose bleeding disorders.

## Textbook Correlations

Before performing this exercise, you should study the introductory material presented here. Further information relating to this exercise can be found in these pages of *Human Physiology,* eighth edition, by Stuart I. Fox:

- *Blood Clotting,* Chapter 13, pp. 374–375.

Damage to a blood vessel initiates a series of events that, if successful, culminate in *hemostasis* (the arrest of bleeding).

1. The first event is **vasoconstriction,** which decreases the flow of the blood in the damaged vessel.
2. The next event is the formation of a **platelet plug.** This response occurs in two steps:
   a. In the first step, platelets adhere to the exposed collagen (connective tissue protein) of the damaged vessel and then release *adenosine diphosphate (ADP).*
   b. The second step occurs when the ADP, by making the adherent platelets sticky, causes other platelets to cling at this site, forming a platelet clump.
3. The third event is the sequential activation of **clotting factors** in the plasma, resulting in the formation of an insoluble fibrous protein, *fibrin,* around the platelet clump. This produces a blood clot.

The formation of fibrin from its precursor, *fibrinogen*, requires the presence of the enzyme *thrombin*.

$$\text{fibronogen} \xrightarrow{\text{thrombin}} \text{fibrin}$$

The insoluble fibrin is formed instantly whenever the enzyme thrombin is present, and thus the formation of thrombin must be a carefully regulated event in the body. The formation of thrombin from its precursor, *prothrombin*, requires the sequential activation of a number of other clotting factors.

When the sequence of events leading to the formation of thrombin is initiated by the release of *tissue thromboplastin* from damaged tissue cells, fibrin is rapidly formed (10–15 seconds). These events constitute the **extrinsic system** of blood clotting, since the sequence is initiated by a factor extrinsic to the blood (fig. 6.9). Alternatively, a blood clot may be initiated in the absence of tissue thromboplastin through the activation of *Hageman factor* (factor XII) by the exposure of plasma to glass, crystals, or collagen. This **intrinsic system** of blood clotting is slower (28–45 seconds) than the extrinsic system (fig. 6.9).

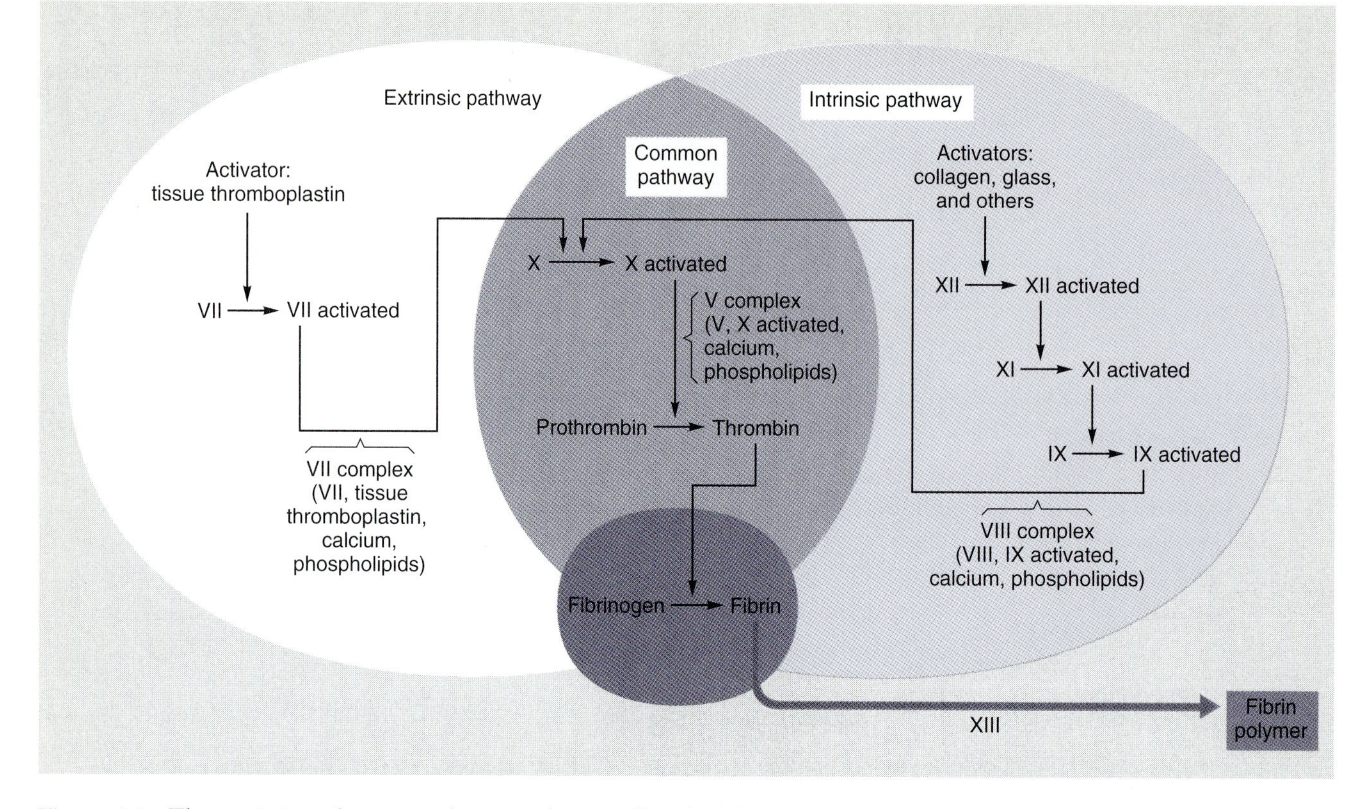


**Figure 6.9 The extrinsic and intrinsic clotting pathways.** These both lead to a common sequence of events that results in the formation of insoluble threads of fibrin polymers.

**(For a full-color version of this figure, see fig. 13.8 in *Human Physiology,* eighth edition, by Stuart I. Fox.)**

Many people have either an acquired or an inherited inability to form fibrin threads within the normal time interval. This inability may be due to a **vitamin K deficiency,** since vitamin K is necessary for the formation of four of the clotting factors, including prothrombin. Vitamin K deficiency may occur in the newborn who has an inadequate intake of milk, in a person with fat malabsorption due to inadequate absorption of this vitamin (bile salts facilitate absorption), as a result of antibiotic therapy destroying intestinal flora (a source of vitamin K), or as a result of oral anticoagulants such as *dicumarol.*

There are many hereditary conditions in which a clotting factor is either missing or defective. The best known of these conditions, classical **hemophilia,** is due to the genetic inability to synthesize normal factor VIII. This condition, as well as *Christmas disease* (defective factor IX), is inherited as a sex-linked recessive trait. Other genetic defects inherited as autosomal traits include those associated with factors II, VII, X, XI, and XII.

In this exercise, two tests will be performed to screen for defective clotting factors. The formation of thrombin in the plasma samples will be inhibited by an anticoagulant. The anticoagulant used, either *citric acid* or *oxalic acid,* removes calcium ion ($Ca^{2+}$) from the plasma. The removal of $Ca^{2+}$ has an anticoagulant effect because calcium is a necessary cofactor in the activation of a number of the clotting factors. This inhibition can be easily reversed by adding calcium ion during the clotting tests.

The test for **prothrombin time** is used to determine deficiencies in the *extrinsic* clotting system and is prolonged when factor V, VII, or X is defective. The test for **activated partial thromboplastin time (APTT)** is used to determine deficiencies in the *intrinsic* clotting system and is sensitive to all defective factors except factor VII (tissue thromboplastin). A person with classical hemophilia (defective factor VIII), for example, would have a normal prothrombin time but an abnormal APTT. In this way, these two tests complement each other and can be used, together with other tests, to determine the exact cause of prolonged clotting time (see table 6.2).

**Table 6.2** Test Results for Clotting Factors

| Defective Factor | Prothrombin Time | APTT |
|---|---|---|
| V | Abnormal | Abnormal |
| VII | Abnormal | Normal |
| VIII | Normal | Abnormal |
| IX | Normal | Abnormal |
| X | Abnormal | Abnormal |
| XI | Normal | Abnormal |
| XII | Normal | Abnormal |

Laboratory

A. Prothrombin
1. Enter your pro

B. Activated Part
1. Enter your AP

2. Compare your

*Test Your Knowl*

1. The factor that
2. Which clotting
3. The factor that
4. The factor name
5. Which vitamin
6. Citric acid (citr
7. The ________ system, whereas intrinsic clotting
8. In the formation makes other plat
9. The general cate ________

## A. Test for Prothrombin Time

### PROCEDURE

1. Pipette 0.10 mL of activated thromboplastin and 0.10 mL of 0.02 M $CaCl_2$ into a test tube. Place the tube in a 37° C water bath and allow it to warm for at least 1 minute.
2. Warm a sample of plasma in the water bath for at least 1 minute. Then use a pipette to forcibly expel 0.10 mL of plasma into the warmed tube containing the thromboplastin–$CaCl_2$ mixture. Start timing at this point.
3. Agitate this tube mixture continuously in the water bath for 10 seconds.
4. Remove the tube, quickly wipe it, and hold it in front of a bright light. Tilt the tube gently back and forth and stop timing when the first fibrin threads appear (the solution will change from a fluid to a semigel). Enter the time in your laboratory report.

The normal prothrombin time is 11 ± 1 seconds.

## B. Test for Activated Partial Thromboplastin Time (APTT)

### PROCEDURE

1. Warm a tube of 0.02 M $CaCl_2$ by placing it in a 37° C water bath.
2. Pipette 0.10 mL of activated cephaloplastin and 0.10 mL of plasma into a test tube and allow it to incubate at 37° C for 3 minutes.
3. Using a pipette, forcibly expel 0.10 mL of warmed $CaCl_2$ into the cephaloplastin-plasma mixture. Start timing at this point.
4. Agitate this tube mixture continuously in the 37° C water bath for 30 seconds. Then remove the tube, quickly wipe it, and hold it against a bright light while rocking the tube back and forth.
5. Stop timing when the first fibrin threads appear. Enter the time in your laboratory report.

The normal APTT is less than 40 seconds.

## *Test Your Understanding of Concepts*

10. Which four factors are common to both the intrinsic and extrinsic clotting pathway? Which clotting test(s) would be abnormally prolonged if a person had a deficiency in factor VII? Explain.

11. Which factor(s) would be defective if a person had a prolonged prothrombin time but a normal partial thromboplastin time (APTT)? Explain.

## *Test Your Ability to Analyze and Apply Your Knowledge*

12. Heparin is a mucopolysaccharide extracted from beef lung and liver that inhibits the action of thrombin. What effect would thrombin have on the prothrombin time and APTT? Why was citric acid or oxalic acid used as an anticoagulant instead of heparin in these tests?

13. Why might a person with an abnormally slow clotting time be given vitamin K? Would treatment with vitamin K immediately improve the clotting time? Explain.

14. You might expect that almost all people with hemophilia due to factor VIII or IX deficiency would be males, whereas those with hemophilia due to factor XI or XII deficiency would be equally like to be males or females. Explain.

# The Cardiovascular System

# Section 7

**T**he blood transports glucose, amino acids, fatty acids, and other monomers from the digestive tract, liver, and adipose tissue to all the cells of the body. The waste products of cellular metabolism are carried by the blood to the kidneys and lungs for elimination. Hormones, secreted by endocrine glands, are carried by the blood to target organs. Blood is thus the major channel of communication between the different specialized organs of the body.

The interchange of molecules between blood and tissue cells occurs across the walls of **capillaries,** which are composed of only a single layer of epithelial cells (endothelium). Blood is delivered to the capillaries in **arterioles,** microscopic vessels with walls of endothelium, smooth muscle, and connective tissue. The arterioles receive their blood from larger, more muscular **arteries.** Blood is drained from the capillaries into microscopic **venules.** The venules drain their blood into larger **veins** that are less muscular and more distensible than arteries.

Since the tissue cells must be located within 0.10 mm of a capillary for molecules to diffuse adequately, the vascular system within an organ is highly branched. The many narrow, muscular arterioles of this *vascular tree* offer great resistance to blood flow *(peripheral resistance)* through the organs. For organs to receive an adequate blood flow *(perfusion),* the arterial blood must be under sufficient pressure to overcome this resistance to blood flow. This arterial pressure is routinely measured with a device known as a *sphygmomanometer.*

The blood pressure required to overcome peripheral resistance and maintain adequate tissue perfusion is generated by a muscular pump—the *heart.* The heart has four chambers: two atria and two ventricles. The *right atrium* receives oxygen-depleted blood returning from body cells in the superior and inferior venae cavae; and the *left atrium* receives oxygen-rich blood from the pulmonary veins. The *right ventricle* pumps blood into the pulmonary arteries to the lungs where oxygen enters and carbon dioxide exits the blood. The *left ventricle* pumps blood into the large *aorta,* which by means of its many branches perfuses all the organs in the body. Since the blood pumped out of the heart by the two ventricles, the **cardiac output,** is carried by arteries to the body's organs, and since blood from the organs is returned by veins to the heart, the cardiovascular system forms a closed circle called the **circulatory system** (fig. 7.1).

The heart's ability to maintain adequate perfusion of the body's organs depends on proper electrical stimulation and muscular contraction, proper functioning of *valves* (directing blood flow within the heart), and integrity of the blood vessels. These functions can be assessed by various techniques, which will be explored in these exercises.

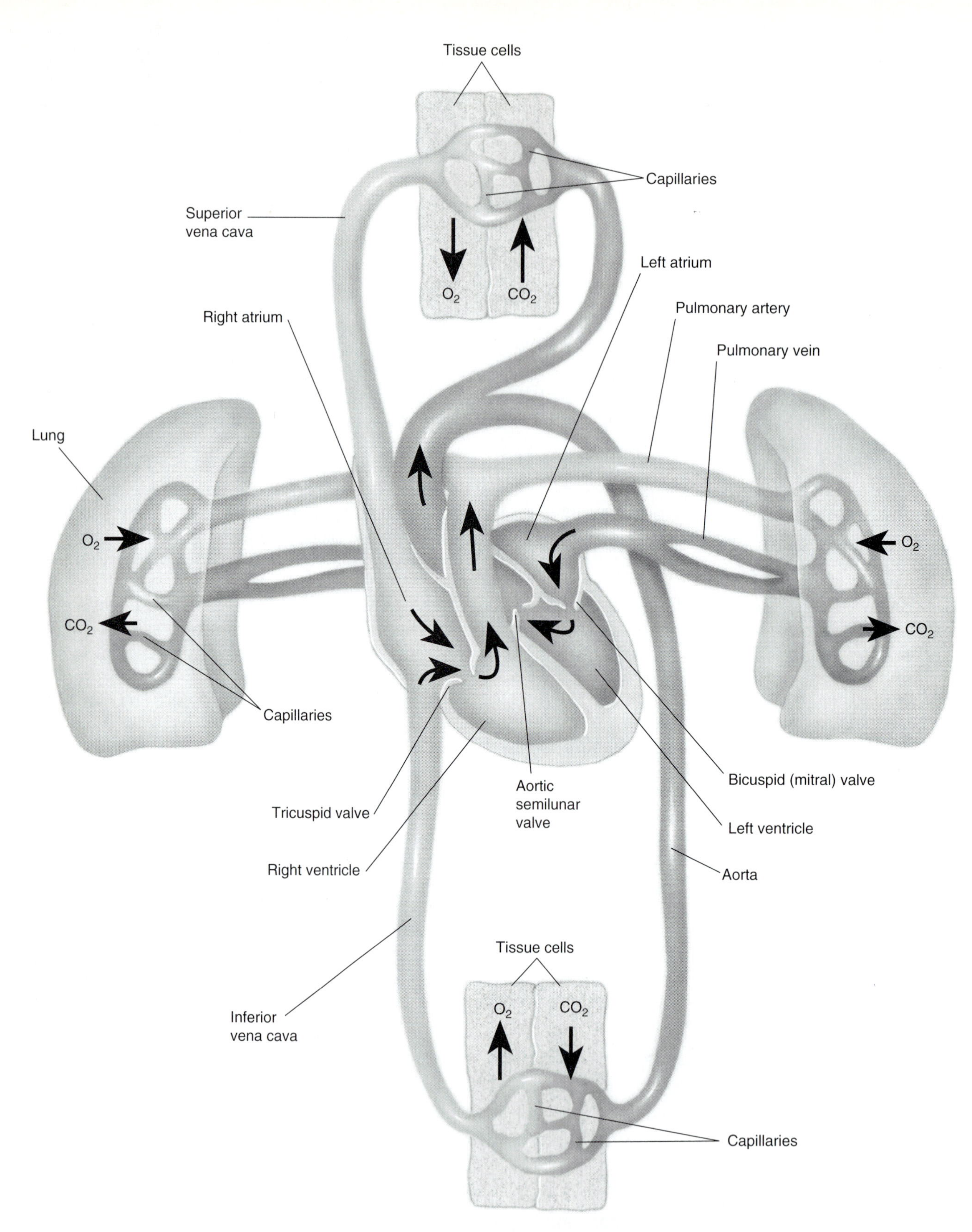


**Figure 7.1 A diagram of the circulatory system.** The pulmonary arteries and veins compose the pulmonary circulation, whereas other arteries and veins are part of the systemic circulation. The right ventricle pumps blood into the pulmonary circulation, while the left ventricle pumps blood into the systemic circulation

**(For a full-color version of this figure, see fig. 13.9 in *Human Physiology,* eighth edition, by Stuart I. Fox.)**

# Effects of Drugs on the Frog Heart

EXERCISE 7.1

## Materials

1. Frogs, dissecting instruments, trays
2. Copper wire or bent pin, thread
3. Recording apparatus: physiograph, transducer coupler, and myograph transducer (Narco); or kymograph, kymograph paper, and kerosene burner
4. As an alternative to pen-and-paper recording equipment, a computerized data acquisition and analysis system, such as those provided by Biopac, iWorx, and others, may be used to perform the exercises in this section.
5. Ringer's solution (see exercise 5.1—all drugs to be prepared using Ringer's solution as a solvent); calcium chloride (2.0 g/100 mL); digitoxin (0.2 g/100 mL); pilocarpine (2.5 g/100 mL); atropine (5.0 g/100 mL); potassium chloride (2.0 g/100 mL); epinephrine (0.01 g/100 mL); caffeine (0.2 g/dL); and nicotine (1.0 g/2L; or 6.16 mL liquid/dL).

The heart of a pithed frog may continue to beat automatically after the frog's central nervous system has been destroyed. By this means, the function of the heart and the effects of various drugs on the heart can be studied.

### OBJECTIVES

1. Describe the pattern of contraction in the frog heart.
2. Describe the effect of various drugs on the heart and explain their mechanisms of action.

A **drug** is a substance that affects some aspects of physiology when given to the body. Drugs may be identical to naturally occurring substances found in the body, such as minerals, vitamins, and hormones, or they may be molecules uniquely produced by particular plants or fungi. Many drugs marketed by pharmaceutical companies are derived from natural products whose chemical structure has been slightly modified to alter the biological activity of the native compounds.

### Textbook Correlations*

Before performing this exercise, you should study the introductory material presented here. Further information relating to this exercise can be found in these pages of *Human Physiology,* eighth edition, by Stuart I. Fox:

- *Cardiac Muscle.* Chapter 12, p. 354–355.
- *Structure of the Heart.* Chapter 13, pp. 378–380.
- *Electrical Activity of the Heart.* Chapter 13, pp. 384–387.

The biological effects of *endogenous* compounds (those compounds normally found in the body) vary with their concentration. A normal blood potassium concentration, for example, is necessary for good health, but too high a concentration can be fatal. Similarly, the actions exhibited by many hormones at abnormally high concentrations may not occur when the hormones are at normal concentrations. It is important, therefore, to distinguish between the **physiological effects** (normal effects) of these substances and their **pharmacological effects** (those that occur when the substances are administered as drugs). A study of the pharmacology of various substances however, can reveal much about the normal physiology of the body.

In this exercise, we will test the effects of various pharmacological agents on the heart of a pithed frog. Although the heart, like skeletal muscle, is striated, it differs from skeletal muscles in several respects. The heartbeat is automatic; it does not have to be stimulated by nerves or electrodes to contract. Action potentials begin spontaneously in the *pacemaker region*—the *sinoatrial* or *SA node* region—of the right atrium and spread through the ventricles in an automatic, rhythmic cycle. As can be seen in the exposed frog heart, this causes the atria to contract before the ventricle. (Note that, unlike mammals, frogs have only one ventricle.)

When the frog heart is connected by a thread to the recording equipment, contractions of the atria and ventricle produce two successive peaks in the recordings. The

*Multimedia Correlations (also see Appendix 3)
- A.D.A.M. *InterActive Physiology* (Cardiovascular System): Cardiac Cycle

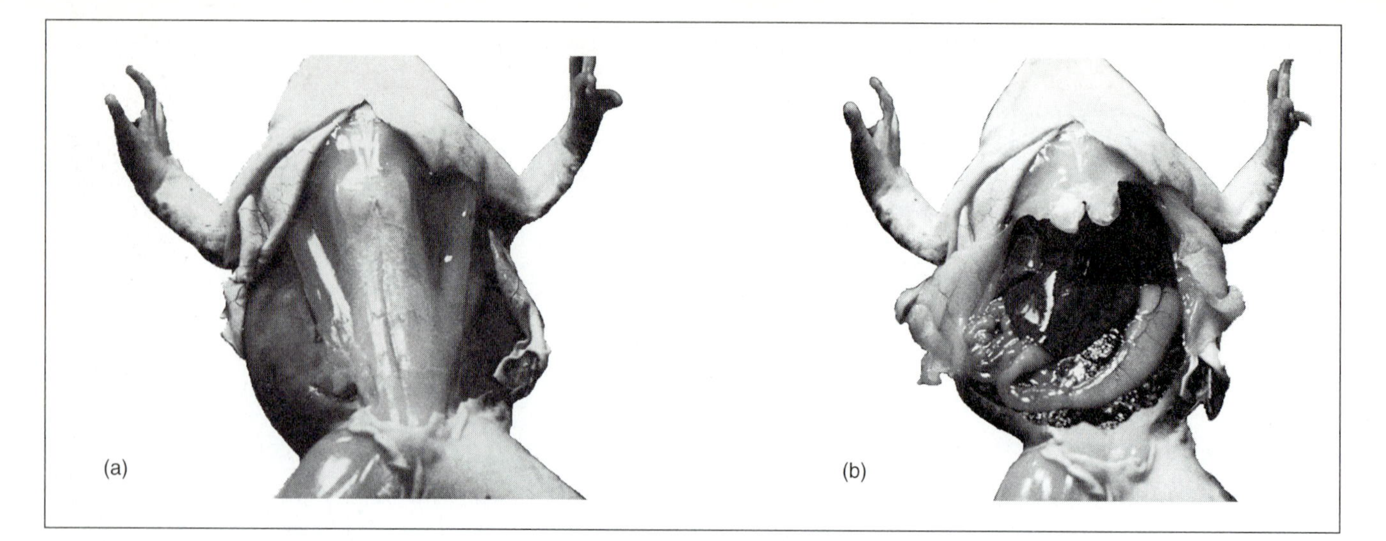


**Figure 7.2 Procedure for exposing the frog heart.** (a) First the skin is cut. (b) The body cavity is exposed by cutting through the muscles to the sternum, which is then split to expose the heart.

strength of contraction is related to the amplitude (height) of these peaks, and the rate of beat can be determined by the distance between the ventricular peaks if the chart speed is known. The rate of impulse conduction between the atria and ventricle is related to the distance between the atrial and ventricular peaks in the recording of each cycle. Therefore, the effects of various drugs on the strength of contraction, rate of contraction, and rate of impulse conduction from the atria to the ventricle can be determined.

## PREPARATION FOR RECORDING

1. Double-pith a frog, and expose its heart (see exercise 5.1 and fig. 7.2). Skewer the apex of the heart muscle with a short length of thin copper wire or a bent pin, being careful not to let the wire enter the chamber of the ventricle. (The frog heart has only one ventricle and two atria.)
2. Bend the copper wire into a loop, and tie one end of cotton thread to this loop or to the head of a bent pin (see the enlarged insert in fig. 7.3).
3. Procedure for **kymograph** recording:
   (a) Tie the other end of the thread to a heart lever. The thread tension should be fairly taut so that contractions of the heart produce movements of the lever.
   (b) Attach kymograph paper (shiny side out) to the kymograph drum, and rotate the drum slowly over a kerosene burner until the paper is uniformly blackened. Arrange the heart lever so that it lightly drags across the smoked paper. Too much pressure of the writing stylus against the kymograph will prevent movement of the heart lever. (See fig. 7.3 for the proper setup.)

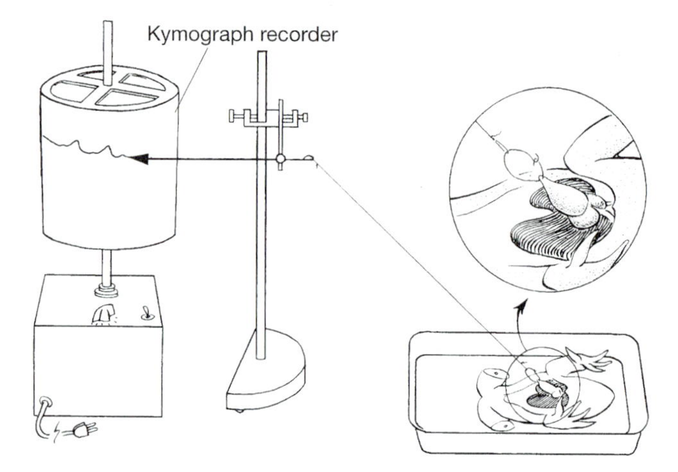


**Figure 7.3 Frog heart setup.** The contractions of the heart pull a lever that writes on a moving chart (kymograph). The setup using electronic recording equipment, such as a physiograph recorder, is similar to that shown here, except that the string from the heart will be connected to a myograph transducer.

4. Procedure for **physiograph** recording (see physiograph, exercise 5.1):
   (a) Tie the other end of the thread to the hook below the myograph transducer. (Make sure the myograph is plugged into the transducer coupler on the physiograph.) The heart should be positioned directly below the myograph. Adjust the height of the myograph on its stand so that the heart is pulled out of the chest cavity (fig. 7.4*b*).
   (b) Make sure the physiograph is properly balanced and set the paper speed at *0.5 cm per second.* Depress the record button, and press the gray paper advance button when ready to record.

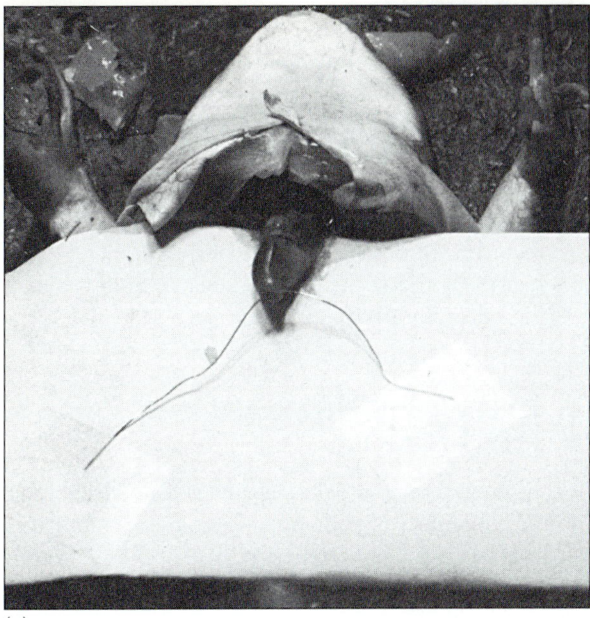

(a)

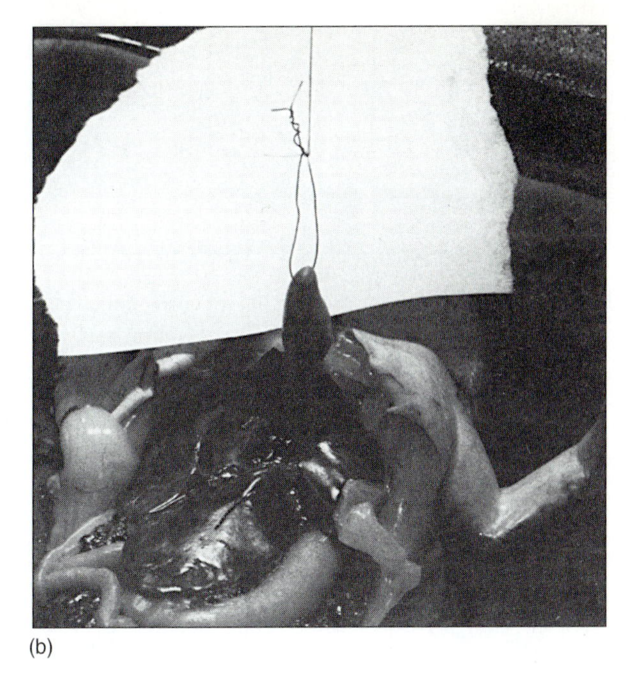

(b)

**Figure 7.4 Procedure for setting up the frog heart to record its contractions.** (a) A small length of thin copper wire is passed through the tip of the ventricle. (b) This wire is then twisted together to form a loop, which is tied by a cotton thread to the hook in the myograph transducer.

5. Procedure for **computerized data acquisition and analysis** equipment.
   a. For the Biopac system, the force transducer assembly (SS12LA) and amplifier (MP30) are connected to each other and to the computer as illustrated in figure 7.5. The procedures in this exercise or in Biopac *PRO* Lesson A04 can then be followed.
   b. For the iWorx system, connect the force transducer to channel 3 of the data acquisition unit. The procedures in this exercise or in iWorx Experiment 10 can then be followed.
6. Observe the pattern of the heartbeat prior to the addition of drugs; this tracing will serve as the normal, or *control,* record (fig. 7.6). Distinguish the atrial from the ventricular beats. Measure the heart rate (in beats per minute), the strength of contraction (in millimeters deflection above baseline), and the distance (in millimeters) between atrial and ventricular peaks of the heartbeat.

   Record this data in the laboratory report.

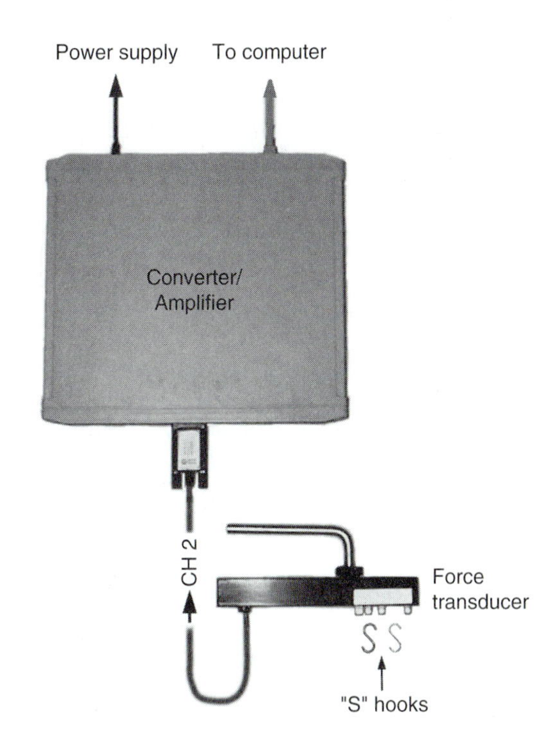


**Figure 7.5 Connection of hardware for use of Biopac system with frog heart exercises.** The details of the setup are described in the Biopac PRO Lesson A04. Once the equipment is set up, it can be used to perform the exercises in this section.

## A. Effect of Calcium Ions on the Heart

In addition to the role of calcium in coupling excitation to contraction, the extracellular $Ca^{2+}/Mg^{2+}$ ratio also affects the permeability of the cell membrane. An increase in the extracellular concentration of calcium (above the normal concentration of 4.5–5.5 mEq/L) affects both the electrical properties and the contraction strength of heart muscle.

The heart is affected in a number of ways by an increase in extracellular calcium. These include (1) an increased force of contraction, (2) a decreased cardiac rate, and (3) the appearance of ectopic pacemakers in the ventricles, producing abnormal rhythms (extrasystoles and idioventricular rhythm—see exercise 7.2).

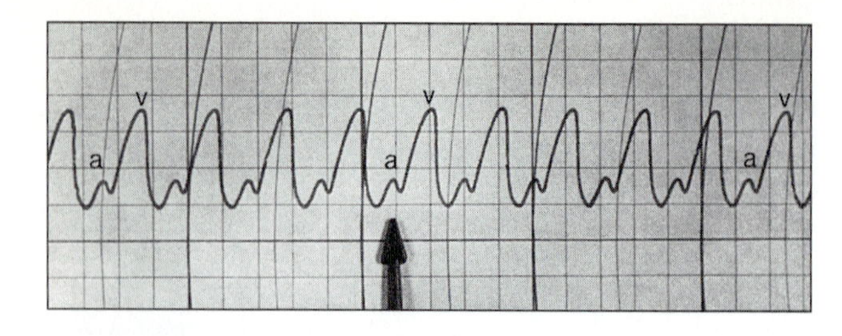


**Figure 7.6 Recordings of the frog heart contractions.** The arrow points to a recording of a smaller atrial contraction, which is followed by a recording of a larger ventricular contraction.

### PROCEDURE

1. Obtain a control record of the normal heartbeat. Then, while the paper continues to run, use a dropper to bathe the heart in a 2.0% solution of calcium chloride ($CaCl_2$). On the moving recording paper, indicate the time at which calcium was added. Observe the effects of the added calcium solution over a period of a few minutes, then stop the recording.
2. Rinse the heart thoroughly with Ringer's solution until the heartbeat returns somewhat to normal; this new rhythm will serve as the next control.
3. In the table of your laboratory report, tape the recording or draw a facsimile of the normal heartbeat and of the changed heartbeat after the calcium solution was added.

## B. Effect of Digitalis on the Heart

The effects of digitalis are believed to be due to its inhibition of the $Na^+/K^+$ (ATPase) pump. This inhibition results in an influx of $Na^+$ and an efflux of $K^+$ and is accompanied by an enhanced uptake of calcium ions. The effects of digitalis and of increased extracellular calcium on the heart are thus very similar.

**Digitalis glycosides,** such as digoxin, or digitalis, are frequently used to treat congestive heart failure, atrial flutter, and atrial fibrillation. Digitalis relieves these conditions by (1) increasing the force of contraction, (2) decreasing the cardiac rate directly by inhibiting the SA node, and (3) by slowing conduction through the bundle of His.

### PROCEDURE

1. Obtain a record of the control heartbeat. Then, bathe the heart in a 2.0% solution of digitalis.
2. Rinse the heart thoroughly with Ringer's solution until the heartbeat returns somewhat to normal; this new rhythm will serve as the next control.
3. In the table of your laboratory report, tape the recording or draw a facsimile of the normal heartbeat and of the changed heartbeat after the digitalis solution was added.

## C. Effect of Pilocarpine on the Heart

Pilocarpine is termed a *parasympathomimetic* drug because it mimics the effect of parasympathetic nerve stimulation. Pilocarpine acts to facilitate the release of the neurotransmitter acetylcholine from the vagus nerve, resulting in a marked decrease in the cardiac rate.

### PROCEDURE

1. Obtain a record of the control heartbeat. Then, bathe the heart in a 2.5% solution of pilocarpine.

**Caution:** *Pilocarpine is very effective; be prepared to add atropine if necessary to counter the effects of pilocarpine (see atropine procedure, next).*

2. Rinse the heart thoroughly with Ringer's solution until the heartbeat returns somewhat to normal; this new rhythm will serve as the next control.
3. In the table of your laboratory report, tape the recording or draw a facsimile of the normal heartbeat and of the changed heartbeat after the pilocarpine solution was added.

## D. Effect of Atropine on the Heart

Atropine is an alkaloid drug derived from the nightshade plant *Atropa belladonna* (the species name, *belladonna*, is often also used as the drug name). Atropine blocks the acetylcholine receptors of postganglionic parasympathetic neurons. Thus, atropine inhibits the effects of parasympathetic activity on the heart, smooth muscles, and glands. If the cardiac rate is decreased as a result of vagal stimulation (or the presence of pilocarpine), the administration of atropine will increase this rate.

The ability of **atropine** to block the effects of parasympathetic nerves is useful clinically. Atropine is used, for example, in ophthalmology to dilate the pupils (parasympathetic nerve activity causes constriction of the pupils) and in surgery to dry the mouth, pharynx, and trachea (parasympathetic nerve activity stimulates glandular secretions that wet these mucous membranes).

#### PROCEDURE

1. Bathe the heart in a 5.0% solution of atropine *while it is still under the influence of pilocarpine*.
2. Record the results; then stop recording and rinse the heart thoroughly with Ringer's solution.
3. In the table of your laboratory report, tape the recording or draw a facsimile of the normal heartbeat and of the changed heartbeat after the atropine solution was added.

## E. Effect of Potassium Ions on the Heart

Since the resting membrane potential of all cells is dependent in large part on the maintenance of a higher concentration of potassium ions ($K^+$) on the inside of the cell than on the outside, an increase in the concentration of extracellular $K^+$ results in a *decrease in the resting membrane potential* (the potential becomes more positive). This, in turn, produces a decrease in the force of contraction and a slower conduction rate of the action potentials. In **hyperkalemia** (high blood potassium), the strength of myocardial contractions is weakened and the cardiac cells become more electrically excitable because the resting potential has risen closer to the threshold required for generating action potentials. In extreme hyperkalemia, the conduction rate may be so depressed that ectopic pacemakers appear in the ventricles and fibrillation may develop.

#### PROCEDURE

1. Obtain a record of the control heartbeat. Then, bathe the heart in a 2.0% solution of potassium chloride (KCl).
2. Record the results; then stop recording and rinse the heart thoroughly with Ringer's solution.
3. In the table of your laboratory report, tape the recording or draw a facsimile of the normal heartbeat and of the changed heartbeat after the potassium solution was added.

## F. Effect of Epinephrine on the Heart

Epinephrine is a hormone secreted by the adrenal medulla. Together with norepinephrine, epinephrine is released in response to sympathetic nerve stimulation. Epinephrine acts to increase both the strength of contraction (contractility) of the heart and the cardiac rate. Exogenous ("from outside") epinephrine is a *sympathomimetic drug*, since it mimics the effect of sympathetic nerve stimulation.

#### PROCEDURE

1. Obtain a record of the control heartbeat. Then, bathe the heart in epinephrine (adrenaline).
2. Record the results; then stop recording and rinse the heart thoroughly with Ringer's solution.
3. In the table of your laboratory report, tape the recording or draw a facsimile of the normal heartbeat and of the changed heartbeat after the epinephrine solution was added.

## G. Effect of Caffeine on the Heart

Caffeine is a mild central nervous system (CNS) stimulant that also acts directly on the myocardium to increase both the strength of contraction and the cardiac rate. Caffeine inhibits activity of the enzyme *phosphodiesterase*, which breaks down a second messenger molecule called *cyclic AMP (cAMP)* that is present in many cells. As a result, the concentration of cAMP rises in heart cells. This duplicates the action of the hormone epinephrine, which utilizes cAMP as a second messenger. Caffeine's usefulness as a central nervous system stimulant is limited because, in high doses, it can promote the formation of ectopic pacemakers (foci), resulting in serious arrhythmias (see exercise 7.2).

#### PROCEDURE

1. Obtain a record of the control heartbeat. Then, bathe the heart with a saturated solution of caffeine.
2. Record the results; then stop recording and rinse the heart thoroughly with Ringer's solution.
3. In the table of your laboratory report, tape the recording or draw a facsimile of the normal heartbeat and of the changed heartbeat after the caffeine solution was added.

## H. Effect of Nicotine on the Heart

Nicotine promotes electrochemical transmission at the autonomic ganglia by stimulating particular nicotinic receptors for acetylcholine in the postganglionic neurons. When applied directly to the heart, the major effect of nicotine will be stimulation of parasympathetic ganglia located within the epicardium. Activation of postganglionic parasympathetic neurons, in turn, will cause slowing of the heart rate. When nicotine is administered into the blood in pharmacological doses (as a drug), it can stimulate sympathetic ganglia and the adrenal medulla (the sympathoadrenal system), resulting in an increase in the heart rate.

### PROCEDURE

1. Obtain a record of the control heartbeat. Then, bathe the heart in a 0.2% solution of nicotine.
2. Record the results; then stop recording and rinse the heart thoroughly with Ringer's solution.
3. In the table of your laboratory report, tape the recording or draw a facsimile of the normal heartbeat and of the changed heartbeat after the nicotine solution was added.
4. Analyze your data and record your results for parts A through H in the Results table of your laboratory report.

# Laboratory Report 7.1

Name ______________________

Date ______________________

Section ______________________

## DATA FROM EXERCISE 7.1

| Condition | Effects (Tape the Recording or Draw a Facsimile) |
|---|---|
| Normal | |
| Calcium ($Ca^{2+}$) | |
| Digitalis | |
| Pilocarpine | |
| Atropine | |
| Potassium ($K^{+}$) | |
| Epinephrine | |
| Caffeine | |
| Nicotine | |

# RESULTS FROM EXERCISE 7.1

| Condition | Rate (beats/min) | Strength (mm above Baseline) | Distance between Atrial and Ventricular Peaks (mm) | Conclusion about Drug Effects |
|---|---|---|---|---|
| Normal | | | | |
| Calcium ($Ca^{2+}$) | | | | |
| Digitalis | | | | |
| Pilocarpine | | | | |
| Atropine | | | | |
| Potassium ($K^{+}$) | | | | |
| Epinephrine | | | | |
| Caffeine | | | | |
| Nicotine | | | | |

# REVIEW ACTIVITIES FOR EXERCISE 7.1

## *Test Your Knowledge of Terms and Facts*

Match these items:

____ 1. endogenous substance that makes the heart beat stronger and faster
____ 2. substance that makes the beat slower and stronger
____ 3. substance that facilitates the release of ACh from parasympathetic nerve endings
____ 4. substance that mimics the action of epinephrine by inhibiting the action of phosphodiesterase
____ 5. substance that stimulates the acetylcholine receptors of autonomic ganglia
____ 6. substance that blocks the acetylcholine receptors for the target cells of postganglionic neurons

(a) digitalis
(b) nicotine
(c) caffeine
(d) epinephrine
(e) atropine
(f) pilocarpine

7. The drug used in this exercise that helps people with atrial fibrillation: ____________________.
8. The drug used in this exercise which is used by ophthalmologists to dilate pupils: ____________________.
9. The term *hyperkalemia* means ____________________.

## *Test Your Understanding of Concepts*

10. What are the effects of hyperkalemia on the heart? How were these effects produced?

11. What is a sympathomimetic drug? What are its effects on the heart?

12. What are the effects of digitalis on the heart? Describe the clinical uses of this drug.

## *Test Your Ability to Analyze and Apply Your Knowledge*

13. What effect did bathing the heart with $Ca^{2+}$ have on the strength of the contraction and the ability of the heart to relax between beats? Provide a physiological explanation of these results.

14. The drug *theophylline* is in the same chemical class as caffeine (they are methylxanthines). Theophylline is sometimes used clinically to dilate the bronchioles in a person suffering from asthma. By analogy with caffeine, provide a physiological explanation for this clinical application.

# Electrocardiogram (ECG)

EXERCISE **7.2**

## Materials

1. Electrocardiograph or other strip chart recorder, such as Physiograph (Narco), or Lafayette Instrument Company, Inc. recorder with EKG module (Alternatively, the Biopac system may be used.)
2. Electrode plates, rubber straps, electrolyte gel or paste; disposable electrodes and electrode clips

### Textbook Correlations*

Before performing this exercise, you should study the introductory material presented here. Further information relating to this exercise can be found in these pages of *Human Physiology,* eighth edition, by Stuart I. Fox:

- *Electrical Activity of the Heart and the Electrocardiogram.* Chapter 13, pp. 383–390.

The regular pattern of electrical impulse production and conduction in the heart results in the mechanical contraction (systole) and relaxation (diastole) of the myocardium—the cardiac cycle. The recording of these electrical events, or electrocardiogram (ECG), may reveal abnormal patterns associated with abnormal cardiac rhythms.

## OBJECTIVES

1. Describe the normal cyclical pattern of electrical impulse production in the heart and conduction along specialized tissues of the heart.
2. Describe the normal electrocardiogram (ECG, or EKG) and explain how it is produced.
3. Obtain an electrocardiogram using the limb leads, identify the waves, determine the P-R interval, and measure the cardiac rate.
4. Describe the common ECG abnormalities or arrhythmias.

Electrical stimulation at any point in the heart musculature (myocardium) results in the almost simultaneous contraction of the individual muscle cells (myocardial cells). This allows the heart to function as an effective pump, with its chambers contracting and relaxing as integrated units. The contraction phase of the cardiac cycle is called **systole,** and the relaxation phase is called **diastole.**

Unlike skeletal muscles, cardiac muscle is able to stimulate itself electrically in the absence of neural input (termed *automaticity*). The sympathetic and parasympathetic nerves that innervate the heart only modulate the ongoing rate of depolarization-contraction and repolarization-relaxation intrinsic to the heart. The intrinsic regulation of systole and diastole, unique to heart muscle, is termed *rhythmicity*.

Although each individual myocardial cell is potentially capable of initiating its own cycle of depolarization-contraction and repolarization-relaxation, a single group of cells usually regulates the cycle of the entire myocardium. This *pacemaker* region establishes its dominance because its cycle is more rapid than other areas, depolarizing the other myocardial cells before they can depolarize themselves (fig. 7.7). A region of the right atrium, the **sinoatrial node (SA node),** serves as the normal pacemaker of the heart. The wave of depolarization initiated by the SA node quickly spreads across the right and left atria as a result of electrical synapses *(gap junctions)* between myocardial cells. However, the depolarization wave cannot easily spread from the myocardial cells of the atria to the myocardial cells of the ventricles. For this to occur, the depolarization wave must be carried from the atria to the ventricles along the specialized conducting tissue of the heart.

The depolarization wave that spreads over the atria stimulates a node called the **atrioventricular node (AV node),** located at the base of the interatrial septum. After a brief delay, the depolarization wave in the AV node is quickly transmitted over a bundle of specialized conducting tissue, the **atrioventricular** (or **AV**) **bundle,** otherwise

*Multimedia Correlations (also see Appendix 3)
- Biopac: Student Lab Lessons 5 and 6
- Intelitool: Cardiocomp
- A.D.A.M. *InterActive Physiology* (Cardiovascular System): Cardiac Action Potential
- *MediaPhys 2.0:* Topics 8.17 and 8.18

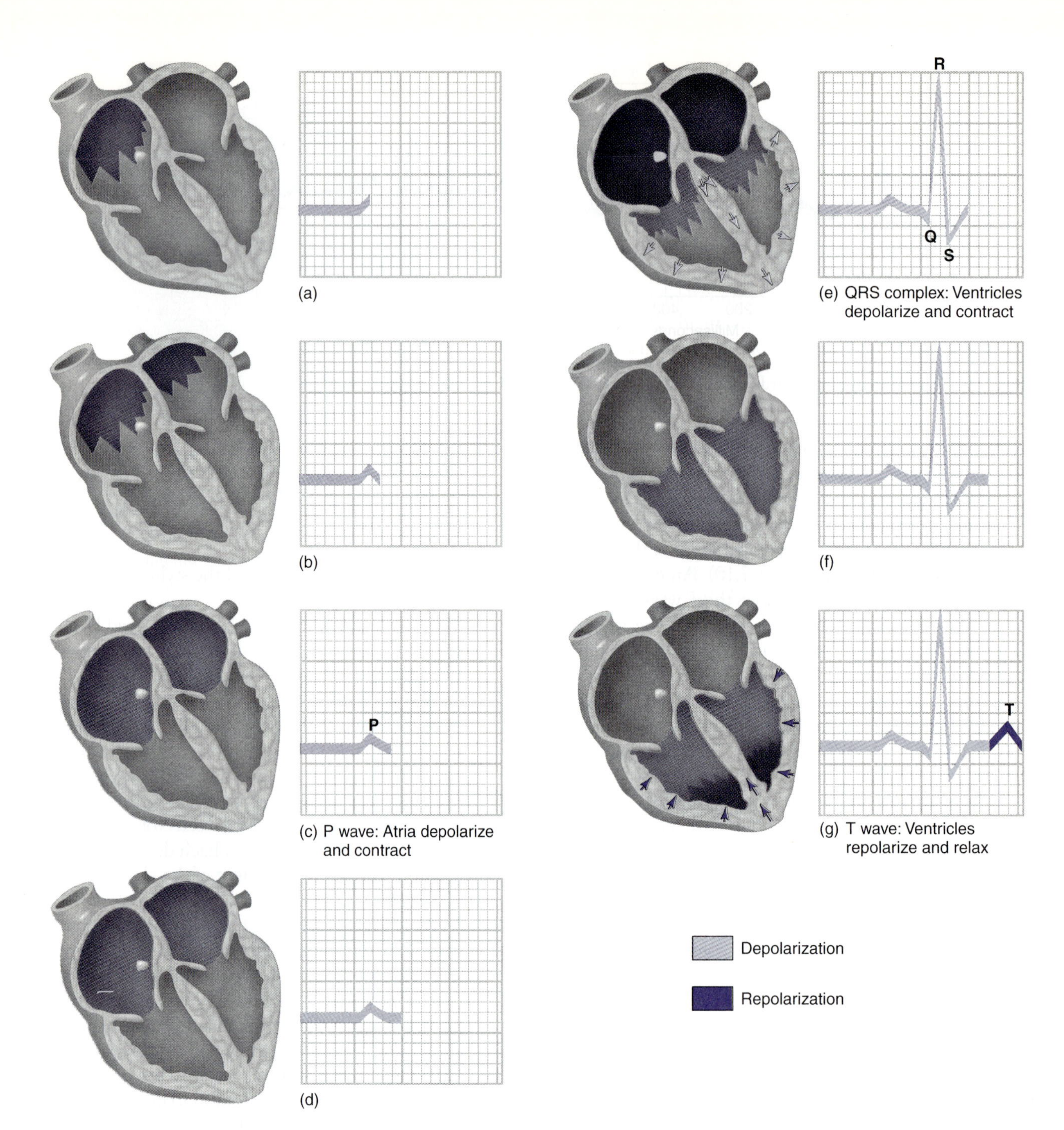


**Figure 7.10 The relationship between impulse conduction in the heart and the ECG.** The direction of the arrows in (e) indicates that depolarization of the ventricles occurs from the inside (endocardium) out (to the epicardium). The arrows in (g), by contrast, indicate that repolarization of the ventricles occurs in the opposite direction.

**(For a full-color version of this figure, see figure 13.23 in *Human Physiology,* eighth edition, by Stuart I. Fox.)**

**Note:** *If you use a multichannel recorder (such as a Physiograph) instead of an electrocardiograph, the recording paper and, consequently, the arithmetic calculation will be different. If you use a Cardiocomp, the P-R interval will be provided automatically.*

P-R interval ________ sec

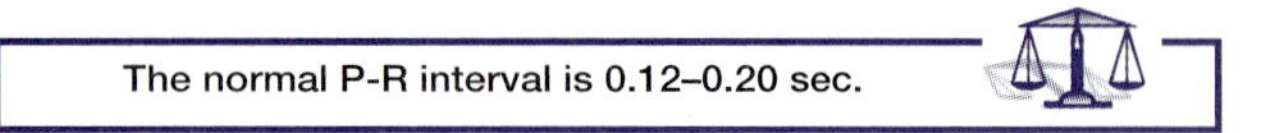


12. Determine the cardiac rate by the following methods:
    (a) Count the number of QRS complexes in a 3-sec interval (the distance between two vertical lines at the top of the ECG paper) and multiply by 20.

**Note:** *If you use a multichannel recorder (such as a Physiograph), the amount of chart paper corresponding to a given time interval must be calculated and will vary with the paper speed. If you use a Cardiocomp, the cardiac rate will be provided automatically.*

Beats per minute = ________

(b) Count the number of QRS complexes in a 6-sec interval, and multiply by 10.
Beats per minute = ________

(c) At a chart speed of 25 mm (2.5 cm) per second, the time interval between one light vertical line and the next is 0.04 sec. The time interval between heavy vertical lines is 0.20 sec. The cardiac rate in beats per minute can be calculated if the time interval between two R waves in two successive QRS complexes is known.

For example, suppose that the time interval from one R wave to the next is exactly 0.60 sec. Therefore,

$$\frac{1 \text{ beat}}{0.60 \text{ sec}} = \frac{x \text{ beats}}{60 \text{ sec}}$$

$$x = \frac{1 \text{ beat} \times 60 \text{ sec}}{0.60 \text{ sec}}$$

$$x = 100 \text{ beats per minute}$$

Beats per minute = ________

(d) The values obtained by *method c* can be approximated by counting the number of heavy vertical lines between one R wave and the next according to the memorized sequence: 300, 150, 100, 75, 60, 50 (at a paper speed of 25 mm per sec):

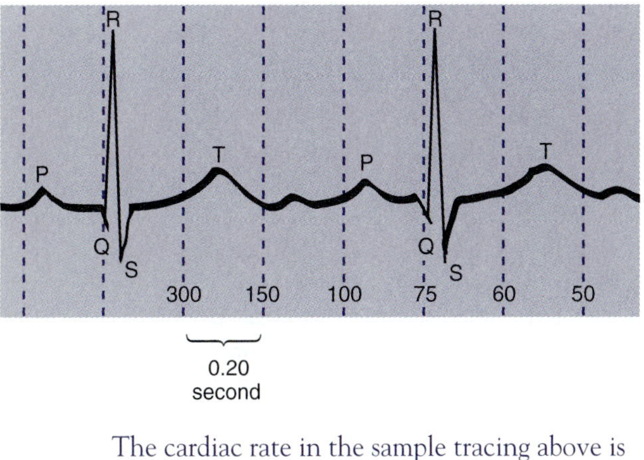


The cardiac rate in the sample tracing above is 75 beats per minute.
Beats per minute = ________

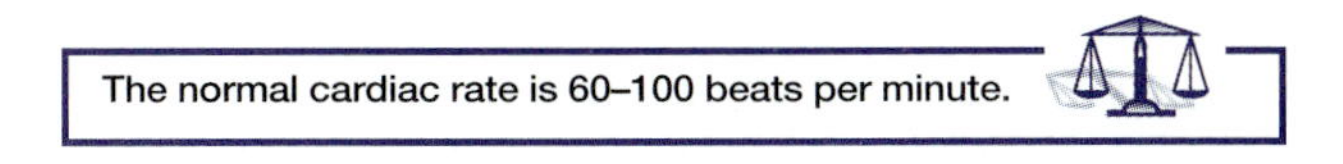


## Abnormal ECG Patterns

Interpretation of the electrocardiogram can provide information about the heart rate and rhythm, as well as possible conditions of *hypertrophy, ischemia* (inadequate blood supply), *necrosis* (death of cells), and other conditions that may produce abnormalities of electrical conduction. According to the standards set by the National Conference on Cardiopulmonary Resuscitation and Emergency Cardiac Care, all health professionals should be able to recognize:

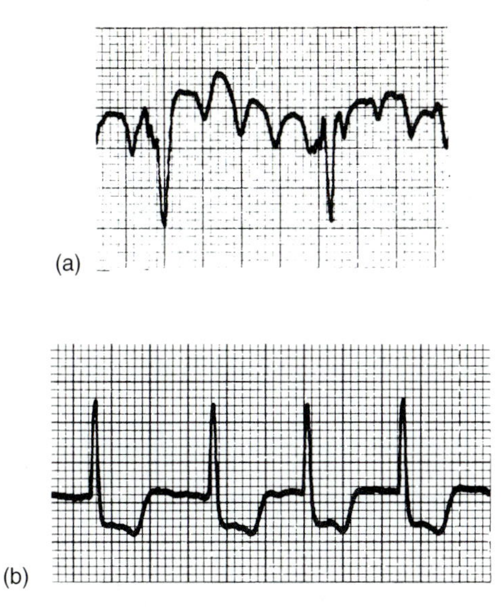


**Figure 7.11 Abnormal atrial rhythms.** (a) Atrial flutter and (b) atrial fibrillation.

1. bradycardia (a ventricular rate slower than 60 beats per minute)
2. the difference between supraventricular and ventricular rhythms
3. premature ventricular contractions
4. ventricular tachycardia
5. atrioventricular block
6. atrial fibrillation and flutter
7. ventricular fibrillation

When *ectopic beats* (beats that are out of place) occur in the atria as a result of the development of an *ectopic pacemaker* (a pacemaker that develops in addition to the normal one in the SA node) or as a result of a derangement in the normal conduction pathway, a condition of atrial flutter or atrial fibrillation may be present. **Atrial flutter** is characterized by very rapid atrial waves (about 300 per minute), producing a sawtoothed baseline. These atrial waves occur with such high frequency that the AV node can beat only to every second, third, or fourth wave it receives (fig. 7.11*a*). A person with atrial flutter may have a normal pulse rate, since the pulse is produced by contraction of the left ventricle.

In **atrial fibrillation,** the depolarization waves occur so rapidly (350–400/min) that the atria no longer function effectively, and the P waves of the ECG are replaced by a wavy baseline (fig. 7.11*b*). Atrial fibrillation is characteristic of atrial enlargement (hypertrophy), as might be produced by *mitral stenosis* (narrowing of the mitral valve), but may occur in all forms of heart disease and occasionally in apparently healthy individuals. *Digitalis* is a drug often used in atrial flutter and fibrillation to decrease the excitability of the AV node, thus maintaining the ventricular rate within the normal range.

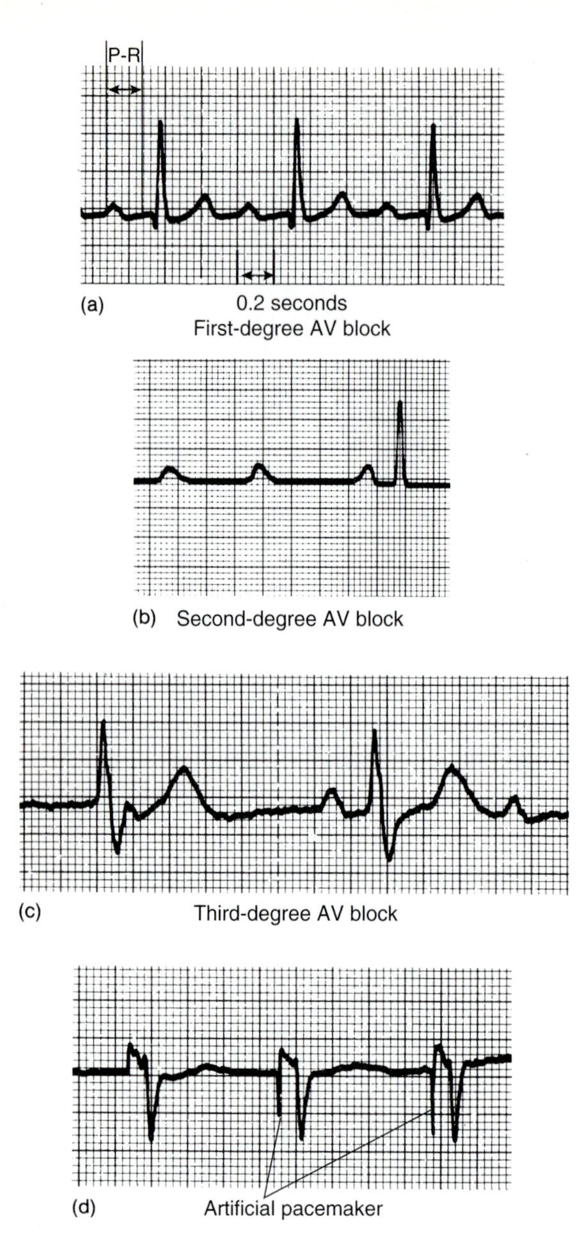


**Figure 7.12 Different stages of AV block.** (a) First-degree AV block, where the P-R interval is greater than 0.2 seconds (one large square); (b) second-degree AV block, where beats (i.e., QRS complexes) are missed (a 3:1 block is shown); and (c) third-degree, or complete, AV block, where a slower than normal heart rate is set by an ectopic focus in the ventricles. (d) An artificial pacemaker, implanted in a person with complete AV node block, produces the sharp downward spikes that precede the inverted QRS waves.

A delay in the conduction of the impulse from the atria to the ventricles is known as **atrioventricular (AV) block.** When the ECG pattern is otherwise normal, a P-R interval greater than 0.20 sec represents a *first-degree AV block* (fig. 7.12*a*). First-degree AV block may be a result of inflammatory states, rheumatic fever, or digitalis treatment.

When the excitability of the AV node is further impaired so that two or more atrial depolarizations are required before the impulse can be transmitted to the

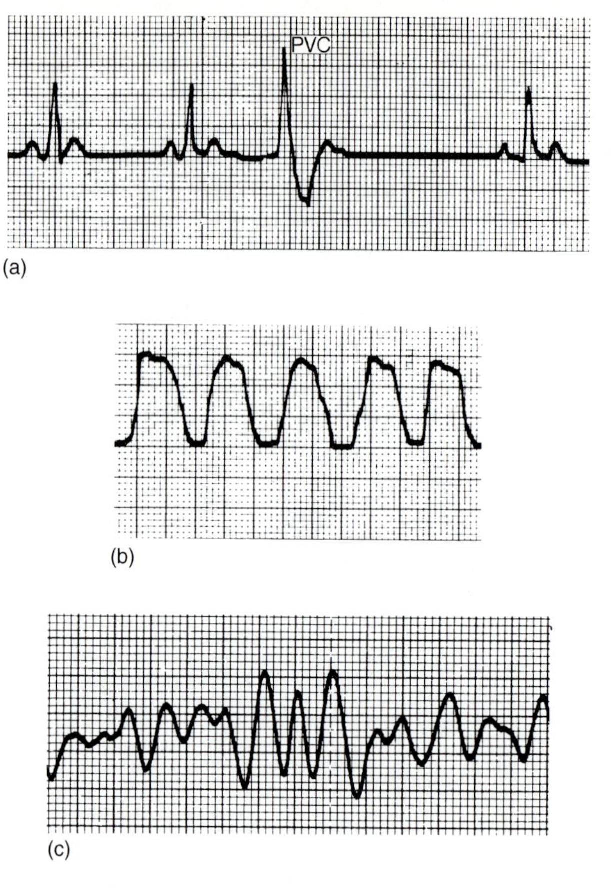


**Figure 7.13 Abnormal ventricular rhythms.** (a) Premature ventricular contraction (PVC), or bigeminy; (b) ventricular tachycardia; and (c) ventricular fibrillation.

ventricles, a *second-degree AV block* is present. This is seen on the ECG as a "dropped" beat (that is, a P wave without an associated QRS complex). In *Wenckebach's phenomenon*, the cycle after the dropped beat is normal, but the P-R interval of successive cycles lengthens until a beat (QRS complex) is again dropped (fig. 7.12*b*).

*Third-degree*, or *complete*, *AV block* occurs when none of the impulses from the atria reach the ventricles (fig. 7.12*c*). In this case, the myocardial cells of the ventricles are freed from their subservience to the SA node, causing one or more ectopic pacemakers to appear in the ventricles without corresponding P waves. The rhythm produced by these ectopic foci in the ventricles is usually very slow (20–45 beats per minute) compared to the normal pace set by the SA node (*sinus rhythm*). An artificial pacemaker may be used to compensate for this condition, as shown by the ECG tracing in figure 7.12*d*.

**Premature ventricular contractions (PVCs)** are produced by ectopic foci in the ventricles when the sinus rhythm is normal. This results in *extrasystoles* (extra beats or QRS complexes without preceding P waves) in addition to the normal cycle, often subjectively described by patients as "palpitations." The ectopic QRS complexes are broad and deformed and may be abnormally coupled to the preceding normal beats. This coupling is termed *bigeminy* and is often seen in digitalis toxicity (fig. 7.13*a*).

When an ectopic focus in the ventricles discharges at a rapid rate, a condition of **ventricular tachycardia** (usually 100–150 beats per minute) develops (fig. 7.13*b*). The ECG shows a widened and distorted QRS complex that often obscures the P wave (although the atria are discharging at their slower, sinus rate). This serious condition should be distinguished from the less serious **supraventricular tachycardia,** in which an ectopic focus above the ventricles results in spontaneous rapid running of the heart (150–250 beats/min) that begins and ends abruptly (paroxysmal). This condition is appropriately called *paroxysmal atrial tachycardia*.

The most serious of all arrhythmias is **ventricular fibrillation** (fig. 7.13*c*). Ventricular fibrillation may develop as ectopic foci emerge and depolarization waves circle the heart *(circus rhythm)*, resulting in an impotent tremor rather than a coordinated pumping action. Under these conditions, the pumping activity of the ventricles ceases, and death occurs within minutes unless emergency measures (including electrical defibrillation) are applied successfully.

# Laboratory Report 7.2

Name ______________________________

Date ______________________________

Section ______________________________

## DATA FROM EXERCISE 7.2

1. Tape your recording in these spaces.
   (a) ***Lead I***

   (b) ***Lead II***

   (c) ***Lead III***

# Effects of Exercise on the Electrocardiogram

EXERCISE 7.3

## MATERIALS

1. Electrocardiograph or multichannel recorder (e.g., Physiograph) with appropriate ECG module
2. ECG plates, straps, gel
3. Alternatively, the Biopac system may be used with the pulse transducer and electrodes for Biopac Student Lab Lesson 7.

## Textbook Correlations*

Before performing this exercise, you should study the introductory material presented here. Further information relating to this exercise can be found in these pages of *Human Physiology,* eighth edition, by Stuart I. Fox:

- *Pressure Changes During the Cardiac Cycle.* Chapter 13, p. 381–382.
- *The Electrocardiogram.* Chapter 13, pp. 387–390.
- *Regulation of Cardiac Rate.* Chapter 14, pp. 408–408.

During exercise, there is a decrease in the activity of parasympathetic innervation to the SA node, conductive tissue, and myocardium and an increase in the activity of sympathetic innervation to the SA node, conductive tissue, and myocardium. More rapid discharge of the SA node, more rapid conduction of impulses, and a faster rate of contraction all result in an increased cardiac rate with exercise.

## OBJECTIVES

1. Describe the effects of the sympathetic and parasympathetic innervation to the heart.
2. Obtain an ECG before, immediately after, and 2 minutes after exercise and explain the differences observed.
3. Determine cardiac rate, P-R interval, and period of ventricular diastole for the ECG tracings obtained in objective 2.

The heart is innervated by both sympathetic and parasympathetic nerve fibers. At the beginning of exercise, the activity of the parasympathetic fibers that innervate the SA node decreases. Since these fibers have an inhibitory effect on the pacemaker, a decrease in their activity results in an increase in cardiac rate. As exercise becomes more intense, the activity of sympathetic fibers that innervate the SA node increases. This has an excitatory effect on the SA node and causes even greater increases in cardiac rate.

Sympathetic fibers also innervate the conducting tissues of the heart and the ventricular muscle fibers. Through these innervations, sympathetic stimulation may increase the velocity of both impulse conduction and ventricular contraction. These effects are most evident at high cardiac rates and contribute only slightly to the increased cardiac rate during exercise. Thus, the increased cardiac rates are mainly due to a shortening of the ventricular diastole (from the peak of the T wave to the beginning of the next QRS complex) and only secondarily due to a shortening of ventricular systole (measured from the QRS peak to the peak of the T wave, fig. 7.14).

## PROCEDURE

1. After the resting ECG has been recorded (from exercise 7.2), unplug the electrode leads from the electrocardiograph.
2. Have the subject exercise while holding the lead wires—for example, by walking up and down stairs, using a stationary bicycle, hopping, doing sit-ups or leg lifts.

**Caution:** *The intensity of exercise should be monitored so that 80% of the maximum cardiac rate is not exceeded. Students who are not in good health should not serve as subjects.*

*Multimedia Correlations (also see Appendix 3)
- Biopac: Student Lab Lesson 7
- Intelitool: Cardiocomp
- A.D.A.M. *InterActive Physiology* (Cardiovascular System): Cardiac Output

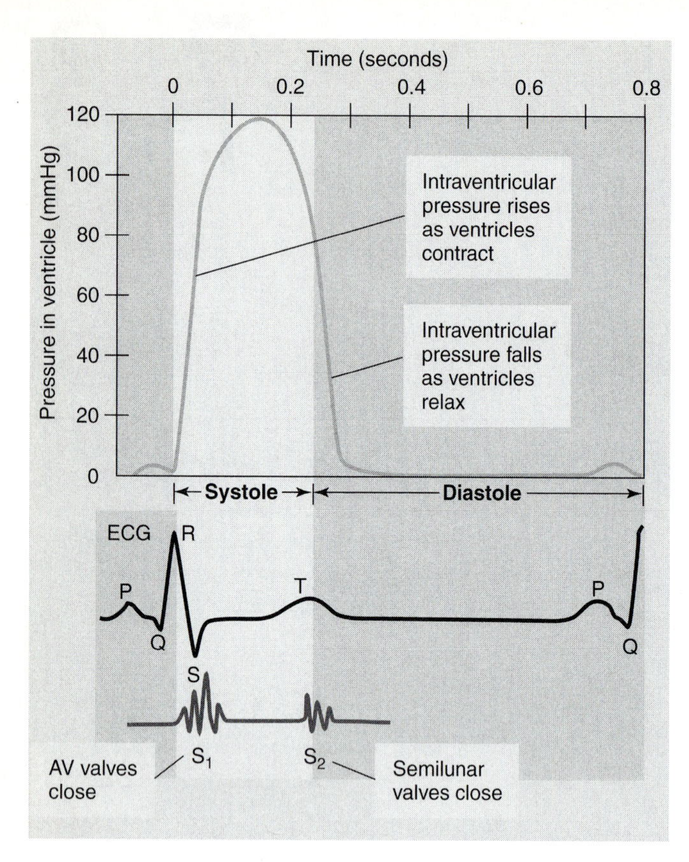


**Figure 7.14 The relationship between changes in intraventricular pressure and the ECG.** The QRS wave (representing depolarization of the ventricles) occurs at the beginning of systole, whereas the T wave (representing repolarization of the ventricles) occurs at the beginning of diastole.

**(For a full-color version of this figure, see fig. 13.24 in *Human Physiology*, eighth edition, by Stuart I. Fox.)**

The rate of blood flowing through the coronary circulation may be adequate to meet the aerobic requirements of the heart at rest, but may be inadequate for the increased metabolic energy demands of the heart during exercise. A portion of the cardiac muscle may receive insufficient blood flow because of a clot in a coronary artery (*coronary thrombosis*) or because of narrowing of the vessel due to *atherosclerosis.* This insufficient blood flow to the heart—**myocardial ischemia**—may, however, be relative to the aerobic demand. Supervised exercise tests such as those on a treadmill with varying levels of aerobic demand help diagnose this kind of ischemia. During these tests, changes in the *S-T segment* of the ECG (fig. 7.15) may indicate ischemia.

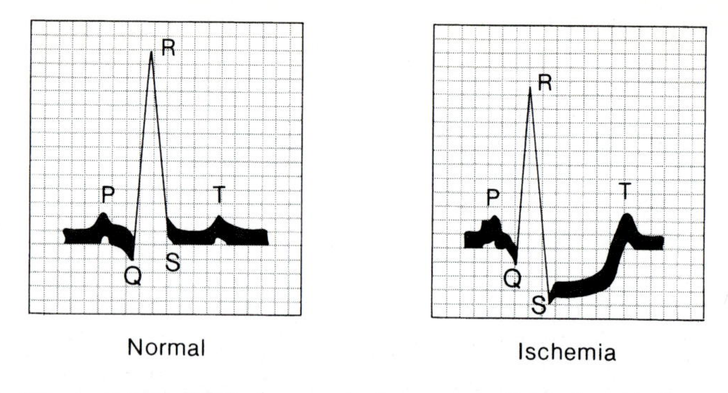


**Figure 7.15 ECG changes during myocardial ischemia.** In myocardial ischemia, the S-T segment of the electrocardiogram may be depressed, as illustrated in this figure.

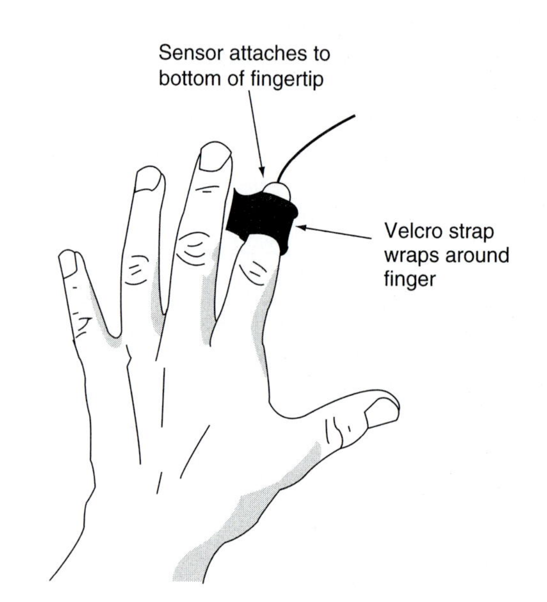


**Figure 7.16 Pulse transducer for Biopac system.**

3. Immediately after exercise, the subject should lie down and the electrode leads are plugged into the electrocardiograph. Record lead II *only.* Wait 2 minutes and record lead II again.
4. Determine the cardiac rate, period of ventricular diastole, and the duration of the QRS complex for the resting ECG and for the two postexercise ECG recordings.

   Enter these data in the table in your laboratory report.
5. Alternatively, the ECG may be obtained using the Biopac system (as set up for Biopac Student Lab Lessons 5 and 6). If the pulse transducer is available (fig. 7.16), as used in Biopac Student Lab Lesson 7, the pulse can be followed directly during rest, exercise, and postexercise.

# Laboratory Report 7.3

Name ______________________

Date ______________________

Section ______________________

## DATA FROM EXERCISE 7.3

1. **Enter your data in this table.**

| ECG Tracing | Cardiac Rate (beats/min) | P-R Interval (sec) (beginning of P to Q) | Ventricular Diastole (sec) (middle of T to next Q) |
|---|---|---|---|
| Resting ECG | | | |
| Immediate postexercise ECG | | | |
| Two-minute postexercise ECG | | | |

(a) Which measurement changed the most as you went from resting to exercise, and then to the two-minute postexercise conditions? ______________________

(b) Which changed the least? What conclusions can you draw regarding the changes in impulse conduction and the cardiac cycle as a result of exercise?

2. **Tape your ECG recordings in this space.**

# REVIEW ACTIVITIES FOR EXERCISE 7.3

## *Test Your Knowledge of Terms and Facts*

1. The ECG wave that occurs at the beginning of ventricular systole is the __________ wave.
2. The ECG wave that occurs at the end of systole and the beginning of diastole is the __________ wave.
3. The ECG wave that is completed just before the end of ventricular diastole is the __________ wave.
4. The nerve that increases the rate of discharge of the SA node is a ______________________________ nerve.
5. The specific nerve that, when stimulated, causes a decrease in the cardiac rate is the ______________________.
6. The scientific term for insufficient blood flow to the heart muscle is ______________________________.

## *Test Your Understanding of Concepts*

7. Describe the regulatory mechanisms that produce an increase in cardiac rate during exercise. Explain how these changes affect the electrocardiogram (ECG).

8. Describe the pressure changes that occur within the ventricles as each ECG wave is produced, and explain how these pressure changes are related to the ECG waves.

## *Test Your Ability to Analyze and Apply Your Knowledge*

9. Explain how a person could have a normal electrocardiogram at rest, but show evidence of myocardial ischemia during exercise. Do you think all people should have regular treadmill (stress) electrocardiogram tests, or only certain people? Explain.

10. Provide a cause-and-effect explanation as to how the pulse rate relates to the ECG rate. How would the relationship change if pulse were measured by palpation of the carotid artery instead of by a pulse transducer in the finger?

# Mean Electrical Axis of the Ventricles

EXERCISE 7.4

## Materials

1. Electrocardiograph, or multichannel recorder (e.g., Physiograph) with appropriate ECG module
2. ECG plates, straps, gel

The voltage changes in the ECG measured by two different leads can be compared and used to determine the mean electrical axis, which corresponds to the average direction of depolarization as the impulses spread into the ventricles. Significant deviations from the normal axis may be produced by specific heart disorders.

## OBJECTIVES

1. Describe the electrical changes in the heart that produce the ECG waves.
2. Determine the mean electrical axis of the ventricles in a test subject, and explain the clinical significance of this measurement.

Depolarization waves spread through the heart in a characteristic pattern. Depolarization waves begin at the SA node and spread from the pacemaker to the entire mass of both atria, producing the *P wave* in an electrocardiogram. After the AV node is excited, the interventricular septum becomes depolarized as the impulses spread through the bundle of His. Since at this point the septum is depolarized while the lateral walls of the ventricles still have their original polarity, there is a potential difference (voltage) between the septum and the ventricular walls. This produces the *R wave*. When the entire mass of the ventricles is depolarized, there is no longer a potential difference within the ventricles, and the voltage returns to zero (completing the *QRS complex*).

The direction of the depolarization waves depends on the orientation of the heart in the chest and on the particular instant of the cardiac cycle being considered. It is clinically useful, however, to determine the **mean axis** (average direction) **of depolarization** during the cardiac cycle. This can be done by observing the voltages of the QRS complex from two different perspectives using two different leads.

## Textbook Correlations*

Before performing this exercise, you should study the introductory material presented here. Further information relating to this exercise can be found in these pages of *Human Physiology,* eighth edition, by Stuart I. Fox:

- The Electrocardiogram. Chapter 13, pp. 387–390.

Lead I provides a horizontal axis of observation (from left arm to right arm); lead III has an axis of about 120° (from left arm to left leg). Using the recordings from leads I and III, the normal mean electrical axis of the ventricles is found to be about 59°, as shown in figure 7.17.

**Hypertrophy** (enlargement) of one ventricle shifts the mean axis of depolarization toward the hypertrophied ventricle because it takes longer to depolarize the larger ventricle. Therefore, a left axis deviation occurs when the left ventricle is hypertrophied (as a result of hypertension or narrowing of the aortic semilunar valve). A right axis deviation occurs when the right ventricle hypertrophies. The latter condition may be secondary to narrowing of the pulmonary semilunar valve or to such congenital conditions as a septal defect or the tetralogy of Fallot.

The depolarization wave normally spreads through both the right and left ventricles at the same time. However, if there is a conduction block in one of the branches of the bundle of His—a **bundle-branch block**—depolarization will be much slower in the blocked ventricle. In left bundle-branch block, for example, depolarization will occur more slowly in the left ventricle than in the right ventricle, and the mean electrical axis will deviate to the left. In right bundle-branch block, there will be a right axis deviation. Deviations of the electrical axis also occur to varying degrees as a result of myocardial infarction.

*Multimedia Correlations (also see Appendix 3)
- Biopac: Student Lab Lesson 6
- Intelitool: Cardiocomp

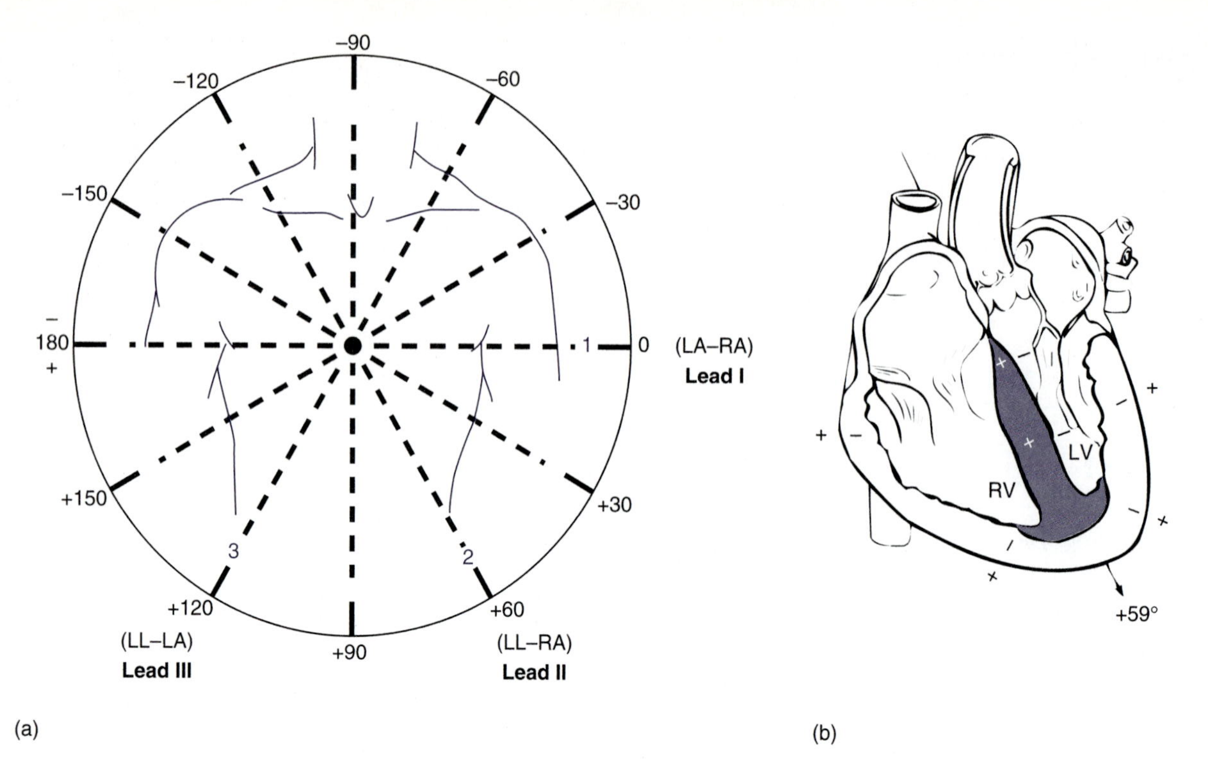


**Figure 7.17 Mean electrical axis of the heart.** (a) The convention by which the axis of depolarization is measured. The bottom half of the circle (with the heart at the center) is considered the positive role (LA = left arm, RA = right arm, LL = left leg). (b) When the interventricular septum is depolarized, the surface of the septum is electrically negative compared to the walls of the ventricles, which have not yet become depolarized. The average normal direction of depolarization, or mean electrical axis of the ventricles, is about 59° (RV = right ventricle, LV = left ventricle).

## PROCEDURE

**Note:** *If a Cardiocomp-7 or –12 is used, the mean electrical axis of the ventricles can be determined by examining the QRS loop of the vectorgram displayed on the computer screen. Alternatively, this procedure can be used by examining leads I and III that were previously obtained (using any equipment) in exercise 7.2.*

1. Analysis of lead I:
   (a) Find a QRS complex and count the number of millimeters (small boxes) that it projects above the upper edge of the baseline.
   Enter this value here:
   + __________ mm
   (b) Count the number of millimeters the Q and S waves (or the R wave, if it is inverted) project below the upper edge of the baseline. Add these measurements of downward deflections.
   Enter this sum here:
   __________ mm
   (c) Algebraically, add the two values from steps (a) and (b), keeping the negative sign if the sum is negative.
   Enter this sum in the laboratory report.
2. Analysis of lead III:
   (a) Find a QRS complex and count the number of millimeters (small boxes) it projects above the upper edge of the baseline.
   Enter this value here:
   + __________ mm
   (b) Count the number of millimeters the Q and S waves (or the R wave, if it is inverted) project below the upper edge of the baseline. Add these measurements of downward deflections.
   Enter this sum here:
   __________ mm
   (c) Algebraically, add the two values from steps (a) and (b), keeping the negative sign if the sum is negative.
   Enter this sum in the laboratory report.
3. On the blank grid chart in your laboratory report, use a straight edge to make a line on the axis of *lead I* that corresponds to the sum you obtained in step 1c (see example grid, fig. 7.19).
4. Use a straight edge to make a line on the axis of *lead III* that corresponds to the sum you obtained in step 2c.
5. Use a straight edge to draw an arrow from the center of the grid chart to the intersection of the

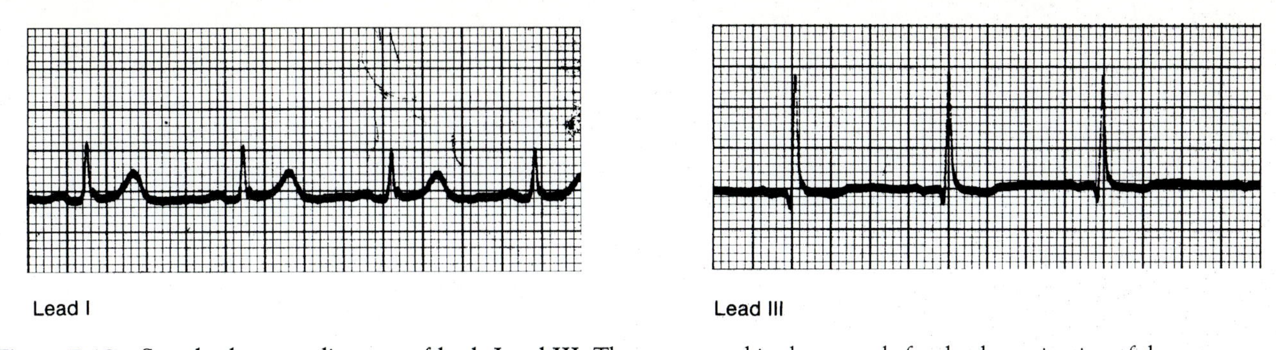


**Figure 7.18 Sample electrocardiograms of leads I and III.** These were used in the example for the determination of the mean electrical axis of the heart. (Note: Each small square is 1 mm on a side.)

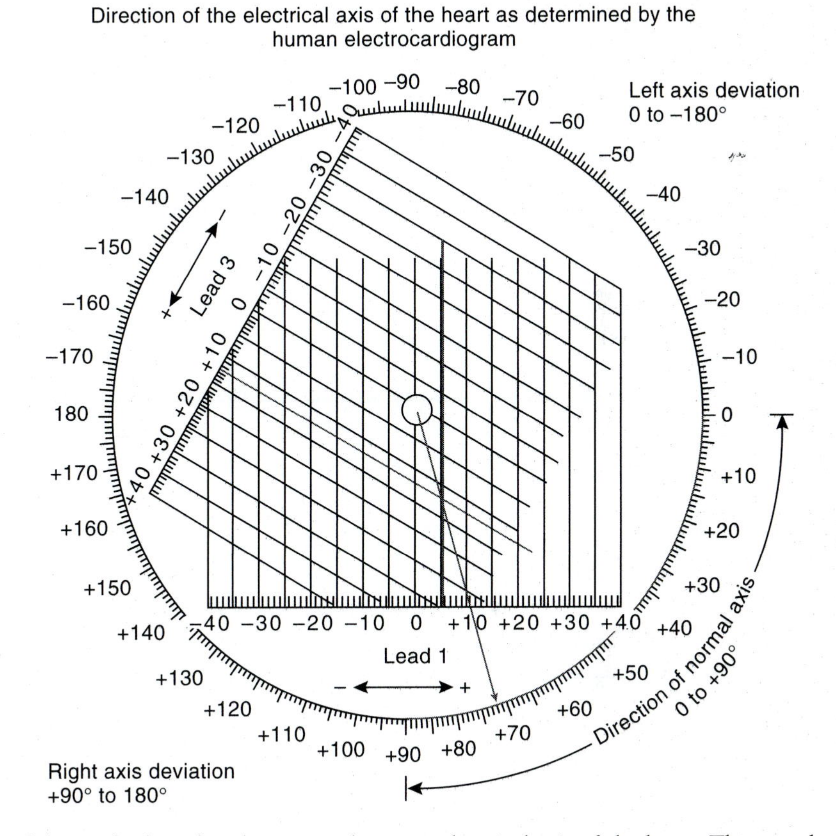


**Figure 7.19 Example of the method used to determine the mean electrical axis of the heart.** This uses the data from leads I and III in the sample ECG shown in figure 7.18. In this example, the mean electrical axis is 72°.

two lines drawn in steps 3 and 4. Extend this arrow to the edge of the grid chart and record the mean electrical axis of the ventricles.

## *Example (fig. 7.18)*

Lead I of sample ECG:

| | |
|---|---|
| Upward deflection: | + 7 mm |
| Downward deflections: | – 1 mm |
| | 6mm |

Lead III of sample ECG:

| | |
|---|---|
| Upward deflection: | + 14 mm |
| Downward deflections: | – 2 mm |
| | 12 mm |

A straight line is drawn perpendicular to the horizontal axis of lead I corresponding to position 6 on the scale. Similarly, a straight line is drawn perpendicular to the axis of lead III that corresponds to position 12 on the scale. An arrow is then drawn from the center of the circle through the intersection of the two lines previously drawn (fig. 7.19).

In this example, the mean electrical axis of the ventricles is +71°.

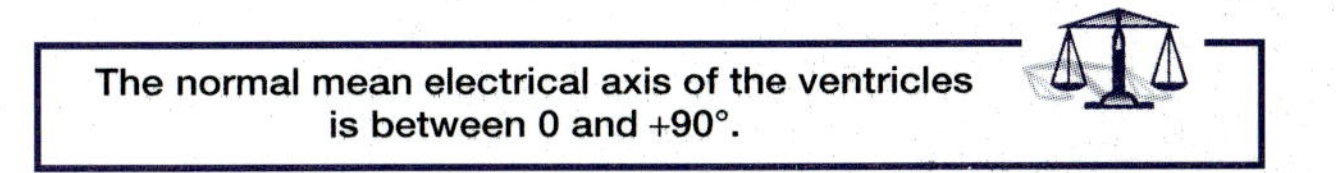

# Laboratory Report 7.4

Name ______________________

Date ______________________

Section ______________________

## DATA FROM EXERCISE 7.4

1. Enter the value for the sum of steps 1a and 1b in the space provided here. Be sure to indicate whether it is a positive or negative number. ______________________
2. Enter the value for the sum of steps 2a and 2b in the space provided here. Be sure to indicate whether it is a positive or negative number. ______________________
3. Use this grid chart to determine the mean electrical axis of the ventricles from your data, as described in steps 3, 4, and 5 of the procedure.
   Mean electrical axis of the ventricles: ______________________

Grid chart

Direction of the electrical axis of the heart as determined by the human electrocardiogram

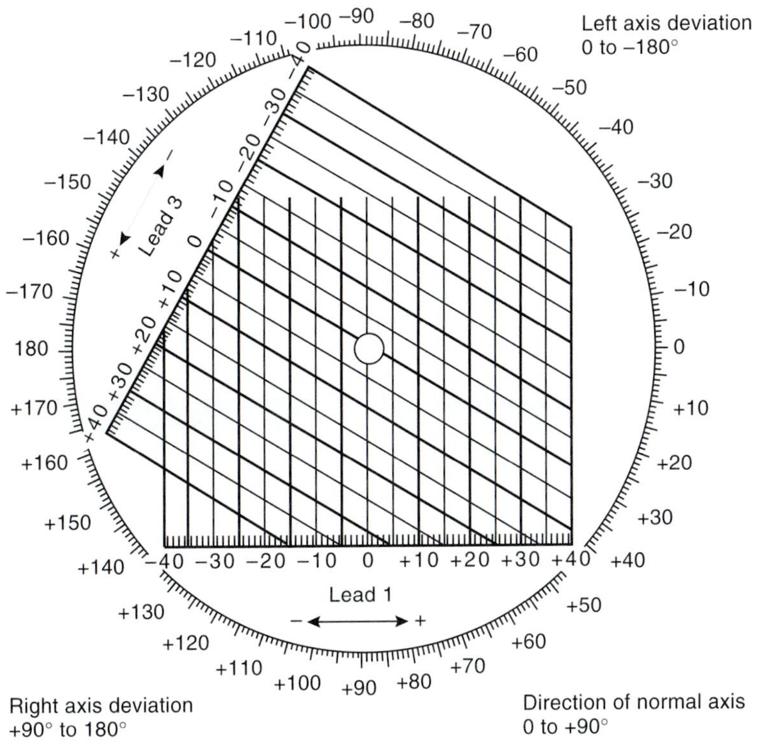

Left axis deviation
0 to −180°

Lead 3

Lead 1

Right axis deviation
+90° to 180°

Direction of normal axis
0 to +90°

# REVIEW ACTIVITIES FOR EXERCISE 7.4

## *Test Your Knowledge of Terms and Facts*

1. Depolarization of the atria produces the ________________________ wave.
2. The difference in polarity between the interventricular septum and the lateral walls of the ventricles produces the ________________________ wave.
3. The mean electrical axis is determined using two bipolar limb leads, lead ________________________ and lead ________________________.
4. Hypertrophy of the left ventricle would shift the mean axis of depolarization to the ________________________.
5. A blockage in conduction in the right branch of the bundle of ________________________ would cause the mean axis of depolarization to shift to the ________________________.

## *Test Your Understanding of Concepts*

6. Explain how the normal pattern of depolarization is affected by ventricular hypertrophy. What factors may be responsible for hypertrophy of the right or left ventricle?

7. Explain how the normal pattern of depolarization is affected by bundle-branch block. What might cause bundle-branch block?

## *Test Your Ability to Analyze and Apply Your Knowledge*

8. Explain the electrical events in the heart that produce the QRS wave. Why does the tracing go up from Q to R, and then back to baseline from R to S?

9. How do you think obesity or pregnancy might influence the results of this laboratory exercise? Explain.

# Heart Sounds

EXERCISE **7.5**

## Materials

1. Stethoscopes
2. Physiograph, high-gain couplers, microphone for heart sounds (Narco), ultrasonic flowmeter (such as Doppler)
3. Alternatively, the Biopac system may be used as per Student Lab Lesson 17, employing the Biopac electrode lead set and the amplified stethoscope.

### Textbook Correlations*

Before performing this exercise, you should study the introductory material presented here. Further information relating to this exercise can be found in these pages of *Human Physiology,* eighth edition, by Stuart I. Fox:

- *Cardiac Cycle and Heart Sounds.* Chapter 13, pp. 381–383.
- *The Electrocardiogram.* Chapter 13, pp. 387–390.

Contraction and relaxation of the ventricles is accompanied by pressure changes that cause the one-way heart valve to close. Closing valves produce sounds that aid in the diagnosis of structural abnormalities of the heart.

### OBJECTIVES

1. Describe the causes of the normal heart sounds.
2. List some of the causes of abnormal heart sounds.
3. Correlate the heart sounds with the waves of the ECG and the events of the cardiac cycle.

## A. Auscultation of Heart Sounds with the Stethoscope

The cycle of mechanical contraction (**systole**) and relaxation (**diastole**) of the ventricles can be followed by listening to the heart sounds with a **stethoscope.** The contraction of the ventricles produces a rise in intraventricular pressure, resulting in the vibration of the surrounding structures as the atrioventricular valves slam shut. The valves closing produces the *first sound* of the heart, usually verbalized as "**lub.**" At the end of the contraction phase, the blood in the aorta and pulmonary arteries pushes the one-way semilunar valves shut, and the resulting vibration of these structures produces the *second sound* of the heart, which is verbalized as "**dub**" (see fig. 7.14).

Careful **auscultation** (listening) to the two heart sounds may reveal two further sounds. This *splitting* of the heart sounds into four components is more evident during inhalation than it is during exhalation. During deep inhalation, the first heart sound may be split into two sounds because the tricuspid and mitral valves close at different times. The second heart sound may also be split into two components because the pulmonary and aortic semilunar valves close at different times.

Auscultation of the chest aids in the diagnosis of many cardiac conditions, including **heart murmurs.** A murmur may be caused by an irregularity in a valve, a septal defect, or the persistent fetal opening *(foramen ovale)* between the right and left atria after birth, resulting in the audible regurgitation of blood in the reverse direction of normal flow. Abnormal splitting of the first and second heart sounds may be due to heart block, septal defects, aortic stenosis, hypertension, or other abnormalities.

*Multimedia Correlations (also see Appendix 3)
- Biopac: Student Lab Lesson 17
- A.D.A.M. *InterActive Physiology* (Cardiovascular System): Cardiac Cycle
- *MediaPhys 2.0*: Topics 8.19–8.23

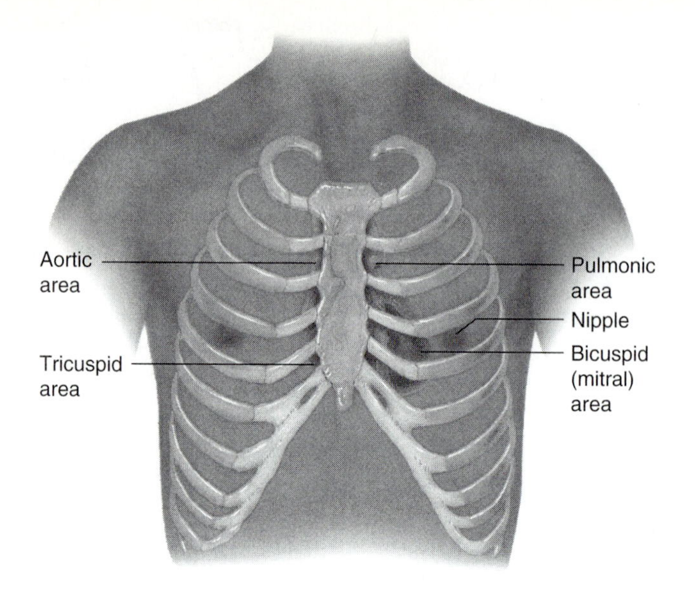


**Figure 7.20 Routine stethoscope positions for listening to the heart sounds.** The first heart sound is caused by closing of the AV valves; the second by closing of the semilunar valves.

**(For a full-color version of this figure, see fig. 13.14 in *Human Physiology*, eighth edition, by Stuart I. Fox.)**

## PROCEDURE

1. Clean the earpieces of the stethoscope with an alcohol swab. To best hear the first heart sound, auscultate the apex beat of the heart by placing the diaphragm of the stethoscope in the *fifth left intercostal space* (the bicuspid area in fig. 7.20).
2. To best hear the second heart sound, place the stethoscope to the right or left of the sternum in the *second intercostal space* (the aortic or pulmonic area in fig. 7.20).
3. Compare the heart sounds in the three stethoscope positions described during quiet breathing, slow and deep inhalation, and slow exhalation.

## B. Correlation of the Phonocardiogram with the Electrocardiogram

If the heart sounds are monitored with a device known as a **phonocardiograph** in conjunction with the monitoring of the electrical patterns of the heart with an electrocardiograph, a correlation of the two recordings will show that the first heart sound occurs at the end of the QRS complex of the ECG, and the second heart sound occurs at the end of the T wave (fig. 7.21; also see fig. 7.14). The arterial pulse is palpated (felt) in either the radial artery or carotid artery in the time interval between the two heart sounds.

## PROCEDURE

1. The following steps relate to the procedure for Physiograph recording. A different procedure for steps 2–7 should be followed if different equipment is used to perform this exercise.
2. Insert two *high-gain couplers* into the physiograph. Plug the cable from the ECG lead selector box into the coupler for channel 1, and the cable from the phonocardiograph microphone into the coupler for channel 2.
3. Attach the ECG electrode plates to the subject in the standard limb lead positions and plug the ECG cable into the lead selector box (see exercise 7.2).
4. For the high-gain coupler in channel 1:
   (a) Turn the time constant knob to the *3.2* position.
   (b) Turn the gain knob to the *× 2* position.
   (c) Turn the knob on the ECG lead selector box to the calibrate position. Turn the outer sensitivity knob on the amplifier to the *10* position. The inner knob should be turned all the way to the right until it clicks.
   (d) Lower the pen lift lever, raise the inkwells, and squeeze the rubber bulbs on the inkwells until ink flows freely. Release the paper-advance button, and position the pen for channel 1 so that it writes on the appropriate heavy horizontal line.
5. For the high-gain coupler in channel 2 adjust the time constant and gain knobs as described in step 3, and position the pen to write on the appropriate heavy horizontal line. The sensitivity knob can be adjusted for the individual subject once recording begins.
6. Place the microphone to the fifth left intercostal space (fig. 7.20) and start the recording. Continue recording during normal breathing and deep inhalation. Move the microphone to the second right (or left) intercostal space and continue recording during normal breathing and deep inhalation. Repeat the procedure on the other side of the sternum.
7. Place the ultrasonic flowmeter crystal on either the radial artery or carotid artery with a dab of electrode paste. Plug the flowmeter into another high-gain coupler (in multichannel recorders) and record the arterial pulse simultaneously with the ECG and phonocardiogram.

Tape your recordings or draw facsimiles in the laboratory report.

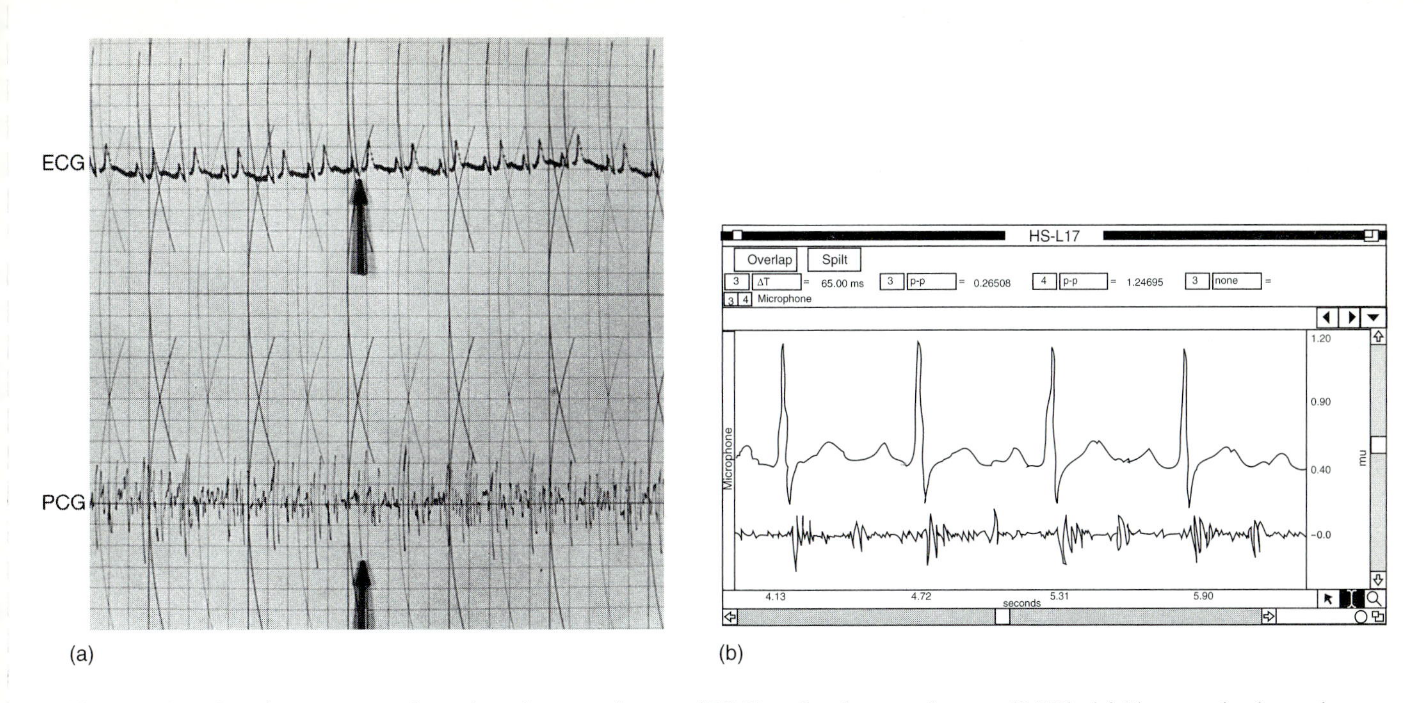


(a) (b)

**Figure 7.21 Simultaneous recording of an electrocardiogram (ECG) and a phonocardiogram (PCG).** (a) Photograph of recordings taken with a Narco physiograph. (b) Idealized computer image of recordings taken with the Biopac system ECG electrodes and amplified stethoscope.

# Laboratory Report 7.5

Name ______________________

Date ______________________

Section ______________________

## DATA FROM EXERCISE 7.5

1. In this table, tape your phonocardiogram recordings or draw facsimiles.

| Microphone Position | Normal Breathing | Deep Inhalation |
|---|---|---|
| Fifth left intercostal space | | |
| Second left intercostal space | | |
| Second right intercostal space | | |

## REVIEW ACTIVITIES FOR EXERCISE 7.5

### *Test Your Knowledge of Terms and Facts*

1. The scientific term for listening carefully (as with a stethoscope) is ______________________________.
2. The first heart sound (lub) is caused by ______________________________.
3. During which phase of the cardiac cycle (systole or diastole) does the first heart sound occur? ______________
4. The second heart sound is caused by ______________________________.
5. The second heart sound occurs at the (beginning or end) ____________________ of (systole or diastole) ____________________.
6. Abnormal heart sounds are called ______________________________.
7. The first heart sound (lub) is correlated with which ECG wave? ______________________________
8. The second heart sound (dub) is correlated with which ECG wave? ______________________________

### *Test Your Understanding of Concepts*

9. Using a complete cause-and-effect sequence, explain the correlation of the heart sounds with the ECG waves.

10. Explain how the heart sounds are normally produced, and describe some of the conditions that can cause heart murmurs.

11. What is meant by the "splitting" of the heart sounds? What can cause this splitting?

## *Test Your Ability to Analyze and Apply Your Knowledge*

12. Can a defective valve be detected by electrocardiography, by auscultation, by palpation of the radial artery, or by all three techniques? Explain your answer.

13. Can the beating of the atria be detected by electrocardiography, by auscultation, by palpation of the radial artery, or by all three techniques? Explain your answer.

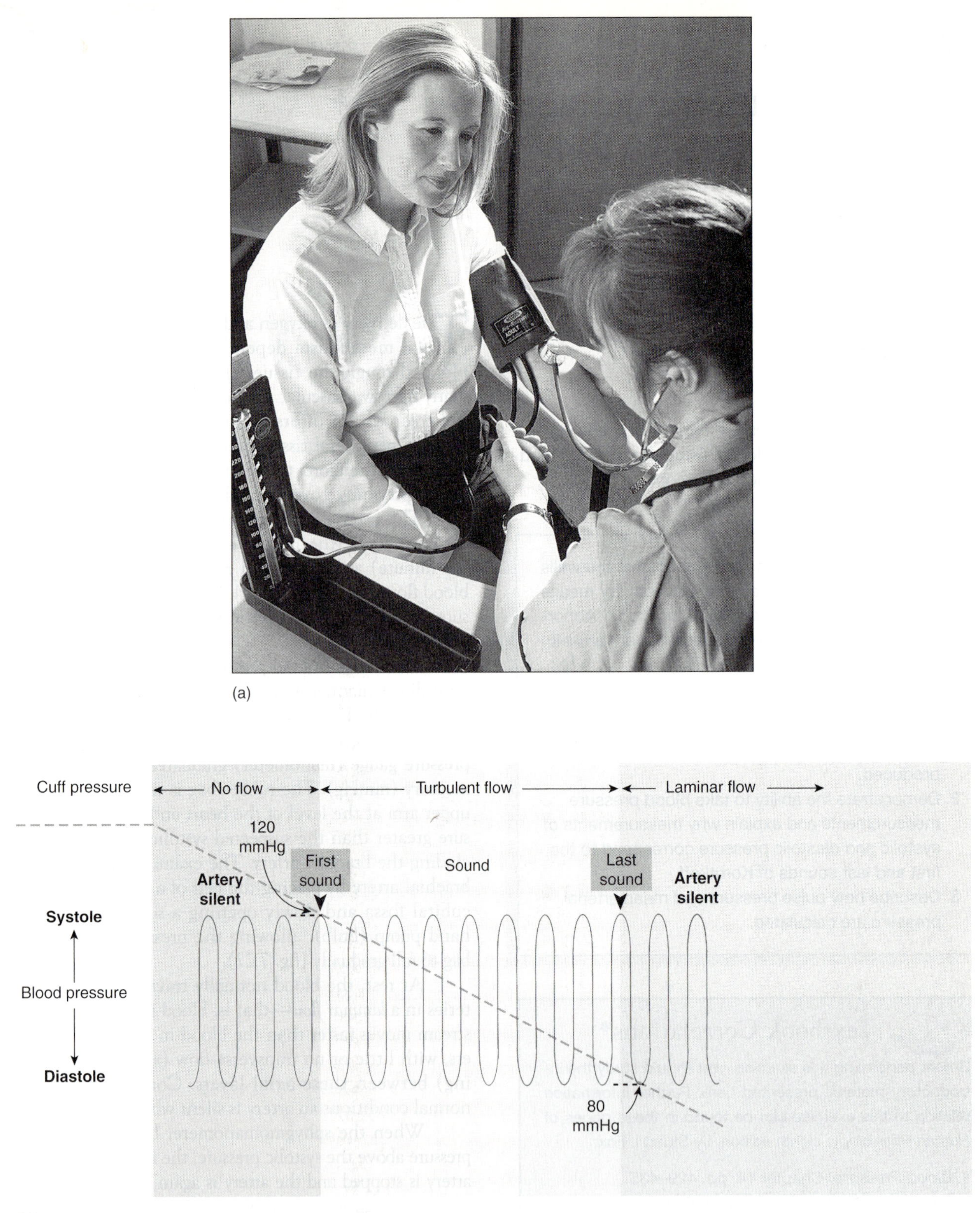

**Figure 7.22 The use of a stethoscope and a sphygmomanometer to measure blood pressure.** (a) The locations where the arm cuff and stethoscope bell are placed. (b) Korotkoff sounds are heard when there is turbulent blood flow through a constriction in the brachial artery. The systolic pressure corresponds to the cuff pressure when the first sound is heard, and the diastolic pressure corresponds to the cuff pressure when the last sound is heard.

**(For a full-color version of this figure, see figs. 14.28 and 14.30 in *Human Physiology*, eighth edition, by Stuart I. Fox.)**

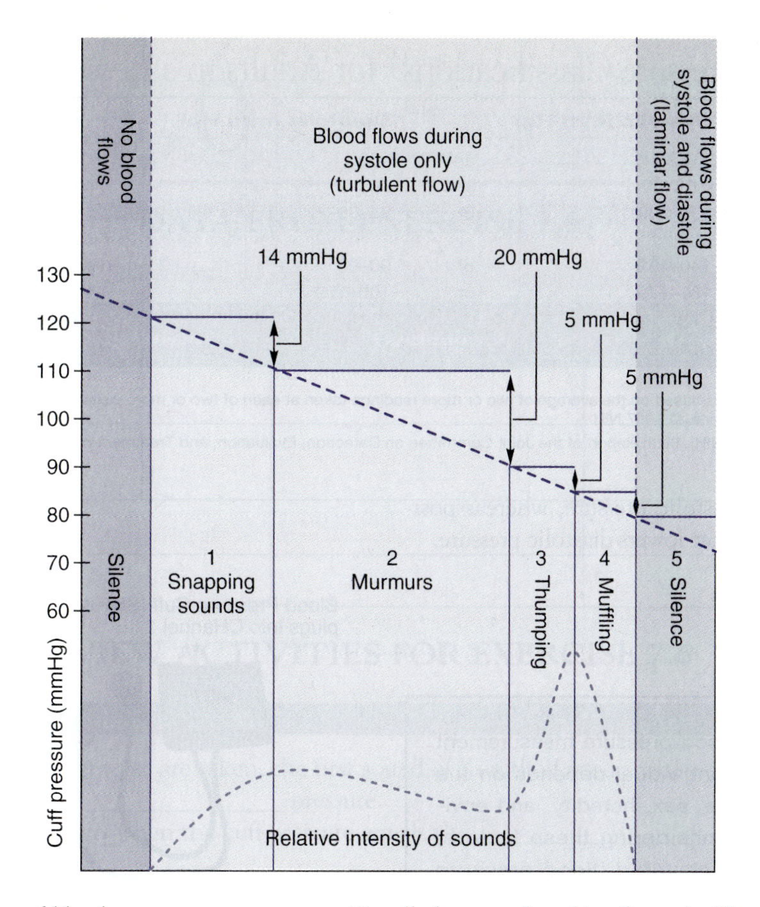


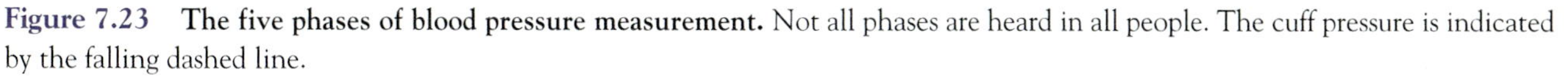

**Figure 7.23 The five phases of blood pressure measurement.** Not all phases are heard in all people. The cuff pressure is indicated by the falling dashed line.

**(For a full-color version of this figure, see fig. 14.31 in *Human Physiology,* eighth edition, by Stuart I. Fox.)**

The sounds of Korotkoff are divided into *five phases* on the basis of the loudness and quality of the sounds, as shown in figure 7.23. Start on the left of the figure with a cuff pressure at approximately 130 millimeters of mercury (mm Hg), and falling.

**Phase 1.** A loud, clear *tapping* (or snapping) sound is evident, which increases in intensity as the cuff is deflated. This phase begins at a cuff pressure of 120 mm Hg and ends at a pressure of 106 mm Hg.

**Phase 2.** A succession of *murmurs* can be heard. Sometimes the sounds seem to disappear during this time (auscultatory gap), perhaps a result of inflating or deflating the cuff too slowly. This phase begins at a cuff pressure of 106 mm Hg and ends at a pressure of 86 mm Hg.

**Phase 3.** A loud, *thumping* sound, similar to phase 1 but less clear, replaces the murmurs. This phase begins at a cuff pressure of 86 mm Hg and ends at a pressure of 81 mm Hg.

**Phase 4.** A *muffled* sound abruptly replaces the thumping sounds of phase 3. This phase begins at a cuff pressure of 81 mm Hg and ends at a pressure of 76 mm Hg.

**Phase 5.** Silence again as all sounds disappear. This phase is absent in some people.

The cuff pressure at which the first sound is heard (the beginning of phase 1) is the **systolic pressure.** The cuff pressure at which the sound becomes muffled (the beginning of phase 4) and the pressure at which the sound disappears (the beginning of phase 5) are taken as measurements of the **diastolic pressure.** Although the phase 5 measurement is closer to the true diastolic pressure than the phase 4 measurement, the beginning of phase 4 is easier to detect and the results are more reproducible. It is often recommended that both measurements of diastolic pressure be recorded. In the example shown in figure 7.23, the pressure would be indicated as 120/81/76 mm Hg. Frequently, however, the blood pressure would simply be recorded in this example as 120/76 mm Hg, or spoken as "120 over 76". The **pulse pressure** is calculated as the difference in these two pressures; and the **mean arterial pressure** is equal to the diastolic pressure plus one-third of the pulse pressure.

Different examiners can record different values for systolic and diastolic pressures, even on the same subject. Blood pressure measurements can vary with the instruments used, the anxiety of the subject, and even body position. Although the systolic pressure (caused by the left ventricle peak ejection pressure) tends to remain fairly constant with changes in body position, the diastolic pressure is affected by gravity. The arm positioned below the heart people, blood levels of lactic acid (lactate) rise significantly when exercise is performed at about 50–70% of their aerobic capacity. This is called the **lactate** (or **anaerobic**) **threshold.** Endurance trained athletes have a higher aerobic capacity and may not reach their lactate threshold until they exercise at about 80% of their maximal oxygen uptake.

The primary cause of the higher aerobic capacity in endurance trained athletes is their higher maximum cardiac outputs, and thus their higher rates of oxygen delivery to the muscles. Endurance training increases the cardiac output through a stronger contraction of the ventricles (thus ejecting more blood per beat), and an increase in blood volume.

The higher stroke volume allows the same cardiac output to be achieved at a slower heart rate, as previously mentioned. The slower resting heart rate of endurance trained athletes—a condition called *athlete's bradycardia*—results from higher levels of inhibitory activity by the vagus nerve innervation to the SA node.

Exercise testing using clinical protocols have proved extremely useful in the diagnosis of heart disease, particularly coronary artery disease and **myocardial ischemia** (inadequate blood flow to the heart). People who appear to be healthy and record normal electrocardiograms at rest may develop *angina pectoris* and abnormal ECG patterns during or after exercise. During these tests and under the watchful eye of a physician, the patient follows a standardized exercise protocol (such as the Balke, Stanford, or Naughton) while the ECG is continuously recorded and blood pressures are taken automatically at regular intervals. Most cardiologists prefer a treadmill test, during which the speed and incline are automatically varied from easy to difficult, with the patient reaching target heart rates as high as 95% of maximum for the patient's age. During the test, irregularities in the ECG or blood pressure, such as depressed or elevated S-T segments, can be observed, which may indicate myocardial ischemia.

## PROCEDURE

1. Measure your reclining (lying down) pulse by placing your fingertips (not the thumb) on the radial artery in the ventrolateral region of the wrist.* Count the number of pulses in 30 seconds and multiply by two. Score points as indicated.

**Reclining Pulse**

| Rate | Points |
|---|---|
| 50–60 | 3 |
| 61–70 | 3 |
| 71–80 | 2 |
| 81–90 | 1 |
| 91–100 | 0 |
| 101–110 | –1 |

Score: ______________________

2. After the reclining pulse has been measured, stand up and measure the pulse rate *immediately* upon standing.

| Rate Points | |
|---|---|
| 60–70 | 3 |
| 71–80 | 3 |
| 81–90 | 2 |
| 91–100 | 1 |
| 101–110 | 1 |
| 111–120 | 0 |
| 121–130 | 0 |
| 131–140 | –1 |

Score: ______________________

*The procedure for the exercise is from "A Cardiovascular Rating as a Measure of Physical Fatigue and Efficiency" by E. C. Schneider. JAMA 74(1920):1507. Copyright 1920 American Medical Association.

3. Subtract the pulse rate of step 1 from the pulse rate of step 2 to obtain the pulse rate increase on standing.

**Pulse Rate Increase on Standing**

| Reclining Pulse | 0–10 Beats | 11–18 Beats | 19–26 Beats | 27–34 Beats | 35–43 Beats |
|---|---|---|---|---|---|
| 50–60 | 3 | 3 | 2 | 1 | 0 |
| 61–70 | 3 | 2 | 1 | 0 | –1 |
| 71–80 | 3 | 2 | 0 | –1 | –2 |
| 81–90 | 2 | 1 | –1 | –2 | –3 |
| 91–100 | 1 | 0 | –2 | –3 | –3 |
| 101–110 | 0 | –1 | –3 | –3 | –3 |

Score: ______________________

4. Place your right foot on a chair or stair that is 18 inches high. Raise your body so that your left foot comes to rest by your right foot. Return your left foot to the original position, followed by the right foot. Repeat this exercise five times, allowing 3 sec for each step up. Immediately upon completion of this exercise, measure the pulse for 15 sec and multiply by four. Record this pulse rate. Pulse: ____________
Measure the pulse as described for 30, 60, 90, and 120 sec after completion of the exercise. Record the time it takes for the pulse to return to normal rate while standing (step 2). Score points as indicated.

**Time for Return of Pulse to Normal, Standing Post-exercise**

| Seconds | Points |
|---|---|
| 0–30 | 4 |
| 31–60 | 3 |
| 61–90 | 2 |
| 91–120 | 1 |
| After 120 | 0 |

Score: ______________________

5. Subtract your normal standing pulse rate (step 2) from your pulse rate immediately after exercise (step 4).

**Pulse Rate Increase Immediately Post-exercise**

| Standing Pulse | 0–10 Beats | 11–20 Beats | 21–30 Beats | 31–40 Beats | Over 41 Beats |
|---|---|---|---|---|---|
| 60–70 | 3 | 3 | 2 | 1 | 0 |
| 71–80 | 3 | 2 | 1 | 0 | –1 |
| 81–90 | 3 | 2 | 1 | –1 | –2 |
| 91–100 | 2 | 1 | 0 | –2 | –3 |
| 101–110 | 1 | 0 | –1 | –3 | –3 |
| 111–120 | 1 | –1 | –2 | –3 | –3 |
| 121–130 | 0 | –2 | –3 | –3 | –3 |
| 131–140 | 0 | –3 | –3 | –3 | –3 |

Score: ______________________

6. Calculate the change in systolic blood pressure as you go from a reclining (lying down position) to a standing position (refer to data in exercise 7.6 or take new measurements).

**Change in Systolic Pressure from Reclining to Standing**

| Change (mm Hg) | Points |
|---|---|
| Rise of 8 or More | 3 |
| Rise of 2–7 | 2 |
| No rise | 1 |
| Fall of 2–5 | 0 |
| Fall of 6 or more | –1 |

Score: ______________________

7. Determine your total score for all the tests and evaluate this score on this basis:
Excellent 18–17
Good 16–14
Fair 13–8
Poor 7 or less

Enter your score and rating in the laboratory report.

# Laboratory Report 7.7

Name ______________________________

Date ______________________________

Section ______________________________

## DATA FOR EXERCISE 7.7

1. Write your total score in the space provided here, and indicate whether this score is excellent, good, fair, or poor according to the rating scale in the procedure.
   Total Score: ____________
   Overall Rating: __________________________

## REVIEW ACTIVITIES FOR EXERCISE 7.7

### *Test Your Knowledge of Terms and Facts*

1. As a person gets older, the maximum cardiac rate ______________________________.
2. If a person has athlete's bradycardia, the resting heart rate is ______________________ than the average.
3. The condition described in question 2 is caused by ______________________________.
4. Define the *aerobic capacity*. ______________________________
5. Define the *lactate threshold*. ______________________________
6. The primary cause of the higher aerobic capacity of endurance trained athletes is ______________________________
   ______________________________.

### *Test Your Understanding of Concepts*

7. What cardiovascular adaptations are associated with endurance training? How do these changes help to improve performance?

8. How does the increase in blood pressure and pulse rate after exercise, and the return of these values to baseline following exercise, compare in people who are and who are not physically fit?

# Measurements of Pulmonary Function

EXERCISE **8.1**

## MATERIALS

1. Collins, Inc. 9-L respirometer or Spirocomp program and equipment
2. Disposable mouthpieces and nose clamp
3. Alternatively, the Biopac system may be used with the set up for Biopac Lesson 12 (for part A of this exercise) and Lesson 13 (for part B of this exercise).
4. Alternatively, the iWorx system may be used with the setup as described in iWorx Exercise 10. Once set up, part A of this exercise can be performed.

## Textbook Correlations*

Before performing this exercise, you should study the introductory material presented here. Further information relating to this exercise can be found in these pages of *Human Physiology*, eighth edition, by Stuart I. Fox:

- *Physical Aspects of Ventilation.* Chapter 16, pp. 483–487.
- *Mechanics of Breathing.* Chapter 16, pp. 488–493.

Spirometry is used to measure lung volumes and capacities and to measure ventilation as a function of time. Such measurements are clinically useful in the diagnosis of restrictive and obstructive pulmonary disorders.

## OBJECTIVES

1. Identify the major muscles involved in inspiration and expiration and explain the mechanics of breathing.
2. Define the different lung volumes and capacities.
3. Perform a normal spirogram, and determine measurements of the different lung volumes and capacities.
4. Describe and perform the forced expiratory volume (FEV) test; and determine FEV from the spirogram.
5. Explain how pulmonary function tests are used in the diagnosis of restrictive and obstructive pulmonary disorders.

**Spirometry** is a technique for measuring lung volumes and capacities. A Collins respirometer can be used for this purpose (fig. 8.1). As the subject exhales into a mouthpiece, the oxygen bell rises, causing a pen to move downward on a moving chart (kymograph). Since this is a closed system, soda lime is provided in the system to remove $CO_2$ from the exhaled air. As the subject inhales, the oxygen bell falls, causing the pen to move upward on the moving chart. The **y** axis of the chart is graduated in milliliters, and the **x** axis is graduated in millimeters. Because the speed with which the chart moves is known, a graph of air volume (mL) moved into and out of the lungs in a given time interval can be obtained.

Alternatively, a computerized setup using a Phipps and Bird spirometer and a program for analyzing the data (Spirocomp) may be used (fig. 8.2). In this case, the treated data are displayed on a computer screen. Or, if the Biopac system is employed, a wet spirometer is avoided because the flow of air is transduced and fed into the Biopac MP30 unit (fig. 8.3), which then displays the data on a computer screen. By means of spirometry, many important aspects of pulmonary function can be visualized and measured, as described below and shown in figure 8.4.

The **total lung capacity (TLC)** is the total volume of gas in the lungs after a maximum (forced) inhalation.

The **vital capacity (VC)** is the maximum volume of gas that can be exhaled after a maximum inhalation.

*Multimedia Correlations (also see Appendix 3)
- Biopac: Student Lab Lessons 12 and 13
- Intelitool: Spirocomp
- A.D.A.M. *InterActive Physiology* (Respiratory System): Pulmonary Ventilation; Gas Exchange
- *MediaPhys 2.0*: Topics 10.10–10.13; Topics 10.22–10.26

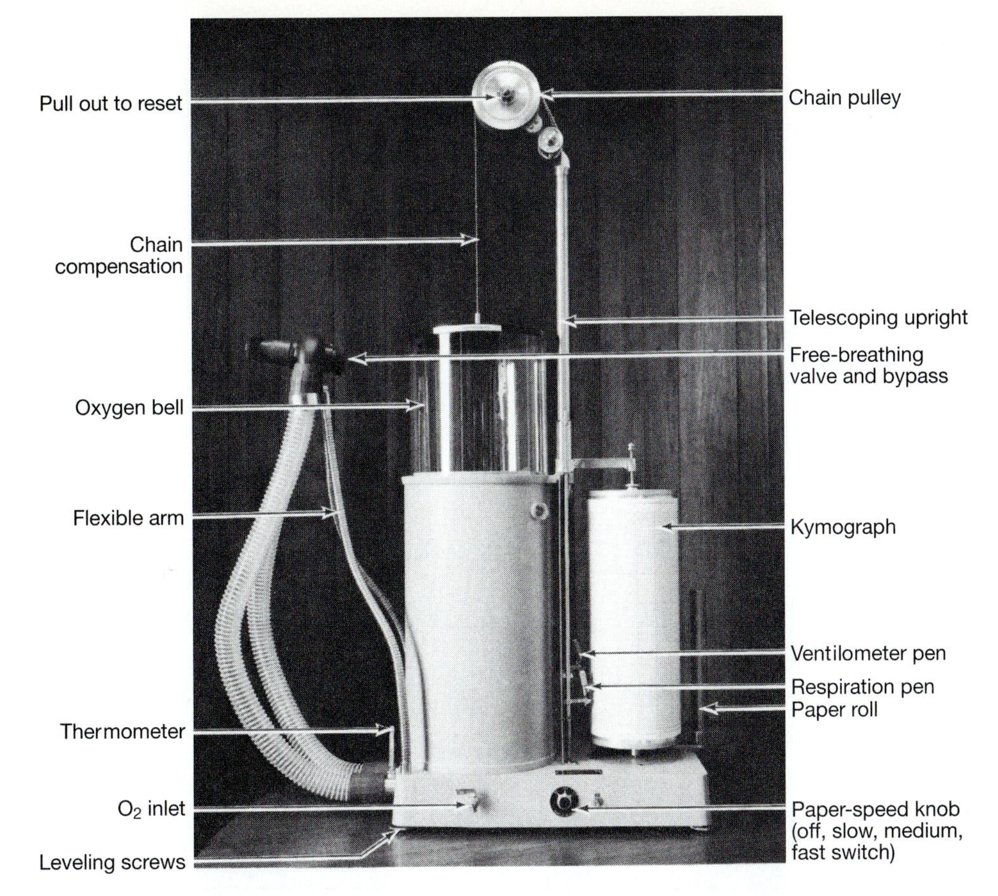


**Figure 8.1** The Collins 9-liter respirometer.

**Figure 8.2** Equipment involved in using the Spirocomp program.

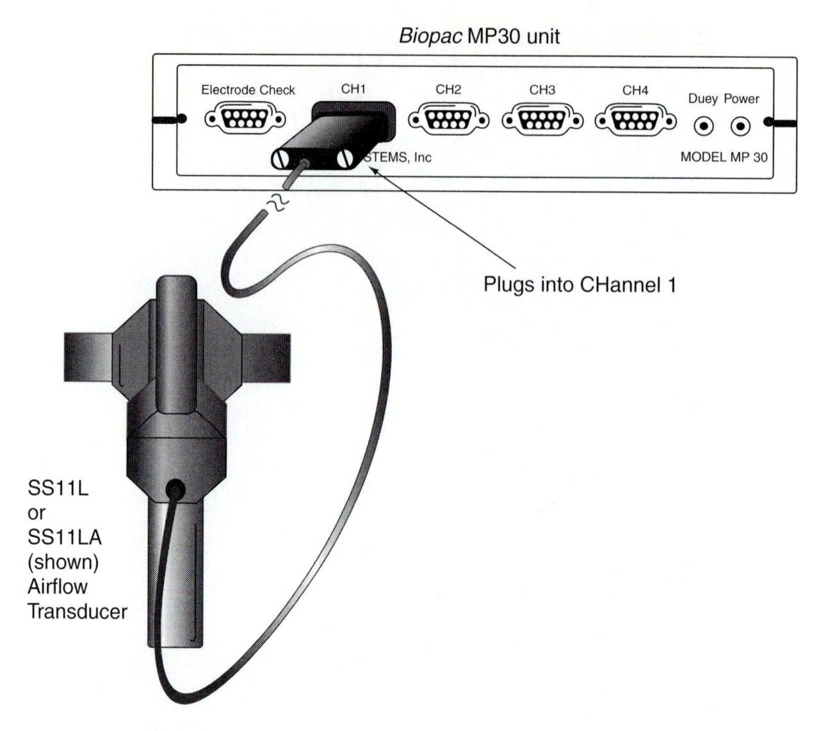


**Figure 8.3** Biopac system equipment for spirometry.

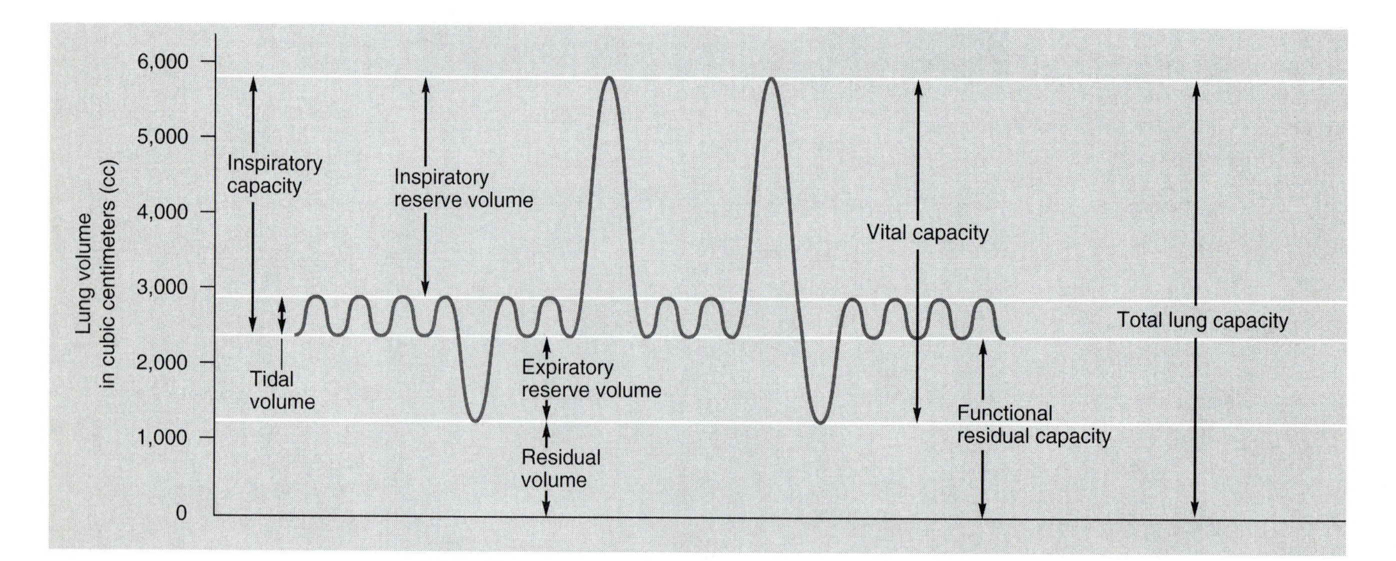


**Figure 8.4** Spirogram recording of lung volumes and capacities.

The **tidal volume (TV)** is the volume of gas inspired or expired during each normal (unforced) ventilation cycle.

The **inspiratory capacity (IC)** is the maximum volume of gas that can be inhaled after a normal (unforced) exhalation.

The **inspiratory reserve volume (IRV)** is the maximum volume of gas that can be forcefully inhaled after a normal (tidal) inhalation.

The **expiratory reserve volume (ERV)** is the maximum volume of gas that can be forcefully exhaled after a normal (tidal) exhalation.

The **functional residual capacity (FRC)** is the volume of gas remaining in the lungs after a normal (unforced) exhalation.

The **residual volume (RV)** is the volume of gas remaining in the lungs after a maximum (forced) exhalation.

The movement of air into and out of the lungs (ventilation) results from a pressure difference between the pulmonary air and the atmosphere. This pressure difference is created by a change in the volume of the thoracic cavity. Since, according to **Boyle's law,** the *pressure*

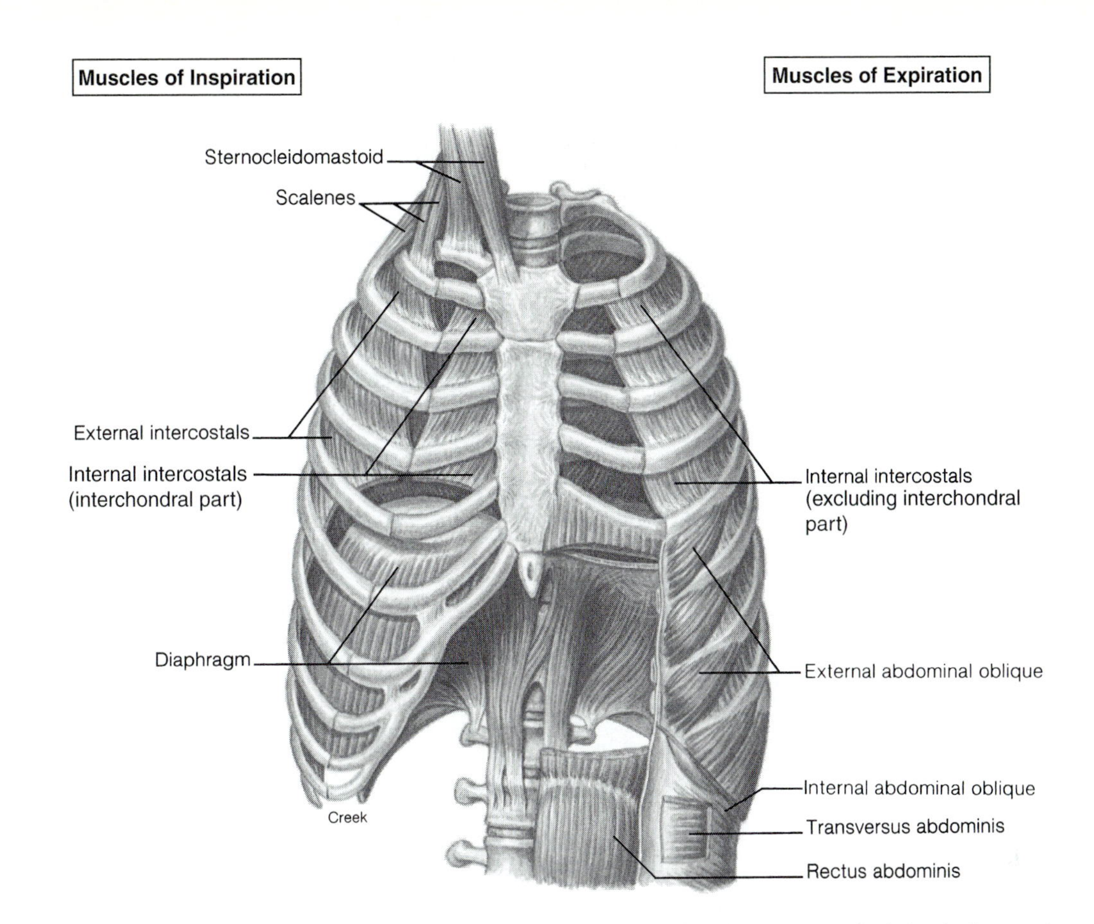


**Figure 8.5 Muscles of respiration.** The right side (of the trunk) shows the muscles of inspiration; the left side depicts the muscles involved in forced expiration.

**(For a full-color version of this figure, see fig. 16.14 in *Human Physiology,* eighth edition, by Stuart I. Fox.)**

*of a gas is inversely proportional to its volume*, an increase in thoracic volume results in a decrease in intrapulmonary pressure. During inhalation, therefore, air is pushed into the lungs by the greater pressure of the atmosphere. When the thoracic volume decreases during exhalation, the intrapulmonary pressure rises above the atmospheric pressure and air is pushed out of the lungs.

In normal (unforced) ventilation, the thoracic volume is regulated by action of the *diaphragm* and the *external intercostal muscles* (fig. 8.5). At rest, the diaphragm forms the convex floor to the thoracic cavity. During inhalation, the diaphragm contracts and pulls itself into a more flattened form. This lowers the floor of the thorax and increases the thoracic volume (and pushes down on the viscera, causing the abdomen to protrude during inhalation). At the same time, the contraction of the external intercostal muscles increases the volume of the thorax by rotating the ribs upward and outward. At the end of inhalation, the diaphragm and external intercostal muscles relax, causing the thorax to resume its original volume and the air inside the lungs to be exhaled. The amount of air inhaled or exhaled in this manner is the tidal volume.

Inhalation can become difficult if the air passages are obstructed or if the lungs lose their normal elasticity. In these cases, the affected person relies increasingly on muscles not usually used in normal (tidal) ventilation: the *scalenus*, *sternocleidomastoid*, and *pectoralis major muscles*. These muscles are also used in healthy people during forced inhalation to obtain the inspiratory reserve volume (IRV).

During forced exhalation, the *internal intercostal muscles* contract, depressing the rib cage, and the *abdominal muscles* contract, pushing the viscera up against the diaphragm. The push of the viscera increases the convexity of the diaphragm and decreases the thoracic volume to a lower level than that achieved in normal exhalation. The amount of air forcefully exhaled by contraction of both groups of muscles is the expiratory reserve volume (ERV).

Even after a maximum forced exhalation, there is still some air left in the lungs. This residual volume of air makes it easier to inflate the lungs during the next inhalation and oxygenates the blood between ventilation cycles.

## A. Measurement of Simple Lung Volumes and Capacities

### PROCEDURE (FOR SPIROCOMP)

1. Press the **"T"** key on the computer and the words "Breathe Normal Cycles" will appear on the computer screen. After three normal tidal volume cycles, the data will appear on the screen.
2. Press the **"E"** key on the computer and the words "Breathe Normal Cycles" will appear on the screen. At the third breathing cycle, the words "Stop After Normal Exhale" will appear.
3. After the pause in breathing, the words "Exhale Forcefully" will appear on the screen. Forcefully exhale all you can at this point.
4. Press the **"V"** key and the words "Inhale Max Then Press V Exhale Fully" will appear on the screen. After a maximal inhalation, press "V" and forcefully exhale all the air you can as fast as possible.
5. Record the data displayed on the screen and use it to complete your laboratory report.

### PROCEDURE (FOR COLLINS RESPIROMETER)

1. Raise and lower the oxygen bell (fig. 8.1) several times to get fresh air into the spirometer. Notice that as the bell moves up and down, one of the pens moves a corresponding distance down and up on a shaft. By adjusting the height of the oxygen bell, position this pen so that it will begin writing in the middle of the chart paper. This pen (the *ventilometer* pen) usually has black ink; whereas the other *(respiration)* pen usually has red ink and will not be used for this exercise. (It should be covered and rocked away from the paper.)
2. With the free-breathing valve set to the *open position*, place the mouthpiece in the buccal cavity (as in breathing through a snorkel), and go through several ventilation cycles to become accustomed to the apparatus. (When the free-breathing valve is open, you will breathe room air.) If a disposable cardboard mouthpiece is used, be particularly careful to prevent air leakage from the corners of the mouth. Breathing through the nostrils can be prevented by means of a nose clamp or by pinching the nose tightly with the thumb and forefinger.
3. Turn the respirometer to the *slow* position (32 mm/min) and close the free-breathing valve so that the oxygen bell and the pen go up and down with each ventilation cycle.
4. Breathe in a normal, relaxed manner for 1-2 min. The breaths should appear relatively uniform, and the slope should go upward (see fig. 8.6). A downward slope indicates that there is an air leak; in this event, tighten the grip of the mouth on the mouthpiece and the nose clamp on the nose, and begin again.

**Note:** *At this speed (32 mm/min), the distance between heavy vertical lines on the chart is traversed in 1 min.*

This procedure measures tidal volume—the amount of air inhaled or exhaled during each resting ventilation cycle.

5. With the drum still turning, perform the test for vital capacity (the maximum amount of air that can be exhaled after a maximum inhalation). At the end of a normal exhalation, inhale as much as possible, and then exhale completely to the fullest possible extent, and stop the recording. At the completion of this exercise, the chart should resemble the one shown in figure 8.6.
6. Remove the chart from the kymograph drum. Notice that the chart is marked horizontally in milliliters.

**Note:** *Since the temperature and pressure of the respirometer are different from those existing in the body, the volume the air occupies in the respirometer will be subject to changes in ambient (room) conditions. To standardize the volumes measured in spirometry, multiply these measured volumes by a correction factor known as the* ***BTPS factor*** *(body temperature, atmospheric pressure, saturated with water vapor). Since the BTPS factor is very close to* ***1.1*** *at normal room temperatures, we will use this figure in the calculations.*

### Calculations

1. Obtain the measured tidal volume from the chart by subtracting the milliliters corresponding to the trough from the milliliters corresponding to the peak of a typical resting ventilation cycle.

#### *Example (from fig. 8.7)*

Step 1
  3,700 mL (inhalation)
− 3,250 mL (exhalation)
  450 mL

Step 2
  450 mL (measured tidal volume)
× 1.1 (BTPS factor)
  495 mL

Enter the corrected measured tidal volume (TV) in the *Measured* column of the table in your laboratory report.

2. Obtain the measured inspiratory capacity from the chart. To do this, subtract the milliliters corresponding to the last normal exhalation before performing the vital capacity maneuver from the milliliters corresponding to the maximum inhalation peak.

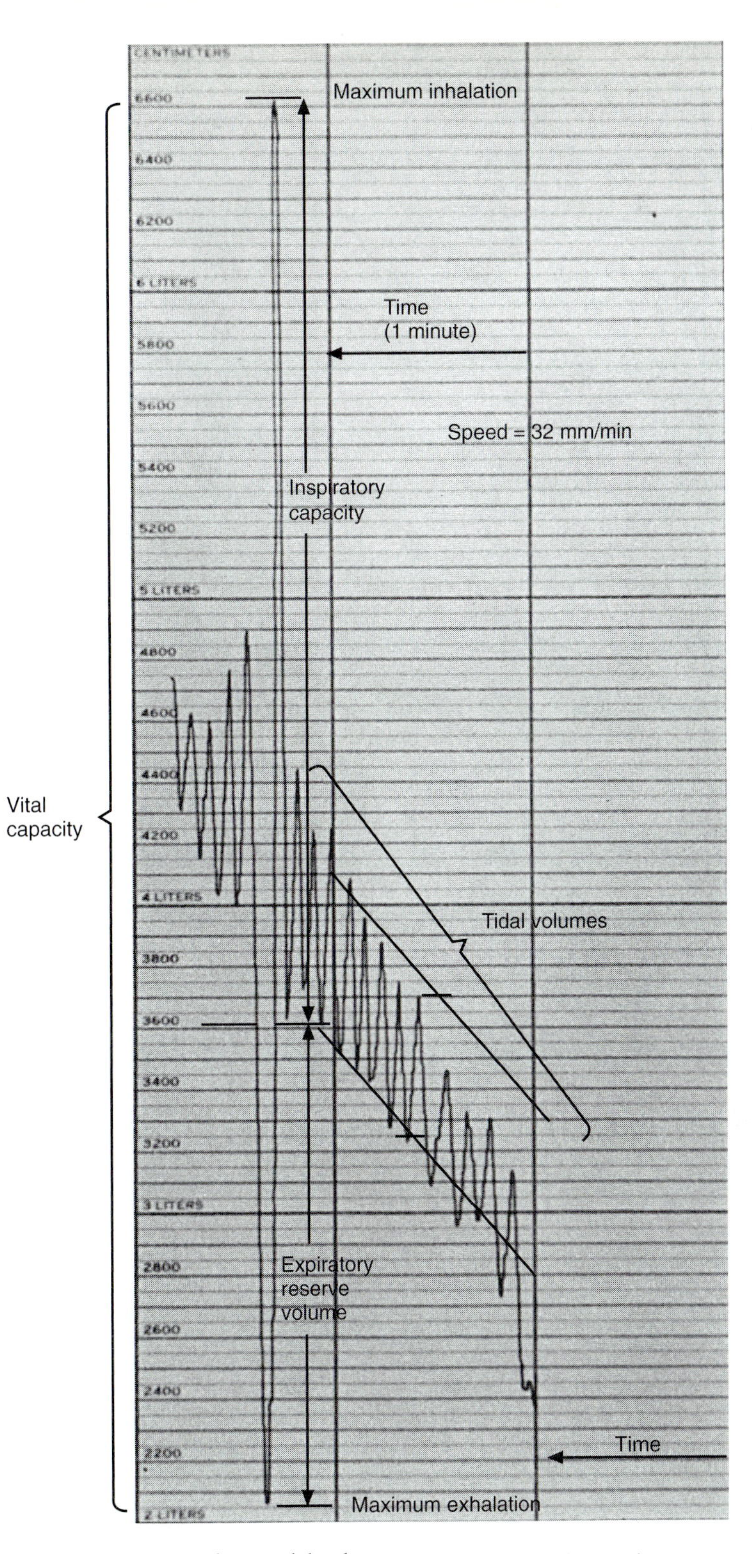


**Figure 8.6 A spirometry recording.** This chart shows tidal volume, inspiratory capacity, expiratory reserve volume, and vital capacity.

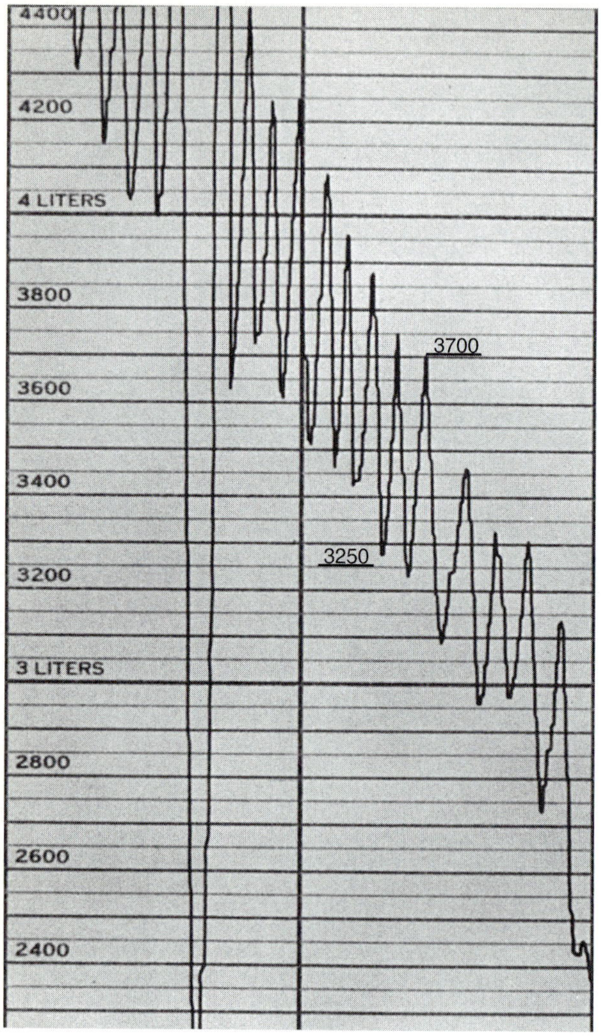

Figure 8.7 A close-up of tidal volume measurements from figure 8.6.

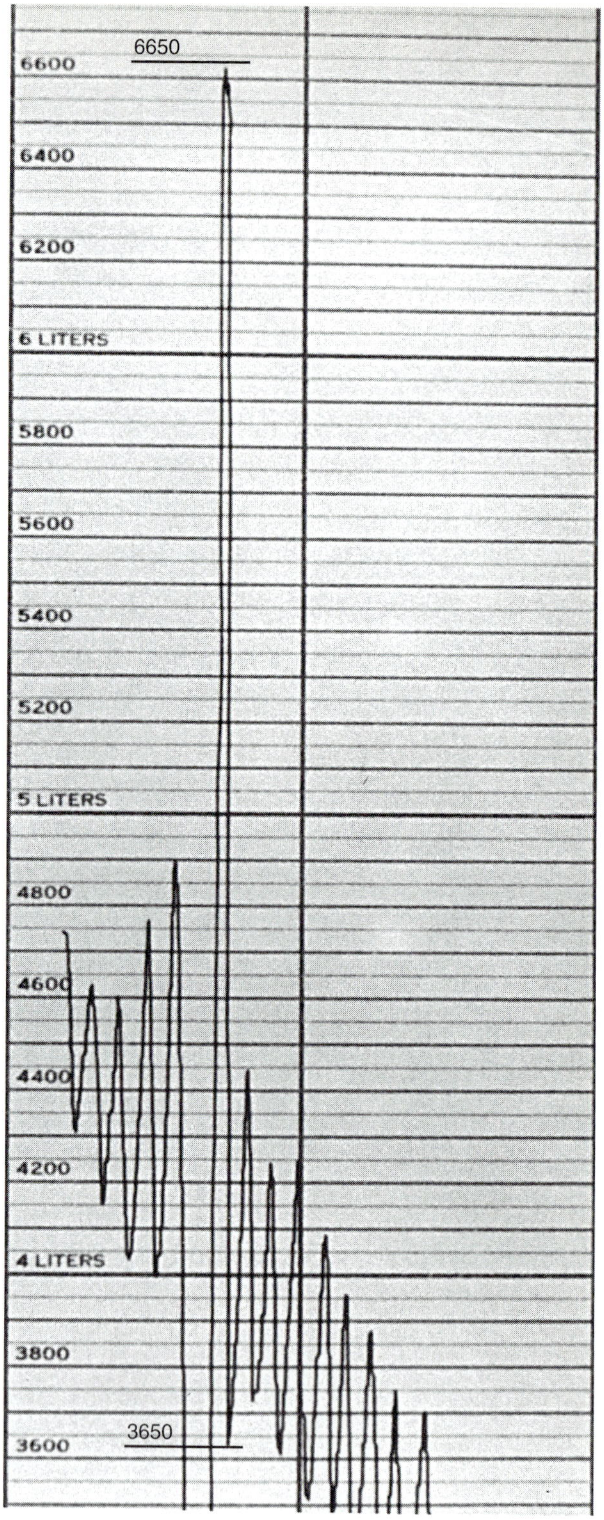

Figure 8.8 A close-up of the inspiratory capacity measurement from figure 8.6.

### *Example (from fig. 8.8)*

Step 1    6,650 mL (maximum inhalation)
          –3,650 mL (normal exhalation)
          3,000 mL

Step 2    3,000 mL (measured inspiratory capacity)
          × 1.1 (BTPS factor)
          3,300 mL

Enter the corrected measured inspiratory capacity (IC) in the *Measured* column of the table in your laboratory report.

3. Obtain the measured expiratory reserve volume by subtracting the milliliters corresponding to the trough for maximum exhalation from the milliliters corresponding to the last normal exhalation before the vital capacity maneuver (same value used for step 2).

### *Example (from fig. 8.9)*

Step 1 3,650 mL (normal exhalation)
−2,050 mL (maximum exhalation)
1,600 mL

Step 2 1,600 mL (measured expiratory reserve)
× 1.1 1,600 mL (volume)
1,760 mL (BTPS factor)

Enter the corrected expiratory reserve volume (ERV) in the *Measured* column of the table in your laboratory report.

4. Obtain the measured vital capacity by either (1) adding the corrected inspiratory capacity (from step 2) and the corrected expiratory reserve volume (step 3) (since these values have already been BTPS standardized, an additional correction step is unnecessary); or (2) subtracting the milliliters corresponding to maximum exhalation from the milliliters corresponding to maximum inhalation. This value must then be multiplied by the BTPS factor.

### Method 1

3,300 mL (corrected inspiratory capacity)
+ 1,760 mL (corrected expiratory reserve volume)
5,060 mL (corrected vital capacity)

### Method 2

6,650 mL (maximum inhalation)
− 2,050 mL (maximum exhalation)
4,600 mL

4,600 mL (measured vital capacity)
× 1.1 (BTPS factor)
5,060 mL (corrected capacity)

Enter the corrected vital capacity (VC) in the *Measured* column of the table in your laboratory report.

5. Obtain the *predicted vital capacity* for the subject's sex, age, and height from tables 8.1 and 8.2. The height in centimeters can be most conveniently obtained by referring to the height conversion scale in figure 8.10.

### *Example*

Sex: male
Age: 34
Height: 174 cm
Predicted vital capacity: 4,140 mL

Enter the subject's predicted vital capacity from the tables of normal values in the *Predicted* column in your laboratory report.

6. To obtain an estimate of the predicted residual volume and the predicted total lung capacity, refer to table 8.3.

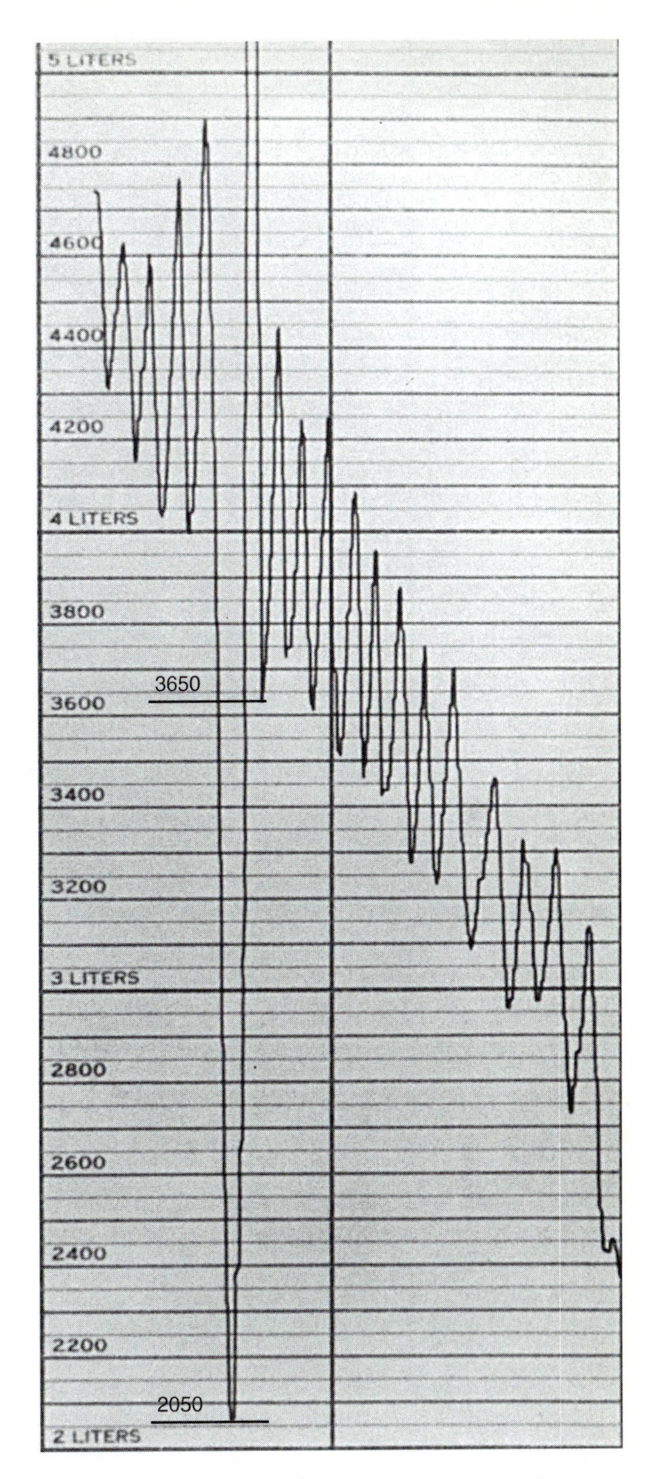


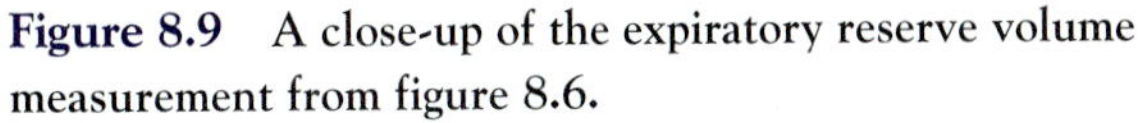

**Figure 8.9** A close-up of the expiratory reserve volume measurement from figure 8.6.

**Note:** *These values cannot be measured by spirometry because residual volume cannot be exhaled; total lung capacity equals the vital capacity plus residual volume.*

7. Obtain the percent predicted value for all the measurements in this way:

$$\text{Percent predicted} = \frac{\text{corrected measured value}}{\text{predicted value}} \times 100\%$$

## Table 8.1 Predicted Vital Capacities, Females (mL)

**Height in Centimeters**

| Age | 146 | 148 | 150 | 152 | 154 | 156 | 158 | 160 | 162 | 164 | 166 | 168 | 170 | 172 | 174 | 176 | 178 | 180 | 182 | 184 | 186 | 188 | 190 | 192 | 194 |
|---|---|---|---|---|---|---|---|---|---|---|---|---|---|---|---|---|---|---|---|---|---|---|---|---|---|
| 16 | 2950 | 2990 | 3030 | 3070 | 3110 | 3150 | 3190 | 3230 | 3270 | 3310 | 3350 | 3390 | 3430 | 3470 | 3510 | 3550 | 3690 | 3630 | 3570 | 3715 | 3755 | 3800 | 3840 | 3880 | 3920 |
| 17 | 2935 | 2975 | 3015 | 3055 | 3095 | 3135 | 3175 | 3215 | 3255 | 3295 | 3335 | 3375 | 3415 | 3455 | 3495 | 3535 | 3575 | 3615 | 3655 | 3695 | 3740 | 3780 | 3820 | 3860 | 3900 |
| 18 | 2920 | 2960 | 3000 | 3040 | 3080 | 3120 | 3160 | 3200 | 3240 | 3280 | 3320 | 3360 | 3400 | 3440 | 3480 | 3520 | 3560 | 3600 | 3640 | 3680 | 3720 | 3760 | 3800 | 3840 | 3880 |
| 20 | 2890 | 2930 | 2970 | 3010 | 3050 | 3090 | 3130 | 3170 | 3210 | 3250 | 3290 | 3330 | 3370 | 3410 | 3450 | 3490 | 3525 | 3565 | 3605 | 3645 | 3695 | 3720 | 3760 | 3800 | 3840 |
| 22 | 2860 | 2900 | 2940 | 2980 | 3020 | 3060 | 3095 | 3135 | 3175 | 3215 | 3255 | 3290 | 3330 | 3370 | 3410 | 3450 | 3490 | 3530 | 3570 | 3610 | 3650 | 3685 | 3725 | 3765 | 3800 |
| 24 | 2830 | 2870 | 2910 | 2950 | 2985 | 3025 | 3065 | 3100 | 3140 | 3180 | 3220 | 3260 | 3300 | 3335 | 3375 | 3415 | 3455 | 3490 | 3530 | 3570 | 3610 | 3650 | 3685 | 3725 | 3765 |
| 26 | 2800 | 2840 | 2880 | 2920 | 2960 | 3000 | 3035 | 3070 | 3110 | 3150 | 3190 | 3230 | 3265 | 3300 | 3340 | 3380 | 3420 | 3455 | 3495 | 3530 | 3570 | 3610 | 3650 | 3685 | 3725 |
| 28 | 2775 | 2810 | 2850 | 2890 | 2930 | 2965 | 3000 | 3040 | 3070 | 3115 | 3155 | 3190 | 3230 | 3270 | 3305 | 3345 | 3380 | 3420 | 3460 | 3495 | 3535 | 3570 | 3610 | 3650 | 3685 |
| 30 | 2745 | 2780 | 2820 | 2860 | 2895 | 2935 | 2970 | 3010 | 3045 | 3085 | 3120 | 3160 | 3195 | 3235 | 3270 | 3310 | 3345 | 3385 | 3420 | 3460 | 3495 | 3535 | 3570 | 3610 | 3645 |
| 32 | 2715 | 2750 | 2790 | 2825 | 2865 | 2900 | 2940 | 2975 | 3015 | 3050 | 3090 | 3125 | 3160 | 3200 | 3235 | 3275 | 3310 | 3350 | 3385 | 3425 | 3460 | 3495 | 3535 | 3570 | 3610 |
| 34 | 2685 | 2725 | 2760 | 2795 | 2835 | 2870 | 2910 | 2945 | 2980 | 3020 | 3055 | 3090 | 3130 | 3165 | 3200 | 3240 | 3275 | 3310 | 3350 | 3385 | 3425 | 3460 | 3495 | 3535 | 3570 |
| 36 | 2655 | 2695 | 2730 | 2765 | 2805 | 2840 | 2875 | 2910 | 2950 | 2985 | 3020 | 3060 | 3095 | 3130 | 3165 | 3205 | 3240 | 3275 | 3310 | 3350 | 3385 | 3420 | 3460 | 3495 | 3530 |
| 38 | 2630 | 2665 | 2700 | 2735 | 2770 | 2810 | 2845 | 2880 | 2915 | 2950 | 2990 | 3025 | 3060 | 3095 | 3130 | 3170 | 3205 | 3240 | 3275 | 3310 | 3350 | 3385 | 3420 | 3455 | 3490 |
| 40 | 2600 | 2635 | 2670 | 2705 | 2740 | 2775 | 2810 | 2850 | 2885 | 2920 | 2955 | 2990 | 3025 | 3060 | 3095 | 3135 | 3170 | 3205 | 3240 | 3275 | 3310 | 3345 | 3380 | 3420 | 3455 |
| 42 | 2570 | 2605 | 2640 | 2675 | 2710 | 2745 | 2780 | 2815 | 2850 | 2885 | 2920 | 2955 | 2990 | 3025 | 3060 | 3100 | 3135 | 3170 | 3205 | 3240 | 3275 | 3310 | 3345 | 3380 | 3415 |
| 44 | 2540 | 2575 | 2610 | 2645 | 2680 | 2715 | 2750 | 2785 | 2820 | 2855 | 2890 | 2925 | 2960 | 2995 | 3030 | 3060 | 3095 | 3130 | 3165 | 3200 | 3235 | 3270 | 3305 | 3340 | 3375 |
| 46 | 2510 | 2545 | 2580 | 2615 | 2650 | 2685 | 2715 | 2750 | 2785 | 2820 | 2855 | 2890 | 2925 | 2960 | 2995 | 3030 | 3060 | 3095 | 3130 | 3165 | 3200 | 3235 | 3270 | 3305 | 3340 |
| 48 | 2480 | 2515 | 2550 | 2585 | 2620 | 2650 | 2685 | 2715 | 2750 | 2785 | 2820 | 2855 | 2890 | 2925 | 2960 | 2995 | 3030 | 3060 | 3095 | 3130 | 3160 | 3195 | 3230 | 3265 | 3300 |
| 50 | 2455 | 2485 | 2520 | 2555 | 2590 | 2625 | 2655 | 2690 | 2720 | 2755 | 2785 | 2820 | 2855 | 2890 | 2925 | 2955 | 2990 | 3025 | 3060 | 3090 | 3125 | 3155 | 3190 | 3225 | 3260 |
| 52 | 2425 | 2455 | 2490 | 2525 | 2555 | 2590 | 2625 | 2655 | 2690 | 2720 | 2755 | 2790 | 2820 | 2855 | 2890 | 2925 | 2955 | 2990 | 3020 | 3055 | 3090 | 3125 | 3155 | 3190 | 3220 |
| 54 | 2395 | 2425 | 2460 | 2495 | 2530 | 2560 | 2590 | 2625 | 2655 | 2690 | 2720 | 2755 | 2790 | 2820 | 2855 | 2885 | 2920 | 2950 | 2985 | 3020 | 3050 | 3085 | 3115 | 3150 | 3180 |
| 56 | 2365 | 2400 | 2430 | 2460 | 2495 | 2525 | 2560 | 2590 | 2625 | 2655 | 2690 | 2720 | 2755 | 2790 | 2820 | 2855 | 2885 | 2920 | 2950 | 2980 | 3015 | 3045 | 3080 | 3110 | 3145 |
| 58 | 2335 | 2370 | 2400 | 2430 | 2460 | 2495 | 2525 | 2560 | 2590 | 2625 | 2655 | 2690 | 2720 | 2750 | 2785 | 2815 | 2850 | 2880 | 2920 | 2945 | 2975 | 3010 | 3040 | 3075 | 3105 |
| 60 | 2305 | 2340 | 2370 | 2400 | 2430 | 2460 | 2495 | 2525 | 2560 | 2590 | 2625 | 2655 | 2685 | 2720 | 2750 | 2780 | 2810 | 2845 | 2875 | 2915 | 2940 | 2970 | 3000 | 3035 | 3065 |
| 62 | 2280 | 2310 | 2340 | 2370 | 2405 | 2435 | 2465 | 2495 | 2525 | 2560 | 2590 | 2620 | 2655 | 2685 | 2715 | 2745 | 2775 | 2810 | 2840 | 2870 | 2900 | 2935 | 2965 | 2995 | 3025 |
| 64 | 2250 | 2280 | 2310 | 2340 | 2370 | 2400 | 2430 | 2465 | 2495 | 2525 | 2555 | 2585 | 2620 | 2650 | 2680 | 2710 | 2740 | 2770 | 2805 | 2835 | 2865 | 2895 | 2925 | 2955 | 2990 |
| 66 | 2220 | 2250 | 2280 | 2310 | 2340 | 2370 | 2400 | 2430 | 2460 | 2495 | 2525 | 2555 | 2585 | 2615 | 2645 | 2675 | 2705 | 2735 | 2765 | 2800 | 2825 | 2860 | 2890 | 2920 | 2950 |
| 68 | 2190 | 2220 | 2250 | 2280 | 2310 | 2340 | 2370 | 2400 | 2430 | 2460 | 2490 | 2520 | 2550 | 2580 | 2610 | 2640 | 2670 | 2700 | 2730 | 2760 | 2795 | 2820 | 2850 | 2880 | 2910 |
| 70 | 2160 | 2190 | 2220 | 2250 | 2280 | 2310 | 2340 | 2370 | 2400 | 2425 | 2455 | 2485 | 2515 | 2545 | 2575 | 2605 | 2635 | 2665 | 2695 | 2725 | 2755 | 2780 | 2810 | 2840 | 2870 |
| 72 | 2130 | 2160 | 2190 | 2220 | 2250 | 2280 | 2310 | 2335 | 2365 | 2395 | 2425 | 2455 | 2480 | 2510 | 2540 | 2570 | 2600 | 2630 | 2660 | 2685 | 2715 | 2745 | 2775 | 2805 | 2830 |
| 74 | 2100 | 2130 | 2160 | 2190 | 2220 | 2245 | 2275 | 2305 | 2335 | 2360 | 2390 | 2420 | 2450 | 2475 | 2505 | 2535 | 2565 | 2590 | 2620 | 2650 | 2680 | 2710 | 2740 | 2765 | 2795 |

Courtesy of Warren E. Collins, Inc., Braintree, MA.

## Table 8.2 Predicted Vital Capacities, Males (mL)

**Height in Centimeters**

| Age | 146 | 148 | 150 | 152 | 154 | 156 | 158 | 160 | 162 | 164 | 166 | 168 | 170 | 172 | 174 | 176 | 178 | 180 | 182 | 184 | 186 | 188 | 190 | 192 | 194 |
|---|---|---|---|---|---|---|---|---|---|---|---|---|---|---|---|---|---|---|---|---|---|---|---|---|---|
| 16 | 3765 | 3820 | 3870 | 3920 | 3975 | 4025 | 4075 | 4130 | 4180 | 4230 | 4285 | 4335 | 4385 | 4440 | 4490 | 4540 | 4590 | 4645 | 4695 | 4745 | 4800 | 4850 | 4900 | 4955 | 5005 |
| 18 | 3740 | 3790 | 3840 | 3890 | 3940 | 3995 | 4045 | 4095 | 4145 | 4200 | 4250 | 4300 | 4350 | 4405 | 4455 | 4505 | 4555 | 4610 | 4660 | 4710 | 4760 | 4815 | 4865 | 4915 | 4965 |
| 20 | 3710 | 3760 | 3810 | 3860 | 3910 | 3960 | 4015 | 4065 | 4115 | 4165 | 4215 | 4265 | 4320 | 4370 | 4420 | 4470 | 4520 | 4570 | 4625 | 4675 | 4725 | 4775 | 4825 | 4875 | 4930 |
| 22 | 3680 | 3730 | 3780 | 3830 | 3880 | 3930 | 3980 | 4030 | 4080 | 4135 | 4185 | 4235 | 4285 | 4335 | 4385 | 4435 | 4485 | 4535 | 4585 | 4635 | 4685 | 4735 | 4790 | 4840 | 4890 |
| 24 | 3635 | 3685 | 3735 | 3785 | 3835 | 3885 | 3935 | 3985 | 4035 | 4085 | 4135 | 4185 | 4235 | 4285 | 4330 | 4380 | 4430 | 4480 | 4530 | 4580 | 4630 | 4680 | 4730 | 4780 | 4830 |
| 26 | 3605 | 3655 | 3705 | 3755 | 3805 | 3855 | 3905 | 3955 | 4000 | 4050 | 4100 | 4150 | 4200 | 4250 | 4300 | 4350 | 4395 | 4445 | 4495 | 4545 | 4595 | 4645 | 4695 | 4740 | 4790 |
| 28 | 3575 | 3625 | 3675 | 3725 | 3775 | 3820 | 3870 | 3920 | 3970 | 4020 | 4070 | 4115 | 4165 | 4215 | 4265 | 4310 | 4360 | 4410 | 4460 | 4510 | 4555 | 4605 | 4655 | 4705 | 4755 |
| 30 | 3550 | 3595 | 3645 | 3695 | 3740 | 3790 | 3840 | 3890 | 3935 | 3985 | 4035 | 4080 | 4130 | 4180 | 4230 | 4275 | 4325 | 4375 | 4425 | 4470 | 4520 | 4570 | 4615 | 4665 | 4715 |
| 32 | 3520 | 3565 | 3615 | 3665 | 3710 | 3760 | 3810 | 3855 | 3905 | 3950 | 4000 | 4050 | 4095 | 4145 | 4195 | 4240 | 4290 | 4340 | 4385 | 4435 | 4485 | 4530 | 4580 | 4625 | 4675 |
| 34 | 3475 | 3525 | 3570 | 3620 | 3665 | 3715 | 3760 | 3810 | 3855 | 3905 | 3950 | 4000 | 4045 | 4095 | 4140 | 4190 | 4225 | 4285 | 4330 | 4380 | 4425 | 4475 | 4520 | 4570 | 4615 |
| 36 | 3445 | 3495 | 3540 | 3585 | 3635 | 3680 | 3730 | 3775 | 3825 | 3870 | 3920 | 3965 | 4010 | 4060 | 4105 | 4155 | 4200 | 4250 | 4295 | 4340 | 4390 | 4435 | 4485 | 4530 | 4580 |
| 38 | 3415 | 3465 | 3510 | 3555 | 3605 | 3650 | 3695 | 3745 | 3790 | 3840 | 3885 | 3930 | 3980 | 4025 | 4070 | 4120 | 4165 | 4210 | 4260 | 4305 | 4350 | 4400 | 4445 | 4495 | 4540 |
| 40 | 3385 | 3435 | 3480 | 3525 | 3575 | 3620 | 3665 | 3710 | 3760 | 3805 | 3850 | 3900 | 3945 | 3990 | 4035 | 4085 | 4130 | 4175 | 4220 | 4270 | 4315 | 4360 | 4410 | 4455 | 4500 |
| 42 | 3360 | 3405 | 3450 | 3495 | 3540 | 3590 | 3635 | 3680 | 3725 | 3770 | 3820 | 3865 | 3910 | 3955 | 4000 | 4050 | 4095 | 4140 | 4185 | 4230 | 4280 | 4325 | 4370 | 4415 | 4460 |
| 44 | 3315 | 3360 | 3405 | 3450 | 3495 | 3540 | 3585 | 3630 | 3675 | 3725 | 3770 | 3815 | 3860 | 3905 | 3950 | 3995 | 4040 | 4085 | 4130 | 4175 | 4220 | 4270 | 4315 | 4360 | 4405 |
| 46 | 3285 | 3330 | 3375 | 3420 | 3465 | 3510 | 3555 | 3600 | 3645 | 3690 | 3735 | 3780 | 3825 | 3870 | 3915 | 3960 | 4005 | 4050 | 4095 | 4140 | 4185 | 4230 | 4275 | 4320 | 4365 |
| 48 | 3255 | 3300 | 3345 | 3390 | 3435 | 3480 | 3525 | 3570 | 3615 | 3655 | 3700 | 3745 | 3790 | 3835 | 3880 | 3925 | 3970 | 4015 | 4060 | 4105 | 4150 | 4190 | 4235 | 4280 | 4325 |
| 50 | 3210 | 3255 | 3300 | 3345 | 3390 | 3430 | 3475 | 3520 | 3565 | 3610 | 3650 | 3695 | 3740 | 3785 | 3830 | 3870 | 3915 | 3960 | 4005 | 4050 | 4090 | 4135 | 4180 | 4225 | 4270 |
| 52 | 3185 | 3225 | 3270 | 3315 | 3355 | 3400 | 3445 | 3490 | 3530 | 3575 | 3620 | 3660 | 3705 | 3750 | 3795 | 3835 | 3880 | 3925 | 3970 | 4010 | 4055 | 4100 | 4140 | 4185 | 4230 |
| 54 | 3155 | 3195 | 3240 | 3285 | 3325 | 3370 | 3415 | 3455 | 3500 | 3540 | 3585 | 3630 | 3670 | 3715 | 3760 | 3800 | 3845 | 3890 | 3930 | 3975 | 4020 | 4060 | 4105 | 4145 | 4190 |
| 56 | 3125 | 3165 | 3210 | 3255 | 3295 | 3340 | 3380 | 3425 | 3465 | 3510 | 3550 | 3595 | 3640 | 3680 | 3725 | 3765 | 3810 | 3850 | 3895 | 3940 | 3980 | 4025 | 4065 | 4110 | 4150 |
| 58 | 3080 | 3125 | 3165 | 3210 | 3250 | 3290 | 3335 | 3375 | 3420 | 3460 | 3500 | 3545 | 3585 | 3630 | 3670 | 3715 | 3755 | 3800 | 3840 | 3880 | 3925 | 3965 | 4010 | 4050 | 4095 |
| 60 | 3050 | 3095 | 3135 | 3175 | 3220 | 3260 | 3300 | 3345 | 3385 | 3430 | 3470 | 3500 | 3555 | 3595 | 3635 | 3680 | 3720 | 3760 | 3805 | 3845 | 3885 | 3930 | 3970 | 4015 | 4055 |
| 62 | 3020 | 3060 | 3110 | 3150 | 3190 | 3230 | 3270 | 3310 | 3350 | 3390 | 3440 | 3480 | 3520 | 3560 | 3600 | 3640 | 3680 | 3730 | 3770 | 3810 | 3850 | 3890 | 3930 | 3970 | 4020 |
| 64 | 2990 | 3030 | 3080 | 3120 | 3160 | 3200 | 3240 | 3280 | 3320 | 3360 | 3400 | 3440 | 3490 | 3530 | 3570 | 3610 | 3650 | 3690 | 3730 | 3770 | 3810 | 3850 | 3900 | 3940 | 3980 |
| 66 | 2950 | 2990 | 3030 | 3070 | 3110 | 3150 | 3190 | 3230 | 3270 | 3310 | 3350 | 3390 | 3430 | 3470 | 3510 | 3550 | 3600 | 3640 | 3680 | 3720 | 3760 | 3800 | 3840 | 3880 | 3920 |
| 68 | 2920 | 2960 | 3000 | 3040 | 3080 | 3120 | 3160 | 3200 | 3240 | 3280 | 3320 | 3360 | 3400 | 3440 | 3480 | 3520 | 3560 | 3600 | 3640 | 3680 | 3720 | 3760 | 3800 | 3840 | 3880 |
| 70 | 2890 | 2930 | 2970 | 3010 | 3050 | 3090 | 3130 | 3170 | 3210 | 3250 | 3290 | 3330 | 3370 | 3410 | 3450 | 3480 | 3520 | 3560 | 3600 | 3640 | 3680 | 3720 | 3760 | 3800 | 3840 |
| 72 | 2860 | 2900 | 2940 | 2980 | 3020 | 3060 | 3100 | 3140 | 3180 | 3210 | 3250 | 3290 | 3330 | 3370 | 3410 | 3450 | 3490 | 3530 | 3570 | 3610 | 3650 | 3680 | 3720 | 3760 | 3800 |
| 74 | 2820 | 2860 | 2900 | 2930 | 2970 | 3010 | 3050 | 3090 | 3130 | 3170 | 3200 | 3240 | 3280 | 3320 | 3360 | 3400 | 3440 | 3470 | 3510 | 3550 | 3590 | 3630 | 3670 | 3710 | 3740 |

Courtesy of Warren E. Collins, Inc., Braintree, MA.

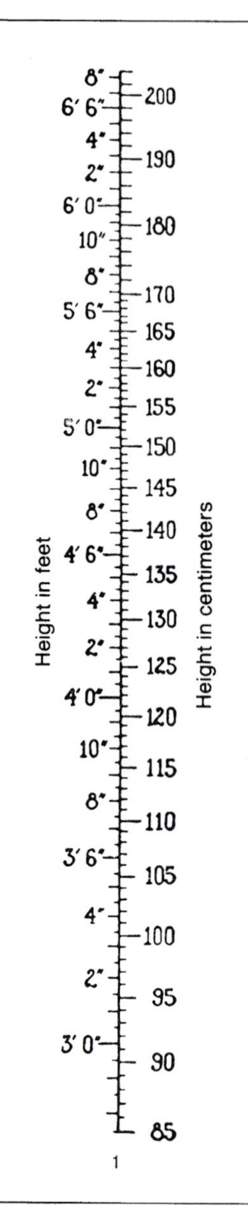


**Figure 8.10** A scale for converting height in feet and inches to height in centimeters.

| Age | Residual Volume: Vital Capacity × Factor | Total Lung Capacity: Vital Capacity × Factor |
|---|---|---|
| 16–34 | 0.250 | 1,250 |
| 35–49 | 0.305 | 1,305 |
| 50–69 | 0.445 | 1,445 |

Table 8.3 Factors for Obtaining the Predicted Residual Volume and Total Lung Capacity

### *Example*

Measured vital capacity 5,060 mL (corrected to BTPS)
Predicted vital capacity 4,140 mL (from table 8.1 or 8.2)

$$\% \text{ predicted} = \frac{5{,}060 \text{ mL}}{4{,}140 \text{ mL}} \times 100\% = 122\%$$

Enter the percent predicted values in the appropriate places in the table in your laboratory report.

> Measurements of vital capacity that are consistently below 80% of the predicted value on repeated tests suggest the presence of a restrictive lung disease, such as emphysema.

## B. Measurement of Forced Expiratory Volume

The ability to ventilate the lungs in a given amount of time is often of greater diagnostic value than measurements of simple lung volumes and capacities. One measurement that considers time intervals is the **forced expiratory volume (FEV),** otherwise known as the *timed vital capacity*.

In the forced expiratory volume test, the subject performs a vital capacity maneuver by inhaling maximally, holding, and then exhaling maximally. While holding at the point of peak inhalation, the respirometer kymograph drum speed is set to its fastest setting (1,920 mm/min); then the subject is instructed to exhale forcefully and maximally. This fast speed stretches out the exhalation tracing and the distance between heavy vertical lines is now traversed in 1 sec. From the recording, the percentage of the total vital capacity that is exhaled in the *first* second ($FEV_1$), the *second* second ($FEV_2$), and the *third* second ($FEV_3$) can be determined. A sample record of the forced expiratory volume is shown in figure 8.11.

### *Pulmonary Disorders*

Chronic (long-term) pulmonary dysfunctions can be divided into two general categories: obstructive disorders and restrictive disorders. These two categories can be distinguished, in part, by the use of the spirometry tests performed in this exercise.

Since the flow of air through a tube is proportional to the fourth power of its radius ($r^4$), a small obstruction in the pulmonary airways results in a greatly magnified resistance to airflow. **Obstructive disorders** of the bronchioles, for example, are characteristic of *emphysema, bronchitis,* and *asthma*. This obstruction can result from bronchiolar secretions, inflammation and edema, or contraction of bronchiolar smooth muscle. These conditions

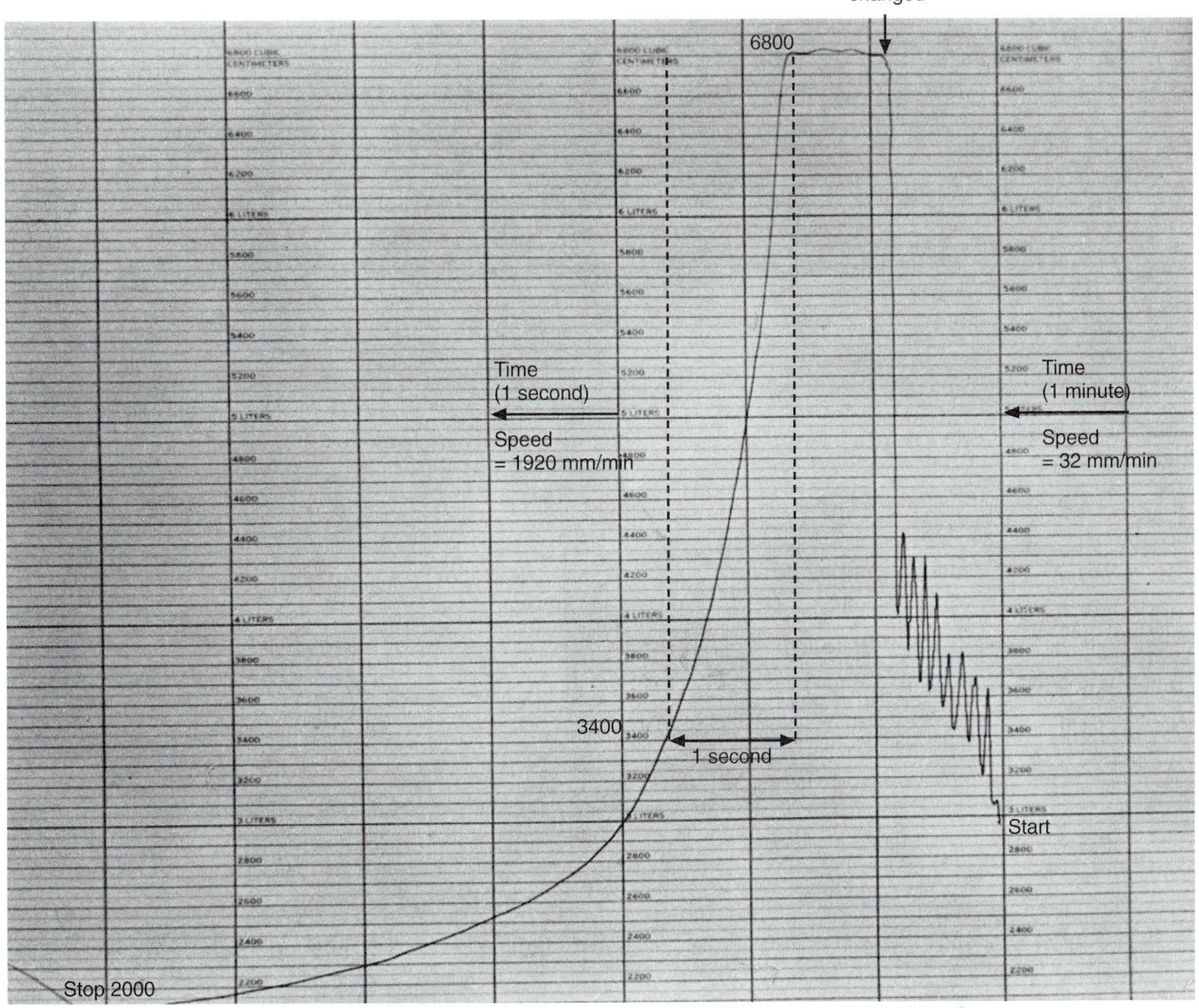


**Figure 8.11** A recording of the forced expiratory volume (FEV) measurement.

make it difficult for sufferers to move air rapidly into and out of the lungs. The bronchioles may be weakened to such a degree that they collapse during exhalation before all the air has been emptied from the lungs. This condition, known as *air trapping,* is revealed by an increase in the functional residual capacity.

In **restrictive disorders,** actual damage to the lung tissue results in an abnormal vital capacity test. However, if the disease is purely restrictive (as in *pulmonary fibrosis*), the airways may be clear, resulting in a normal forced expiratory volume (FEV) test. The vital capacity is reduced, but it can be quickly exhaled. This is not true of emphysema, which is both a restrictive and an obstructive disease.

Caused primarily by cigarette smoking and aggravated by air pollution, emphysema reduces the number of alveoli in the lungs. This results in an abnormally low vital capacity, indicative of restrictive disease. However, since the elastic alveolar tissue normally helps to keep the thin-walled bronchioles open during exhalation, the loss of alveoli in emphysema also results in a reduction in the elastic support of the bronchioles. During exhalation, such bronchioles may narrow (increasing the resistance to airflow) and even collapse. This adds an obstructive component to the disease, resulting in an abnormal forced expiratory volume test.

The $FEV_1$ test detects increased airway resistance, as occurs in emphysema, bronchitis, and asthma. Also used preoperatively, this test is used extensively to predict a patient's response to general anesthesia and to estimate the length of time the patient must be kept on a respirator postoperatively. The $FEV_1$ test is also valuable in research on the effects of air pollutants—such as cigarette smoke and ozone—on pulmonary function.

## PROCEDURE

**Note:** *If the Spirocomp was used, the $FEV_1$ has already been computed and was displayed on the computer screen when the "V" test in part A was performed. The following procedure is for the Collins respirometer.*

1. Start the kymograph slowly rotating (32 mm/min), and have the subject breathe normally into the respirometer for a few breaths, until comfortable.
2. After a normal (unforced) exhalation, instruct the subject to take a deep, forceful inhalation and to hold this inhalation momentarily.
3. Switch the kymograph to the fast speed (1,920 mm/min), and instruct the subject to exhale as rapidly and as forcefully as possible.

**Note:** *At a speed of 1,920 mm/min, the time interval between two heavy vertical lines on the chart is 1 sec.*

### Calculations: Forced Expiratory Volume ($FEV_1$)

1. Measure the vital capacity from the chart by subtracting the exhalation trough (which is flat, because all air has been expelled) from the inhalation peak (also flat, because the subject's breath is held). Do not multiply this value by the BTPS factor.

#### *Example (from fig. 8.11)*

6,800 mL (maximum inhalation)
– 2,000 mL (maximum exhalation)
4,800 mL (vital capacity)

Enter the *uncorrected* vital capacity here:
__________ mL

2. Measure the amount of air exhaled in the first second by subtracting the milliliters corresponding to the exhalation line after 1 sec (3,400 mL, in fig. 8.11) from the milliliters of the inhalation peak at the moment of exhale (6,800 mL). Remember that here the distance between heavy vertical lines is passed in 1 sec. If the subject does not begin to exhale exactly on a vertical line, use a ruler to measure 3.2 cm horizontally from the start of exhalation. (This is the distance between vertical lines and is equivalent to 1 sec at this chart speed.)

#### *Example (from fig. 8.11)*

6,800 mL (maximum inhalation)
– 3,400 mL (exhalation line after 1 sec)
3,400 mL (amount exhaled in first second)

Enter the amount exhaled in the first second here:
__________ mL

3. Calculate the percentage of the vital capacity exhaled in the first second (the $FEV_1$).

#### *Example*

$$FEV_1 = \frac{3,400 \text{ mL (step 2)}}{4,800 \text{ mL (step 1)}} \times 100\% = 70.8\%$$

Enter the $FEV_1$ in the laboratory report. Refer to table 8.4, and enter the predicted percentage for the $FEV_1$ in the laboratory report.

**Table 8.4** Predicted Percentage of the Vital Capacity (VC) Exhaled during the First Second ($FEV_1$)

| Age | Predicted Percent VC ($FEV_1$) |
|---|---|
| 18–29 | 82–80 % |
| 30–39 | 78–77 % |
| 40–44 | 75.5 % |
| 45–49 | 74.5 % |
| 50–54 | 73.5 % |
| 55–64 | 72–70 % |

From *Archives of Environmental Health,* Volume 12, p. 146, 1966. Reprinted with permission of the Helen Dwight Reid Educational Foundation. Published by Heidref Publications, 1319 Eighteenth St., N.W., Washington, D.C. 20036-1802. 

# Laboratory Report 8.1

Name ____________________

Date ____________________

Section ____________________

## DATA FROM EXERCISE 8.1

**A. Measurement of Simple Lung Volumes and Capacities**

1. Enter your data (corrected to BTPS) under the *Measured* column, and enter your calculated *Percent Predicted,* in this table.

| Volume/Capacity | Measured | Predicted | Percent Predicted |
|---|---|---|---|
| Tidal volume | | 500 mL (avg. normal) | |
| Inspiratory capacity | | 2,800 mL (avg. normal) | |
| Expiratory reserve volume | | 1,200 mL (avg. normal) | |
| Vital capacity | | (from tables) | |
| Residual volume | not measured | (vital capacity × factor) | cannot calculate |
| Total lung capacity | not measured | (vital capacity × factor) | cannot calculate |

**B. Measurement of Forced Expiratory Volume**

1. Enter your measured and predicted $FEV_1$: measured = ______________%; predicted range = ______________%.

2. Compare your values to the normal range and enter your conclusions here.

# REVIEW ACTIVITIES FOR EXERCISE 8.1

## *Test Your Knowledge of Terms and Facts*

Identify these lung volumes and capacities:

1. Maximum amount of air that can be expired after a maximum inspiration: ______________________________.
2. Maximum amount of air that can be expired after a normal expiration: ______________________________.
3. Maximum amount of air that can be inspired after a normal expiration: ______________________________.
4. Amount of air left in the lungs after a maximum expiration: ______________________________.
5. Category of pulmonary disorders in which the alveoli are normal but there is an abnormally high resistance to air flow: ______________________________.
6. An example of a disorder in the category described in question 5 is ______________________________.
7. A pulmonary function test for the category of disorders named in question 5 is the ____________________ test.

## *Test Your Understanding of Concepts*

8. Calculate these values for the spirogram shown here. Be sure to correct your values to BTPS (use a BTPS value of 1.1).

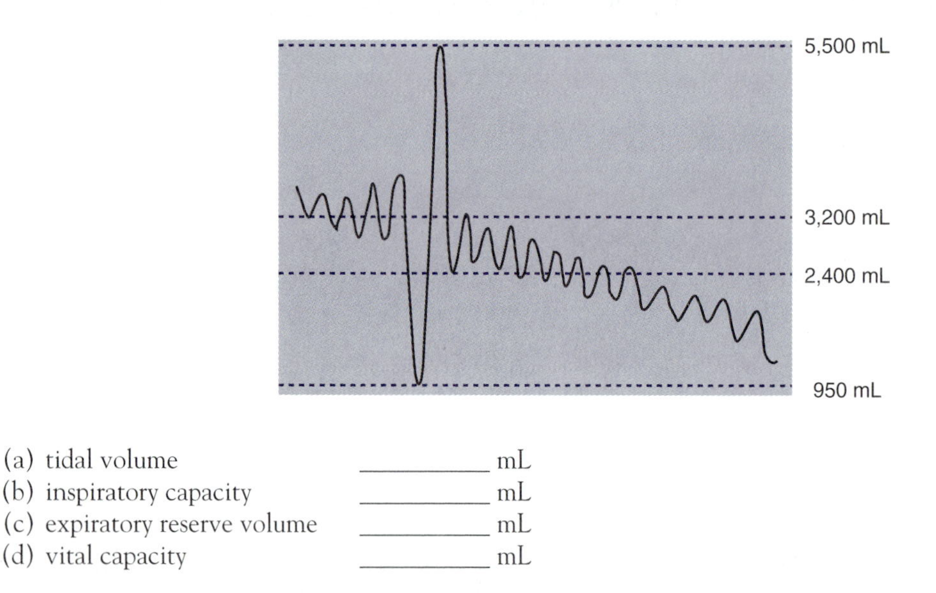


(a) tidal volume ____________ mL
(b) inspiratory capacity ____________ mL
(c) expiratory reserve volume ____________ mL
(d) vital capacity ____________ mL

9. Calculate the $FEV_1$ value for the spirogram shown here.

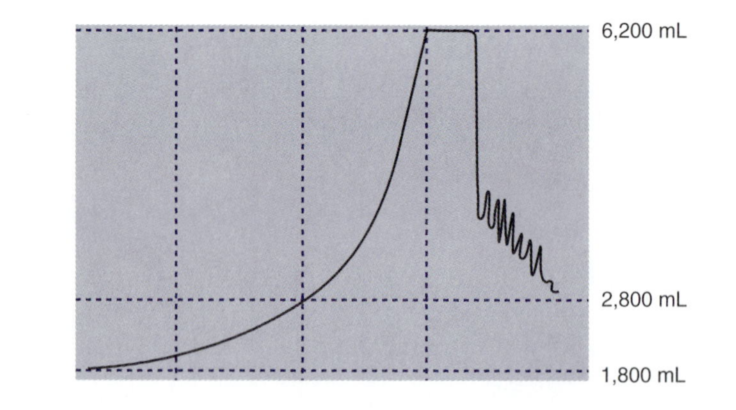

$FEV_1$ = ____________ %

10. Distinguish between obstructive and restrictive disorders. How does spirometry aid in their diagnosis?

## *Test Your Ability to Analyze and Apply Your Knowledge*

11. Does your chest expand because your lungs inflate, or do your lungs inflate because your chest expands? Explain.

12. When a person's lungs are ventilated by a tank of gas, the technique is known as *intermittent positive pressure breathing (IPPB)*. Analyzing this term, explain how this technique inflates the lungs for inspiration, and deflates the lungs for expiration.

# Effect of Exercise on the Respiratory System

EXERCISE 8.2

## Materials

1. Collins 9-L respirometer or Spirocomp program and equipment
2. Nose clamps and disposable mouthpieces
3. Alternatively, the Biopac system may be used with the set up for Biopac Lesson 12.
4. Alternatively, the iWorx system may be used with the setup as described in iWorx Exercise 10. Once set up, the measurements for this exercise can be obtained.

## Textbook Correlations*

Before performing this exercise, you should study the introductory material presented here. Further information relating to this exercise can be found in these pages of *Human Physiology,* eighth edition, by Stuart I. Fox:

- *Gas Exchange in the Lungs.* Chapter 16, pp. 493–499.
- *Regulation of Breathing.* Chapter 16, pp. 499–504.
- *Ventilation During Exercise.* Chapter 16, pp. 513–514.

Total minute volume is the product of the rate and depth of breathing per minute. Oxygen consumption per minute is a measure of the metabolic rate. Total minute volume is adjusted by physiological mechanisms to compensate for changes in metabolic rate.

## OBJECTIVES

1. Define the term *total minute volume* and explain how this measurement is obtained.
2. Describe how the rate of oxygen consumption is measured and explain how it is used as a measure of the metabolic rate.
3. Describe the relationship between the total minute volume and the rate of oxygen consumption; and explain how, and why, these measurements are changed during exercise.
4. Explain why oxygen consumption and total minute volume remain elevated after exercise has ceased.

The volume of air exhaled in a minute of unforced breathing is known as the **total minute volume** and is equal to the product of tidal volume (milliliters per breath) and the frequency of breathing (breaths per minute). Only about two-thirds of this volume actually reaches the alveoli (this is known as the *alveolar minute volume*). The remaining one-third stays within the dead space of the lungs and is not involved in gas exchange.

**Gas exchange** occurs in the alveoli of the lungs. Oxygen diffuses from the alveolar air into the blood, while carbon dioxide diffuses from the blood into the alveolar air. Thus, the blood leaving the lungs is rich in oxygen and reduced in carbon dioxide. When the blood reaches the tissue capillaries, oxygen diffuses from the blood into the tissues, where it can be used by the cells in aerobic respiration. Meanwhile carbon dioxide, formed as a waste product of aerobic respiration, diffuses from the tissues into the capillary blood.

Since oxygen is consumed by the body's cells in aerobic respiration, the initial volume of air within the oxygen bell of the respirometer decreases as the subject breathes through the mouthpiece. (In addition, the exhaled carbon dioxide is removed by soda lime within the respirometer.) Thus, oxygen consumption results in the removal of air from the bell and on a spirogram is seen as an upward slope of the tidal volume tracing. The amount of oxygen consumed per minute can be calculated as the difference between milliliter levels before and after 1 minute of resting ventilation (fig. 8.12).

*Multimedia Correlations (also see Appendix 3)
- Intelitool: Spirocomp
- A.D.A.M. InterActive Physiology (Respiratory System): Control of Respiration

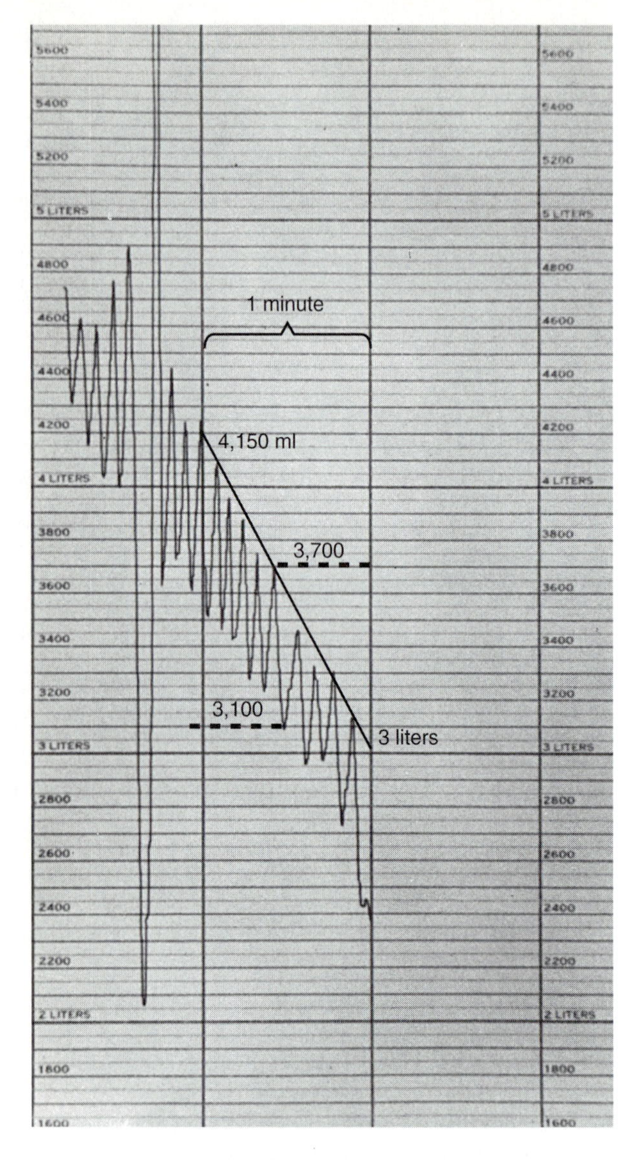


**Figure 8.12 Spirogram showing tidal volume measurements over a 1 minute interval at rest.** The rising slope of the tidal volume measurements indicates oxygen consumption.

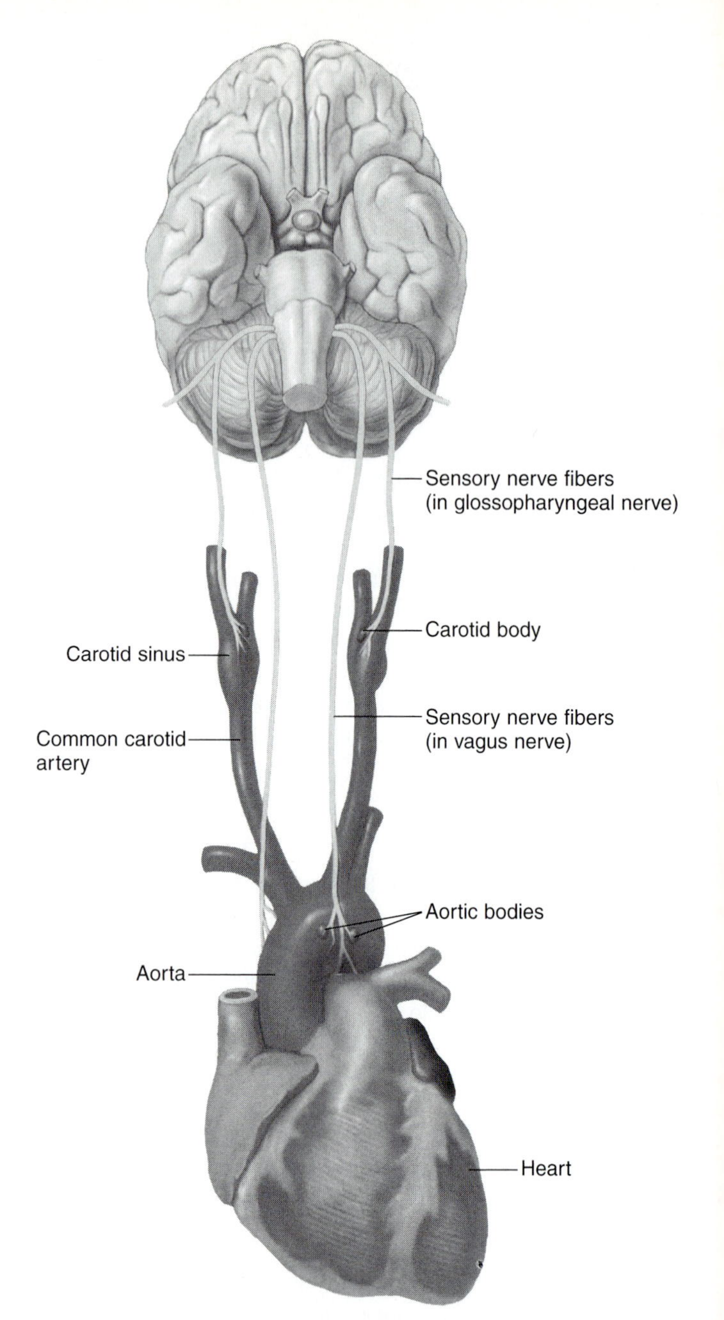


**Figure 8.13 Sensory input from the aortic and coratid bodies.** The peripheral chemoreceptors (aortic and carotid bodies) regulate the brain stem respiratory centers by means of sensory nerve stimulation.

**(For a full-color version of this figure, see fig. 16.26 in *Human Physiology*, eighth edition, by Stuart I. Fox.)**

The rate at which oxygen is consumed and carbon dioxide is produced by the body cells during aerobic respiration is related to the *metabolic rate* of the person. When the individual is relaxed and comfortable and has not eaten for 12 to 15 hours, the metabolic rate is lowest. This rate is referred to as the **basal metabolic rate (BMR).** Under these conditions, the metabolic rate is set primarily by the activity of the thyroid gland; in the past, measurements of BMR were used to assess thyroid function.

When a person exercises, however, the metabolic rate increases greatly (the metabolism of muscles can increase as much as sixtyfold during strenuous exercise). As a result, oxygen is consumed and carbon dioxide produced at much more rapid rates than during resting conditions. The respiratory system keeps pace with this increased demand by increasing the total minute volume.

The **respiratory control center** in the *medulla oblongata* contains inspiratory and expiratory neurons that regulate breathing via motor nerves to the respiratory

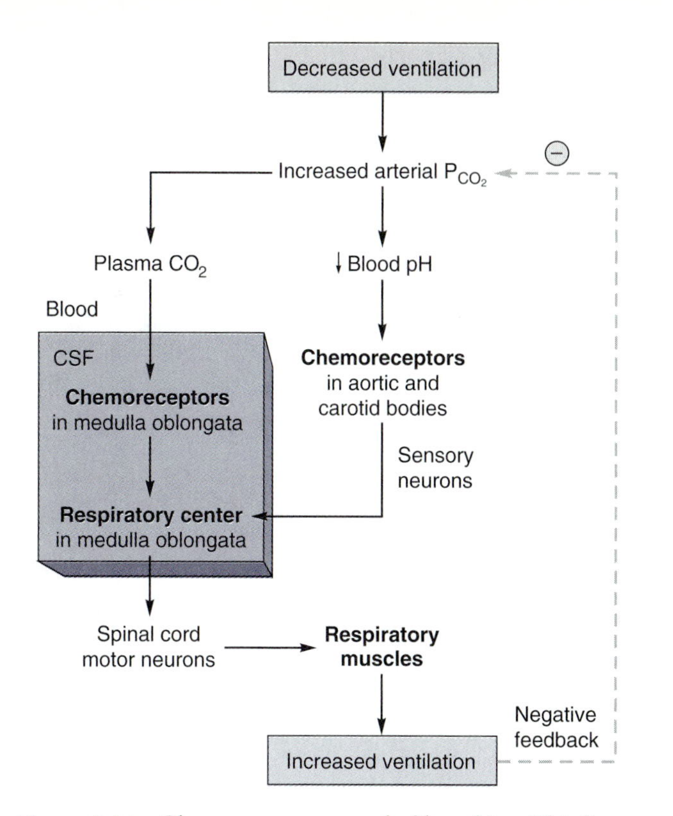


**Figure 8.14** **Chemoreceptor control of breathing.** This figure depicts the negative feedback control of ventilation through changes in blood $P_{CO_2}$ and pH. The blood-brain barrier, represented by the box, allows $CO_2$ to pass into the cerebrospinal fluid but prevents the passage of H+.

**(For a full-color version of this figure, see fig. 16.29 in *Human Physiology,* eighth edition, by Stuart I. Fox.)**

muscles. The respiratory control center's activity, however, is influenced by neurons in the pons, and also by chemical changes in the blood. Specifically, an increase in plasma $CO_2$, acting through its lowering of blood pH and cerebrospinal fluid pH, stimulates **chemoreceptors** (fig. 8.13). The *peripheral chemoreceptors* are located in the **aortic** and **carotid bodies;** the *central chemoreceptors* are in the medulla oblongata. These chemoreceptors modify the activity of the respiratory control center in the medulla, such that a rise in plasma $CO_2$ (and consequent fall in pH) will stimulate breathing (fig. 8.14).

The increase in total minute volume that occurs during exercise may be due in part to an increase in $CO_2$ production, although concentrations of arterial $CO_2$ during exercise are not usually increased. Anticipation and excitement coming from conscious brain areas and sensory feedback from the exercising muscles may also contribute to the **hyperpnea** (increased breathing) of exercise.

Oxygen consumption and the total minute volume remain elevated immediately after exercise. This extra oxygen consumption (over resting levels) following exercise is called the **oxygen debt.** The extra oxygen is used to oxidize lactic acid produced as a result of anaerobic respiration in the exercising muscles and to support an increased metabolism within the warmed muscles.

**Hypoventilation** occurs when the alveoli are inadequately ventilated (alveolar minute volume is reduced). Due to a reduction in the total minute volume where either the tidal volume or the frequency of breathing is depressed, hypoventilation can also result from a pathological increase in dead air space in the lungs. The latter may be caused by any condition that affects lung tissue (such as emphysema) or by inadequate blood flow to well-ventilated alveoli. Such ventilated alveoli that are lacking the appropriate perfusion of capillary blood (producing an abnormal *ventilation/perfusion ratio*) are incapable of fully oxygenating blood. As a result of hypoventilation, there is inadequate elimination of carbon dioxide from the blood. Since the plasma carbon dioxide levels are directly affected by ventilation, hypoventilation may be operationally defined as an *abnormally increased plasma carbon dioxide level* (above 40 Torr, or mm Hg).

## PROCEDURE

1. Set the Collins respirometer to the slow speed of 32 mm/min. (At this speed the distance between two heavy vertical lines is traversed in 1 minute.) To avoid air leakage, position the mouthpiece securely in the mouth with the lips tightly sealed around it, and clamp the nostrils closed.
2. Under resting conditions, breathe normally into the respirometer for 1 minute (that is, perform the procedure for measuring tidal volume—see exercise 8.1).

**Note:** *The tidal volume measurements must have an upward slope, indicating that the oxygen being consumed is from the air trapped in the oxygen bell. If the slope is downward, air is leaking into the system, usually through the corners of the mouth or the nose. In this event, reposition the mouthpiece, check the nose clamp, and begin the measurements again.*

3. After 1 minute, stop the respirometer, remove the chart from the kymograph drum and determine the frequency of ventilation (number of breaths per minute) and the tidal volume (corrected to BTPS).

### *Example (from fig. 8.12)*

$$\begin{array}{rl} 3{,}700 \text{ mL} & \text{(inhalation peak of tidal volume)} \\ -\ 3{,}100 \text{ mL} & \text{(exhalation trough of tidal volume)} \\ \hline 600 \text{ mL} & (\textit{uncorrected} \text{ tidal volume}) \end{array}$$

$$\begin{array}{rl} 600 \text{ mL} & \\ \times\ 1.1 & \text{(BTPS factor)} \\ \hline 660 \text{ mL} & (\textit{corrected} \text{ tidal volume}) \end{array}$$

From figure 8.12, frequency = 10 breaths/min

Enter the subject's corrected tidal volume here:
________ mL
Enter the subject's frequency of ventilation here:
________ breaths/min

| The average ventilation frequency is 14 breaths/min. |
|---|

4. Determine the subject's total minute volume at rest by multiplying the frequency of ventilation by the tidal volume.
   Enter this value in the data table in your laboratory report.

| The average total minute volume is 6,750 mL/min. |
|---|

### *Example (from fig. 8.12)*

660 mL/breath × 10 breaths/min
= 6,600 mL/min total minute volume

5. Use a straight edge to draw a line that touches either the peaks or the troughs of the tidal volume measurements. Determine the oxygen consumption per minute by subtracting the milliliters where this straight line intersects the heavy vertical chart line at the beginning of one minute from the milliliters where the straight line intersects the next heavy vertical line at the end of one minute.

### *Example (from fig. 8.12)*

Using a line that averages the peaks:

$$\begin{array}{rl} 4{,}150 \text{ mL} & \text{(at end of 1 minute)} \\ -\ 3{,}000 \text{ mL} & \text{(at beginning of 1 minute)} \\ \hline 1{,}150 \text{ mL} & \text{(oxygen consumption)} \end{array}$$

Enter the resting oxygen consumption in the data table in your laboratory report.

6. Now, have the subject perform light exercise, such as 5 to 10 jumping jacks, and then repeat the respirometer measurements and data calculation described in steps 1–5. Alternatively, if a bicycle ergometer is available, the total minute volume may be determined while the student pedals lightly for one minute.

**Caution:** *If breathing into the respirometer becomes difficult after exercise, the subject should stop the procedure. Results obtained in less than a minute can then be extrapolated to 1 minute. Alternatively, the bell can be filled with 100% oxygen to prevent the possible occurrence of hypoxia.*

Enter the corrected tidal volume after exercise in this space:
__________ mL

Enter the frequency of ventilation after exercise in this space:
__________ breaths/min

Enter the total minute volume and the oxygen consumption per minute after exercise in the data table of the laboratory report.

7. Calculate the percent increase after exercise for total minute volume and for oxygen consumption. This is the difference between the exercise and resting measurements, divided by the resting measurement and multiplied by 100%. Enter these values in the data table in your laboratory report.

# Laboratory Report 8.2

Name ______________________________

Date ______________________________

Section ______________________________

## DATA FOR EXERCISE 8.2

1. Enter the total minute volume and rate of oxygen consumption during rest and after exercise in this table.
2. Calculate and enter the percent increases in your measurements after exercise, and enter these values in this table.

| Measurement | Resting | During or After Exercise | Percent Increase |
|---|---|---|---|
| Total minute volume | | | |
| Oxygen consumption | | | |

## REVIEW ACTIVITIES FOR EXERCISE 8.2

### *Test Your Knowledge of Terms and Facts*

1. The net diffusion of oxygen and carbon dioxide in opposite directions across the wall of alveoli is known as ______________________________.
2. Define the term *tidal volume*. ______________________________
3. The total minute volume is obtained by multiplying the ______________________________ times the ______________________________.
4. The peripheral chemoreceptors include the ____________________ and the ____________________.
   ______________________________
5. The stimulus that directly activates the peripheral chemoreceptors is ______________________________.
6. Define the term *hyperpnea*. ______________________________
7. The term *hypoventilation* may be operationally defined as ______________________________.

### *Test Your Understanding of Concepts*

8. Why did the rate of oxygen consumption increase during exercise? What happened to the plasma levels of carbon dioxide during exercise? Explain.

9. What happened to the total minute volume during exercise? What happened to the total minute volume right after exercise was completed? What physiological mechanisms might account for these changes?

10. Define the oxygen debt, and explain the functions of the oxygen debt.

## *Test Your Ability to Analyze and Apply Your Knowledge*

11. High total minute volume during mild to moderate exercise is more accurately termed hyperpnea than hyperventilation. Distinguish between these two terms and explain why this statement is true.

12. How might the plasma $CO_2$ levels differ during the performance of a particular exercise in a trained versus an untrained person? Explain your answer.

# Oxyhemoglobin Saturation

EXERCISE 8.3

## Materials

1. Graduated cylinder and a 1-cc syringe
2. Test tube and distilled water
3. Colorimeter and cuvettes
4. Sodium dithionite (hydrosulfite), 1.0 g per 100 mL
5. Alcohol swabs and lancets for preparing fingertip blood. Alternatively, dog or cat blood (obtained from a veterinarian) may be used.

The iron atoms within the heme groups of hemoglobin may be free (unbound), or they may be bonded to oxygen or carbon monoxide gas. Each of these different forms of hemoglobin has a slightly different color, which allows the percentage of each type in a mixture to be measured using the absorption spectrum of each hemoglobin type. Measurement of the percent oxyhemoglobin is used clinically to assess lung function and the capacity of blood to transport oxygen.

## OBJECTIVES

1. Define the term percent saturation.
2. Explain the clinical significance of the percent oxyhemoglobin measurement.
3. Explain the clinical significance of the percent carboxyhemoglobin measurement.
4. Describe how an absorption spectrum is obtained, and explain how the absorption spectra of the different forms of hemoglobin are used to determine the percent saturation.

### Textbook Correlations*

Before performing this exercise, you should study the introductory material presented here. Further information relating to this exercise can be found in these pages of *Human Physiology,* eighth edition, by Stuart I. Fox:

- *Partial Pressure of Gases in Blood* Chapter 16, pp. 494–496.
- *Hemoglobin and Oxygen Transport,* Chapter 16, pp. 504–510.

The ability of the blood to carry oxygen depends on (1) ventilation; (2) gas exchange across the alveoli of the lungs; (3) the red blood cell count and hemoglobin concentration; and (3) the chemical form of the hemoglobin.

There are two chemical forms of normal hemoglobin. Normal hemoglobin without oxygen is called **deoxyhemoglobin;** after it binds to oxygen, it is called **oxyhemoglobin.** If a person suffers from *carbon monoxide poisoning,* however, the abnormal hemoglobin called **carboxyhemoglobin** (hemoglobin bound to carbon monoxide) causes the blood to carry a lower amount of oxygen. This is because the carbon monoxide displaces oxygen and binds to hemoglobin with a higher affinity than does oxygen. Therefore, when health professionals need to determine the oxygen carrying capacity of the blood, they need to learn the relative proportion of each hemoglobin type as well as the total hemoglobin concentration.

The relative proportion of each type of hemoglobin is given as its **percent saturation.** The percent oxyhemoglobin saturation, for example, is the proportion of hemoglobin bound to oxygen. Normally, this value is approximately 97% in arterial blood and 75% in venous blood.

$$\text{\% oxyhemoglobin saturation} = \frac{\text{oxyhemoglobin}}{\text{total blood hemoglobin}} \times 100$$

The determination of percent oxyhemoglobin saturation is a very sensitive means of assessing the effectiveness of pulmonary function. When pulmonary function and blood hemoglobin are normal, the arterial blood has a percent oxyhemoglobin saturation of about 97%. Even when pulmonary function is normal, the percent oxyhemoglobin saturation can be decreased by carbon monoxide poisoning. When a person's blood has a carboxyhemoglobin

*Multimedia Correlations (also see Appendix 3)

- A.D.A.M. *InterActive Physiology* (Respiratory System): Gas Transport; Gas Exchange; Control of Repiration

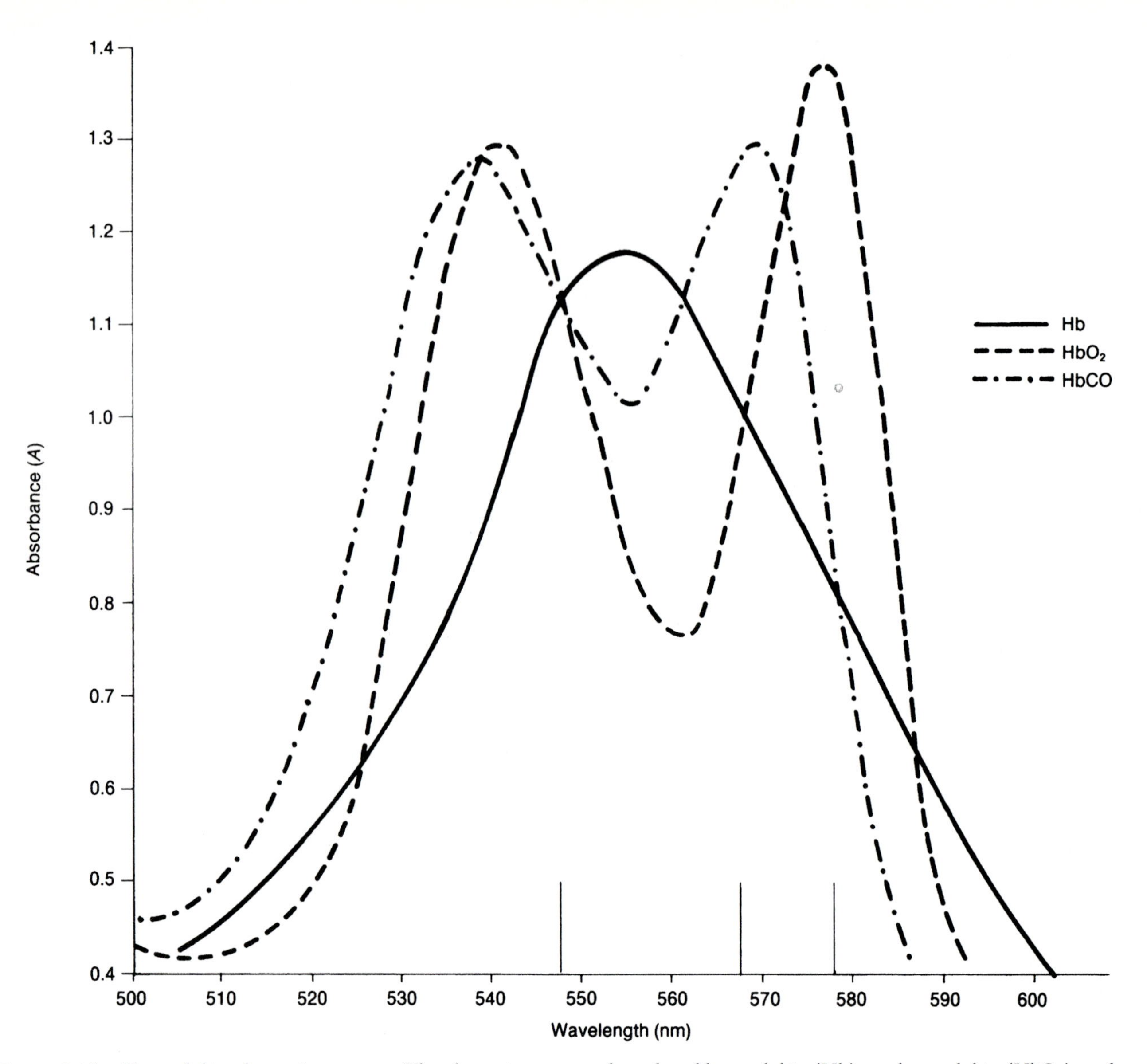


**Figure 8.15** **Hemoglobin absorption spectra.** The absorption spectra for reduced hemoglobin (Hb), oxyhemoglobin ($HbO_2$), and carboxyhemoglobin (HbCO) are shown. Source: *Instrumentation Lab, Inc.*

saturation of 6.9%, for example (obtained in one study using cigarette-smoking New York taxicab drivers), the percent oxyhemoglobin saturation is decreased accordingly. Another abnormal form of hemoglobin is **methemoglobin,** which is oxidized hemoglobin that lacks the electron needed to bond with oxygen, and therefore cannot participate in oxygen transport. Notice that these impairments in oxygen transport cannot be detected by standard measurement of red blood cell count, hematocrit, or total blood hemoglobin.

The percent saturation of the different types of hemoglobins is measured by comparing the absorption spectrum of an unknown sample of blood with the absorption spectra of pure oxyhemoglobin, pure reduced hemoglobin, and pure carboxyhemoglobin. Since these hemoglobins have different colors, they absorb different amounts of light at each wavelength. A graph of absorbance versus wavelength (where the concentration is constant) is called an **absorption spectrum** (fig. 8.15).

The absorption spectrum of an unknown sample of blood will display some combination of these three absorption spectra, since the blood contains all three types of hemoglobins. The relative contribution of each hemoglobin type to the absorption spectrum is proportional to the relative amount of each type in the blood. This analysis, which is obviously complex, is usually performed by a laboratory instrument specifically manufactured for this purpose.

In this exercise, you will construct an absorption spectrum for 100% oxyhemoglobin and 100% reduced hemoglobin. By bubbling air into a flask containing blood until the blood is in equilibrium with the air, 100% oxyhemoglobin may be obtained. This process essentially duplicates the process that occurs in the capillaries surrounding the lung alveoli. A simpler (but less accurate) method is to obtain a sample of blood from the fingertip, which has a high percent oxyhemoglobin saturation. The sample of 100% reduced hemoglobin may be obtained by adding sodium hydrosulfite ($Na_2S_2O_4$) to a second sample of blood. The sodium hydrosulfite (also called sodium dithionite) removes oxygen from oxyhemoglobin.

A person may have normal hemoglobin, hematocrit, and red blood cell counts and still not be delivering adequate amounts of oxygen to the body cells. This may be due to inadequate lung function, resulting in poor oxygenation of the blood, or due to an abnormally high blood concentration of carboxyhemoglobin or methemoglobin. In this event, the hemoglobin cannot become fully saturated with oxygen, and therefore the percent oxyhemoglobin saturation of arterial blood may drop below normal. These effects are similar to the effects of anemia, since in both cases the amount of oxygen carried by the blood is reduced. In this sense, carbon monoxide poisoning may be thought of as a functional anemia. The hemoglobin and red blood cell counts are normal, but the red blood cells are not transporting the normal amount of oxygen.

## PROCEDURE

1. Add 8.0 mL of distilled water to a test tube. Obtain a large drop of blood by wiping the fingertip with 70% alcohol, let dry, and puncturing it with a sterile lancet. Then, mix this blood with the distilled water by inverting the test tube over the punctured finger.
2. Transfer half the contents of the test tube (4.0 mL) to a second tube.
3. Add 0.20 mL of 1.0% sodium dithionite solution to the second test tube and mix thoroughly.

**Note:** *The dithionite solution should be freshly prepared just prior to use, and the absorbance values of the two tubes should be determined within 5 minutes of the time the dithionite is added to the second tube.*

4. Transfer the two solutions to two cuvettes. Fill a third cuvette with distilled water and use it as a blank to standardize the spectrophotometer at 500 nm. (See exercise 2.1 for a description of the spectrophotometer, standardizing procedures, and Beer's law.)
5. Record the absorbances of solutions 1 and 2. Continue using the blank to standardize the spectrophotometer at each of the successive wavelengths from 510 to 600 nm, and then record the absorbances of the two solutions in the laboratory report.
6. Graph the absorption spectra of oxyhemoglobin and reduced hemoglobin on the graph provided in the laboratory report.

# Laboratory Report 8.3

Name ______________________

Date ______________________

Section ______________________

## DATA FOR EXERCISE 8.3

| Wavelength | Oxyhemoglobin Absorbance | Reduced Hemoglobin Absorbance | Wavelength | Oxyhemoglobin Absorbance | Reduced Hemoglobin Absorbance |
|---|---|---|---|---|---|
| 500 nm | | | 560 nm | | |
| 510 nm | | | 570 nm | | |
| 520 nm | | | 580 nm | | |
| 530 nm | | | 590 nm | | |
| 540 nm | | | 600 nm | | |
| 550 nm | | | | | |

# REVIEW ACTIVITIES FOR EXERCISE 8.3

## *Test Your Knowledge of Terms and Facts*

1. In the lungs, normal hemoglobin without oxygen, or ______________________________,
   binds to oxygen to become ______________________________.
2. The type of hemoglobin bound to carbon monoxide: ______________________________.
3. The type of hemoglobin where the heme iron is in the oxidized $Fe^{2+}$ state: ______________________________.
4. What is the normal percent oxyhemoglobin saturation of arterial blood? __________%
5. A graph of the absorbance of light as a function of the wavelength of light is called a(n) ____________________
   ______________________________.
6. The different forms of hemoglobin can be distinguished visually because they have different ____________________
   ______________________________.

## *Test Your Understanding of Concepts*

7. What blood measurement would be abnormally increased in a person with carbon monoxide poisoning? What are the dangers of carbon monoxide poisoning?

8. In what way are carbon monoxide poisoning and anemia different? In what way are they similar? Explain.

## *Test Your Ability to Analyze and Apply Your Knowledge*

9. Suppose a person's percent oxyhemoglobin saturation in the venous blood was 75% at rest, but decreased to 35% during moderate exercise. (a) Explain why the venous percent oxyhemoglobin saturation decreased during the exercise; and (b) predict how the arterial percent oxyhemoglobin saturation might change during exercise. Explain your answer.

10. The percent oxyhemoglobin saturation is measured in babies under treatment for *respiratory distress syndrome (RDS)* and in patients under general anesthesia. What information would these measurements provide under these conditions? Explain.

the red blood cells. This reaction is catalyzed by an enzyme called *carbonic anhydrase* (fig. 8.16).

$$CO_2 + H_2O \xrightarrow{\text{carbonic anhydrase}} H_2CO_3$$

Some of the carbonic acid formed can immediately dissociate to yield $H^+$ and *bicarbonate ion* ($HCO_3^-$). The $H^+$ derived from carbonic acid and other acids in the blood gives normal arterial blood a pH of 7.40 ± 0.05 (table 8.6 and fig. 8.16).

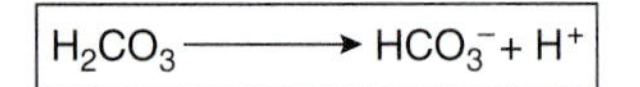

$$H_2CO_3 \longrightarrow HCO_3^- + H^+$$

## A. Ability of Buffers to Stabilize the pH of Solutions

Plasma has a particular concentration of bicarbonate as a result of the dissociation of carbonic acid. Bicarbonate serves as the major *buffer* of the blood, helping to stabilize the pH of plasma despite the continuous influx of $H^+$ from molecules of lactic acid, fatty acids, ketone bodies, and other metabolic products. The $H^+$ released by these acids is prevented from lowering the blood pH because it is combined with bicarbonate. Although a new acid molecule (carbonic acid) is formed, this reaction prevents a rise in the free $H^+$ concentration (fig. 8.16).

$$H^+ + HCO_3^- \longrightarrow H_2CO_3$$

Carbonic acid formed in this way can provide a source of new $H^+$ if the blood pH should begin to rise (from a loss of blood $H^+$) beyond normal levels. The carbonic acid/bicarbonate buffer system helps to stabilize the blood pH under normal conditions. Disease states, however, may cause the blood pH to fall below 7.35 or to rise above 7.45. These conditions are called *acidosis* and *alkalosis*, respectively.

Normally, the rate of ventilation is matched to the rate of $CO_2$ production by the tissues, so that the carbonic acid, bicarbonate, and $H^+$ concentrations in the blood remain within the normal range. If **hypoventilation** occurs, however, the carbonic acid levels will rise above normal and the pH will fall below 7.35. This condition is called **respiratory acidosis** (table 8.6). **Hyperventilation,** conversely, causes an abnormal decrease in carbonic acid and a corresponding rise in blood pH. This condition is called **respiratory alkalosis.** Thus, respiratory acidosis or alkalosis occurs when the blood $CO_2$ level (as measured by its partial pressure or $P_{CO_2}$, in millimeters of mercury) is different from the normal value (40 mm Hg) as a result of abnormal breathing patterns (fig. 8.17).

### PROCEDURE

1. Allow the pH meter to warm up by setting the selector switch to the *standby* position. Be sure that the pH electrodes are immersed in buffer and are not allowed to dry. Verify that the temperature selector switch is set at the current room temperature.
2. Turn the selector switch to *pH* and take a reading of the buffer. Use the calibration knob to set the pH meter to the correct pH of the buffer (7.000). Now, turn the selector switch back to the *standby* position

**Table 8.5** The pH Scale

| | $H^+$ Concentration (Molar) | pH | $OH^-$ Concentration (Molar) |
|---|---|---|---|
| | 1.0 | 0 | $10^{-14}$ |
| | 0.1 | 1 | $10^{-13}$ |
| | 0.01 | 2 | $10^{-12}$ |
| Acids | 0.001 | 3 | $10^{-11}$ |
| | 0.0001 | 4 | $10^{-10}$ |
| | $10^{-5}$ | 5 | $10^{-9}$ |
| | $10^{-6}$ | 6 | $10^{-8}$ |
| Neutral | $10^{-7}$ | 7 | $10^{-7}$ |
| | $10^{-8}$ | 8 | $10^{-6}$ |
| | $10^{-9}$ | 9 | $10^{-5}$ |
| | $10^{-10}$ | 10 | 0.0001 |
| Bases | $10^{-11}$ | 11 | 0.001 |
| | $10^{-12}$ | 12 | 0.01 |
| | $10^{-13}$ | 13 | 0.1 |
| | $10^{-14}$ | 14 | 1.0 |

**Table 8.6** The Effect of Respiration on Blood pH

| $P_{CO_2}$ (mm Hg) | $H_2CO_3$ (mEq/L)[1] | $HCO_3^-$ (mEq/L)[1] | $HCO_3^-/H_2CO_3$ Ratio | Blood pH | Condition |
|---|---|---|---|---|---|
| 20 | 0.6 | 24 | 40/1 | 7.70 | Respiratory alkalosis |
| 30 | 0.9 | 24 | 26.7/1 | 7.53 | Respiratory alkalosis |
| 40 | 1.2 | 24 | 20/1 | 7.40 | Normal |
| 50 | 1.5 | 24 | 16/1 | 7.30 | Respiratory acidosis |
| 60 | 1.8 | 24 | 13.3/1 | 7.22 | Respiratory acidosis |

1. Ion concentrations are commonly measured in milliequivalents (mEq) per liter. This measurement is equal to the millimolar concentration of the ion multiplied by its number of charges.

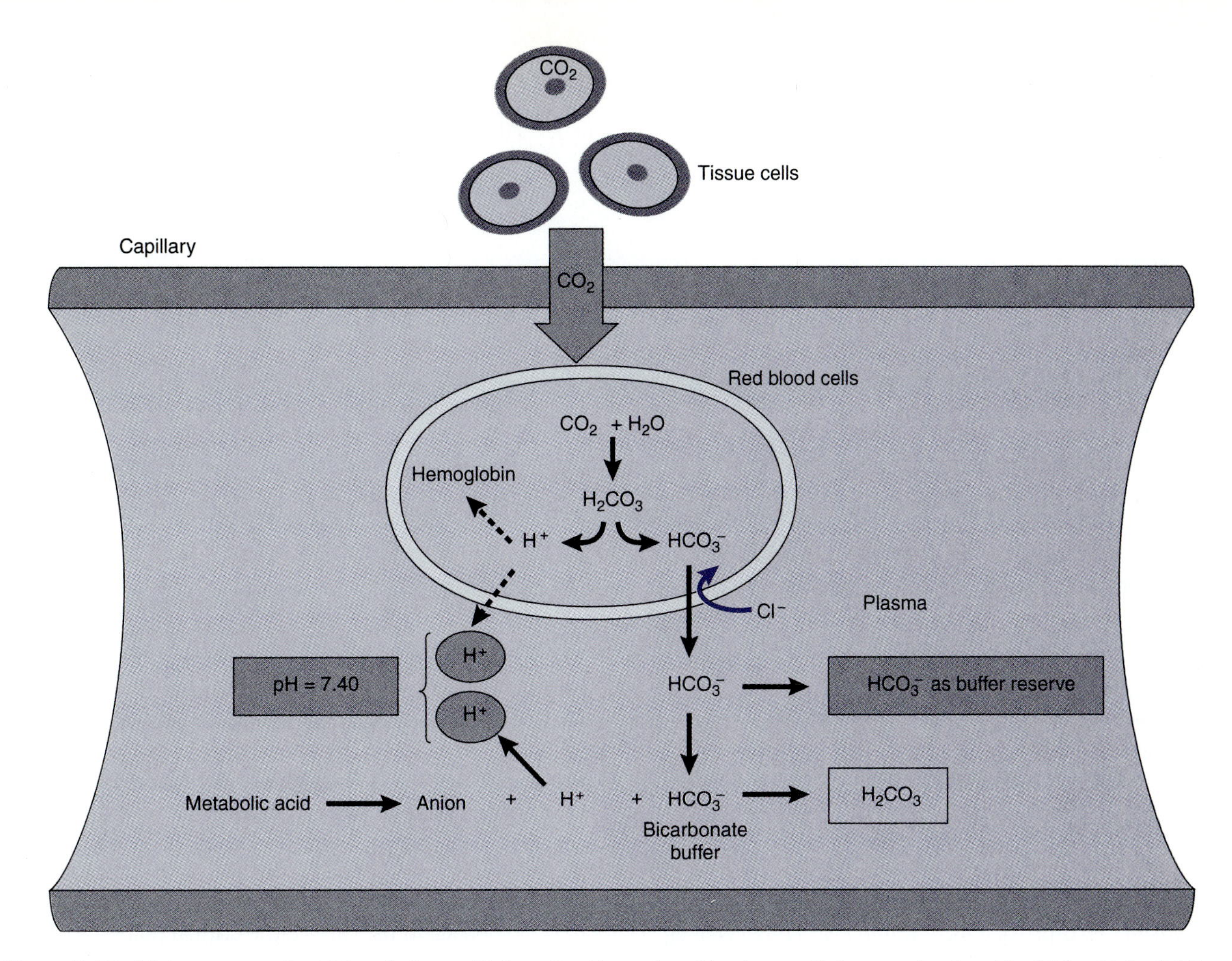


**Figure 8.16 Maintentance of acid-base balance.** Carbon dioxide produced by tissue cells forms carbonic acid, which adds both $H^+$ and bicarbonate ($HCO_3^-$) to the plasma. The bicarbonate released from red blood cells buffers the $H^+$ produced by ionization of metabolic (nonvolatile)acids, such as lactic acid and ketone bodies.

**(For a full-color version of this figure, see fig. 16.49 in *Human Physiology,* eighth edition, by Stuart I. Fox.)**

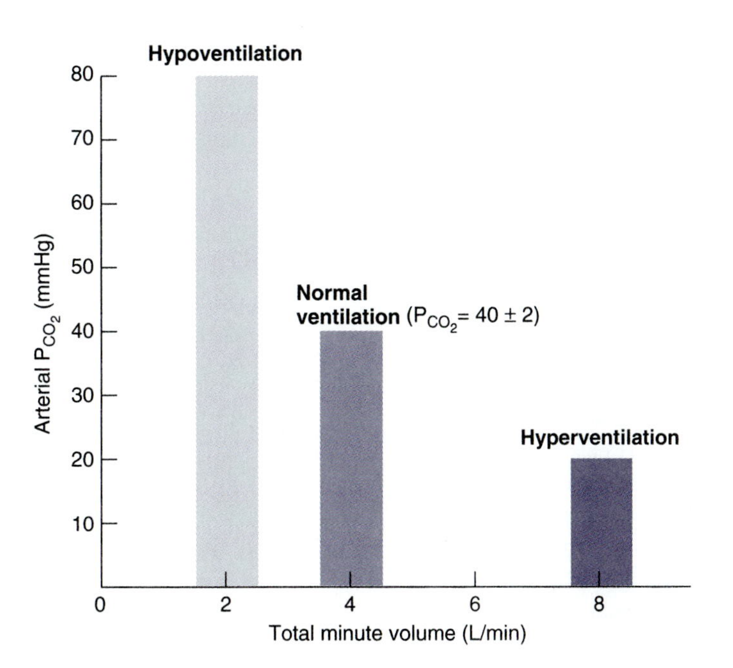


**Figure 8.17 The relationship between total minute volume and arterial $P_{CO_2}$.** Notice that these are inversely related when the total minute volume increases by a factor of 2, the arterial $P_{CO_2}$ decreases by half. The total minute volume measures breathing, and is equal to the amount of air in each breath (the tidal volume) multiplied by the number of breaths per minute. The $P_{CO_2}$ measures the $CO_2$ concentration of arterial blood plasma.

**(For a full-color version of this figure, see fig. 16.28 in *Human Physiology,* eighth edition, by Stuart I. Fox.)**

and transfer the pH electrodes to a beaker of distilled water. Turn the selector switch to *pH* and record the pH of distilled water. Return the selector switch to the *standby* position and the electrodes back to the buffer.

3. Add one drop of concentrated hydrochloric acid (HCl) to the beaker of distilled water and mix thoroughly. Transfer the electrodes to this solution, turn the selector switch to the *pH* position, and record the pH of the solution in your laboratory report.

**Note:** *After recording the pH of a solution, always turn the selector switch to standby and use a squeeze bottle of distilled water to rinse the electrodes thoroughly. Wipe the electrodes with lint-free paper and return them to the buffer solution. Check the pH of the buffer solution after the cleaning procedure to be sure you have adequately cleaned the electrodes.*

4. Add one drop of concentrated NaOH to a fresh beaker of distilled water and record the pH of the water before and after adding the NaOH.
5. Add one drop of concentrated HCl to a beaker containing standard buffer solution (pH = 7.000).
6. Add one drop of concentrated NaOH to a fresh beaker of standard buffer solution (pH = 7.000).
7. Add two drops of concentrated HCl to a beaker of fresh standard buffer solution (pH = 7.000).
8. Add two drops of concentrated NaOH to a beaker of fresh standard buffer solution (pH = 7.000).

## B. Effect of Exercise on the Rate of $CO_2$ Production

Increased muscle metabolism during exercise results in an increase in $CO_2$ production. Despite this, the $CO_2$ levels and pH of arterial blood do not normally change significantly during exercise. This is because the increased rate of $CO_2$ production is matched by an increase in the rate of its elimination through ventilation. The mechanisms responsible for exercise *hyperpnea* (increased breathing) are complex and incompletely understood.

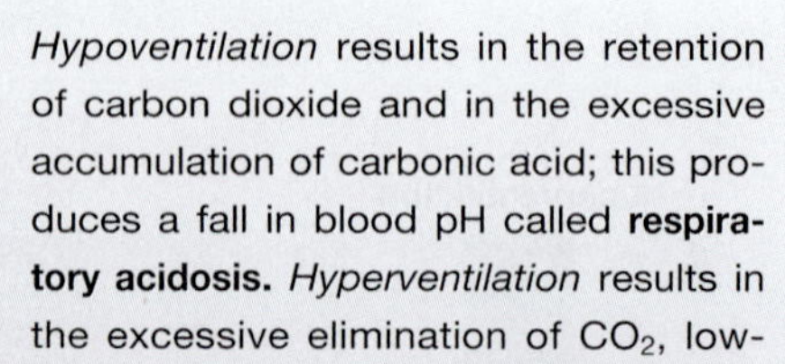

*Hypoventilation* results in the retention of carbon dioxide and in the excessive accumulation of carbonic acid; this produces a fall in blood pH called **respiratory acidosis.** *Hyperventilation* results in the excessive elimination of $CO_2$, lowered carbonic acid, and a rise in pH, causing **respiratory alkalosis.** This differs from the normal hyperpnea (increased total minute volume) that occurs during exercise, where increased respiration matches increased $CO_2$ production so that the arterial $CO_2$ levels and pH remain in the normal range.

### PROCEDURE

1. Fill a beaker with 200 mL of distilled water and add 5.0 mL of 0.10N NaOH and a few drops of phenolphthalein indicator. This indicator is pink in alkaline solutions and clear in neutral or acidic solutions. Divide this solution into two beakers.
2. While sitting quietly, exhale through a glass tube or straw (or double straws) into the solution in the first beaker. Carefully record the time required to turn the solution from pink to clear in your laboratory report.
3. Exercise vigorously for 2 to 5 minutes by running up and down stairs or by doing jumping jacks. Exhale through a glass tube or straw (or double straws) into the second beaker, and again record the time it takes to clear the pink solution.

## C. Role of Carbon Dioxide in the Regulation of Ventilation

The carbon dioxide concentration of the blood reflects a balance between the rate of its production (by aerobic cell respiration) and the rate of its elimination through the lungs. When a person consciously holds his or her breath for a sufficiently long time, the carbon dioxide level rises (and the pH falls) to such an extent that reflex breathing occurs. On the other hand, during hyperventilation, the pH of the blood can rise to the point that the desire to breathe is eliminated until the amount of carbon dioxide in the blood again rises above the critical point.

### PROCEDURE

1. Count the number of breaths you take in 1 minute of relaxed, unforced breathing. Enter this number in your laboratory report.
2. Force yourself to hyperventilate for about 10 seconds; stop if you begin to feel dizzy.
3. Immediately after hyperventilation, count the number of breaths you take in 1 minute of relaxed, unforced breathing.

# Laboratory Report 8.4

Name ______________________

Date ______________________

Section ______________________

## DATA FROM EXERCISE 8.4

### A. Ability of Buffers to Stabilize the pH of Solutions

1. Enter your data in these spaces:

pH of distilled water: ________

pH of water + 1 drop HCl: ________

pH of water + 1 drop NaOH: ________

pH of buffer: 7.000

pH of buffer +1 drop HCl: ________

pH of buffer +1 drop NaOH: ________

pH of buffer +3 drops HCl:________

pH of buffer +3 drops NaOH: ________

2. Does your data support the statement that "buffers help to stabilize the pH of solutions"? Explain your answer.

### B. Effect of Exercise on the Rate of $CO_2$ Production

1. Enter your data in these spaces:

Time for color change at rest: ______________________

Time for color change after exercise: ______________________

2. Explain your results here:

### C. Role of Carbon Dioxide in the Regulation of Ventilation

1. Enter your data in these spaces:

Rate of breathing at rest: __________ breaths/min

Rate of breathing after hyperventilation: __________ breaths/min

2. Explain your results here.

# REVIEW ACTIVITIES FOR EXERCISE 8.4

## *Test Your Knowledge of Terms and Facts*

1. A solution with a $H^+$ concentration of $10^{-9}$ molar has a pH of ____________________.
2. Hypoventilation produces a condition called respiratory ____________________; hyperventilation produces a condition called respiratory____________________.
3. Define these terms:
   (a) *acid* ____________________
   (b) *base* ____________________
   (c) *acidosis* ____________________
   (d) *alkalosis* ____________________
4. What is the normal measurement of arterial carbon dioxide levels? __________mm Hg
5. The free bicarbonate in the plasma serves as the major ____________________ of the blood.
6. The enzyme in red blood cells that catalyzes the formation of carbonic acid is ____________________.

## *Test Your Understanding of Concepts*

7. Draw equations to show how carbon dioxide affects the blood concentration of $H^+$ (and thus the pH) and the blood concentration of $HCO_3^-$. Indicate the directions of change in these values if blood carbon dioxide levels were to rise.

8. Use the equations shown in question 7 to explain how hyperventilation and hypoventilation affect the blood pH.

9. Describe how your breathing rate changed following 10 seconds of hyperventilation. Explain the physiological mechanisms responsible for the changed breathing pattern.

## *Test Your Ability to Analyze and Apply Your Knowledge*

10. People who hyperventilate may get dizzy (due to cerebral vasoconstriction), causing anxiety and further hyperventilation. Such people are sometimes urged to breathe into a paper bag. What good would this do? Explain the physiological mechanisms involved.

11. Intravenous infusions of sodium bicarbonate are often given to acidotic patients to correct the acidosis and relieve the strain of rapid breathing. Explain why bicarbonate is helpful in this situation. What would happen if too much bicarbonate were given? Explain.

# Renal Function and Homeostasis

# Section 9

**T**he kidneys are responsible for the elimination of most of the waste products of metabolism. These wastes include urea and creatinine (derived from protein catabolism) and ketone bodies (derived from fat catabolism). The kidney must also retain (or *reabsorb*) molecules essential for normal body function, such as glucose, amino acids, and bicarbonate.

Through these actions, the kidneys are involved in maintaining a constant internal environment (homeostasis), including the regulation of *electrolyte concentrations, fluid balance,* and *acid-base balance.* The fluid volume of the blood is maintained by the reabsorption of 98% to 99% of the water that leaves the blood in the initial step of urine formation, while the electrolyte and pH balance of the blood is maintained by the selective reabsorption of such ions as $Na^+$, $K^+$, and $HCO_3^-$.

The kidneys contain approximately 2 million functional units called **nephrons** (fig. 9.1*c*). Each nephron is composed of two parts: (1) the *glomerulus,* which is a tightly woven, highly permeable capillary bed at the end of an arteriole, and (2) the *renal tubule,* which is a bent and convoluted tubule lined by epithelial cells. The mouth of each renal tubule (Bowman's capsule) envelops a glomerulus. The last part of the renal tubule (the collecting duct) empties its contents into the renal pelvis as urine, which is then funneled down the ureters (fig. 9.1).

The formation of urine in the nephron occurs in two stages. (1) The hydrostatic pressure of the blood squeezes fluid out of the capillary wall of the glomerulus, producing an *ultrafiltrate* of blood. Except for proteins, which are usually too large to leave the capillaries, the glomerular filtrate contains the same solute molecules as plasma and is isotonic to plasma. (2) As the glomerular filtrate passes through the renal tubules, the cells of the tubules selectively reabsorb and secrete solute molecules and ions. The solution that emerges at the end of the collecting duct is urine, and thus very different in composition and concentration from the glomerular filtrate that enters the tubule.

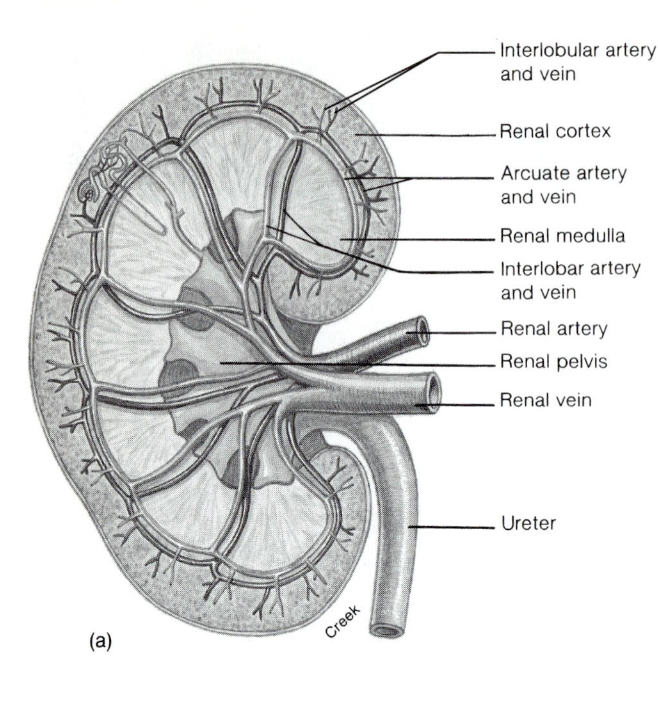


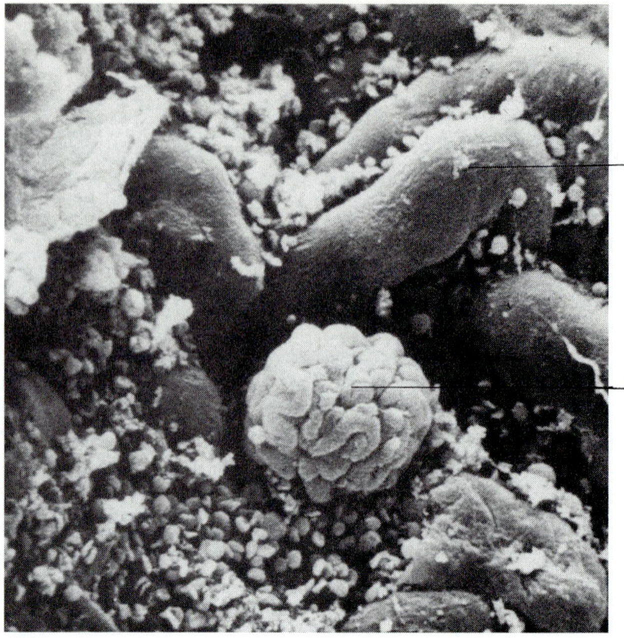


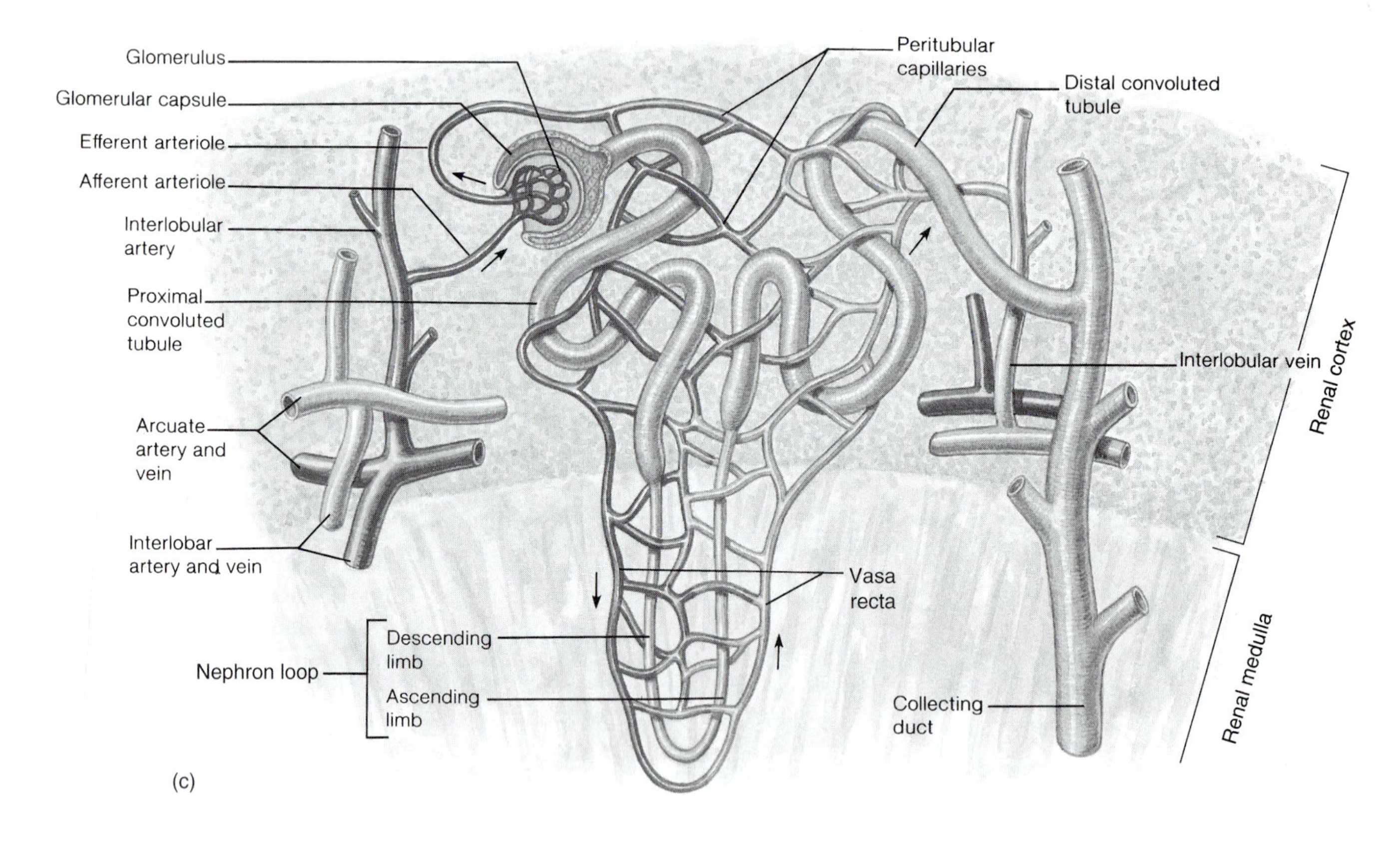


**Figure 9.1 Structure of the kidney.** (a) A diagram of a sectioned kidney, illustrating the arrangement of blood vessels. (b) A scanning electron micrograph of glomeruli and tubules. (c) The tubules and associated blood vessels that compose a nephron.

**(For full-color versions of these figures, see figs. 17.4 and 17.5 in *Human Physiology,* eighth edition, by Stuart I. Fox.)**

# Renal Regulation of Fluid and Electrolyte Balance

EXERCISE **9.1**

## Materials

1. Urine collection cups
2. Urinometers and droppers

**Note:** *Specific gravity can also be measured visually using disposable urine dip strips (Miles Inc., Curtin Matheson Scientific Inc.)*

3. pH paper (pH range 3–9), potassium chromate (20 g per 100 mL), silver nitrate (2.9 g per 100 mL)
4. NaCl crystals or salt tablets
5. Alternatively, normal and abnormal artificial urine is available (Wards Biology). However, modifications must be made by the instructor to simulate the conditions in this exercise.

Urine volume, solute concentration, and electrolyte content are adjusted by the kidneys to maintain homeostasis of the blood. Drinking excess water or eating salty foods results in a rising blood volume, which is followed by compensatory increases in the urinary excretion of the salt and water.

## OBJECTIVES

1. Describe the roles of ADH and aldosterone in the regulation of fluid and electrolyte balance.
2. Calculate the concentration of ions in solution (in milliequivalents per liter).
3. Demonstrate and explain how the kidneys respond to water and salt loading by changes in urinary volume, specific gravity, pH, and electrolyte composition.

### Textbook Correlations*

Before performing this exercise, you should study the introductory material presented here. Further information relating to this exercise can be found in these pages of *Human Physiology,* eighth edition, by Stuart I. Fox:

- *Reabsorption of Salt and Water.* Chapter 17, pp. 532–539.
- *Renal Control of Electrolyte and Acid-Base Balance.* Chapter 17, pp. 544–549.

The reabsorption of fluid and electrolytes (ions) into the blood from the renal tubular filtrate is adjusted to meet the needs of the body by the action of hormones. The major hormones involved in this process are **antidiuretic hormone (ADH),** released by the posterior pituitary gland, and **aldosterone,** secreted by the cortex of the adrenal gland.

The release of antidiuretic hormone by the posterior pituitary is regulated by osmoreceptors in the hypothalamus (fig. 9.2). These receptors are stimulated by an increase in the osmotic pressure of the blood, as might occur in dehydration. The ADH released in response to this stimulus promotes the reabsorption of water from the renal tubules, resulting in (1) the retention of water and therefore a decrease in the osmotic pressure of the blood back to the normal level, and (2) the excretion of a small volume of highly concentrated (hypertonic) urine.

The secretion of aldosterone by the adrenal cortex may be stimulated by an increase in blood $K^+$ or by a decrease in blood $Na^+$ or blood volume. An increase in blood $K^+$ directly stimulates the adrenal cortex to secrete aldosterone. A decrease in blood $Na^+$ or blood volume indirectly affects aldosterone secretion by stimulating the kidneys to secrete the enzyme **renin** into the blood (fig. 9.3). Renin catalyzes the reaction that leads to the formation of a polypeptide known as **angiotensin II,** which has two major effects. Angiotensin II: (1) stimulates vasoconstriction, thus increasing the blood pressure, and (2) stimulates the secretion of aldosterone

*Multimedia Correlations (also see Appendix 3)
- A.D.A.M.*InterActive Physiology*: Glomerular Filtration
- *MediaPhys 2.0*: Topics 11.34–11.46

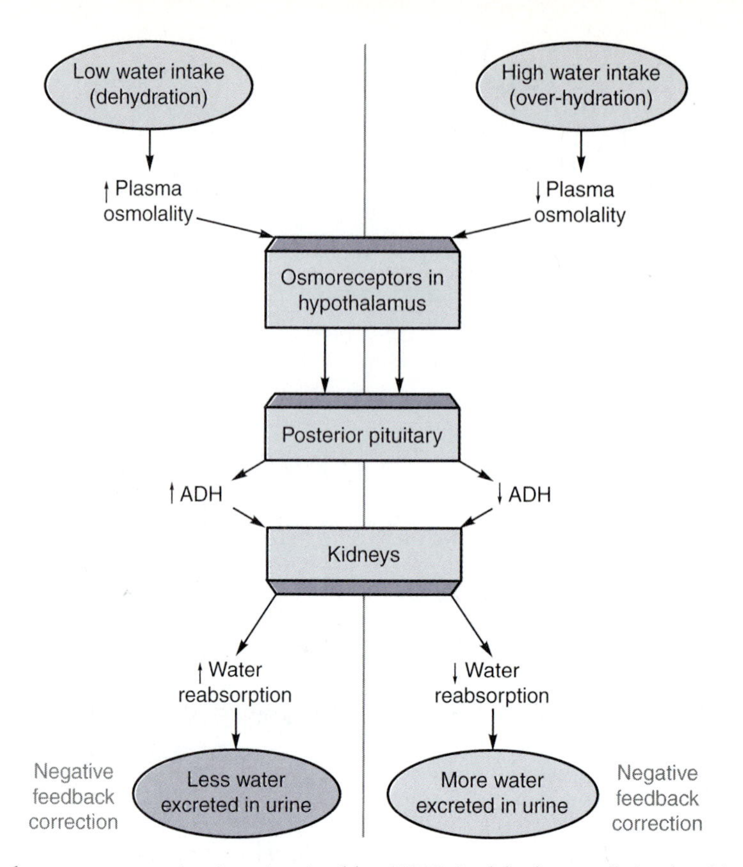


**Figure 9.2 Homeostasis of plasma concentration is maintained by ADH.** In dehydration (left side of figure), a rise in ADH secretion results in a reduction in the excretion of water in the urine. In over-hydration (right side of figure), the excess water is eliminated through a decrease in ADH secretion. These changes provide negative feedback correction, maintaining homeostasis of plasma osmolality and, indirectly, blood volume.

**(For a full-color version of this figure, see fig. 17.20 in *Human Physiology,* eighth edition, by Stuart I. Fox.)**

from the adrenal cortex. This released aldosterone then promotes the reabsorption of $Na^+$ from the glomerular filtrate into the blood, in exchange for $K^+$, which is secreted from the blood into the renal tubules. When $Na^+$ ions are reabsorbed, water follows passively owing to the osmotic gradient that is created. In this way, aldosterone prompts a rise in blood $Na^+$ and volume, correcting the original deviation and maintaining homeostasis.

## Milliequivalents

The concentrations of ions in body fluids (electrolytes) are usually given in terms of milliequivalents (mEq) per liter. To convey the meaning and significance of this unit of measurement, consider the chloride ($Cl^-$) concentration of the urine.

Suppose a urine sample had a chloride concentration of 610 mg per 100 mL. How does this number of ions and this number of charges compare with the number of other ions and charges that are present in the urine? To determine this, we must first convert the chloride concentration from milligrams per 100 mL to millimoles per liter.

### *Example*

The atomic weight of chloride is 35.5. Therefore,

$$\frac{610 \text{ mg of Cl}^-}{100 \text{ mL}} \times \frac{1 \text{ g}}{1{,}000 \text{ mg}} \times \frac{1{,}000 \text{ mL}}{1 \text{ L}} \times \frac{1 \text{ mole}}{35.5 \text{ g}}$$

$$= 0.171 \text{ M} \times \frac{1{,}000 \text{ mM}}{1 \text{ M}} = 171 \text{ mM}$$

One mole of chloride has the same number of ions as 1 mole of $Na^+$ or 1 mole of $Ca^{2+}$ or 1 mole of anything else. One mole of $Ca^{2+}$, however, has twice the number of charges (*valence*) as 1 mole of $Cl^-$. Therefore, 2 moles of $Cl^-$ are required to neutralize 1 mole of $Ca^{2+}$. If charges are taken into account by multiplying the moles by the valence, the product is termed the **equivalent weight** of an ion. One-thousandth of the equivalent weight dissolved in 1 liter of solution gives a concentration of *milliequivalents per liter (mEq/L)*.

### *Example*

$$171 \text{ mM} \times 1 \text{ (the valence of Cl}^-\text{)}$$
$$= 171 \text{ mEq/L of Cl}^-$$

The major advantage of expressing the concentrations of ions in milliequivalents per liter is that the total

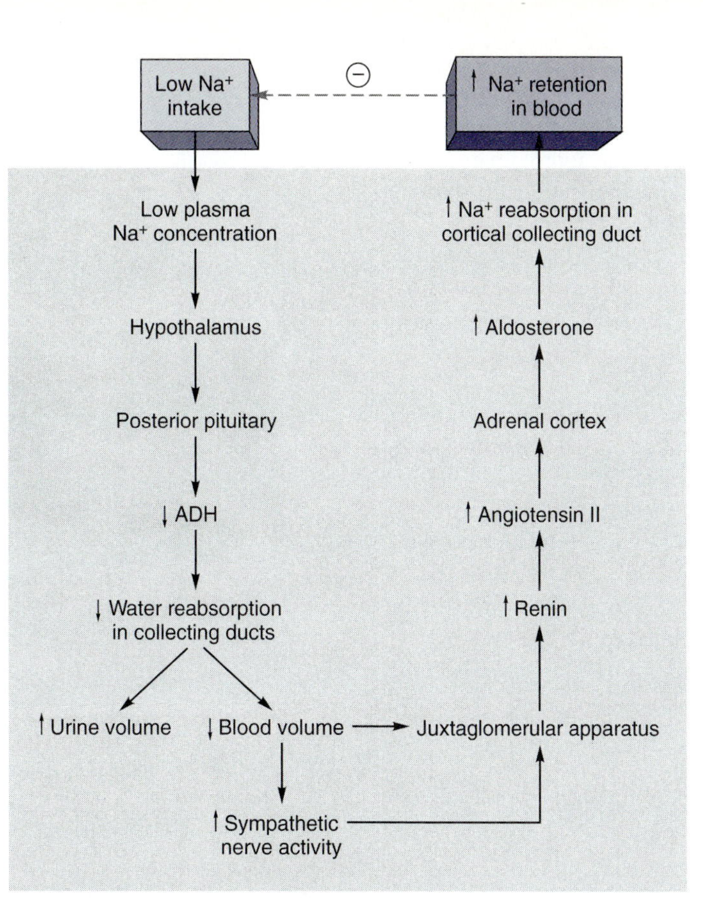


**Figure 9.3 Homeostasis of plasma $Na^+$.** This is the sequence of events by which a low sodium (salt) intake leads to increased sodium reabsorption by the kidneys. The dashed arrow and negative sign indicate the completion of the negative feedback loop.

**(For a full-color version of this figure, see fig. 17.26 in *Human Physiology,* eighth edition, by Stuart I. Fox.)**

concentration of anions can be easily compared with the total concentration of cations. In an average sample of venous plasma, for example, the total anions and the total cations are each equal to 156 mEq/L. Chloride, the major anion, has a plasma concentration of 103 mEq/L, whereas the chloride concentration in the urine is highly variable, ranging from 61 to 310 mEq/L.

Due in large part to the effects of ADH and aldosterone, the kidneys can vary their excretion of water and electrolytes to maintain homeostasis of the blood volume and composition. Abnormally low blood volume can produce **hypotension** (low blood pressure) and may result in circulatory shock; abnormally high blood volume contributes to **hypertension.** Renal regulation of $Na^+$ balance is also critical for health. Changes in blood $Na^+$ cause secondary changes in blood volume, as water follows sodium by osmosis. Changes in blood $K^+$ affect the bioelectrical properties of all cells, but the effects on the heart are particularly serious. **Hyperkalemia** (high blood $K^+$) is usually fatal when the $K^+$ concentration rises from 4 mEq/L (normal) to over 10 mEq/L. This may be caused by a variety of conditions, including inadequate aldosterone secretion (in Addison's disease), or by an excessive intake of potassium.

## PROCEDURE

1. The students void their urine into collection cups at the beginning of the laboratory session. In the analyses done in step 4, this sample will serve as the control (time zero).
2. The students drink 500 mL of water. One group just drinks the water; another group ingests NaCl (salt tablets are easiest to take) in addition to drinking the water. Most people can tolerate up to 4.5 g of salt, but the amount ingested should not be so great as to cause nausea.

**Note:** *Students with hypertension or on sodium-restricted diets should not perform the salt-loading exercise. Interestingly, 4.5 g/500 mL of NaCl is a 0.9 g/100 mL saline solution that is isotonic to plasma (normal saline).*

3. After drinking the solutions described in step 2, the students void their urine every 30 minutes for 2 hours. The urine samples are analyzed as described in step 4.

4. Each of the five urine samples collected are analyzed for pH, specific gravity, and chloride content as follows:
   (a) **Volume (mL).** Measure the approximate volume of urine obtained and enter the data in the table of the laboratory report.
   (b) **pH.** Determine the pH of the urine samples by dipping a strip of pH paper into the urine and matching the color developed with a color chart. The urine normally has a pH between 5.0 and 7.5.
   (c) **Specific gravity.** Determine the specific gravity of the urine samples by floating a urinometer in a cylinder (fig. 9.4) nearly filled with the specimen. Read the specific gravity at the meniscus on the urinometer scale, making sure that the urinometer float is not touching the bottom or the sides of the cylinder. The specific gravity is directly related to the amount of solutes in the urine and ranges from 1.010 to 1.025. (Pure water should have a specific gravity of 1.000.)
   (d) **Chloride concentration.** When $Na^+$ is reabsorbed by the renal tubules, $Cl^-$ follows passively by electrostatic attraction. Follow the steps below to determine the chloride concentration of the urine samples.
      (1) Measure 10 drops of urine into a test tube (1 drop is approximately 0.05 mL).
      (2) To this tube add 1 drop of 20% potassium chromate solution with a second dropper.
      (3) Add 2.9% silver nitrate solution one drop at a time using a third dropper, while shaking the test tube continuously. Count the number of full drops required to cause a permanent change in the color of the solution from yellow to brown.
      (4) Determine the chloride concentration of the urine sample. Because each drop of 2.9% silver nitrate added in step 3 is equivalent to 61 mg of $Cl^-$ per 100 mL of urine, simply multiply the number of drops by 61 to obtain the chloride concentration of the urine in milligrams per 100 mL.

**Figure 9.4 Instruments for determining the specific gravity of urine.** (a) A glass cylinder and (b) a urinometer float.

### *Example*

If 10 drops of 2.9% silver nitrate were required,

$$10 \times 61 \text{ mg of } Cl^-/100 \text{ mL} = 610 \text{ mg } Cl^-/100 \text{ mL}$$

5. Convert the chloride concentration to mEq/L and enter your data in the appropriate table in the laboratory report.

# Laboratory Report 9.1

Name ______________________

Date ______________________

Section ______________________

## DATA FROM EXERCISE 9.1

1. Enter your data in the appropriate table.

   (a) Ingestion of water only

| Time | Volume (mL) | pH | Specific Gravity | Chloride (mEq/L) |
|---|---|---|---|---|
| 0 | | | | |
| 30 | | | | |
| 60 | | | | |
| 90 | | | | |
| 120 | | | | |

   (b) Ingestion of water and NaCl

| Time | Volume (mL) | pH | Specific Gravity | Chloride (mEq/L) |
|---|---|---|---|---|
| 0 | | | | |
| 30 | | | | |
| 60 | | | | |
| 90 | | | | |
| 120 | | | | |

## REVIEW ACTIVITIES FOR EXERCISE 9.1

### *Test Your Knowledge of Terms and Facts*

1. Osmoreceptors in the hypothalamus of the brain are stimulated by a(n) ______________________ (increase/decrease) in the plasma osmotic pressure.
2. As a result of stimulation, the osmoreceptors stimulate the secretion of ______________________ from the ______________________ gland.
3. The hormone named in question 2 specifically stimulates the kidneys to ______________________.
4. The hormone that stimulates the reabsorption of $Na^+$ from the nephron tubules, and also stimulates the secretion of $K^+$ into the tubules, is ______________________.
5. The substance that stimulates vasoconstriction and also stimulates the secretion of the hormone named in question 4: ______________________.
6. The measurement that is 1.000 for pure water and that increases in proportion to the general solute concentration of a solution: ______________________.

## Test Your Understanding of Concepts

7. Calcium is normally present at a concentration of about 0.1 g/L. Calculate the mEq/L of $Ca^{2+}$ in the plasma (the atomic weight of calcium is 40).

8. Describe how the measurement of urine specific gravity changed after drinking water, and explain the physiological mechanisms responsible for this change.

## Test Your Ability to Analyze and Apply Your Knowledge

9. Imagine a dehydrated desert prospector and a champagne-quaffing partygoer, each of whom drinks a liter of water at time zero and voids urine over a period of 3 hours. Using their urine samples, compare the probable differences in volume and composition. (*Hint:* Alcohol inhibits ADH secretion.) Use relative terms, such as *increased* or *decreased*.

| Observation | Prospector | Partygoer |
|---|---|---|
| Urine volume | | |
| Specific gravity | | |
| $Na^+$ and $Cl^-$ Content | | |

(a) Explain the answers you provided in this table.

10. Many diuretic drugs used clinically inhibit $Na^+$ reabsorption in the loop of Henle. Predict the effect of these drugs on the urinary excretion of $Cl^-$ and $K^+$, and explain your answer.

# Renal Plasma Clearance of Urea

EXERCISE **9.2**

## Materials

1. Pipettes with mechanical pipettors (or Repipettes), mechanical microliter pipettes (capacity μL), disposable tips
2. Colorimeter and cuvettes
3. BUN reagents and standard (Stanbio, through Curtin Matheson Scientific, Inc.)
4. Sterile lancets and 70% alcohol
5. File
6. Microhematocrit centrifuge and heparinized capillary tubes
7. Container for the disposal of blood-containing objects
8. As an alternative, normal and abnormal artificial urine is available (Wards Biology). However, modifications must be made by the instructor to simulate the conditions in this exercise.

Urea and other waste products in the plasma are filtered by the kidneys and excreted in the urine. The efficiency of the kidneys in performing these processes for each solute excreted is measured by its renal plasma clearance.

## OBJECTIVES

1. Describe the chemical nature of urea and explain its physiological significance.
2. Define renal plasma clearance and explain how this value is calculated.
3. Perform a renal plasma clearance measurement for urea and explain the physiological significance of this measurement.
4. Explain how the renal plasma clearance for a solute is affected by filtration, reabsorption, and secretion.

## Textbook Correlations*

Before performing this exercise, you should study the introductory material presented here. Further information relating to this exercise can be found in these pages of *Human Physiology,* eighth edition, by Stuart I. Fox:

- *Renal Plasma Clearance.* Chapter 17, pp. 539–544.

When amino acids are broken down in the process of cellular respiration, or when they are converted to glucose (a process known as gluconeogenesis), the amino (—$NH_2$) groups are removed and secreted into the blood in the form of **urea.** This function is performed by the liver.

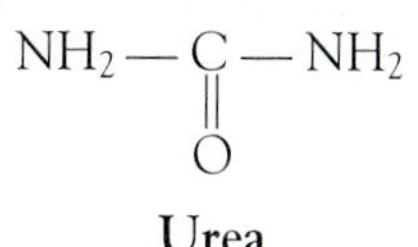


**Urea**

The urea is filtered by the glomeruli and enters the renal tubules. Although urea is a waste product of amino acid metabolism, some of it is transported (through facilitative diffusion) out of the nephron tubules by specific carriers. This increases the urea concentration in the interstitial fluid of the renal medulla, thus contributing to the hypertonicity of the renal medulla. Because urea is reabsorbed after filtration, only 60% of the blood filtered by the glomeruli is cleared of urea. Since the average glomerular filtration rate (GFR) is 125 mL/min (for both kidneys), this amounts to an average of 75 mL of plasma cleared of urea per minute. This value is termed the **renal plasma clearance** of urea (fig. 9.5).

The clearance rate of a substance depends on the size of the kidneys and the rate of urine production, as well as the other factors discussed. In this exercise, it will be assumed that the kidneys are of average size and that urine production is equal to or greater than 2.0 mL/min.

*Multimedia Correlations (also see Appendix 3)

- A.D.A.M. *InterActive Physiology* Early Filtrate Processing: Late Filtrate Processing

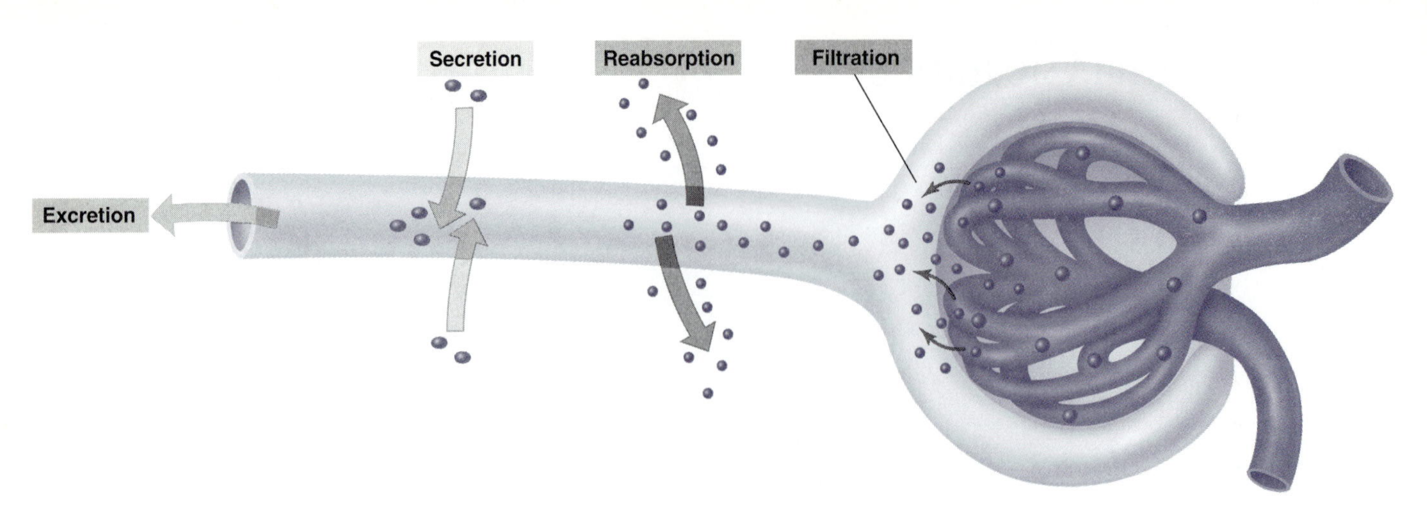


**Figure 9.5 Filtration, reabsorption, and secretion by the nephron.** Urea is filtered and about 40% of the amount filtered is passively reabsorbed (there is no secretion of urea). Therefore, approximately 60% of the amount filtered is excreted in the urine.

**(For a full-color version of this figure, see fig. 17.21 in *Human Physiology*, eighth edition, by Stuart I. Fox.)**

Under these conditions, the clearance of urea can be calculated using the formula

$$\text{Clearance} = \frac{U \times V}{P}$$

where,

$U$ is the concentration of urea in the urine, in milligrams per 100 mL
$P$ is the concentration of urea in the plasma, in milligrams per 100 mL
$V$ is the urine excreted in milliliters per minute

The concentration of a substance in urine, $U$, multiplied by the volume, $V$, of urine produced per minute gives the milligrams of the substance excreted in the urine per minute. When this figure is divided by the concentration, $P$, of that substance in the plasma, the result indicates the volume of plasma that contained the amount of the substance excreted per minute. This is the amount of plasma "cleared" by passage through the kidneys per minute.

If the substance is filtered from the glomeruli but is not reabsorbed or secreted (as with the polysaccharide *inulin*), the renal clearance equals the glomerular filtration rate (GFR) (fig. 9.6). If a substance is filtered but then reabsorbed into the blood (as with glucose, amino acids, urea, and many other substances), the renal plasma clearance must be less than the glomerular filtration rate. (How much less depends on the degree of reabsorption.) If a substance enters the renal tubules both by filtration and by active transport from the capillaries into the nephron tubules (a process called **secretion**), the renal plasma clearance is greater than the glomerular filtration rate. This is the case with the substance *para-aminohippuric acid (PAH)*, which is almost entirely removed or "cleared" from the blood in a single passage through the kidneys.

The plasma concentration of **blood urea nitrogen (BUN)** reflects both the rate of urea formation from protein in the liver and the rate of urea excretion by filtration through the glomeruli of the kidneys. In the absence of liver malfunction and abnormal protein metabolism, a rise in BUN indicates a kidney disorder such as nephritis, pyelonephritis, or kidney stones.

Since the renal plasma clearance for urea is substantially less than the glomerular filtration rate (GFR) due to its reabsorption, the urea clearance is not a particularly good indicator of kidney function. More useful clinically are the renal plasma clearances for exogenously administered *inulin* (a large polysaccharide) and endogenous *creatinine* (a byproduct of creatine, a molecule found primarily in muscle). Since inulin is neither reabsorbed nor secreted by the nephron, its clearance equals the GFR. Creatinine is secreted to a slight degree by the renal nephron, so its clearance rate is 20–25% greater than the true GFR (as defined by the inulin clearance test).

## PROCEDURE

### Collection of Plasma and Urine Samples

1. Empty the urinary bladder. Then, have the subject drink 500 mL of water as quickly as is comfortable and record the time (time zero).
2. About 20 minutes later, cleanse the fingertip with 70% alcohol and, using a sterile lancet, obtain a drop of blood from the subject's fingertip. Fill a heparinized capillary tube at least halfway with blood, plug one end, and centrifuge in a microhematocrit centrifuge

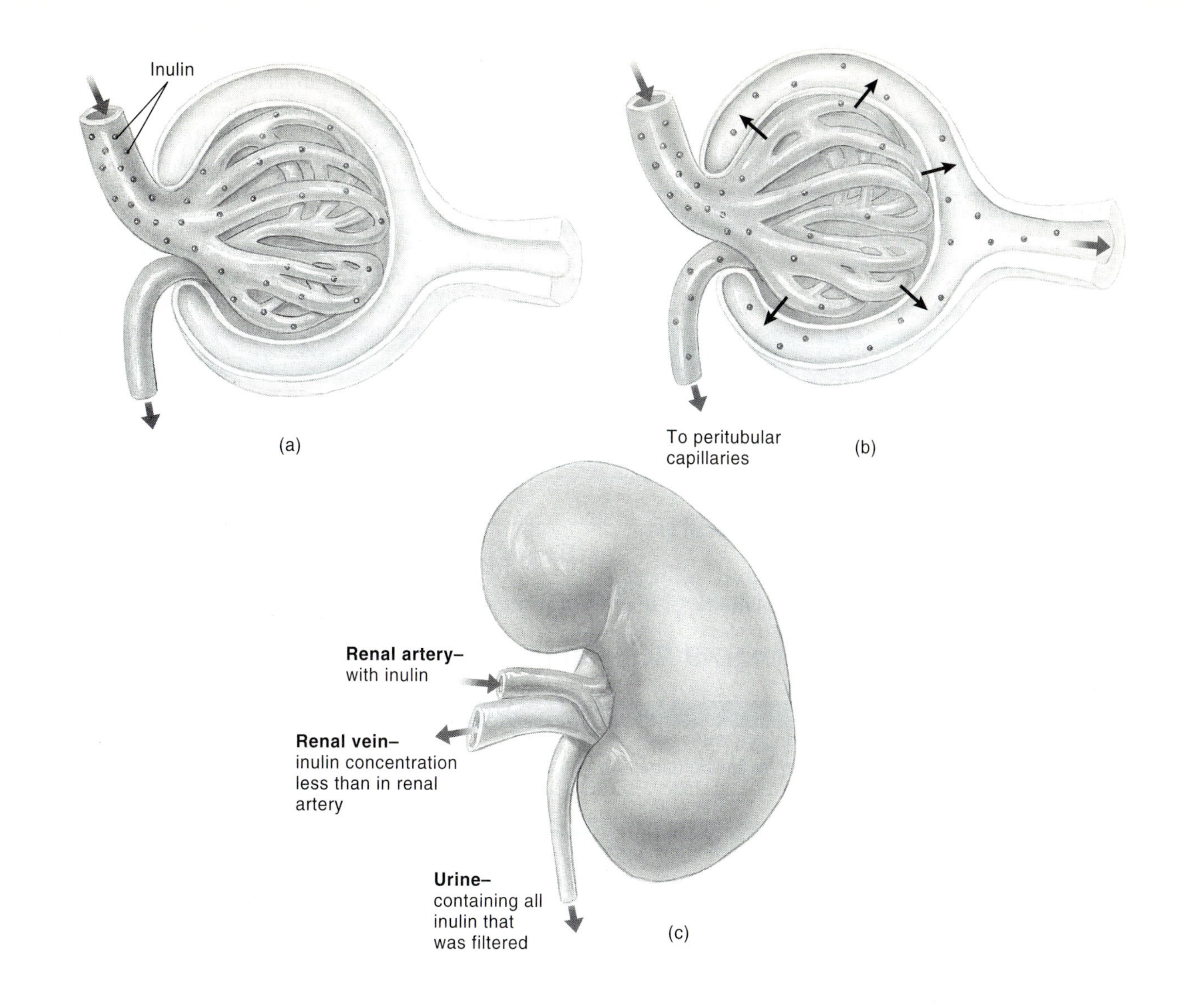


**Figure 9.6 Renal clearance of inulin.** (a) Inulin is present in the blood entering the glomeruli, and (b) some of this blood, together with its dissolved inulin, is filtered. All of this filtered inulin enters the urine, whereas most of the filtered water is returned to the vascular system (is reabsorbed). (c) The blood leaving the kidneys in the renal vein, therefore, contains less inulin than the blood that entered the kidneys in the renal artery. Since inulin is filtered but neither reabsorbed nor secreted, the inulin clearance equals the glomerular filtration rate (GFR).

**(For a full-color version of this figure, see fig. 17.22 in *Human Physiology,* eighth edition, by Stuart I. Fox.)**

for 3 minutes (as in performing a hematocrit measurement—see exercise 6.1).

**Caution:** *Always handle only your own blood and place all objects that have been in contact with blood in the container indicated by the instructor.*

3. Score the capillary tube lightly with a file at the plasma-cell junction and break the tube at the scored mark. Expel the plasma into a labeled small beaker or test tube.
4. Collect a complete urine sample 30 minutes after drinking the 500 mL of water. Measure the milliliters of water produced in 30 minutes, divide by 30, and enter the volume of urine produced per minute in the laboratory report.
5. Dilute the urine 1:20 with water. This can be done by adding 19 mL of water to 1 mL of urine, or by adding 1.9 mL of water to 0.10 mL (100 μL) of urine. Mix the diluted urine solution.

### Measurement of Urea

6. Label four test tubes or cuvettes: *B* (blank), *S* (standard), *P* (plasma), and *U* (urine).

10. What is the significance of the glomerular filtration rate (GFR) measurement? Describe two ways that this measurement may be obtained.

## *Test Your Ability to Analyze and Apply Your Knowledge*

11. Suppose substance "A" is reabsorbed about 30%, whereas substance "B" is secreted about 30%. Explain how their renal plasma clearance values relate to each other and to the glomerular filtration rate (GFR).

12. Explain the mechanism by which each of the following conditions might produce an increase in the plasma concentration of urea: (a) increased protein catabolism; (b) decreased blood pressure, as in circulatory shock; and (c) kidney failure.

# Clinical Examination of Urine

EXERCISE **9.3**

## Materials

1. Microscopes
2. Urine collection cups, test tubes, microscope slides, and coverslips
3. Albustix, Clinitest tablets, Ketostix, Hemastix, Ictotest tablets, or Multistix (Ames Laboratories; Ictotest can be obtained from Hardy Diagnostics).
4. Sediment stain (such as Sternheimer-Mablin stain)
5. Centrifuge and centrifuge tubes
6. Transfer pipettes (droppers)
7. Alternatively, normal and abnormal artificial urine is available with a few test strips (Wards Biology).

The presence of abnormally large amounts of proteins and casts in the urine can indicate damage to the glomeruli of the kidney. The presence of bacteria and a large number of white blood cells in the urine sediment indicates urinary tract infection. Abnormal concentrations of glucose, ketone bodies, bilirubin, and other plasma solutes in the urine may reflect abnormally high concentrations in the plasma.

### OBJECTIVES

1. Describe the physiological processes responsible for normal urinary concentrations of protein, glucose, ketone bodies, and bilirubin in the urine.
2. Describe the pathological processes that may produce abnormal solute concentrations and explain the clinical significance of this information.
3. Describe the normal constituents of urine sediment and explain how the microscopic examination of urine sediment can be clinically useful.

A clinical examination of urine may provide evidence of urinary tract infection (UTI) or kidney disease. Additionally, since urine is derived from plasma, an examination of the urine provides a convenient, nonintrusive means of assessing the composition of plasma and of detecting a variety of systemic diseases. A clinical examination of the urine includes an observation of its appearance (table 9.1), tests of its chemical composition, and a microscopic examination of urine sediment.

### Textbook Correlations

Before performing this exercise, you should study the introductory material presented here. Further information relating to this exercise can be found in these pages of *Human Physiology,* eighth edition, by Stuart I. Fox:

- *Glomerular Filtration.* Chapter 17, pp. 529–531.
- *Reabsorption of Glucose.* Chapter 17, pp. 543–544.

When the kidneys are inflamed the permeability of the glomerular capillaries may be increased, resulting in the leakage of proteins into the urine **(proteinuria)** and the appearance of casts in the urine sediment. Since this leakage represents a continuous loss of the protein solutes responsible for the colloid osmotic pressure of plasma, fluid may accumulate in the tissues, resulting in *edema.*

The appearance of glucose **(glycosuria)** in the urine suggests the presence of *diabetes mellitus.* If diabetes is suspected, however, the test for glucose in the urine alone is not sufficient because a person may have hyperglycemia without glycosuria. In such a case, the glucose concentration in the plasma at the time of the test may not be high enough to exceed the ability of the nephron tubules to completely reabsorb glucose from the filtrate (known as the *transport maximum* ($T_m$) for glucose). Because tubular reabsorption is so efficient (the $T_m$ is high), the urine will be free of glucose until the concentrations of glucose in the blood exceed its *renal plasma threshold.* A more conclusive test for diabetes is the *oral glucose tolerance test.*

## A. Test for Proteinuria

Since proteins are very large molecules (macromolecules), they are not normally present in measurable amounts in the glomerular ultrafiltrate or the urine. The presence of proteins in the urine may therefore indicate an abnormal increase in the permeability of the kidney glomeruli. Such permeability changes may be caused by renal infections

**Table 9.1** Appearance of Urine and Cause

| Color | Cause |
|---|---|
| Yellow-orange to brownish green | Bilirubin from obstructive jaundice |
| Red to red-brown | Hemoglobinuria |
| Smoky red | Unhemolyzed RBCs from urinary tract |
| Dark wine color | Hemolytic jaundice |
| Brown-black | Melanin pigment from melanoma |
| Dark brown | Liver infections, pernicious anemia, malaria |
| Green | Bacterial infection (*Pseudomonas aeruginosa*) |

*Note:* Certain foods (e.g., beets, rhubarb) and some commonly prescribed drugs and vitamin supplements may also alter the color of urine.

(glomerulonephritis) or by other diseases that have secondarily affected the kidneys, such as diabetes mellitus, jaundice, or hyperthyroidism.

### PROCEDURE

Dip the yellow end of a disposable Albustix strip into a urine sample and compare the color developed with the chart provided. Record the albumin (protein) concentration in the data table in your laboratory report.

## B. Test for Glycosuria

Although glucose is easily filtered by the glomerulus, it is not normally present in the urine. All of the filtered glucose is normally reabsorbed from the renal tubules into the blood. This reabsorption process is carrier mediated—that is, the filtered glucose is transported across the wall of the renal tubule by a protein carrier.

When the glucose concentration both in the plasma and in the glomerular ultrafiltrate is within the normal limits (70–110 mg per 100 mL), there is a sufficient number of carrier molecules along the renal tubules to transport all the glucose back into the blood. However, if the blood glucose level exceeds a certain limit, called the **renal plasma threshold** for glucose, (about 180 mg per 100 mL), the number of glucose molecules in the glomerular ultrafiltrate will be greater than the number of available carrier molecules, and the nontransported glucose will "spill over" into the urine.

The chief cause of glycosuria is diabetes mellitus, although other conditions, such as hyperthyroidism, hyperpituitarism, and liver disease may also have this effect. Glycosuria, therefore, is not a renal disease but a symptom of other systemic diseases that raise the blood sugar level.

### PROCEDURE

1. Place 10 drops of water and 5 drops of urine in a test tube.
2. Add a Clinitest tablet.
3. Wait 15 seconds and compare the color developed with the color chart provided.
   Or: Dip the end of a disposable Clinistix strip into a urine sample, wait the required amount of time, then compare the color developed with the chart provided on the bottle. Record the glucose concentration in the data table of the laboratory report.

## C. Test for Ketonuria

When there is carbohydrate deprivation, such as in starvation or high-protein diets, the body relies increasingly on the metabolism of fats for energy. This pattern is also seen in people with diabetes mellitus, where lack of the hormone insulin prevents the body cells from utilizing the large amounts of glucose available in the blood. This occurs because insulin is necessary for the transport of glucose from the blood into the body cells.

The metabolism of fat proceeds in a stepwise manner:

1. triglycerides are hydrolyzed to fatty acids and glycerol;
2. fatty acids are converted into smaller intermediate compounds—*acetoacetic acid*, *β-hydroxybutyric acid*, and *acetone* (collectively known as **ketone bodies**);
3. the ketone bodies are broken down in aerobic cellular respiration, releasing energy. When the production of ketone bodies from fatty acid metabolism exceeds the ability of the body to metabolize these compounds, they accumulate in the blood *(ketonemia)* and spill over into the urine *(ketonuria)*.

### PROCEDURE

Dip a disposable Ketostix strip into a urine sample; 15 seconds later, compare the color developed with the color chart. Record the ketone concentration in the data table of the laboratory report.

## D. Test for Hemoglobinuria

Hemoglobin may appear in the urine in the event of hemolysis in the systemic blood vessels (e.g., in transfusion reactions), of rupture in the capillaries of the glomerulus, or of hemorrhage in the urinary system. In the latter condition, whole red blood cells may be found in the urine *(hematuria)*, although the low osmotic pressure of the urine may cause hemolysis and the release of hemoglobin *(hemoglobinuria)* from these red cells. Hemoglobinuria is normally found in the samples from menstruating women.

## PROCEDURE

Dip the test end of a disposable Hemastix strip into the urine sample, wait 30 seconds, and compare the color developed with the color chart. Enter the hemoglobin content in the data table of the laboratory report.

## E. Test for Bilirubinuria

The fixed phagocytic cells of the spleen and bone marrow *(reticuloendothelial system)* destroy old red blood cells and convert the *heme* groups of hemoglobin into the pigment **bilirubin.** The bilirubin is secreted into the blood and carried to the liver, where it is bonded to (*conjugated* with) glucuronic acid, a derivative of glucose. Most of this conjugated bilirubin is secreted into the bile as bile pigment; the rest is released into the blood.

The blood normally contains a small amount of free and conjugated bilirubin. An abnormally high level of blood bilirubin *(hyperbilirubinemia)* may result from (1) an increased rate of red blood cell destruction, seen in hemolytic anemia, for example; (2) liver damage, as in hepatitis and cirrhosis; or (3) obstruction of the common bile duct, as might occur because of a gallstone. The increase in blood bilirubin results in **jaundice,** a condition characterized by a brownish yellow pigmentation of the skin, sclera of the eye, and mucous membranes.

Normally, the kidneys can excrete only the more water-soluble bilirubin that is conjugated with glucuronic acid. An increase in the urine bilirubin *(bilirubinuria)*, therefore, may be associated with jaundice due to liver disease or bile duct obstruction, but is not normally observed in jaundice due to hemolytic anemia. In the latter case, the excess bilirubin is in a free, nonpolar state where it binds to plasma proteins, and thus is not filtered into the nephron tubules.

## PROCEDURE

Bilirubin can be quickly tested using the Multistix strip. If a more accurate determination is desired, the Ictotest procedure may be employed as described in these steps.

1. Place 5 drops of urine on a square of the test mat.
2. Place an Ictotest tablet in the center of the mat.
3. Place 2 drops of water on the tablet.
4. Interpret the test as follows:
   (a) Negative: Mat has no color or a slight pink-to-red color.
   (b) Positive: Mat turns blue to purple. The speed and intensity of color development is proportional to the amount of bilirubin present.
5. Record your observations in the data table in your laboratory report.

## F. Microscopic Examination of Urine Sediment

Microscopic examination of the urine sediment may reveal the presence of various cells, crystals, bacteria, and casts (figs. 9.7, 9.8, and 9.9). **Casts** are cylindrical structures formed by the precipitation of protein molecules within the renal tubules. Although a small number of casts are found in normal urine, a large number indicate renal disease, such as *glomerulonephritis* or *nephrosis*. The casts may lack cells, or contain cells such as leukocytes, erythrocytes, or epithelial cells (fig. 9.7). The presence of large numbers of erythrocytes, leukocytes, or certain epithelial cells in the urine is indicative of renal disease.

Like casts, a small number of crystals are present in normal urine. Their presence in large numbers, however may suggest a tendency to form kidney stones. Additionally, large numbers of *uric acid* crystals occur in *gout*, a form of arthritis (fig. 9.8).

## PROCEDURE

1. Fill a conical centrifuge tube three-quarters full of urine, and centrifuge at a moderate speed for 5 minutes.
2. Discard the supernatant and place one drop of Sternheimer-Mablin stain (or similar sediment stain) on the sediment. Mix by aspiration with a transfer pipette.
3. Place 1 drop of the stained sediment on a clean slide and cover with a coverslip.
4. Scan the slide with the low-power objective under reduced illumination (close the diaphragm) and identify the components of the sediment.

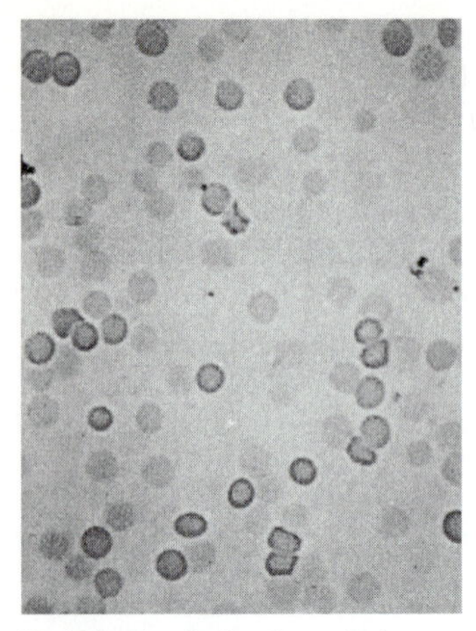

Red blood cells (erythrocytes)

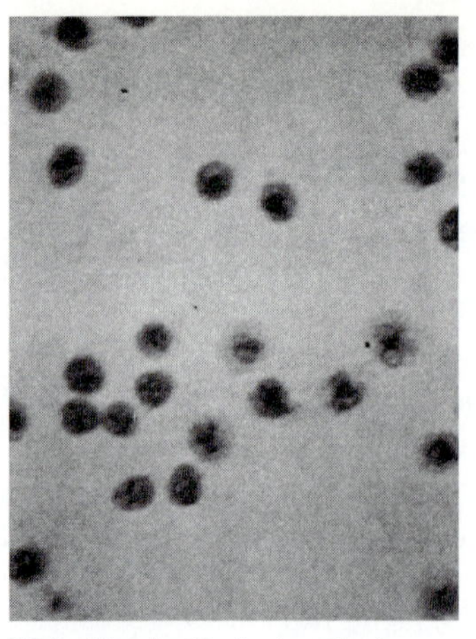

White blood cells (leukocytes)

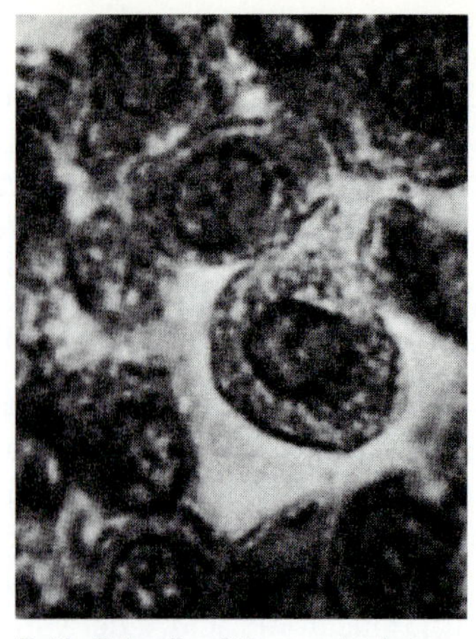

Epithelial cells of the renal tubules

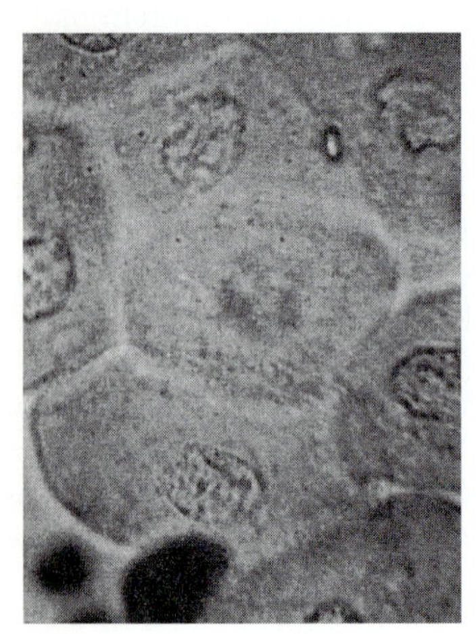

Epithelial cells of the bladder

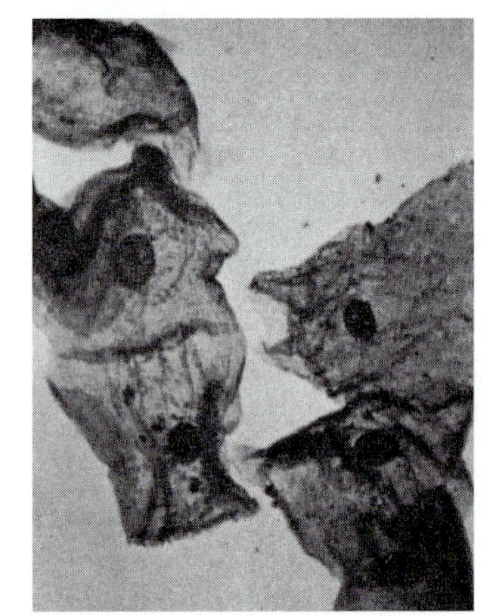

Epithelial cells of the urethra

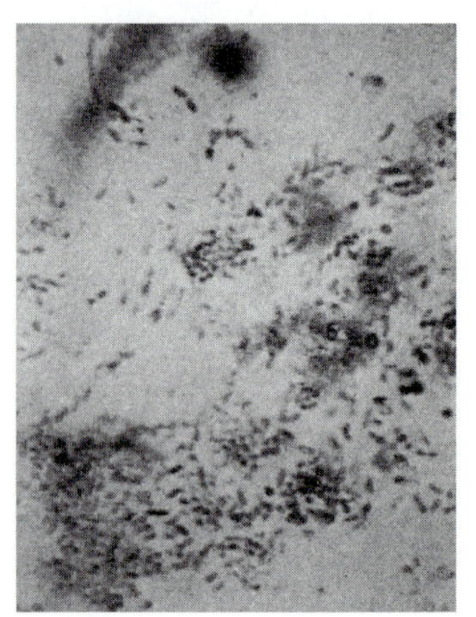

Bacteria

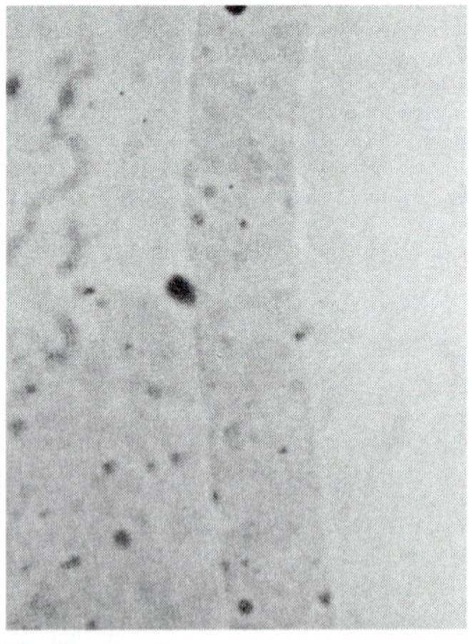

Hyaline cast

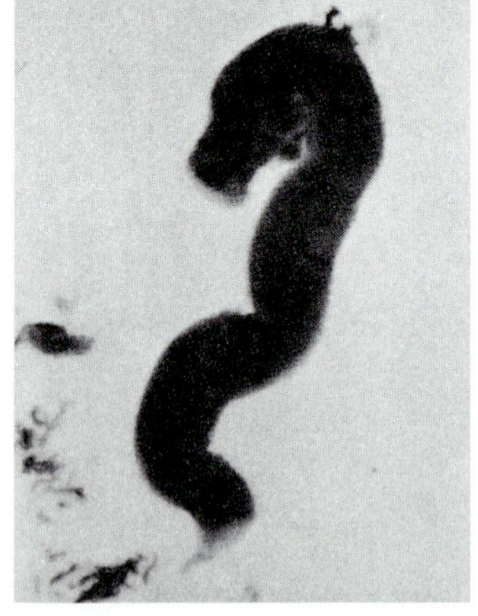

Waxy cast

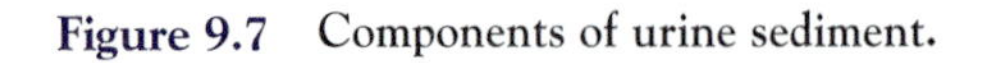

**Figure 9.7** Components of urine sediment.

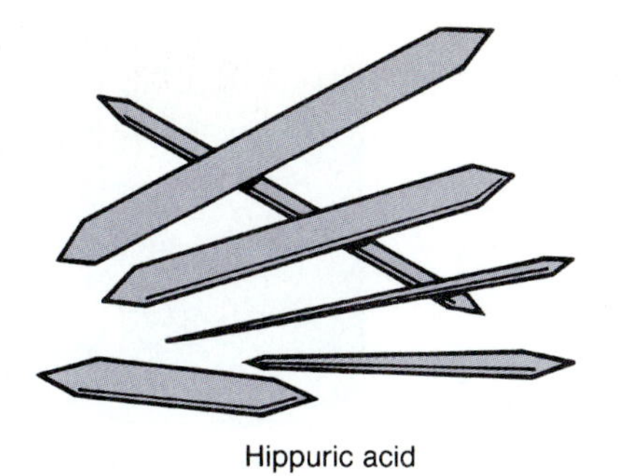


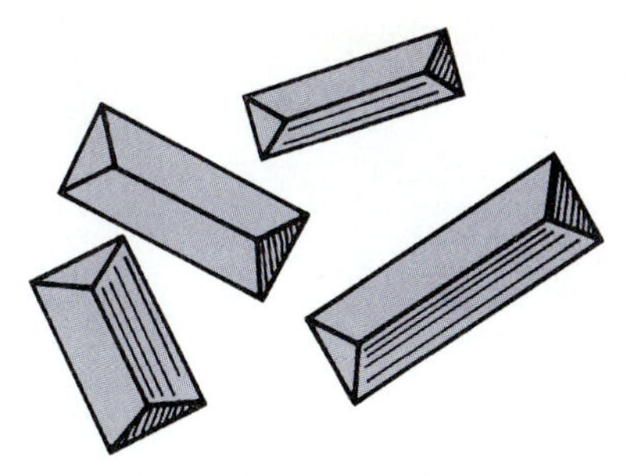


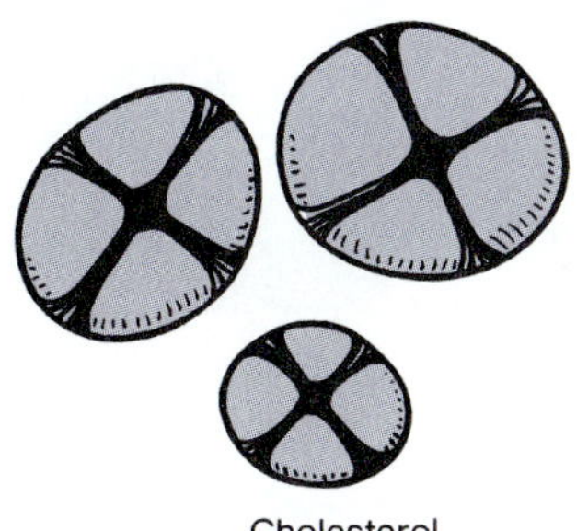


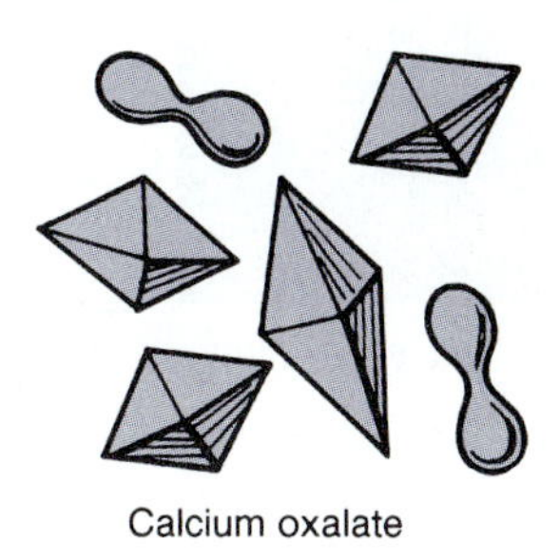


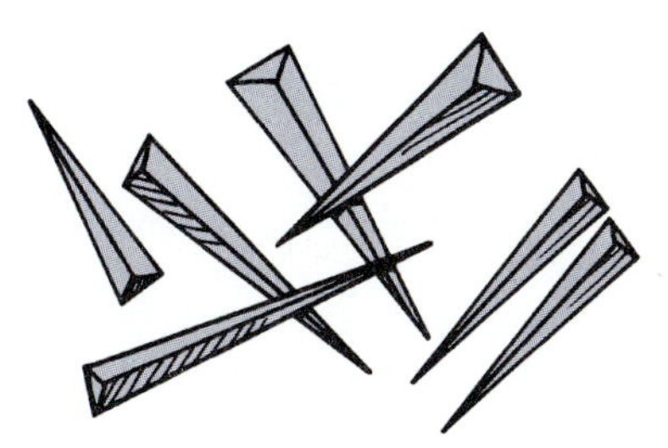

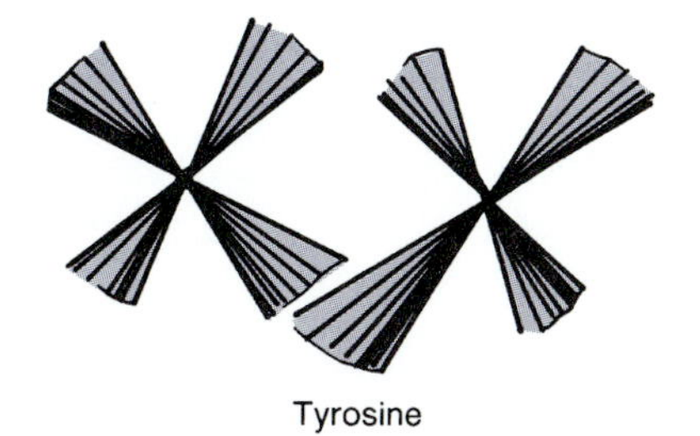


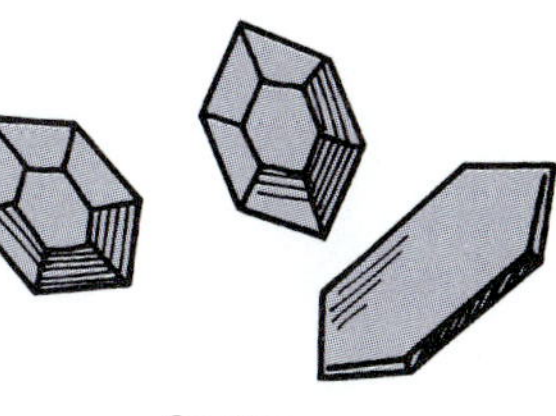


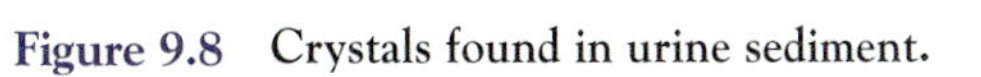

**Figure 9.8** Crystals found in urine sediment.

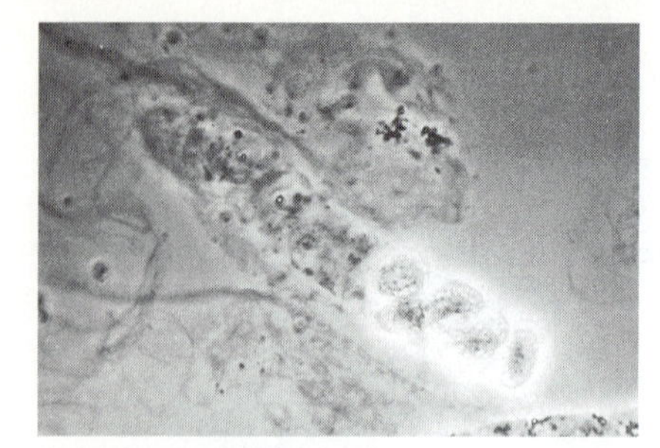

Hyaline casts; phase contrast. (400×)

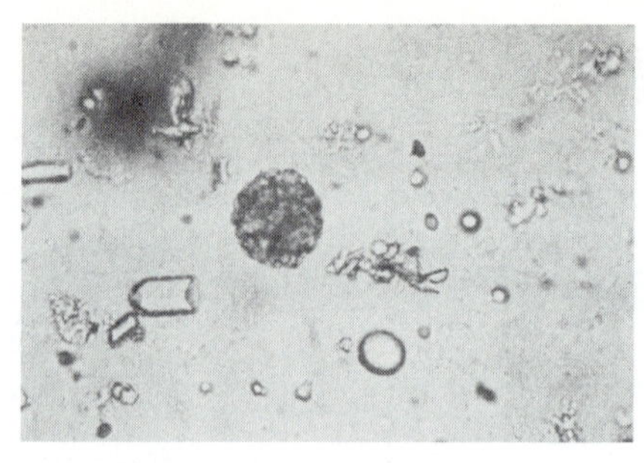

Oval fat bodies; "mulberry cell." Urine from a diabetic rat. (400×)

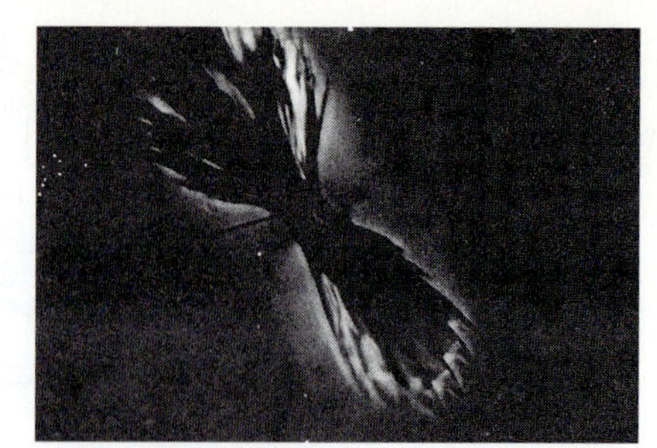

"Drug" crystal (sulfa); polarized light. Note sharp needle-like structure that may cause damage to the tubules. (400×)

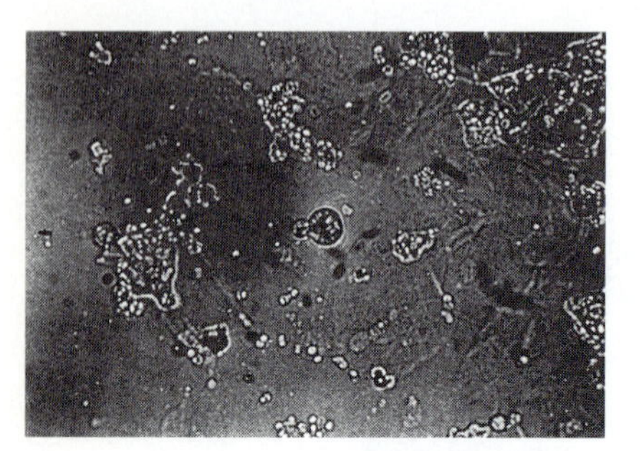

Leucine crystals; note size variation. (400×)

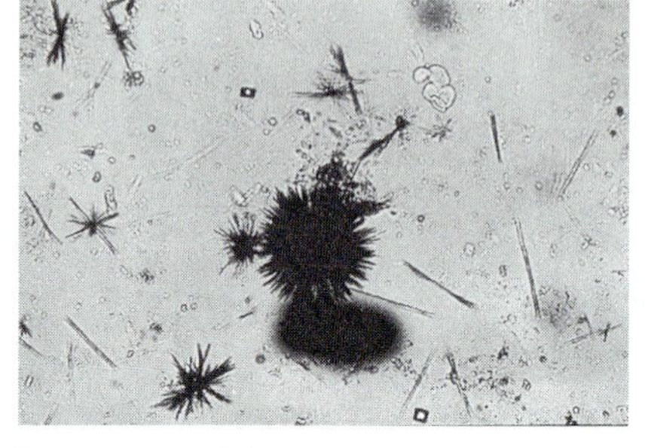

Tyrosine crystals (rosette needles). (100×)

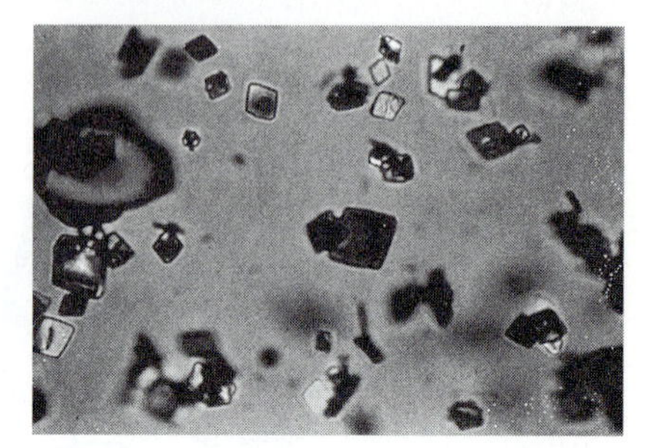

Uric acid crystals (plates); polarized light. (400×)

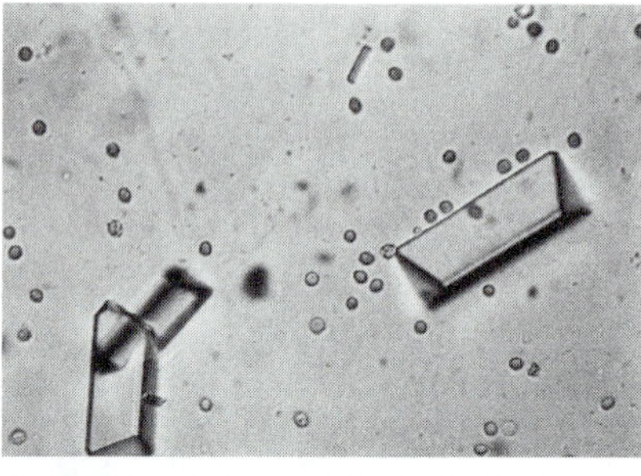

Triple phosphate crystals; "coffin-lid" form. (400×)

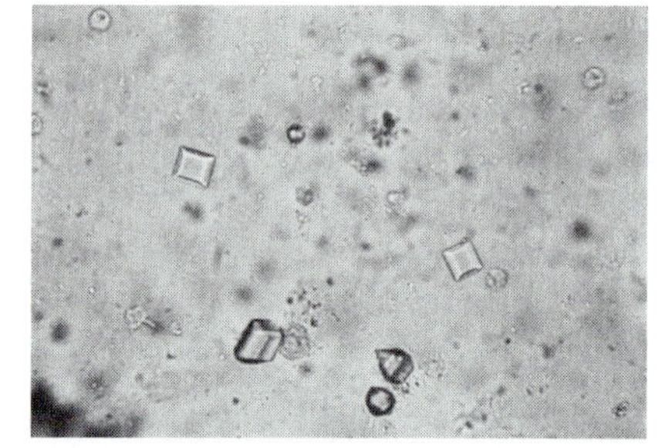

Calcium Oxalate crystals. (400×)

**Figure 9.9** Photographs of urine sediment.

# Laboratory Report 9.3

Name ____________________

Date ____________________

Section ____________________

## DATA FROM EXERCISE 9.3

1. Record your data and the interpretations of your data in this table.

| Urine Test | Result of Exercise (Positive or Negative) | Physiological Reason for Negative Result | Clinical Significance of Positive Result |
|---|---|---|---|
| Proteinuria | | | |
| Glycosuria | | | |
| Ketonuria | | | |
| Hemoglobinuria | | | |
| Bilirubinuria | | | |

(a) Describe the appearance (such as color, intensity, turbidity) of the urine sample and explain the possible causes of these observations.

(b) List the casts, cells, or crystals of the urine sediment identified under low-power; and suggest the possible clinical significance if these components were found in abnormally high concentrations.

## REVIEW ACTIVITIES FOR EXERCISE 9.3

### *Test Your Knowledge of Terms and Facts*

Match these molecules with their descriptions:

| | |
|---|---|
| _____1. normally not filtered | (a) ketone bodies |
| _____2. filtered, then normally completely reabsorbed | (b) bilirubin |
| _____3. derived from fat breakdown | (c) glucose |
| _____4. derived from heme groups of hemoglobin | (d) protein |

5. Common cells found in normal urine sediment ______________________________.
6. An abnormal component of urine sediment, formed from protein ______________________________.
7. Indicate a possible cause for each of these conditions:
   (a) proteinuria ______________________________
   (b) glycosuria ______________________________
   (c) ketonuria ______________________________
   (d) bilirubinuria ______________________________

### *Test Your Understanding of Concepts*

8. Describe the composition of urinary casts and explain how they can get into the urine.

9. Which component of the urine would be raised if a person were on a very low carbohydrate weight-reducing diet? Explain the process involved.

## *Test Your Ability to Analyze and Apply Your Knowledge*

10. Is it possible for someone to have an abnormally high plasma glucose concentration and yet not have glycosuria? Explain your answer. Relate this answer to a person who eats a couple of sugar-coated doughnuts compared to a person with uncontrolled diabetes mellitus.

11. Proteinuria and the presence of numerous casts in the urine sediment are often accompanied by edema. What is the relationship between these symptoms? (*Hint:* Review exercise 2.1—particularly the clinical applications box regarding plasma protein concentration.)

# Digestion and Nutrition

# Section 10

The digestive tract is a continuous tube that is open to the external environment at both ends, by way of the mouth and the anus (fig. 10.1). Material inside this digestive tract is outside the body in the sense that it can contact only the epithelial cells that line the tract. For this material to reach the inner cells of the body, it must pass through the epithelial cells of the tract (a process known as **absorption**) into the blood. Before nutritive material can be absorbed, however, it must first be broken down by physical processes, such as chewing (mastication), and by enzymatic hydrolysis into its monomers. The process of hydrolyzing larger food molecules (polymers) into absorbable monomers is known as **digestion.**

The embryonic digestive system consists of a hollow tube only one cell layer thick. As the embryo develops, different regions of the digestive tract become specialized for different functions. Some epithelial cells that line the tract become secretory, forming exocrine glands that secrete mucus, HCl, or particular hydrolytic enzymes characteristic of a certain region of the digestive tract. The liver, which produces bile, and the pancreas, which produces the many digestive enzymes found in *pancreatic juice,* develop as outpouchings *(diverticula)* of the embryonic small intestine and maintain their connections with the intestine by means of the hepatic and pancreatic ducts. Other regions of the small intestine become specialized for absorption through an increase in surface area. This increased surface area is produced by epithelial folds known as **villi** and minute foldings of the epithelial cell membranes called **microvilli.**

The digestive tract may be visualized as a "disassembly" line, where the food is conveyed, by means of muscular movements of the tract *(peristalsis)* and the opening and closing of sphincter muscles, from one stage of processing to the next. Coordination of these processes is achieved by neural reflexes and by hormones secreted by the gastrointestinal tract (table 10.1).

Diet maintains the consistent supply of nutrients to the body cells as energy sources for fuel, for growth, and for the replacement of cellular parts. *Carbohydrates, fats, proteins, vitamins, minerals,* and *water* are six nutrient classes recommended for daily consumption. Only carbohydrates, fats, and proteins can provide energy, measured in **kilocalories** *(kcals).* The **basal metabolic rate (BMR)** is the minimum amount of energy required by the body at rest.

| | |
|---|---|
| **Exercise 10.1** | Histology of the Gastrointestinal Tract, Liver, and Pancreas |
| **Exercise 10.2** | Digestion of Carbohydrate, Protein, and Fat |
| **Exercise 10.3** | Nutrient Assessment, BMR, and Body Composition |

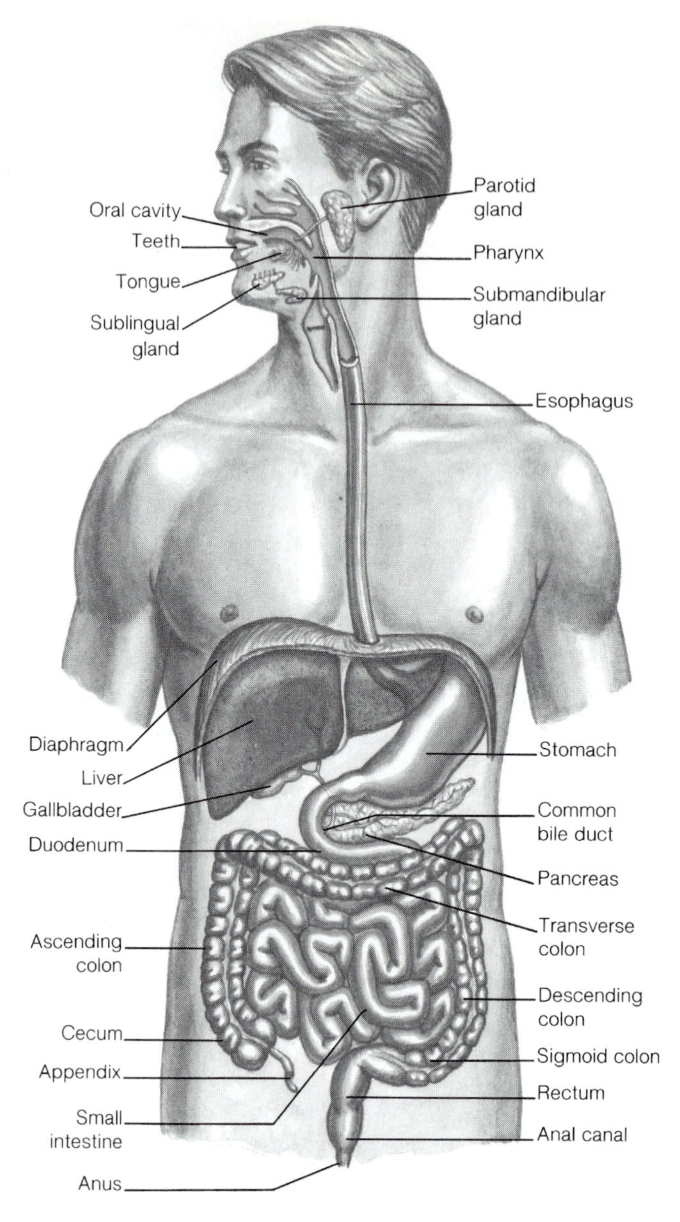


**Figure 10.1** Organs of the digestive system.

**(For a full-color version of this figure, see fig. 18.2 in *Human Physiology,* eighth edition, by Stuart I. Fox.)**

## Table 10.1 Physiological Effects of Gastrointestinal Hormones

| Secreted by | Hormone | Effects |
|---|---|---|
| Stomach | Gastrin | Stimulates parietal cells to secrete HCl<br>Stimulates chief cells to secrete pepsinogen<br>Maintains structure of gastric mucosa |
| Small intestine | Secretin | Stimulates water and bicarbonate secretion in pancreatic juice<br>Potentiates actions of cholecystokinin on pancreas |
| Small intestine | Cholecystokinin (CCK) | Stimulates contraction of gallbladder<br>Stimulates secretion of pancreatic juice enzymes<br>Inhibits gastric motility and secretion<br>Maintains structure of exocrine pancreas (acini) |
| Small intestine | Gastric inhibitory peptide (GIP) | Inhibits gastric motility and secretion<br>Stimulates secretion of insulin from pancreatic islets |
| Ileum and colon | Glucagon-like peptide-1 (GLP-1) | Inhibits gastric motility and secretion<br>Stimulates secretion of insulin from pancreatic islets |
| | Guanylin | Stimulates intestinal secretion of $Cl^-$, causing excretion of NaCl and water in the feces |

# Histology of the Gastrointestinal Tract, Liver, and Pancreas

EXERCISE **10.1**

**MATERIALS**

1. Microscopes
2. Prepared tissue slides of the digestive system

All regions of the gastrointestinal tract have a mucosa, submucosa, muscularis, and serosa, but these layers of the wall display different specializations in different regions of the tract. The histology of the liver and pancreas provides insights into the functions of these organs.

**OBJECTIVES**

1. Identify the mucosa, submucosa, muscularis, and serosa layers of different regions of the gastrointestinal tract.
2. Describe the structure and function of the layers of the esophagus, stomach, small intestine, and large intestine.
3. Describe the microscopic anatomy of the liver and explain its functional significance.
4. Describe the microscopic anatomy of the pancreas and distinguish the parts involved in the endocrine and exocrine functions of the pancreas.

**Textbook Correlations***

Before performing this exercise, you should study the introductory material presented here. Further information relating to this exercise can be found in these pages of *Human Physiology,* eighth edition, by Stuart I. Fox:

- *Esophagus and Stomach.* Chapter 18, pp. 564–566.
- *Small Intestine.* Chapter 18, pp. 568–572.
- *Large Intestine.* Chapter 18, pp. 572–574.
- *Live, Gallbladder, and Pancreas.* Chapter 18, pp. 575–582.

The tubular digestive tract, including the esophagus, stomach, small intestine, and large intestine, consists of four major layers, or *tunics* (fig. 10.2). From the innermost layer outward, they are as listed here:

1. The **mucosa,** or mucous membrane, consists of an inner epithelium spread over a thin layer of connective tissue, the *lamina propria,* which is bordered by a ribbon of smooth muscle, the *muscularis mucosa.* The epithelium is stratified squamous in the esophagus and anal canal and simple columnar in the stomach, small intestine, and large intestine.
2. The **submucosa** is connective tissue and therefore has abundant extracellular space for blood vessels, nerves, and mucus-secreting glands. Parasympathetic fibers and ganglia can be seen as the *submucosal (Meissner's) plexus* in the submucosa.
3. The **muscularis externa** consists of smooth muscle, arranged in an inner circular and outer longitudinal layer throughout most of the digestive tract. Parasympathetic fibers and ganglia can be seen as the *myenteric (Auerbach's) plexus* in this layer.
4. The **serosa** consists of a *simple squamous* epithelium and connective tissue, and is the outermost covering of the digestive tract.

## A. Esophagus and Stomach

The mucosa layer of the esophagus is lined with a stratified squamous epithelium (fig. 10.3). The muscles of the first third of the esophagus, like those of the pharynx and mouth, are striated to provide voluntary control of swallowing. The middle third contains a mixture of striated and smooth muscle, and the last third of the esophagus contains only involuntary smooth muscle.

The submucosa of the stomach is thrown into large folds, or *rugae,* which can be seen with the unaided eye. Microscopic examination of the mucosa shows that it, too, is folded. The openings of these folds into the stomach lumen are called *gastric pits.* The cells that line the

*Multimedia Correlations (also see Appendix 3)
- *MediaPhys 2.0:* Topics 14.9–14.55

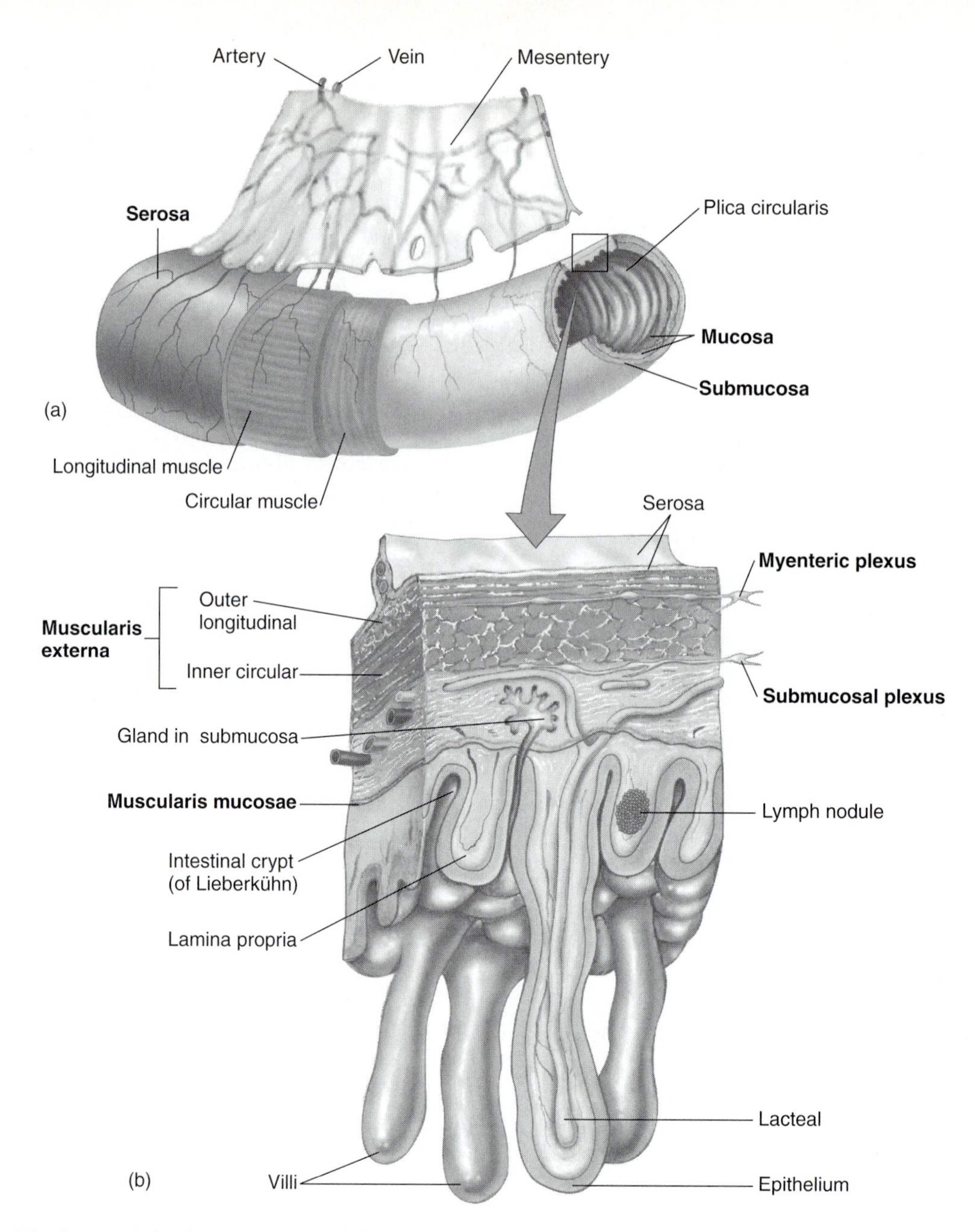


**Figure 10.2 The layers of the digestive tract.** (a) An illustration of the major tunics, or layers, of the small intestine. The insert shows how folds of mucosa form projections called villi in the small intestine. (b) An illustration of a cross section of the small intestine showing layers and glands.

**(For a full-color version of this figure, see fig. 18.3 in *Human Physiology*, eighth edition, by Stuart I. Fox.)**

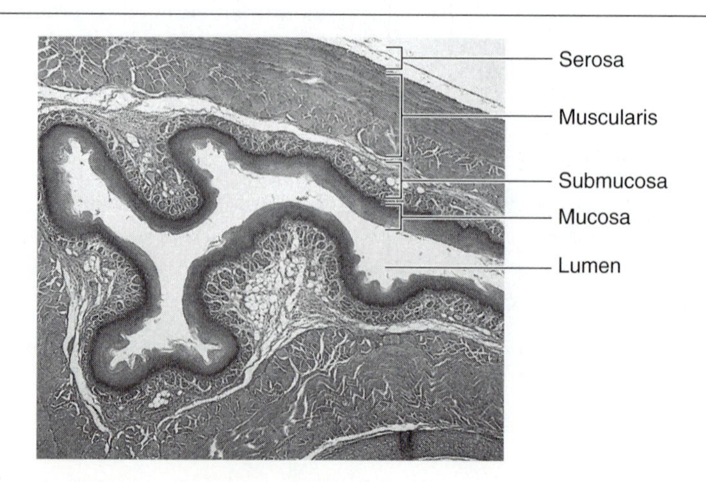


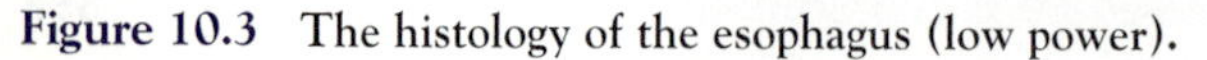

**Figure 10.3** The histology of the esophagus (low power).

folds of mucosa are secretory and form the **gastric glands** (figs. 10.4 and 10.5).

The gastric glands include: (1) *goblet cells,* which secrete mucus; (2) *parietal cells,* which secrete hydrochloric acid (HCl); (3) *chief cells,* which secrete pepsinogen (the inactive precursor of pepsin, a protein-digesting enzyme); (4) *enterochromaffin-like cells (ECL)* which secrete histamine; (5) *G cells,* which secrete the hormone gastrin into the blood; and (6) *D cells,* which secrete the hormone somatostatin. The gastric mucosa also secretes a polypeptide called *intrinsic factor,* which aids in the absorption of vitamin $B_{12}$ in the intestine.

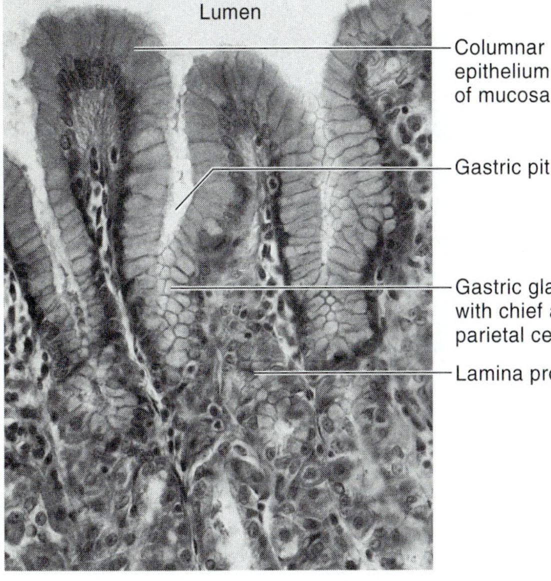


**Figure 10.4 The microscopic structure of the stomach.**

## PROCEDURE

1. Observe a cross section of the esophagus under 100× (using the 10× objective lens) and note the four major layers, or tunics.
2. Hold a slide of a stomach section up to a light source and observe a fold, or ruga. Now, place the slide on a microscope and, under 100×, observe the gastric pits and glands in the mucosa, the submucosa, and the muscularis externa.
3. Using the high-dry objective lens (45×), observe the gastric glands in the mucosa under a total magnification of 450×. Identify goblet cells near the surface of the gastric pits. These mucus-secreting cells are numerous and clear in appearance. Near the base of the glands, identify parietal cells (with red-staining cytoplasm) and chief cells (with blue-staining cytoplasm). ECL cells and G cells cannot be easily identified without specially stained slides.

## B. Small Intestine and Large Intestine

The small intestine is approximately 21 feet long (in a cadaver) and divided into three regions. The first region, approximately 12 inches long, is called the **duodenum.** The next region, the **jejunum,** is about 8 feet long and constitutes two-fifths of the entire length of the intestine. The **ileum,** about 12 feet long (constituting three-fifths of the intestine), is the terminal region.

The mucosa and submucosa of the small intestine form large folds called the *plicae circulares.* The surface area of the mucosa is further increased by microscopic

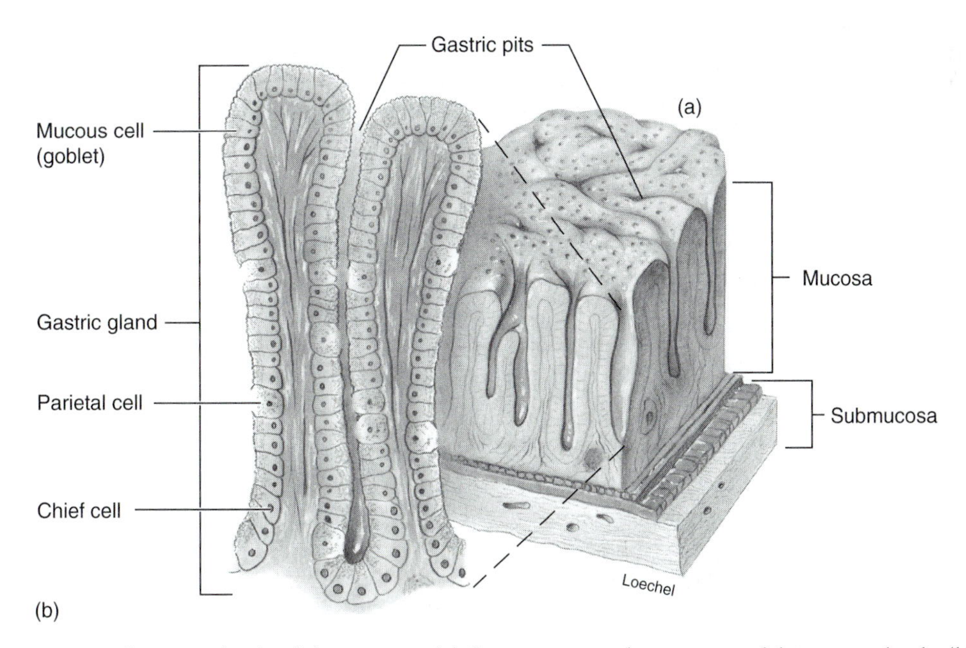


**Figure 10.5 Gastric pits and gastric glands of the mucosa.** (a) Gastric pits are the openings of the gastric glands. (b) Gastric glands consist of mucous cells, chief cells, and parietal cells, each of which produces a specific secretion.

**(For a full-color version of this figure, see fig. 18.7 in *Human Physiology,* eighth edition, by Stuart I. Fox.)**

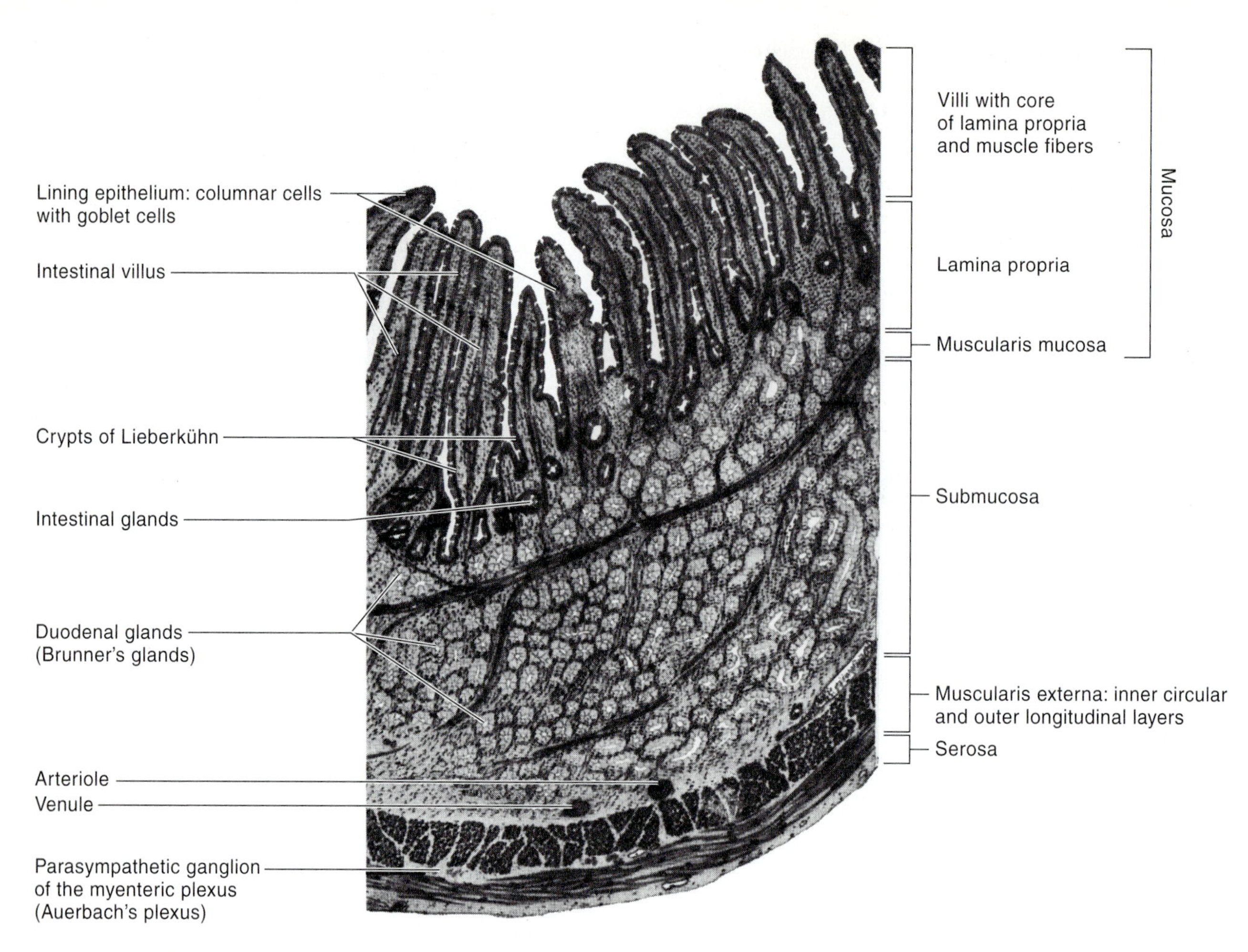


**Figure 10.6** The microscopic structure of the duodenum.

## Table 10.2 Characteristics of the Major Digestive Enzymes

| Enzyme | Site of Production | Source | Substrate | Optimum pH | Product(s) |
|---|---|---|---|---|---|
| Salivary amylase | Mouth | Saliva | Starch | 6.7 | Maltose |
| Pepsin | Stomach | Gastric glands | Protein | 1.6–2.4 | Shorter polypeptides |
| Pancreatic amylase | Duodenum | Pancreatic juice | Starch | 6.7–7.0 | Maltose, maltriose, and oligosaccharides |
| Trypsin, chymotrypsin, carboxypeptidase | | | Polypeptides | 8.0 | Amino acids, dipeptides, and tripeptides |
| Pancreatic lipase | | | Triglycerides | 8.0 | Fatty acids and monoglycerides |
| Maltase | | Epithelial membranes | Maltose | 5.0–7.0 | Glucose |
| Sucrase | | | Sucrose | 5.0–7.0 | Glucose + fructose |
| Lactase | | | Lactose | 5.8–6.2 | Glucose + galactose |
| Aminopeptidase | | | Polypeptides | 8.0 | Amino acids, dipeptides, tripeptides |

folds that form fingerlike projections called *villi* (fig. 10.6). Each villus has a core of connective tissue (the lamina propria) covered with a simple columnar epithelium. The apical surface (facing the lumen) of each epithelial cell has a slightly blurred, "brush border" appearance because of numerous projections of its cell membrane in the form of *microvilli*. Microvilli can be clearly seen only with an electron microscope.

The microvilli, villi, and plicae circulares increase the surface area of the small intestine tremendously, thus maximizing the rate at which the products of digestion can be absorbed by transport through the epithelium into the blood. Various digestive enzymes—called **brush-border enzymes**—are fixed to the cell membranes of the microvilli and act together with enzymes from pancreatic juice to catalyze hydrolysis reactions of food molecules. The small intestine epithelium is also the source of important gastrointestinal hormones (table 10.2).

The epithelium at the base of the villi invaginates to form pouches called *crypts of Lieberkühn*, or intestinal

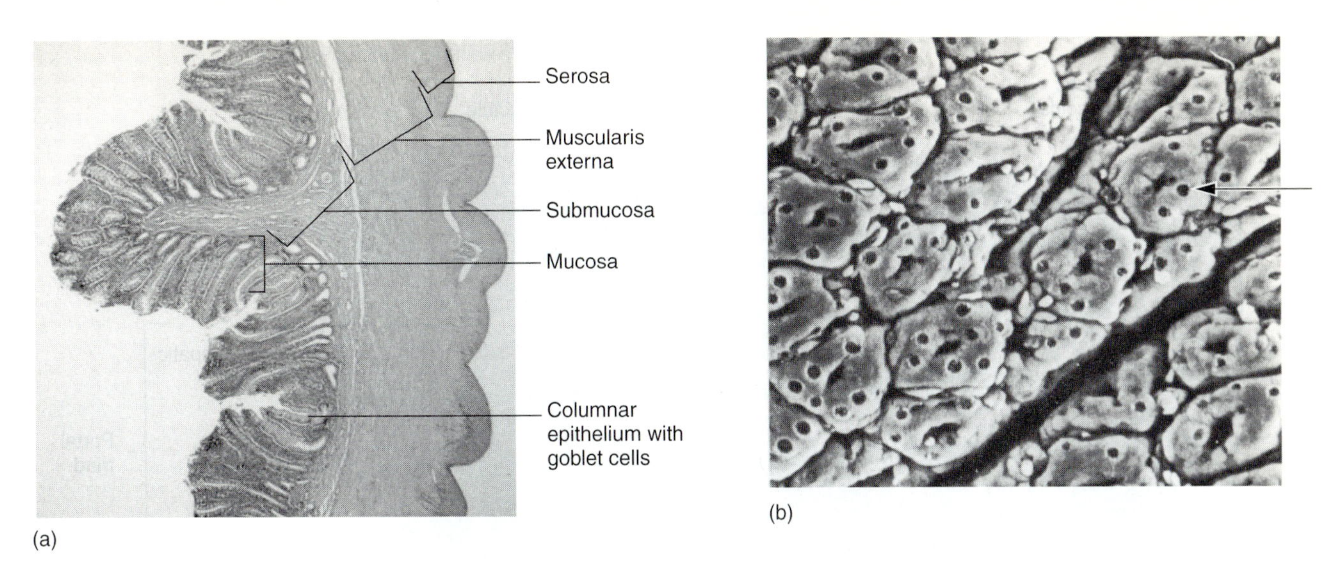


**Figure 10.7 The microscopic structure of the ileum and large intestine.** (a) A photomicrograph of a cross section of the ileum. (b) A scanning electron micrograph of the luminal surface of the large intestine (the arrow points to the opening of a goblet cell).

(b is from R. G. Kessel and R. H. Kardon, *Tissues and Organs: A Text-Atlas of Scanning Electron Microscopy* © 1979 W. H. Freeman and Company.)

crypts. Although somewhat similar in appearance to gastric glands of the stomach, these intestinal crypts are not secretory. Instead, the cells within the crypts undergo mitotic division and push upward to replace those cells that are continuously shed into the lumen from the tips of the villi. Within the submucosa of the duodenum are alkaline mucus-secreting *Brunner's glands* (fig. 10.6). The jejunum and ileum lack these glands but are otherwise similar in structure (fig. 10.7).

Waste products from the small intestine pass into the **colon** of the large intestine where water, sodium, and potassium are absorbed. The mucosa of the large intestine contains crypts of Lieberkühn but not villi, so its surface has a flat appearance. As with the small intestine, numerous lymphocytes can be seen in the lamina propria, and large lymphatic nodules appear at the junction of the mucosa and submucosa. Lymphatic nodules are clearly evident in a section of the *appendix*, a short outpouching from the cecum.

## PROCEDURE

1. Use the lowest power available on the microscope to observe the layers of a section of small intestine. Identify the villi, submucosa, and muscularis externa.
2. Observe the villi using the 45× objective lens. Identify the goblet cells in the epithelium and the numerous lymphocytes (small, blue-staining cells) in the lamina propria within each villus. Also within the lamina propria, note the *central lacteal*—a lymphatic vessel that transports absorbed fat from the intestine.
3. Observe a slide of the large intestine using the 10× objective lens (fig. 10.7). Note the four layers, absence of villi, and goblet cells in the columnar epithelium of the mucosa.

## C. Liver

The liver aids digestion by producing and secreting **bile,** which emulsifies fat. Bile leaves the liver in the *common hepatic duct*, which branches to form the *cystic duct* and the *common bile duct*. The cystic duct channels bile to the gallbladder where it is stored and concentrated. The common bile duct joins the pancreatic duct and, together, they empty into the duodenum.

The liver also serves to modify the composition of the blood that drains from capillaries of the intestine through the *hepatic portal vein*. Before this rich venous blood can return to the heart and circulation, it must pass through *sinusoids* in the liver tissue (figs. 10.8 and 10.9). Liver sinusoids, however, are wider than most types of capillaries and are lined with phagocytic cells called *Kupffer cells*. Blood is drained from these sinusoids by small *central veins* that ultimately merge to form the large *hepatic vein*, which carries blood away from the liver. Liver sinusoids also receive arterial blood from branches of the *hepatic artery*. Arterial blood mixes with blood from the hepatic portal vein, and together they pass through the sinusoids to the central vein (fig. 10.9).

Bile is produced and secreted by the liver cells *(hepatocytes)*, but does not mix with blood because bile never enters the sinusoids. Instead, the hepatocytes secrete bile into *bile canaliculi* that are located between adjacent hepatocytes (fig. 10.9). Bile is drained from the canaliculi into *bile ducts*, located near the entry of the portal vein and hepatic artery into the sinusoid. The grouping of the portal vein, hepatic artery, and bile duct that one sees in a microscopic view of the liver is called a *portal*, or *hepatic*, *triad* (fig. 10.8).

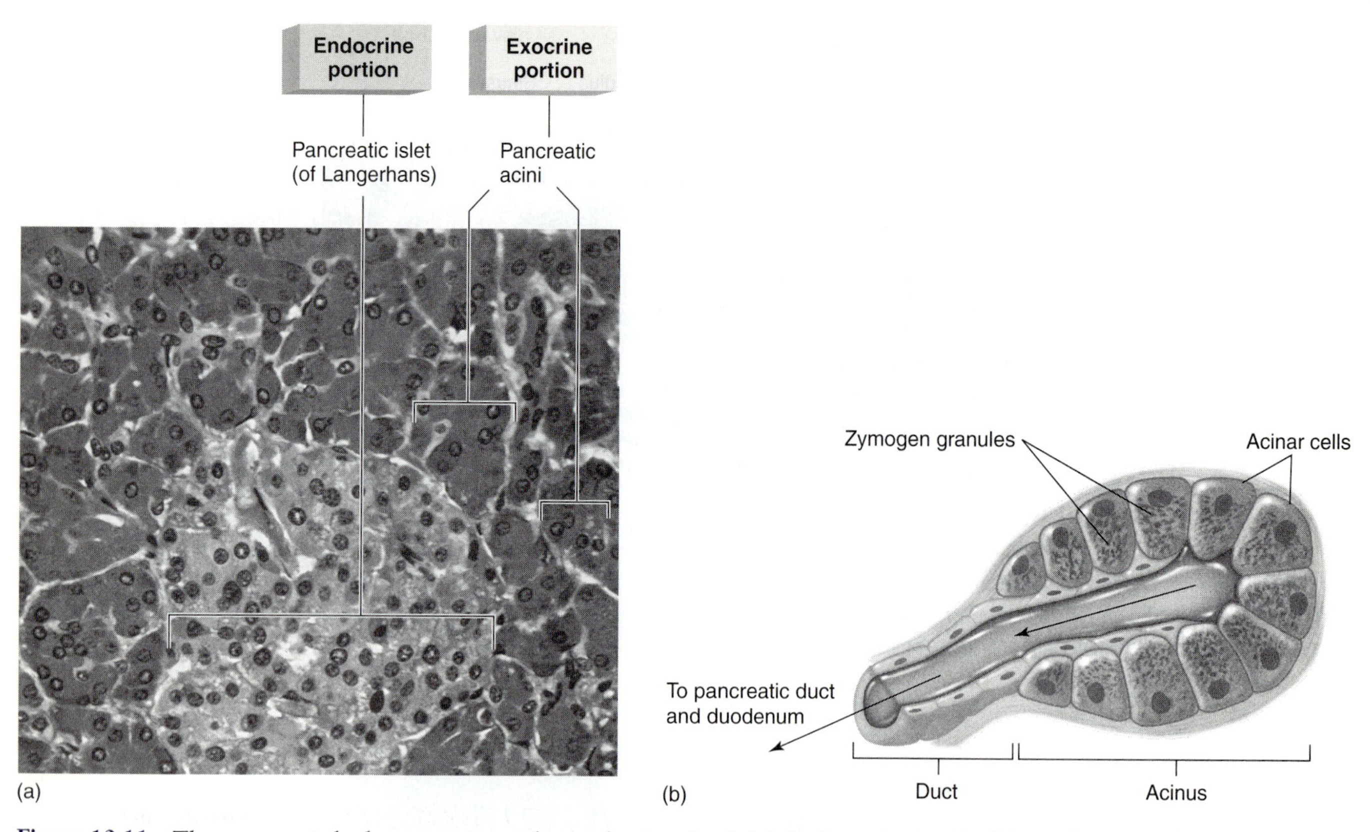


**Figure 10.11 The pancreas is both an exocrine and an endocrine gland.** (a) A photomicrograph of the endocrine and exocrine portions of the pancreas. (b) An illustration depicting the exocrine pancreatic acini, where the acinar cells produce inactive enzymes stored in zymogen granules. The inactive enzymes are secreted by way of a duct system into the duodenum.

**(For a full-color version of this figure, see fig. 18.28 in *Human Physiology*, eighth edition, by Stuart I. Fox.)**

Inflammation of the liver, or **hepatitis,** may be caused by bacterial or viral infections, alcohol abuse, allergy, or drugs. This condition is usually reversible. In **cirrhosis,** however, large areas of liver tissue are destroyed and replaced with permanent connective tissue and "regenerative nodules" of hepatocytes that lack the platelike structure of normal liver tissue. Since, among its many functions, the liver produces plasma *albumin* and converts ammonia to *urea,* liver disease may be accompanied by a decrease in the plasma albumin concentration and in the appearance of ammonia in the blood.

Inflammation of the pancreas, or **pancreatitis,** can result from the action of digestive enzymes on pancreatic tissue. The digestive enzymes produced within the pancreas are normally inactive until they enter the duodenum, but activated enzymes may reflux from the duodenum into the pancreatic duct. This produces an inflammation reaction accompanied by a "leakage" of enzymes into the blood. This may be detected clinically by a rise in the concentration of pancreatic amylase in the plasma.

## PROCEDURE

1. Examine the pancreas under low magnification. The numerous small clusters of cells are the pancreatic acini. Occasional larger groupings of cells less intensely stained are pancreatic islets of Langerhans (fig. 10.11).
2. Scan the slide for a pancreatic duct in cross section. When one has been found, change to high magnification and observe its simple columnar epithelium.

# Laboratory Report 10.1

Name ______________________

Date ______________________

Section ______________________

## REVIEW ACTIVITIES FOR EXERCISE 10.1

### *Test Your Knowledge of Terms and Facts*

1. Identify the cells of the gastric mucosa that secrete:
   (a) HCl ______________________
   (b) pepsinogen ______________________
   (c) histamine ______________________
2. The three components of the mucosa layer of the digestive tract are the ______________________, the ______________________, and the ______________________.
3. Microscopic fingerlike projections of mucosa in the small intestine are called ______________________.
4. The foldings of the plasma membrane of intestinal epithelial cells that produce the "brush border" are called ______________________.
5. Blood is transported from the intestine to the liver in a large vessel known as the ______________________.
6. Once blood has reached the liver, it travels through large capillaries called ______________________.
7. The microscopic exocrine units of the pancreas are called ______________________; the endocrine structures are known as the ______________________.
8. The glands in the duodenum that secrete an alkaline mucus: ______________________.

### *Test Your Understanding of Concepts*

9. Describe the structural adaptations of the small intestine that help increase the surface area and the rate at which digestion products can be absorbed.

In this exercise, the effects of pH and temperature on the activity of amylase will be tested by checking for the disappearance of substrate (starch) and the appearance of product (maltose) at the end of an incubation period. The appearance of a reducing sugar (maltose) in the incubation medium will be determined by the *Benedict's test*, where an alkaline solution of cupric ions ($Cu^{2+}$) is reduced to cuprous ions ($Cu^{+}$), forming a yellow-colored precipitate of cuprous oxide ($Cu_2O$).

Although starch digestion begins in the mouth with the action of salivary amylase, this is usually of minor importance in digestion (unless one chews excessively). Most of the digestion of polysaccharides and complex sugars to monosaccharides occurs in the small intestine when exposed to pancreatic and fixed brush-border enzymes. The hydrolytic action of amylase in combination with the buffering action of saliva may help prevent the accumulation of fermentable carbohydrates between the gums (gingiva) and the teeth, thus serving to protect against the growth of harmful bacteria that result in dental cavities (**caries**).

## PROCEDURE (SEE FIG. 10.12)

**Step 1:**

(1) Label four clean test tubes 1–4.

(2) Obtain 10 mL of saliva in a small, graduated cylinder. (Salivation can be aided by chewing a piece of paraffin.) If only 5 mL of saliva is obtained, dilute the saliva with an equal volume of distilled water.

**Note:** *As an alternative to saliva, the instructor may provide a solution of amylase derived from the pig or human pancreas.*

**Step 2:**

(1) Add 3.0 mL of distilled water to tube 1.

(2) Add 3.0 mL of saliva to tubes 2 and 3.

(3) Add 3 drops of concentrated HCl to tube 3.

(4) Boil the remaining saliva in a separate Pyrex test tube by passing the tube through the flame of a Bunsen burner. Use a test-tube clamp and keep the tube at an angle, pointed away from yourself and others. When cool, add 3.0 mL of this boiled saliva to tube 4.

**Step 3:** Add 5.0 mL of cooked starch (provided by the instructor) to each of the four tubes.

**Step 4:** Incubate all tubes for 1–1 1/2 hours in a 37° C water bath.

**Step 5:** Divide the contents of each sample by pouring half into four new test tubes.

**Step 6:** Test one set of four solutions for starch by adding a few drops of iodine solution *(Lugol's reagent)*. A positive test is indicated by the development of a purplish black color.

**Step 7:** Test the other set of four solutions for reducing sugars in the following way:

(1) Add 5.0 mL of Benedict's reagent to each of the four test tubes and immerse them in a rapidly boiling water bath for 2 minutes.

(2) Remove the tubes from the boiling water with a test-tube clamp and rate the amount of reducing sugar present according to the following scale:

| | |
|---|---|
| Blue (no maltose) | – |
| Green | + |
| Yellow | ++ |
| Orange | +++ |
| Red (most maltose) | ++ ++ |

Enter your results in the data table in your laboratory report.

## B. Digestion of Protein (Egg Albumin) by Pepsin

Although amylase is most active at the pH of saliva (pH 6–7), the enzyme **pepsin** has a pH optimum that is adapted to the normal pH of the stomach (pH less than 2—see table 10.2). The low pH of the stomach is due to the secretion of hydrochloric acid (HCl) by parietal cells in the gastric glands (fig. 10.13). The strong acidity of the stomach coagulates proteins, thus facilitating their digestion by pepsin and, later by other proteolytic enzymes in the small intestine.

**Gastric ulcers** are apparently not due to an increase in stomach acidity, but rather to a breakdown in the normal mucosal barriers to digestion. The barriers are believed to be (1) the tight junctions between adjacent epithelial cells that prevent hydrogen ions from entering the mucosa and (2) the rapid renewal of surface epithelial cells. (The stomach sheds half a million cells a minute, completely renewing the gastric mucosa every three days.) By itself, the thick layer of mucus that covers the gastric epithelium is not an effective barrier to self-digestion. The weakening of the mucosal barrier to hydrogen ions is promoted by alcohol and salicylates, such as aspirin. The bacterium *Helicobacter pylori* resides in the gastrointestinal tract of many people, and this bacterium may be a causative agent in peptic ulcers. Indeed, peptic ulcer can be treated and even cured in some people by antibiotic therapy.

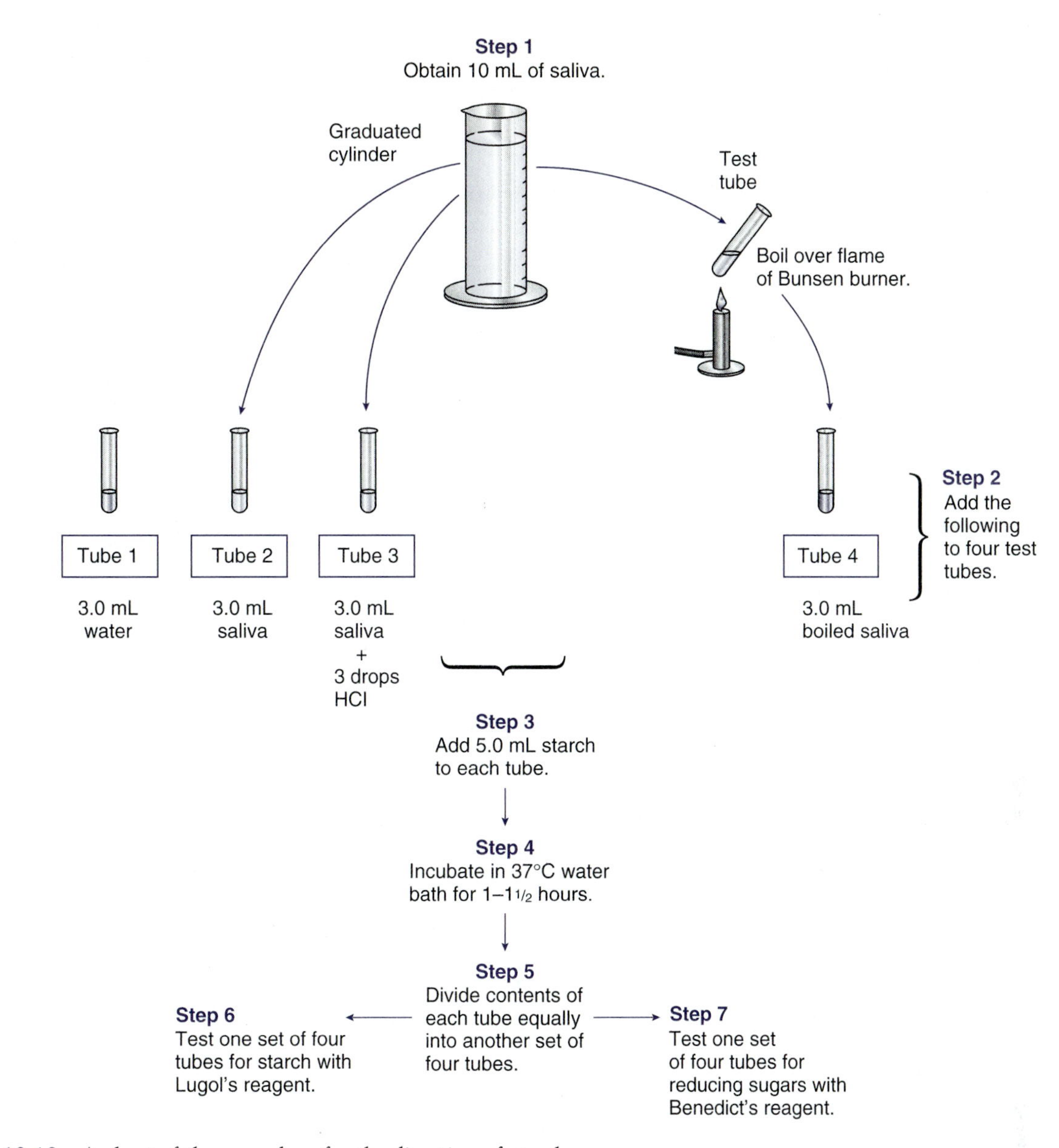


**Figure 10.12** A chart of the procedure for the digestion of starch.

Pepsin, secreted by the *chief cells* of the gastric glands, is responsible for the digestion of less than 15% of ingested protein. Removal of the entire stomach (complete gastrectomy) thus has little effect on protein digestion. The major site of protein digestion is the small intestine, where the enzymes *trypsin* and *chymotrypsin* (secreted by the pancreas) and the *dipeptidases* (fixed in the intestinal brush-border mucosa) hydrolyze proteins and smaller polypeptides into absorbable amino acids (table 10.2).

The stomach does not normally digest itself. A *peptic ulcer* may form when the mucosa of the stomach (gastric ulcer) or duodenum (duodenal ulcer) is digested by the strongly acidic gastric juice. Although the etiology of peptic ulcers is not entirely known, it is believed that ulcers are caused by acid ($H^+$) from the gastric lumen eroding the mucosal surface rather than by the digestive action of pepsin on structural proteins within the mucosa. When the stomach produces an excess of acid, protective mechanisms may not be sufficient to protect the intestinal mucosa. Excessive stomach acid may be produced in susceptible individuals as a result of vagus stimulation and aggravate a duodenal ulcer.

When the acidic products of the stomach (called *chyme*) enter the duodenum, the intestine is stimulated to release the hormone *secretin*, which inhibits the gastric secretion of acid and stimulates the release of alkaline pancreatic juice (see table 10.1). The acidic chyme, therefore, is diluted and neutralized in the small intestine.

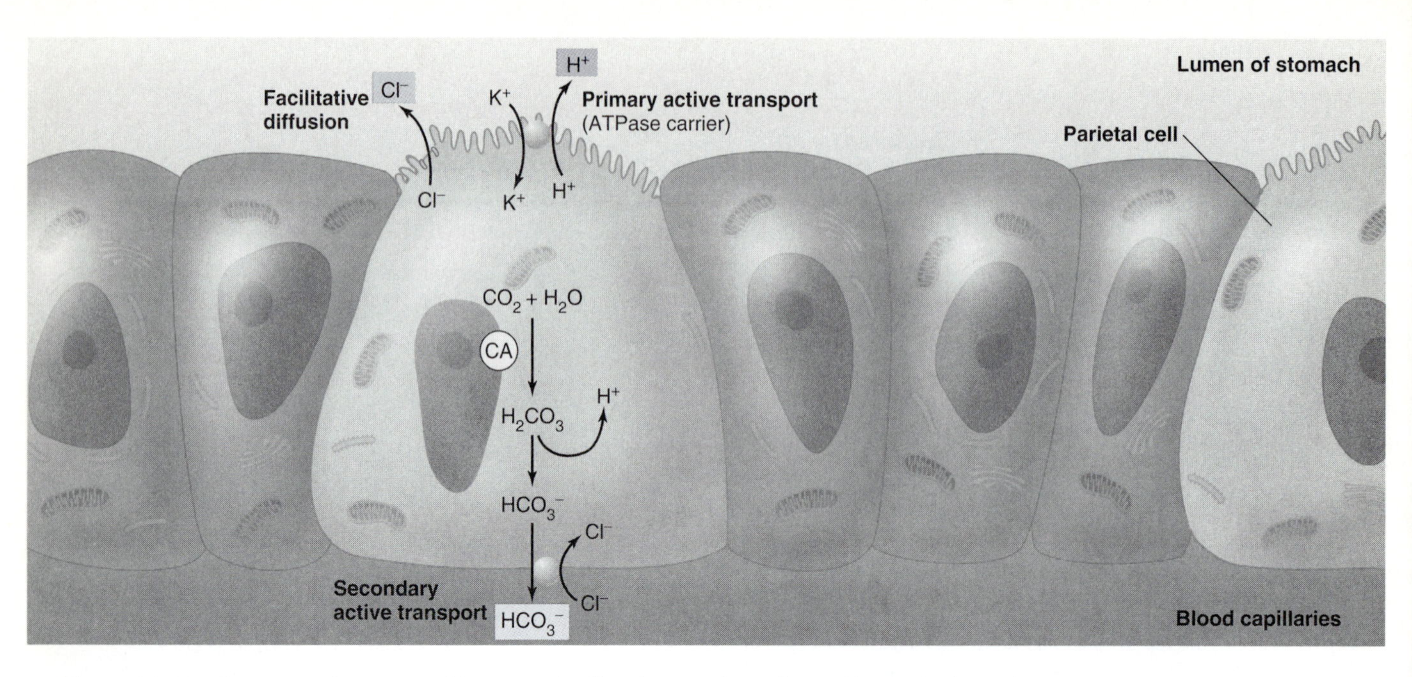


**Figure 10.13 Secretion of gastric acid by parietal cells.** The apical membrane (facing the lumen) secretes $H^+$ in exchange for $K^+$ using a primary active transport carrier that is powered by the hydrolysis of ATP. The basolateral membrane (facing the blood) secretes bicarbonate ($HCO_3^-$) in exchange for $Cl^-$. The $Cl^-$ moves into the cell against its electrochemical gradient, powered by the downhill movement of $HCO_3^-$ out of the cell. This $HCO_3^-$ is produced by the dissociation of carbonic acid ($H_2CO_3$), which is formed from $CO_2$ and $H_2O$ by the action of the enzyme carbonic anhydrase (abbreviated CA). The $Cl^-$ then leaves the apical portion of the membrane by diffusion through a membrane channel. The parietal cells thus secrete HCl into the stomach lumen as they secrete $HCO_3^-$ into the blood.

**(For a full-color version of this figure, see fig. 18.8 in *Human Physiology,* eighth edition, by Stuart I. Fox.)**

## PROCEDURE (SEE FIG. 10.14)

**Step 1:** Label five clean test tubes 1–5. Using a sharp scalpel or razor blade, cut five slices of egg white about the size of a fingernail and as thin as possible. It is essential that the slices be very thin and uniform in size. Place a slice of egg white in each of the five test tubes.

**Step 2:**
(1) Add 1 drop of distilled water to tube 1.
(2) Add 1 drop of concentrated hydrochloric acid (HCl) to tubes 2, 3, and 4.
(3) Add 1 drop of concentrated (10N) NaOH to tube 5.

**Step 3:**
(1) Add 5.0 mL of pepsin solution to tubes 1, 2, 3, and 5.
(2) Add 5.0 mL of distilled water to tube 4.

**Step 4:**
(1) Place tubes 1, 2, 4, and 5 in a 37° C water bath. Place tube 3 in a freezer or ice bath.
(2) Incubate all tubes for 1–1 1/2 hours, remove the tubes (thaw the one that was frozen).

**Step 5:** Remove all incubated tubes and record the appearance of the egg white in the data table in your laboratory report.

## C. Digestion of Triglycerides by Pancreatic Juice and Bile

Although the stomach produces a gastric lipase, the major digestion of triglycerides (fats and oils) occurs in the small intestine through the action of **pancreatic** and **intestinal lipase** (see table 10.2). The digestion of fat in the small intestine is dependent upon the presence of *bile,* which is produced by the liver and transported to the duodenum via the bile duct (fig. 10.15). (The gallbladder serves only to store and concentrate the bile.)

Since fat is not soluble in water, dietary fat enters the duodenum in the form of large fat droplets containing the fat-soluble vitamins A, D, E, and K. The detergent action of bile salts lowers the surface tension of these large droplets, breaking them up into smaller droplets in a process called **emulsification.** Following this process, more surface area is presented to the lipase enzymes, promoting the digestion of fat into monoglycerides and fatty acids, and the release of the fat-soluble vitamins (fig. 10.14).

The absorption of fat is more complicated than that of the water-soluble monomers. The glycerol and fatty acids produced by lipase action aggregate to form spherical structures (*micelles*), which are absorbed by the intestinal

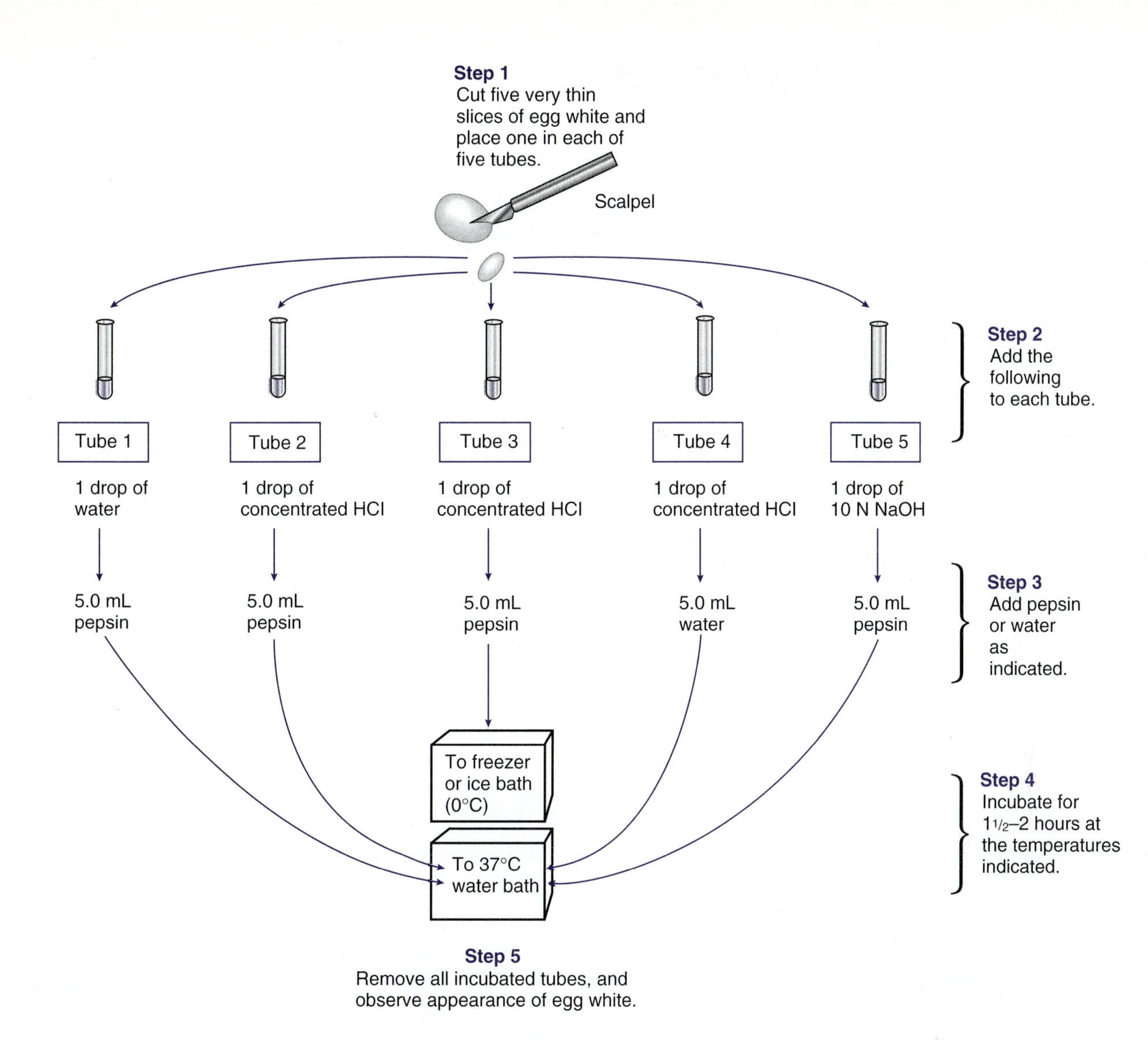


**Figure 10.14** **A chart of the procedure for the digestion of protein.**

epithelium. Once in the epithelial cells, the monomers are resynthesized to form tiny lipid droplets primarily composed of triglycerides *(chylomicrons)*, which are then secreted into lymphatic vessels (lacteals) of intestinal villi (fig. 10.15). From there, chylomicrons are carried via lymph to veins. Unlike the other products of digestion, therefore, lipids enter the blood as polymers rather than monomers. It should be emphasized, however, that all foodstuffs, including fats, must be completely digested into their monomers before they can be absorbed by the digestive epithelium.

In this exercise, we will test the digestion of fat into glycerol and fatty acids by measuring the decrease in pH produced by the liberation of free fatty acids, as the digestion of triglycerides proceeds.

## PROCEDURE (SEE FIG. 10.16)

**Step 1:** Add 3.0 mL of cream or vegetable oil to three test tubes, numbered 1–3.

**Step 2:** Add the following:

(1) To tube 1, add 5.0 mL of water and a few grains of bile salts

(2) To tube 2, add 5.0 mL of pancreatin solution

(3) To tube 3, add 5.0 mL of pancreatin solution and a few grains of bile salts

**Step 3:** Incubate the tubes at 37° C for 1 hour, checking the pH of the solutions at 20-minute intervals with a pH meter or with short-range pH paper.

**Step 4:** Record your data in the table in your laboratory report.

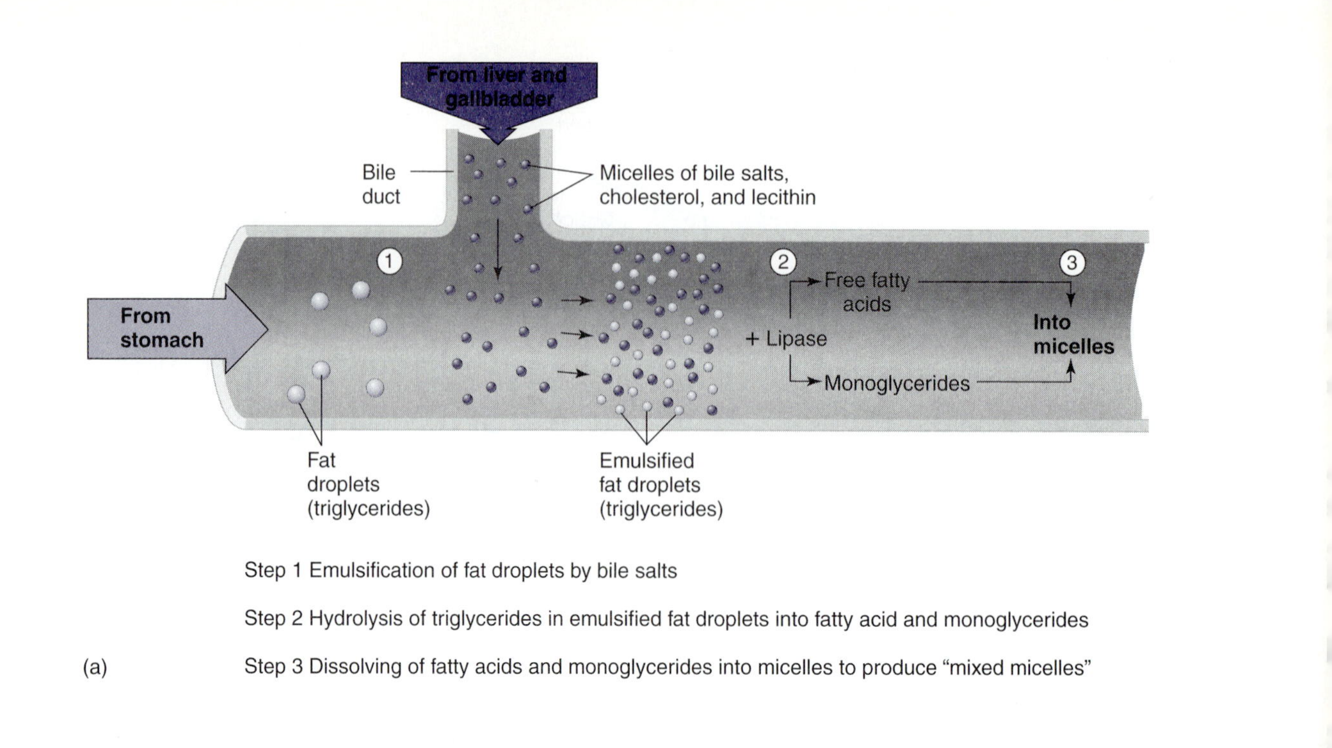


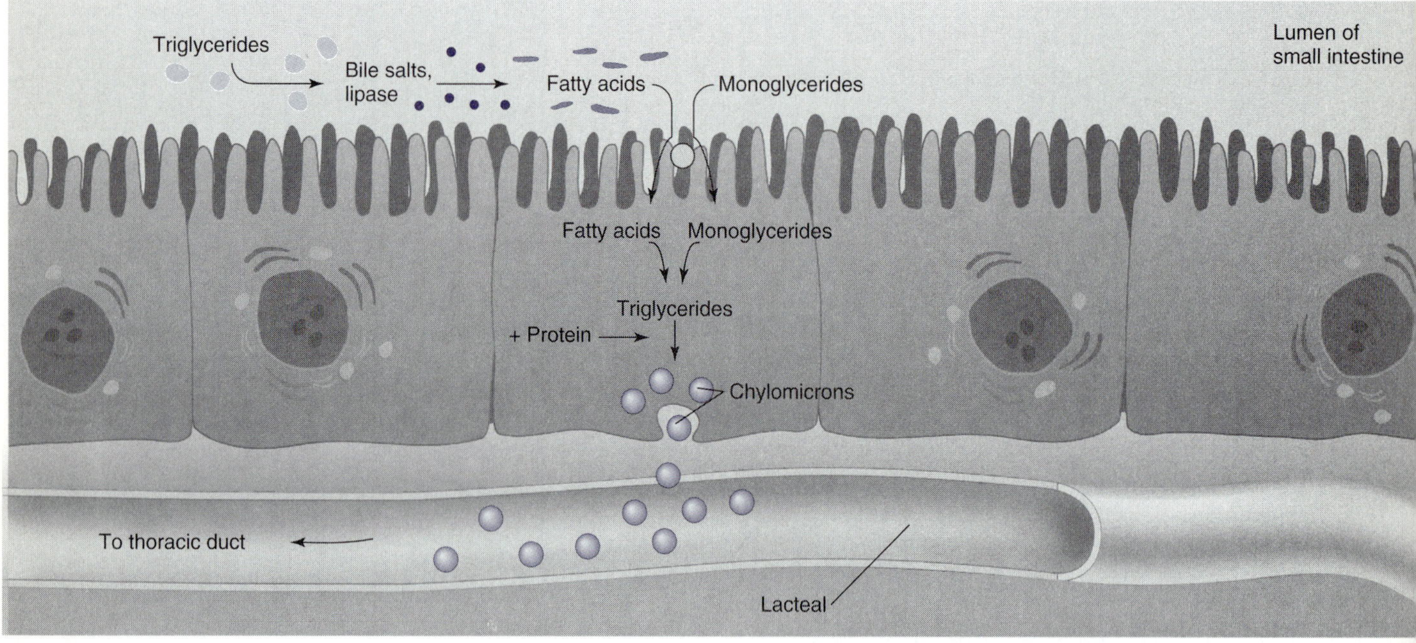


**Figure 10.15** **Fat digestion and absorption.** (a) The steps in the digestion of fat. (b) The process of fat absorption into the intestinal epithelial cells and secretion into lymphatic capillaries (lacteals).

**(For full-color versions of these figures, see figs. 18.35 and 18.36 in *Human Physiology,* eighth edition, by Stuart I. Fox.)**

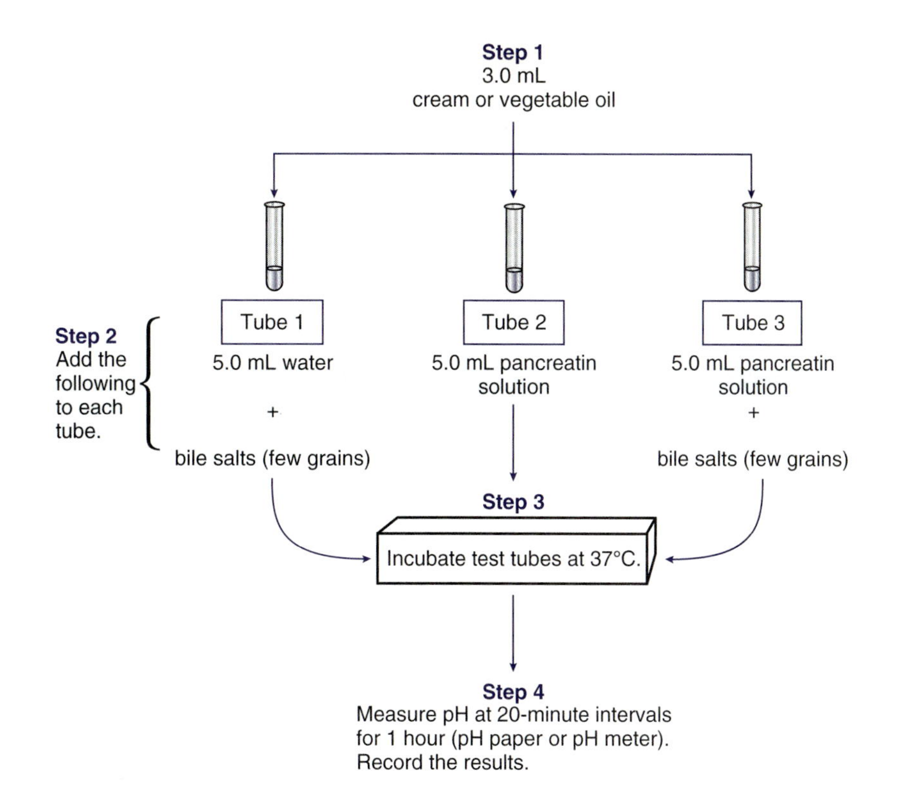


**Figure 10.16** A chart of the procedure for fat digestion.

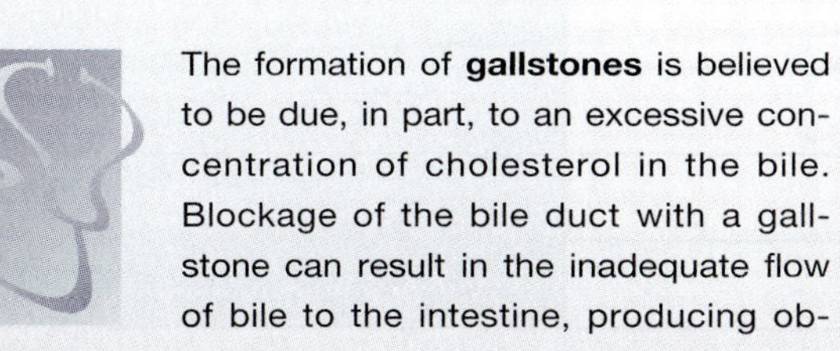

The formation of **gallstones** is believed to be due, in part, to an excessive concentration of cholesterol in the bile. Blockage of the bile duct with a gallstone can result in the inadequate flow of bile to the intestine, producing obstructive jaundice and steatorrhea. *Obstructive jaundice* is an elevation in the blood levels of the bile pigment bilirubin due to blockage of the bile duct. High bilirubin levels produce a yellowish discoloration of the skin, the sclera of the eyes, and the mucous membrane. *Steatorrhea,* the appearance of fat in the feces due to the inadequate digestion and absorption of fat, is associated with a deficiency in the uptake of fat-soluble vitamins A, D, E, and K. Since vitamin K is necessary for normal blood clotting, this condition can be serious. Treatment for gallstones include surgery, dissolution by drugs, and fragmentation by ultrasound.

# REVIEW ACTIVITIES FOR EXERCISE 10.2

## *Test Your Knowledge of Terms and Facts*

1. Starch is partially digested into maltose by the action of ________________.
2. The enzyme in gastric juice that partially digests proteins is ________________; this enzyme has a pH optimum of ________________.
3. Which food group—carbohydrates, lipids, or proteins—is not digested significantly until it reaches the small intestine? ________________
4. Bile is produced by the ________________ and stored in the ________________.
5. What is the function of bile salts? ________________
6. The particles consisting of a combination of triglycerides and protein, which are secreted by intestinal epithelial cells into the central lacteals of the villi, are called ________________.

## *Test Your Understanding of Concepts*

7. In exercise A, which tube(s) contained the most starch following incubation? Which tube(s) contained the most reducing sugars? What conclusion can you draw from these results?

8. In exercise B, which tube showed the most digestion of egg albumin? What can you conclude about the pH optimum of pepsin?

9. Compare the effects of HCl on protein digestion by pepsin and on starch digestion by salivary amylase. Explain the physiological significance of these effects.

10. In part C, which test tube displayed the most rapid fall in pH? Explain the reason for this, including an explanation of how the digestion of fat can cause a fall in the pH of the solution, and the function of bile salts.

11. How does the digestion and absorption of fat differ from the digestion and absorption of glucose and amino acids?

12. Why doesn't the stomach normally digest itself? Why doesn't gastric juice normally digest the duodenum?

## *Test Your Ability to Analyze and Apply Your Knowledge*

13. Suppose, in performing part A, the test for starch and for reducing sugars both came out positive after a 1-hour incubation, but after the tubes incubated for 2 hours, the test for starch came out negative. How could you explain these results?

14. Reviewing your data from part A, what do you think happens to the digestion of a bite of bread after you swallow it?

15. Reviewing your data from part B, explain why frozen food keeps longer than food kept at room temperature.

16. A person with gallstones may have jaundice and an abnormally long clotting time. Explain the possible relationship between gallstones, jaundice, and blood clotting.

# Nutrient Assessment, BMR, and Body Composition[1]

EXERCISE **10.3**

## Materials

1. Home scale or physicians height-weight scale
2. Tape measure, fat calipers (if available)
3. Calorie counting guide such as the U.S. Department of Agriculture Handbook, cookbooks, or popular diet books
4. Alternatively, caloric values of food, and the caloric expenditure of exercise, can be obtained from the Web. One good source is *www.caloriecontrol.org*.

Energy consumed in food and expended by the metabolic activities of the body is measured in kilocalories. Weight is gained when the energy consumption is greater than the energy expenditure, and weight is lost when the reverse is true.

### OBJECTIVES

1. Describe the different nutrient classes; and list the calories per gram for carbohydrates, lipids, and fats.
2. Demonstrate how a dietary record is used to assess food and fluid consumption.
3. Define BMR and demonstrate two different methods for estimating BMR.
4. Define activity factor (AF) and demonstrate how to estimate the number of calories burned for various activities.
5. Calculate and balance the number of calories consumed in the diet and calories expended in activities over three days.
6. Describe how each pound of body weight gain or loss is related to calories consumed or expended.

1. Courtesy of Dr. Lawrence G. Thouin, Jr., Pierce College.

### Textbook Correlations

Before performing this exercise, you should study the introductory material presented here. Further information relating to this exercise can be found in these pages of *Human Physiology,* eighth edition, by Stuart I. Fox:

- *Nutritional Requirement.* Chapter 19, pp. 598–604.
- *Regulatory Functions of Adipose Tissue.* Chapter 19, pp. 605–608.
- *Caloric Expenditures.* Chapter 19, pp.608–609.

## A. Body Composition Analysis

Body composition refers to relative proportion of both the actively metabolizing tissues, or *lean body mass*, and fat tissue in the body. Although some adipose tissue is essential, an excess of fat is detrimental to health. An estimated 25% of the U.S. population is *obese* (more than 20% above the "ideal body weight"), with greater risk of health problems.

People with **apple-shaped bodies** (more fat around the abdominal area) seem to be more likely to develop cardiovascular disease, hypertension, and diabetes mellitus than those with **pear-shaped bodies** (more fat in the hips, buttocks, and thighs). Since males tend to become apple shaped, and females tend to become pear shaped, it appears that sex hormones help to direct the distribution of fat. For reasons still unknown, abdominal fat (stored deep in the body within the greater omentum) seems to pose a greater health risk than the subcutaneous fat stored under the skin in the hips and thighs. For example, the pear-to-apple shift in fat distribution seen in postmenopausal females is accompanied by an increase in the risk of diseases, such as cardiovascular disease (usually more common in males) and breast cancer.

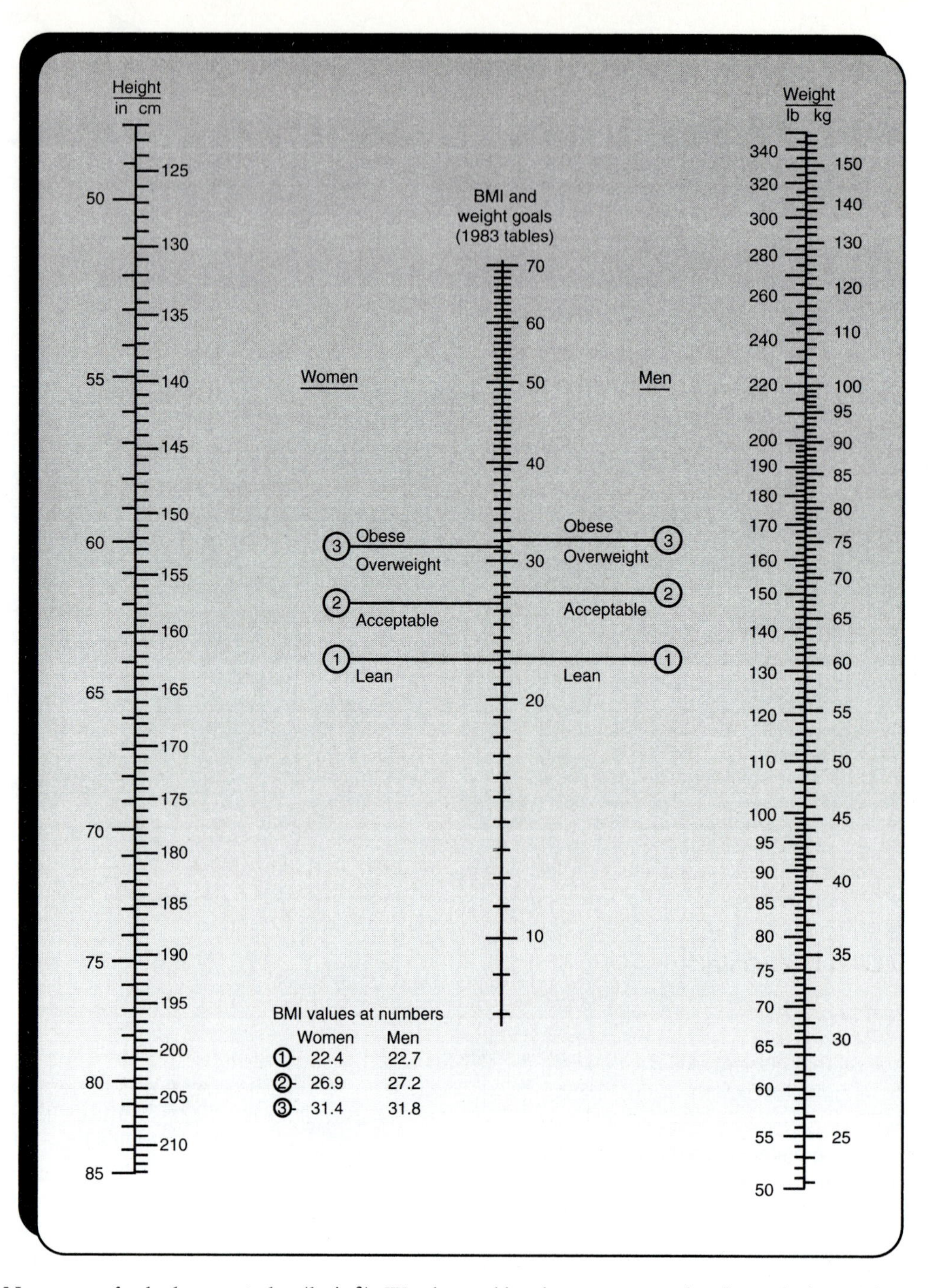


**Figure 10.17 Nomogram for body mass index (kg/in²).** Weights and heights are measured without clothing. The ratio weight/height² (metric units) is read from the central scale after a straight edge is placed between height and body weight.

Body composition can be estimated using a variety of techniques. One estimate of body fat is a height-to-weight ratio called the **body mass index (BMI).** BMI equals the body weight in kilograms divided by height in meters squared (the arithmetic can be avoided using the nomogram in fig. 10.17). In general, a person with a BMI of 27 or greater is obese, with an associated increased risk of health problems. *Skinfold calipers* measure the thickness of the subcutaneous fat layer at particular body sites. These measurements are then compared to norms for total body fat. Underwater weighing, one of the most accurate methods of determining body fat, is also the most expensive and inconvenient. *Bioelectric impedance* instruments send a small current through skin electrodes to measure the resistance in the electrolyte-rich body fluids. This method estimates body water, and then uses a computer data bank to derive body fat indirectly (lean body mass has more water than fat tissue). *Infrared wands* derive the percent body fat from density data taken over the biceps brachii muscle.

## PROCEDURE (BODY MASS INDEX)

1. The body mass index, or BMI, is obtained by the following formula:

$$BMI = \frac{w}{h^2}$$

where,

$w$ = weight in kilograms (pounds divided by 2.2)

$h$ = height in meters (inches divided by 39.4)

2. Alternatively, the BMI may by obtained by use of the nomogram in figure 10.17. Simply draw a straight line between your height in the left column and your weight in the right column. Your BMI will be shown in the middle column.
3. Alternatively, you can simply enter your height and weight in the calculator provided at this website: *www.caloriecontrol.org/bmi.html*
4. Record your BMI here: ________________________.

**A BMI of over 30 places a person at high risk for the diseases of obesity. A BMI of under 27 is considered healthy, and a BMI in the range of 23–25 appears to be optimum for health.**

## PROCEDURE (WAIST-TO-HIP RATIO):

1. Stand and measure your waist at the navel. Record this value (in cm): __________ cm.
2. Measure your hips at the greatest circumference (buttocks). Record this value (in cm): __________ cm.
3. Divide the waist circumference by the hip circumference to get the waist-to-hip ratio. Record this waist-to-hip ratio: __________

**According to the American Heart Association, a waist-to-hip ratio above 1.0 for men and above 0.8 for women is associated with an increased risk of cardiovascular disease.**

# B. Energy Intake and the Three-Day Dietary Record

Both the chemical energy consumed in foods and the metabolic energy expended by the cell are measured in **kilocalories (kcal),** or **Calories (C).** The major sources of food calories are carbohydrates, fats (lipids), and proteins. When allowances are made for inefficiency in the assimilation of each nutrient, one gram *(g)* of each of the three energy nutrients provides the body with approximately the following number of calories:

1 g carbohydrate = 4.0 kcal
1 g fat = 9.0 kcal
1 g protein = 4.0 kcal

The primary **carbohydrates** in food are the *sugars* (such as glucose, fructose, and sucrose) and *complex carbohydrates* (such as starches and dietary fiber). To meet the energy requirements of children and adults, recommended dietary allowances suggest that more than half (about 55%) of the calories consumed per day come from carbohydrate sources in the diet. Emphasis should be on the increased consumption of complex carbohydrates, especially dietary fibers found in fruits, vegetables, legumes, and whole-grain cereals. In addition to providing a source of calories, dietary fiber has been associated with improving overall health by promoting normal stool elimination, enhancing satiety, and lowering plasma cholesterol levels.

**Lipids** are generally divided into *triglycerides*, *phospholipids*, and *sterols* (steroids). The digestion, emulsification, and absorption of lipids also facilitates the absorption of *fat-soluble vitamins A, D, E,* and *K* and the *essential fatty acids*. Two primary unsaturated fatty acids are considered essential and must be present in the diet to maintain health: *linoleic acid* and α-linolenic acid.

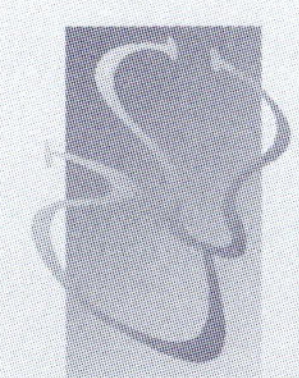

**Triglycerides** are the major lipid components of foods and the most concentrated source of energy (9 kcal/g). The average American currently derives about 36% of the daily total dietary calories from fats. High dietary fat and cholesterol intakes have been associated with an increased risk of cardiovascular disease and cancer. As recommended by the Food and Nutrition Board's Committee on Diet and Health, the fat content of the U.S. diet should be lowered so as not to exceed 30% of the caloric intake (10% of fat calories from *saturated* fatty acids, 10% from *polyunsaturated* fatty acids, and 10% from *monounsaturated* fatty acids); and dietary cholesterol should be less than 300 mg/day (National Research Council, 1989).

According to U.S. Department of Agriculture (USDA) surveys, about 15% of the total food energy intake of the average American is derived from **protein.** Most of this protein, about 65%, is derived from animal sources, primarily meat and dairy products, with only about 20% from cereal grains. Despite increased protein requirements for certain populations, such as growing children, pregnant or lactating females, and the elderly, the typical American diet normally meets or exceeds the requirements. Interestingly, there is little evidence that physical exercise increases the need for protein, other than that required during the initial conditioning period. Therefore, individuals eating a typical American diet need make no adjustment to the recommended allowance for protein.

Nutrients that do not contribute energy to the body but that still are required to maintain body functions are vitamins, minerals, and water. The **fat-soluble vitamins** (A, D, E, and K) are absorbed from the small intestine with other food lipids, and are concentrated to some degree in adipose tissue. The **water-soluble vitamins** are vitamin C, thiamine ($B_1$), riboflavin ($B_1$), niacin ($B_1$), pyridoxine ($B_6$), folate, cyanocobalamin ($B_{12}$), biotin, and pantothenic acid. Most water-soluble vitamins serve as coenzymes that assist enzymes in the regulation of metabolism. The **major minerals,** required in higher quantities, are calcium, phosphorus, magnesium, and the electrolytes (sodium, chloride, and potassium). The **minor minerals,** or **trace elements,** are required in lesser quantities. They include iron, zinc, iodine, selenium, copper, manganese, fluoride, chromium, and molybdenum. Recommended quantities of these nutrients are normally met when a variety of foods are consumed. For the average person eating a typical American diet, therefore, no supplementation is recommended. Indeed, high intakes of certain nutrients in supplement form can be toxic.

**Water** is also an essential dietary nutrient. Although assessments of adequate water intake involve many complex factors, the general recommendation is a minimum of 1.0 mL of water per kilocalorie of energy expended per day. Therefore, an individual expending 2,000 kilocalories per day should consume at least 2 liters of water.

## PROCEDURE

1. Complete the 3-day dietary record (provided in the laboratory report). The 3 days must be consecutive (attempt to include at least one weekend day). At roughly the same time each day, try to weigh yourself under consistent conditions (comparable clothing, same scale etc.). Record your weight in pounds in the space provided for that day.
2. Record all foods and fluids consumed each day in the appropriate columns, noting the approximate time of day. Estimate food quantities by weight (such as ounces) and fluids by volume (e.g., cups, or liters), depending upon the units listed in your calorie guide. Record the total volume of fluids consumed at the bottom of the column.
3. Look up the estimated number of calories for each food or fluid noted in your diet record. The caloric values can be looked up in reference books or on the net, for example at *www.caloriescount.org/calculator.html*.
   Record the total calorie intake at the bottom of the column.
4. Comment on where the food or fluid was consumed. Do you always eat sitting down at a table, in a quiet, relaxing environment? Or are you sometimes in the car, between classes, in front of the TV, or in bed studying?
5. Comment on why you ate or drank. Do you always eat because you are hungry? On occasion, do you eat because you are bored, or because the food is there, or because someone else is paying for it?

## C. Energy Output: Estimates of the BMR and Activity

The total energy expended each day includes the energy required at rest and that expended during physical activity. For most people, the calories consumed at rest make up most of the total daily energy expenditure. This energy is used to pump blood, inflate the lungs, transport ions, and carry on the other functions of life. Measurement of this resting energy expenditure shortly after awakening and at least 12 hours after the last meal is known as the **basal metabolic rate (BMR).** With all other factors equal, BMR is influenced most by the amount of actively metabolizing tissues, or *lean body mass*. The BMR is higher in younger, more muscular people, and in males (who have a higher average muscle mass than females). The BMR is also influenced by **thyroxine.** People who are *hypothyroid* have a low BMR; those who are *hyperthyroid* have a high BMR.

While most of our caloric output is spent at rest, most people are physically active and expend calories beyond the BMR. The additional number of activity calories expended will vary with the individual, and with the duration, intensity, and types of activities performed. This increased caloric output can be estimated by multiplying an *activity factor* (AF) by the BMR. In general, the average sedentary person raises the total number of calories burned per day to about 130% (AF = 1.3) of the estimated BMR. Moderately active people may raise daily expenditures upwards of 150% (AF = 1.5) above the BMR estimates. Top athletes may double the BMR estimates, or more (AF ≥ 2.0). Aerobic activities, such as running, swimming, bicycling, and dancing, normally burn more calories than anaerobic activities, such as weight lifting.

| Activity | Activity Factor (AF) |
|---|---|
| Lying in bed all day—equal to BMR | 1.00–1.29 |
| Mild activity—normal routine, no exercise | 1.30–1.49 |
| Moderate activity—1 hour of aerobic exercise | 1.50–1.69 |
| Heavy activity—2 to 4 hours of aerobic exercise | 1.70–1.99 |
| Rigorous athletic training | 2.0 and above |

## PROCEDURE

**Step 1.** Estimate your basal metabolic rate (BMR) using two different methods:

**Method 1:** Your weight in kilograms (2.2 lb/kg) __________ (kg):

Female BMR = 0.7 kcal/kg/hr

Male BMR = 1.0 kcal/kg/hr

Use the above conversion factors to calculate your kilocalories per hour. Then, multiply this figure by 24 (hours per day) and enter your answer in the space below.

In one day (24 hours), your BMR is approximately __________ kcal.

**Method 2:** Estimate your *Ideal Body Weight (IBW)* in pounds.

Female IBW = 100 lb for the first 5 feet in height + 5 lb per inch above 5 feet in height

Male IBW = 106 lb for the first 5 feet in height + 6 lb per inch above 5 feet in height

Your Ideal Body Weight is approximately __________ lb.

Next, multiply your IBW by 10 for your daily estimated BMR.

In one day (24 hours), your BMR is approximately __________ kcal.

**Note:** *Since these BMRs are only estimates, a difference between the two values is to be expected. Select the one BMR estimate you feel is most accurate, write that number in the space provided on the dietary record in the laboratory report, and use it for calculations.*

**Step 2.** For each day in your 3-day dietary record, select one activity factor (AF) from the chart above that best reflects your total activity for that 24-hour period, with 1.30 typical for the casual college routine (without exercise); and write that decimal for each day in the AF box of the dietary record.

**Note:** *For variety (and more fun) attempt to vary your activities and the AF each day—another reason to include one weekend day in this report.*

**Step 3.** Calculate the total number of calories expended (output) each day as:

Total calories expended (kcal) =

Activity AF (from box) × BMR (from Step 1, Method 1 or 2)

Write this total in the line 3 of the dietary record.

**Step 4.** Subtract total calories expended from total calories consumed to determine the caloric balance lost or gained that day (in your record subtract line 3 from line 1). Write the number of excess calories in line 4 of the dietary record; and circle either calories "gained" or "lost."

**Step 5.** Assuming that 1 lb of body tissue (not just fat) that is gained or lost represents approximately 3,500 kcal, convert the excess calories from line 4 into body weight:

Body weight (lost/gained) = __________ (kcal) ÷3,500 (kcal/lb) = __________ (lb)

Write the pounds lost/gained that day in line 5 of the dietary record.

**Step 6.** Complete the evaluation section in the laboratory report, which follows the 3-day dietary record.

3-Day Dietary Record

Day # ______ Day of Week ______ Your Weight (lb) ______

| Time (A.M./P.M.) | Foods-Units (oz, g, . . .) | Fluid-Units (cup, mL, . . .) | Calories (kcal) | Where ? | Why ? | Activities |
|---|---|---|---|---|---|---|
| | | | | | | |
| | | Total fluid: | Total kcal: | | | |

1. Total calories consumed (intake): ______________ kcal
2. Estimated BMR at rest: ______________ kcal (Step 1, Method 1 or 2)
3. Total calories expended (output): ______________ kcal (Step 3)
4. Today's caloric balance: ______________ kcal (Step 4)
   gained / lost (circle one)
5. Given that 1 lb of body weight (not just fat) is equal to approximately 3,500 kcal gained or lost, how many pounds of body weight were gained or lost today? ______________ lb.
6. Enter values from today's report into the Dietary Record Evaluation form in the Laboratory Report.

AF

## 3-Day Dietary Record

**Day #** ______ **Day of Week** ______ **Your Weight (lb)** ______

| Time (A.M./P.M.) | Foods-Units (oz, g, . . .) | Fluid-Units (cup, mL, . . .) | Calories (kcal) | Where ? | Why ? | Activities |
|---|---|---|---|---|---|---|
| | | | | | | |
| | | Total fluid: | Total kcal: | | | |

1. Total calories consumed (intake): ______________ kcal
2. Estimated BMR at rest: ______________ kcal (Step 1, Method 1 or 2)
3. Total calories expended (output): ______________ kcal (Step 3)
4. Today's caloric balance: ______________ kcal (Step 4)
   gained / lost (circle one)
5. Given that 1 lb of body weight (not just fat) is equal to approximately 3,500 kcal gained or lost, how many pounds of body weight were gained or lost today? ______________ lb.
6. Enter values from today's report into the Dietary Record Evaluation form in the Laboratory Report.

**AF**

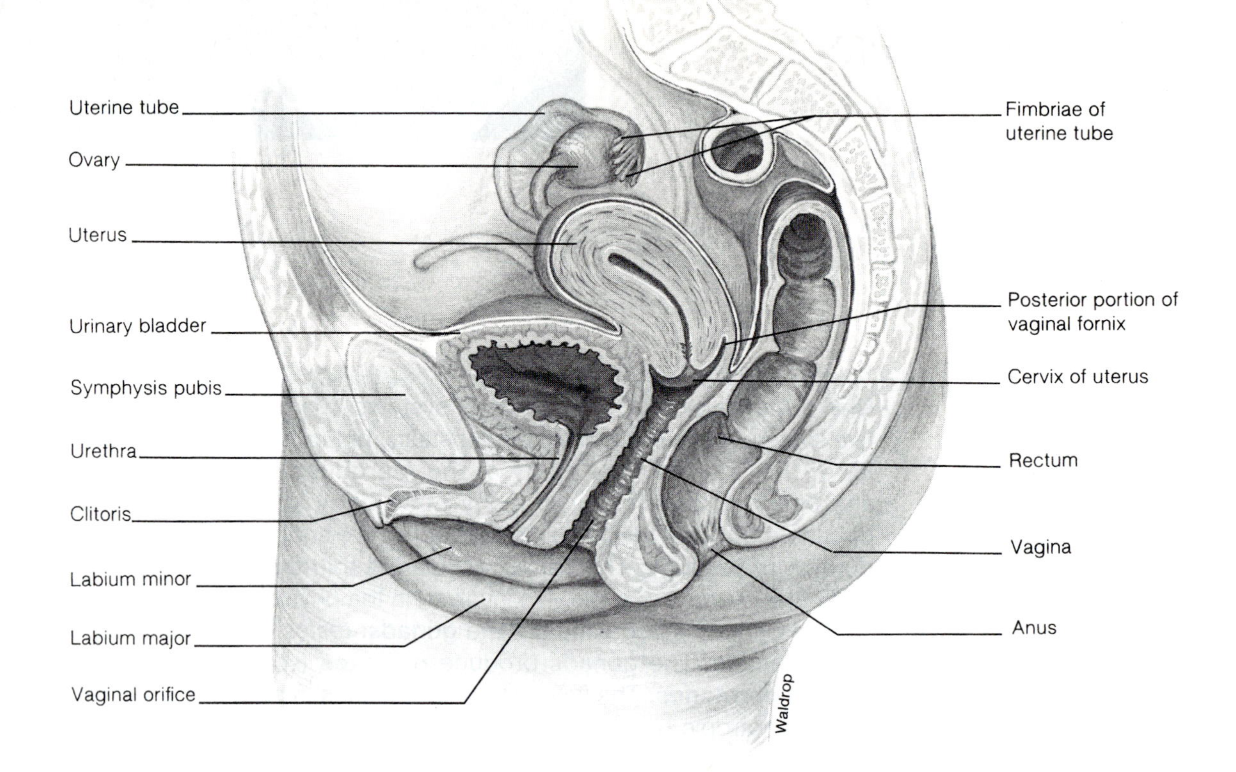


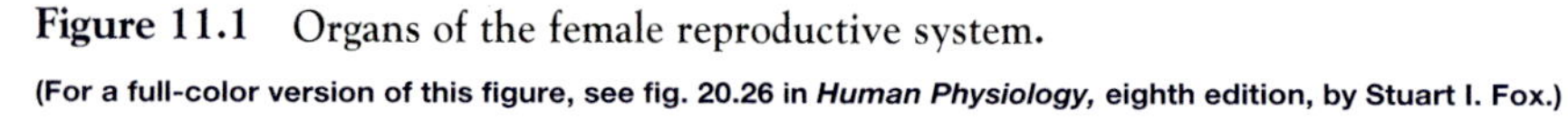

**Figure 11.1** Organs of the female reproductive system.

**(For a full-color version of this figure, see fig. 20.26 in *Human Physiology,* eighth edition, by Stuart I. Fox.)**

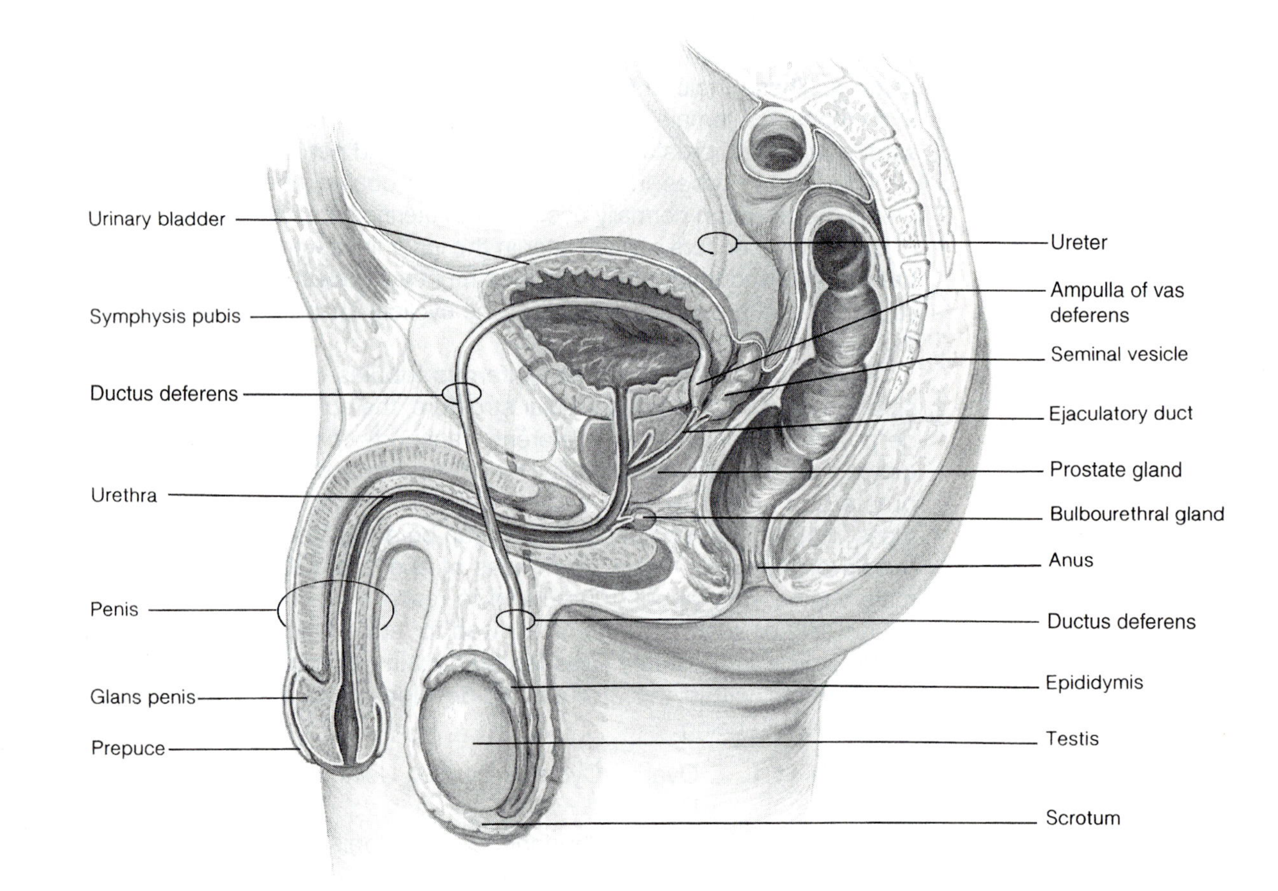


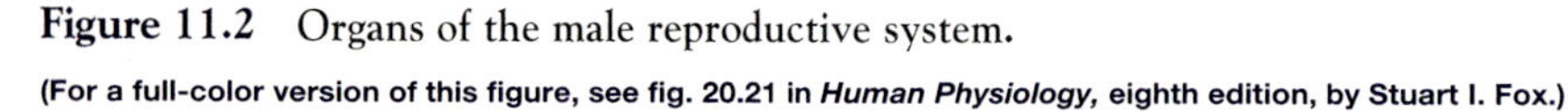

**Figure 11.2** Organs of the male reproductive system.

**(For a full-color version of this figure, see fig. 20.21 in *Human Physiology,* eighth edition, by Stuart I. Fox.)**

# Ovarian Cycle as Studied Using a Vaginal Smear of the Rat

EXERCISE 11.1

## Materials

1. Young female rats
2. Ether jar (large, widemouth jar with close-fitting lid) and ether. This is optional.
3. Isotonic saline and cotton swabs (or, alternatively, eyedroppers may be used)
4. Giemsa's stain (dilute concentrate 1:50) and absolute methyl alcohol in staining jars
5. Microscopes and microscope slides

The cyclic changes in ovarian hormone secretion cause cyclic changes in the epithelium of the female reproductive tract. By observing exfoliated epithelial cells, the stage of the ovarian cycle and the level of ovarian hormone secretion can be determined.

## OBJECTIVES

1. Identify the phases of the ovarian cycle.
2. Describe the changes that occur in the endometrium and correlate these changes with the stages of the ovarian cycle.
3. Describe the appearance of a vaginal smear at different stages of the cycle and explain the clinical usefulness of a vaginal smear.

## Textbook Correlations*

Before performing this exercise, you should study the introductory material presented here. Further information relating to this exercise can be found in these pages of *Human Physiology,* eighth edition, by Stuart I. Fox:

- *Menstrual Cycle.* Chapter 20, pp. 659–665.

*Multimedia Correlations (also see Appendix 3)
- *MediaPhys 2.0:* Topics 13.17–13.24

The amount of gonadotropic hormones (FSH and LH) secreted by the anterior pituitary of females increases and decreases in a cyclical fashion. The secretion of estrogen and progesterone by the ovary will follow the same cycle. In most mammals, sexual receptivity (heat, or estrus) occurs during a specific part of the cycle. This pattern is called an **estrous cycle.** In human and subhuman primates, sexual receptivity occurs throughout the cycle, with monthly bleeding occurring at the beginning of each cycle. These cycles are called **menstrual cycles** (*menses* means "monthly") (fig. 11.3).

The uterus is one of the target organs of the ovarian hormones. As the secretions of estrogen and progesterone increase during the cycle, the inner lining of the uterus (the endometrium) increases in thickness (fig. 11.3). The ovarian hormones are preparing the uterus for the possible implantation of the developing embryo should fertilization occur. If fertilization does not occur, the cyclical decrease in plasma levels of estrogen and progesterone causes the necrosis (cellular death) and sloughing off of the upper two-thirds of the endometrium. The cycle is ready to begin anew (fig. 11.3).

The rise and fall in the secretion of estrogen and progesterone from the ovaries during the course of a menstrual cycle causes cyclic changes in the endometrium, as illustrated in figure 11.3. At the same time, the rise and fall in estrogen and progesterone during a menstrual or estrus cycle produce cyclic changes in the vaginal mucosa. This causes different cells to be shed, or *exfoliated,* at different stages of the cycle and these can be observed by taking a **vaginal smear.**

The estrous cycle of a rat is usually completed in 4 to 5 days. The cycle is roughly divisible into four stages that can be identified by a vaginal smear (fig. 11.4).

1. **Proestrus.** Proestrus is the beginning of a new cycle. The follicles of the ovary start to mature under the influence of the gonadotropic hormones, and the ovary starts to increase its secretion of estrogen. *Vaginal smear.* Nucleated epithelial cells (fig. 11.4*a*,*b*).
2. **Estrus.** The uterus is enlarged and distended because of the accumulation of fluid. Estrogen secretion is at its height. The rat thus becomes sexually receptive as ovulation occurs. *Vaginal smear.* Squamous cornified cells (fig. 11.4*c*).

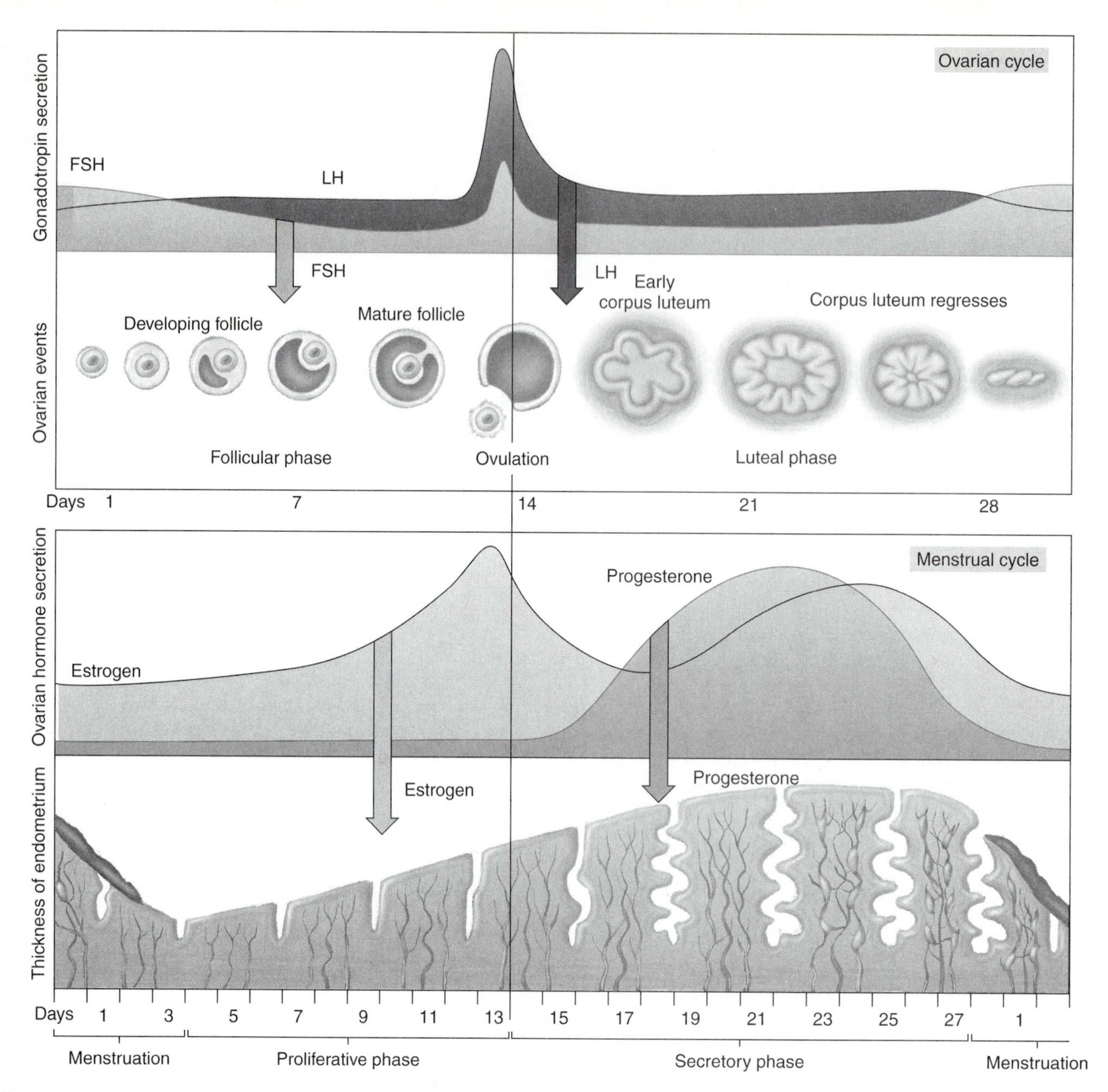


**Figure 11.3** **Cycle of ovulation and menstruation.** The downward arrows indicate the effects of the hormones.

**(For a full-color version of this figure, see fig. 20.35 in *Human Physiology,* eighth edition, by Stuart I. Fox.)**

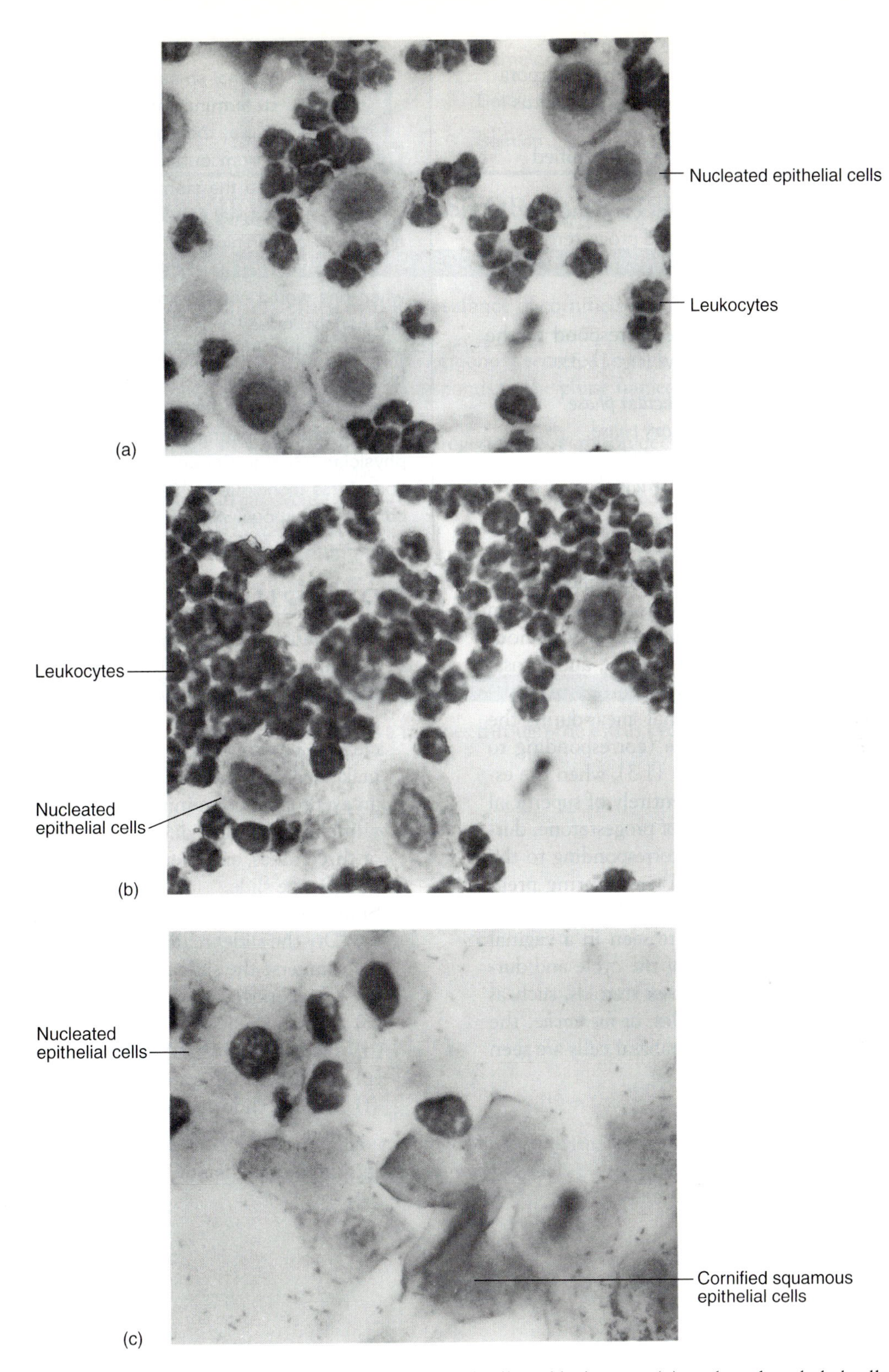


**Figure 11.4 The vaginal smear of a rat.** (a,b) Nucleated epithelial cells and leukocytes; (c) nucleated epithelial cells and cornified cells.

quantitative measurements of hCG that can give information regarding the health of the pregnancy. Most commonly, however, pregnancy tests employ urine samples. Modern over-the-counter home pregnancy tests can detect hCG in urine one to two weeks following conception, and are generally used in the week following the first missed menstrual period.

> Since hCG acts like LH, injections of hCG are sometimes given clinically to trigger ovulation in women. Interestingly, this LH-like action of hCG was the basis of the old **rabbit test** for pregnancy. Beginning in the 1930s, the phrase "the rabbit died" meant that the woman was pregnant. Rabbits are reflex ovulators—if they don't have sexual intercourse, they don't ovulate. In this test, a virgin rabbit was injected with urine from a possibly pregnant woman. If the woman was pregnant, the rabbit ovulated and produced a corpus luteum. The rabbit had to be killed in order to examine its ovaries; thus, the rabbit died in the test whether the woman was pregnant or not. Although it has been more than forty years since the rabbit test was replaced by immunological detection of hCG, the phrase "the rabbit died" may still be heard to announce a positive pregnancy test.

Modern pregnancy tests still use animals (rabbits, goats, or others), but these are used in the production of the test kit, and the animals don't have to be killed in the process. Companies that produce the test kit inject the animals with hCG, which is a foreign antigen that stimulates the animal's immune system to produce antibodies against the hCG. The antibodies circulate in the blood plasma, and blood samples are taken to derive the antibodies for the test. All modern pregnancy tests detect the presence of hCG in a blood or urine sample by means of antigen-antibody bonding. Tests that employ antibodies to detect a specific molecule are known as **immunoassays.**

There are two polypeptide subunits, alpha and beta, in the hCG glycoprotein molecule. The alpha subunit of hCG is also present in TSH, FSH, and LH; it is the beta subunit that is unique to hCG. Thus, antibodies directed against the beta subunit of hCG provide the least amount of cross-reaction with other hormones. Accurate and sensitive immunoassays for hCG in pregnancy tests employ antibodies that are produced by a clone of lymphocytes—termed *monoclonal antibodies*—against the specific beta subunit of hCG.

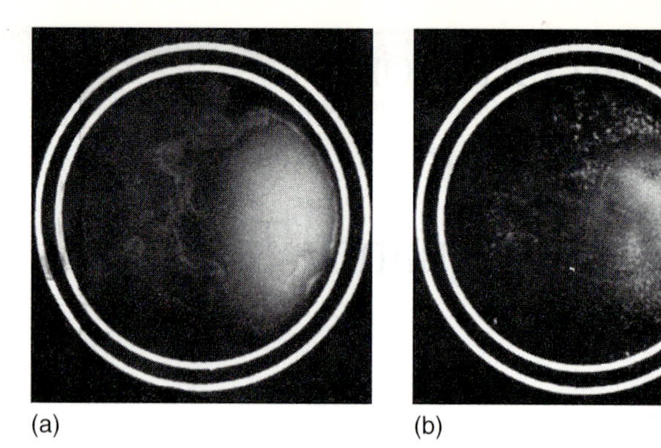


**Figure 11.6 A pregnancy test.** (a) Negative—no agglutination reaction occurs when urine is added to a control solution containing white latex particles with rabbit gamma globulin (from a rabbit not sensitized to hCG). (b) Positive—agglutination of latex particles occurs when urine from a pregnant woman is added to latex coated with antibodies from a rabbit sensitized to hCG.

All pregnancy tests involve the use of antibodies against hCG, but the way these are used varies with the type of test. Although not widely used anymore, the easiest tests to understand involve agglutination reactions. In these tests, antibodies are stuck onto tiny white latex particles to make them visible. If hCG is present in the urine sample, the hCG binds to the anti-hCG antibodies and causes the latex particles to agglutinate (fig. 11.6), much like the agglutination reaction used for blood typing (see exercise 6.3).

Over-the-counter home pregnancy kits are also based on this antigen-antibody reaction. They are easier to perform, but the way these tests work is more complex than with the simpler agglutination reaction. In the common one-step types of tests, the test apparatus has two windows. One window has a visible line; this is a "control" that demonstrates how a positive response in the other window should appear. The other window has a membrane with an invisible line. Antibodies against hCG are attached to the membrane along that line. When urine containing hCG moves along the membrane, antigen-antibody bonds form. Latex particles may be used to help visualize the line in a positive pregnancy test (fig. 11.7), but the exact way that each test works is proprietary information (not available to the public).

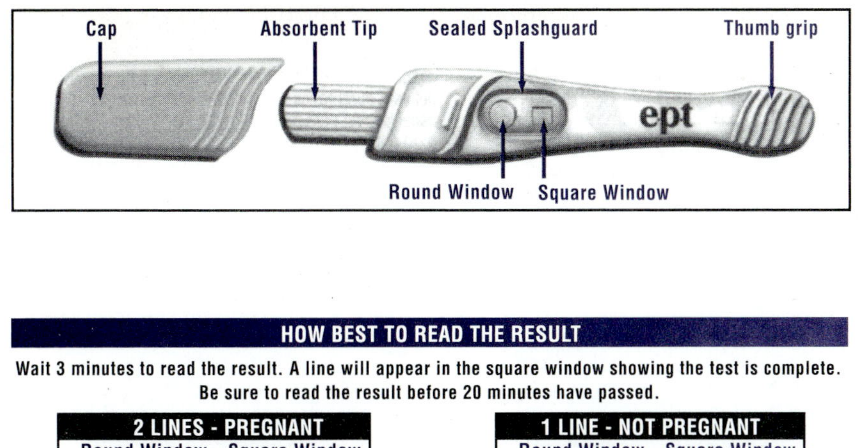


HOW BEST TO READ THE RESULT

Wait 3 minutes to read the result. A line will appear in the square window showing the test is complete. Be sure to read the result before 20 minutes have passed.

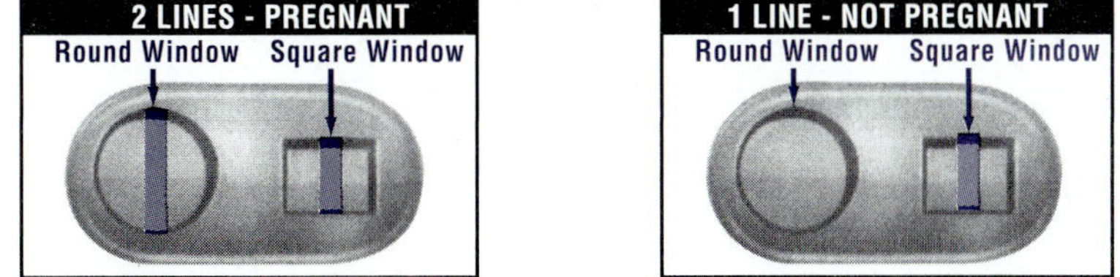


**Figure 11.7** **A typical over-the-counter home pregnancy test.** This pregnancy test (e.p.t., Warner-Lambert Consumer Healthcare) is similar to the over twenty different brands currently available in drugstores, supermarkets, and over the Internet. Follow the specific procedures that come with the specific pregnancy test to obtain accurate results.

## PROCEDURE

### *Agglutination Pregnancy Test*

1. Allow the urine sample to reach room temperature.
2. Fill the plastic reservoir provided in the pregnancy kit with urine and insert the filtering attachment.
3. Expel the urine onto two circles on the disposable slide provided by gently squeezing the reservoir.
4. Shake the latex control (latex particles with gamma globulin antibody) and add 1 drop to the first circle. Mix the urine and reagent with an applicator stick by spreading the mixture over the entire circle.
5. Shake the bottle of reagent (latex particles with antibodies against hCG) and add 1 drop to the second circle. Mix as before.
6. Rock the slide gently for 1 minute; then look for agglutination. If *negative:* the solution will remain milky (fig. 11.6*a*). If *positive:* the solution will appear grainy (fig. 11.6*b*).

### *Home Pregnancy Test*

1. Remove the test apparatus from its foil package and follow the directions as outlined for the specific pregnancy test. This generally involves putting the absorbent tip of the apparatus into a flow of urine, or dipping the tip into a container of urine.
2. Follow the remaining steps of the procedure. This generally involves placing the apparatus on a flat surface for 3 minutes and then reading the results. An example of such results is illustrated in figure 11.7.

# Laboratory Report 11.2

Name ____________________

Date ____________________

Section ____________________

## REVIEW ACTIVITIES FOR EXERCISE 11.2

### *Test Your Knowledge of Terms and Facts*

1. Menstruation is caused by a(n) ____________________ (increase/decrease) in the secretion of estrogen and progesterone.
2. The structure that secretes estrogen and progesterone for the first 10 weeks of pregnancy is the ____________________.
3. The hormone tested for in a pregnancy test is ____________________.
4. The hormone named in question 3 is produced by the ____________________.
5. The hormone named in question 3 has an action similar to which pituitary hormone? ____________________.

### *Test Your Understanding of Concepts*

6. Describe the formation, function, and fate of the corpus luteum during an unfertile menstrual cycle. What happens to the corpus luteum if fertilization occurs?

7. Why is this pregnancy test called an immunoassay? Explain how this test works to detect pregnancy.

8. Why are most pregnancy tests not valid if they are performed too soon after conception?

9. Suppose a man has a tumor that secretes hCG. Would he give a positive result on a pregnancy test? Would hCG have any physiological effect on him? Explain

# Patterns of Heredity

EXERCISE **11.3**

## Materials

1. Phenylthiocarbamide (PTC) paper (VWR Scientific Products, Ward's)
2. Sickle cell turbidity test (Chembio Diagnostic Systems, Inc.); prepared slides of sickle cell anemia and normal blood.
3. Ishihara color-blindness charts

## Textbook Correlations

Before performing this exercise, you should study the introductory material presented here. Further information relating to this exercise can be found in these pages of *Human Physiology,* eighth edition, by Stuart I. Fox:

- *DNA Synthesis and Cell Division.* Chapter 3, pp. 69–78.
- *Cones and Color Vision.* Chapter 10, p. 272.
- *Inherited Defects in Hemoglobin Structure and Function.* Chapter 16, p. 508–509.

The ways in which many aspects of body structure and function are inherited can be understood by applying relatively simple concepts. The patterns of heredity are important in anatomy and physiology because of the numerous developmental and functional disorders that have a genetic basis. The knowledge of which disorders and diseases are inherited finds practical application in the genetic counseling of prospective parents.

## OBJECTIVES

1. Define the terms dominant, recessive, homozygous, and heterozygous.
2. Distinguish between autosomal and sex-linked inheritance.
3. Explain the nature of sickle-cell anemia and describe how it is inherited.
4. Describe how hemophilia and color blindness are inherited.

A person inherits two sets of genes controlling every trait: one from the mother and one from the father (if these genes are *autosomal*—that is, not located on the sex chromosomes). If both genes are identical, the person is said to be **homozygous** for that trait. A person who is homozygous for normal adult hemoglobin A, for example, has the **genotype** *AA*; a person who is homozygous for the sickled hemoglobin S has the genotype *SS*.

If a person inherits the gene for hemoglobin A from one parent and the gene for hemoglobin S from the other parent, this person is said to be **heterozygous** for that trait and has the genotype *AS*. This person is a carrier and has the sickle-cell *trait* but does not have sickle-cell *disease*. The **phenotype** (in this case the absence of sickle cell disease) is the same for the heterozygote as it is for the person who is homozygous normal. Thus, the gene for hemoglobin A is **dominant** to the gene for hemoglobin S (or, stated another way, the gene for hemoglobin S is **recessive** to the gene for hemoglobin A).

Although the heterozygote does not display the phenotype of sickle cell disease, this person is a carrier of the sickle cell trait since one-half of the gametes will contain the gene for hemoglobin A and one-half will contain the gene for hemoglobin S. (In the process of gamete formation, known as *meiosis*, the chromosome number is halved). If this individual mates with one who is homozygous AA, the probability is half the progeny will be homozygous AA, and half will be heterozygous AS.

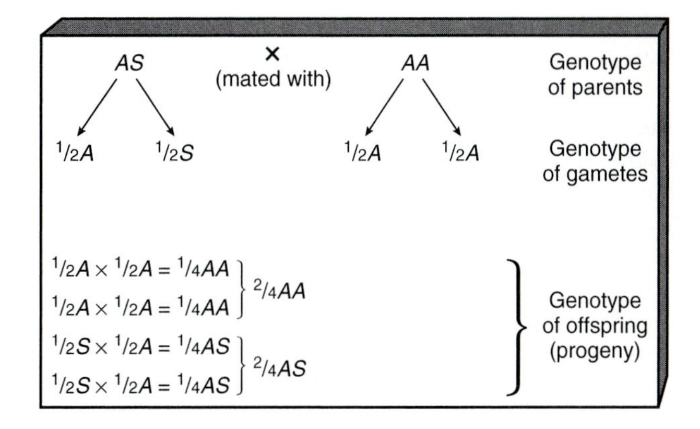

If two individuals who are both heterozygous *AS* mate, one-fourth of the progeny will have the genotype *AA*, one-fourth will have the genotype *SS*, and one-half will have the genotype *AS*. Although individuals with the homozygous genotype *AA* and the heterozygous genotype *AS* are healthy, there is a one-in-four (25%) probability that a child from this mating will have the phenotype of sickle cell disease (genotype *SS*).

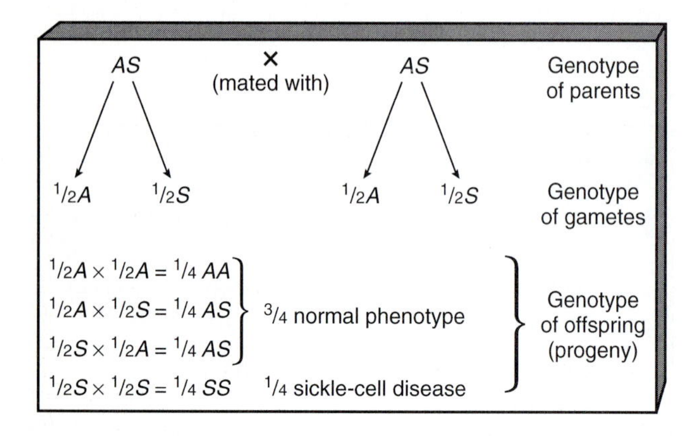


Most of the concepts of heredity discussed in this exercise were discovered in the 1860s by an Austrian monk named Gregor Mendel; consequently, these patterns of heredity are often called *simple Mendelian heredity.* A proper knowledge of these patterns is obviously needed for genetic counseling of carriers of genetic diseases. If both parents are carriers of such diseases as sickle cell anemia, Tay-Sachs disease, phenylketonuria (PKU), and others that are inherited as *autosomal recessive* traits, they should be aware that there is a 25% chance that their children will get the disease. If only one parent is a carrier, they should know that there is no chance of their children getting the disease. Further, couples should be informed that whether they have no children or a dozen, the probability that their next child will get the disease will always remain the same.

## A. Sickle-Cell Anemia

Sickle-cell anemia is an autosomal recessive disease affecting 8–11% of the African-American population. In this disease, a single base change in the DNA, through the mechanisms of transcription and translation, results in the production of an abnormal hemoglobin (hemoglobin S). Hemoglobin S differs from the normal adult hemoglobin (hemoglobin A) by the substitution of one amino acid for another (valine for glutamic acid) in one position of the protein. A quick test for sickle-cell anemia is based on the fact that, under conditions of reduced oxygen tension, hemoglobin S is less soluble than hemoglobin A and tends to make a solution turbid, or cloudy (fig. 11.8*a*).

### PROCEDURE

1. Fill a calibrated capillary tube with blood up to the line. Then, expel the blood into a test tube containing 2.0 mL of test reagent (contains sodium dithionite, which produces low oxygen tension).
2. If the solution does not become cloudy within 5 minutes the test is negative.
3. Place a drop of solution on a slide, add a coverslip and compare your cells with those in figure 11.8*b*. If available, compare your sample to that of a prepared slide of sickle cell anemia.
4. Record your data in the laboratory report.

## B. Inheritance of PTC Taste

The ability to taste PTC paper (phenylthiocarbamide) is inherited as an autosomal dominant trait. Therefore, if *T* is a taster and *t* is a nontaster, tasters have the genotype *TT* or *Tt* and nontasters have the genotype *tt*.

### PROCEDURE

1. Taste the PTC paper by leaving a strip of it on the tongue for a minute or so. If the paper has an unpleasantly bitter taste, you are a taster.
2. Determine the number of tasters and nontasters, calculate the proportion of each in the class, and enter this data in your laboratory report.

## C. Sex-Linked Traits: Inheritance of Color Blindness

The sex of an individual is determined by one pair of the twenty-three pairs of chromosomes inherited from the parents. These are the sex chromosomes, X and Y. The female has the genotype XX and the male has the genotype XY. Traits that are determined by genes located on the X sex chromosome (as opposed to the other, autosomal chromosomes) are called **sex-linked traits** (the Y chromosome apparently carries very few genes).

Unlike the patterns of heredity previously considered, where the genes are carried on autosomal chromosomes, the inheritance of genes carried on the X chromosome follows a different pattern for males than for females. This is because the male inherits only one X chromosome (and only one set of sex-linked traits) from his mother, whereas the female inherits an X chromosome from both parents.

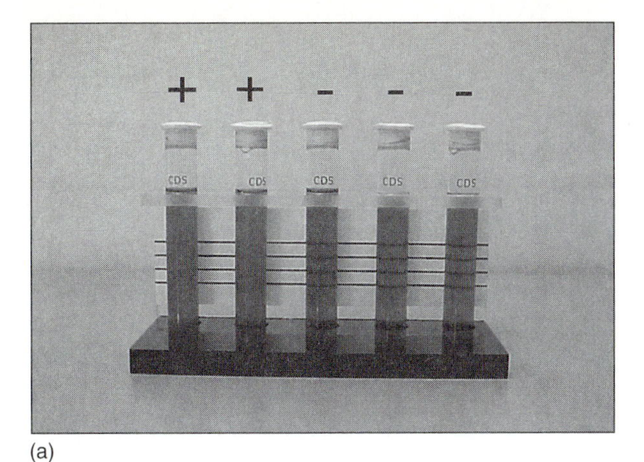


(a)

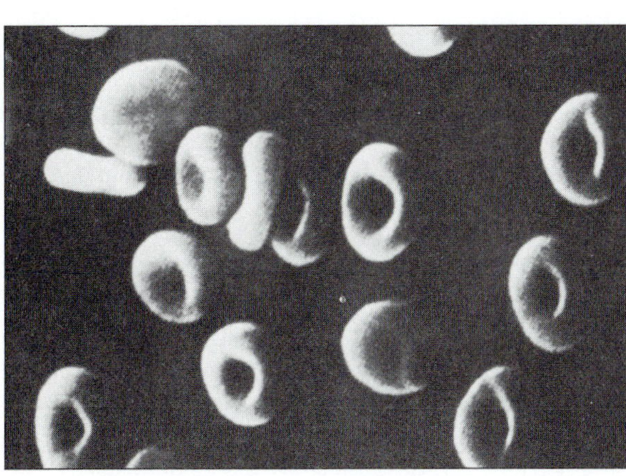

(b)

(c)

**Figure 11.8 Sickle-cell anemia.** (a) A turbidity test for sickle-cell anemia. Tubes indicated with a "+" are too cloudy to see through, indicating the presence of hemoglobin S. (b) Normal red blood cells under a scanning electron microscope. (c) Sickled red blood cells under a scanning electron microscope.

The genes for color vision and for some of the blood-clotting factors are carried on the X chromosome, where the phenotypes for **color blindness** and **hemophilia** are recessive to the normal phenotypes. A normal female may have either the homozygous or the heterozygous ("carrier") genotypes, whereas a male must have either the normal or the affected phenotypes.

Let's suppose a man with the normal phenotype mates with a woman who is a carrier for color blindness (C is normal, *c* is color blind).

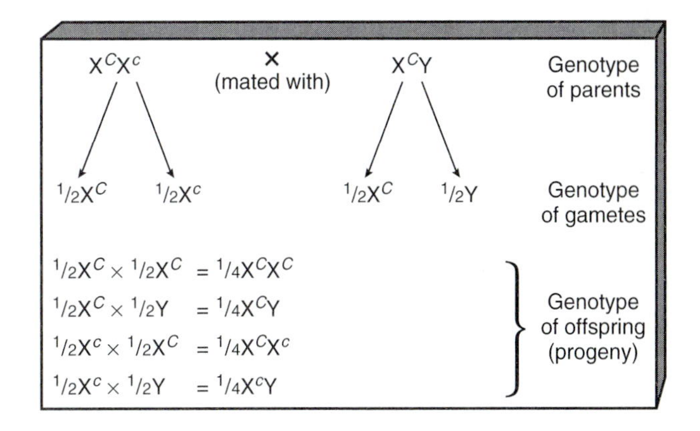


The probability that a child formed from this union will be color blind is one in four (25%). And in the event of a color-blind offspring, that offspring is certain to be male (100% probability). All female children formed from this union will have the normal phenotype, but the probability that a given female child will be a carrier for color blindness is one in two (50%).

The perception of color is due to the action of certain photoreceptor cells, known as **cones,** in the retina of the eye. According to the *Young-Helmholtz theory* of color vision, the perception of all the colors of the visible spectrum is due to the stimulation of only three types of cones—*blue, green,* and *red.* Their names refer to the regions of the wavelength spectrum at which each type of cone is maximally stimulated. When one of these three types of cones is defective owing to the inheritance of a sex-linked recessive trait, the ability to distinguish certain colors is diminished.

## PROCEDURE

In the *Ishihara test,* colored dots are arranged in a series of circles in such a way that a person with normal vision can see a number embedded within each circle. By contrast, a color-blind person will see only an apparently random array of colored dots.

# Laboratory Report 11.3

Name ____________________

Date ____________________

Section ____________________

## DATA FOR EXERCISE 11.3

**A. Sickle-Cell Anemia**

1. Was your test positive or negative? ____________________
2. Describe the appearance of your red blood cells in the microscope; compare to those of a prepared slide of sickle-cell disease.

**B. Inheritance of PTC Taste**

1. Are you a taster? __________
2. Enter the number of tasters and nontasters in your class in the table below.
3. Calculate the proportion of tasters (the number of tasters divided by the total number of students). Enter this value in this table.

| Phenotype | Number in Class | Proportion of Tasters |
|---|---|---|
| Tasters | | |
| Nontasters | | |

**C. Sex-Linked Traits: Inheritance of Color Blindness**

1. Are you color-blind? YES or NO (circle one). If YES, what type of color blindness do you have? ____________

## REVIEW ACTIVITIES FOR EXERCISE 11.3

### *Test Your Knowledge of Terms and Facts*

1. Chromosomes other than the sex chromosomes are called ____________________ chromosomes.
2. If a person has two identical genes for a trait, the person is said to be ____________________ for that trait.
3. If a person inherits a different gene from one parent than the other for a trait, the person is said to be ____________________ for that trait.
4. Genes inherited on the X chromosomes code for ____________________ traits.
5. The physical manifestation of a genotype is called a ____________________.

Oxygen gas is composed of oxygen molecules formed by the covalent bonding of two oxygen atoms. In this case, two pairs of electrons are shared by the two atoms, forming a double bond between them.

$$:\ddot{O}: + :\ddot{O}: \longrightarrow :\ddot{O}::\ddot{O}:, \quad \text{or} \quad O{=}O, \quad \text{or} \quad O_2$$

An atom of nitrogen has seven electrons, two in its inner shell and five in its outer shell. It requires three electrons to complete its outer shell. This requirement may be met by sharing electrons with three hydrogen atoms, forming a molecule of ammonia, or by sharing three pairs of electrons with another atom of nitrogen, forming a molecule of nitrogen gas.

$$:\dot{N}\cdot + 3H\cdot \longrightarrow :N:H \text{ (with H above and below N)}, \quad \text{or} \quad N{-}H \text{ (with H above and below N)}, \quad \text{or} \quad NH_3$$

$$\cdot N: + :N\cdot \longrightarrow N:::N, \quad \text{or} \quad N{\equiv}N, \quad \text{or} \quad N_2$$

When the electrons are not shared equally, but instead are held by only one of the two nuclei, the atom that captures the electron has a negative charge and the atom that loses the electron has a positive charge. These charged atoms (called *ions*) may be held together by a weak electronic attraction known as an **ionic bond.**

# Ions and Electrolytes

When a compound that is held together by weak ionic bonds is dissolved in water, it dissociates into positively charged ions *(cations)* and negatively charged ions *(anions)*. These ions can conduct electricity, and hence the original ionic compound is called an **electrolyte.** The most ubiquitous electrolyte is common table salt (NaCl).

$$\underset{\text{Ionic compound}}{NaCl} \longrightarrow \underset{\text{Cation}}{Na^+} + \underset{\text{Anion}}{Cl^-}$$

Some atoms form ionic bonds as a group with other atoms and remain grouped when the ionic compound dissociates. These groups are called *radicals*. Examples of radicals include sulfate ($SO_4^{2-}$), phosphate ($PO_4^{3-}$), ammonium ($NH_4^+$), and hydroxyl ($OH^-$).

$$\underset{\text{Ammonium sulfate}}{(NH_4)_2SO_4} \longrightarrow \underset{\text{Ammonium}}{2\ NH_4^+} + \underset{\text{Sulfate}}{SO_4^{2-}}$$

Notice that the sulfate radical has two negative charges and that two ammonium radicals are needed to retain electrical neutrality.

| Cation | Symbol | Anion | Symbol |
|---|---|---|---|
| Sodium | $Na^+$ | Chloride | $Cl^-$ |
| Potassium | $K^+$ | Sulfate | $SO_4^{2-}$ |
| Calcium | $Ca^{2+}$ | Bicarbonate | $HCO_3^-$ |
| Magnesium | $Mg^{2+}$ | Phosphate | $PO_4^{3-}$ |
| Hydrogen | $H^+$ | Hydroxyl | $OH^-$ |
| Ammonium | $NH_4^+$ | Carbonate | $CO_3^{2-}$ |

# pH and Buffers

The hydrogen ion concentration of a solution can vary between $10^{-14}$ molar and zero (see exercise 2.6 for a discussion of molarity). Pure water, which has a hydrogen ion concentration of $10^{-7}$ molar, is considered neutral.

$$H{-}O{-}H \longrightarrow H^+ + OH^-$$

Any substance that increases the $H^+$ concentration is called an **acid,** and any substance that decreases the $H^+$ concentration is called a **base.** Bases decrease the $H^+$ concentration by adding $OH^-$ to the solution. The $OH^-$ can combine with free hydrogen ions to form water.

$$\underset{\text{Hydrochloric acid}}{HCl} \longrightarrow H^+ + Cl^-$$

$$\underset{\text{Sodium hydroxide (a base)}}{NaOH} \longrightarrow Na^+ + OH^-$$

When equal amounts of hydrogen cation and hydroxyl anion are added to a solution, the acid and base neutralize each other, forming water and a **salt.**

$$\underset{\text{Acid}}{HCl} + \underset{\text{Base}}{NaOH} \longrightarrow \underset{\text{Salt}}{NaCl} + \underset{\text{Water}}{H_2O}$$

A convenient way of expressing the hydrogen ion concentration of a solution is by means of the symbol **pH,** which is the negative logarithm of the hydrogen ion concentration.

$$pH = \log \frac{1}{[H^+]}$$

Thus, pure water, with $10^{-7}$ moles of hydrogen ions/L, has a pH of 7.000. Since the pH is an inverse function of the $H^+$ concentration, an increase in the hydrogen concentration above that of water (i.e., an *acidic* solution) is indicated by a pH of less than 7.000, whereas a decrease in the $H^+$ concentration (i.e., a *basic* solution) has a pH between 7.000 and 14. A solution that has $10^{-2}$ moles of hydrogen ions/L (pH 2) is acidic, whereas one that has $10^{-12}$ moles of hydrogen ions/L (pH 12) is basic.

| Acid | Symbol | Base | Symbol |
|---|---|---|---|
| Hydrochloric acid | $HCl$ | Sodium hydroxide | $NaOH$ |
| Phosphoric acid | $H_3PO_4$ | Potassium hydroxide | $KOH$ |
| Nitric acid | $HNO_3$ | Calcium hydroxide | $Ca(OH)_2$ |
| Sulfuric acid | $H_2SO_4$ | Ammonium hydroxide | $NH_4OH$ |

A **buffer** is a compound that serves to prevent drastic pH changes when acids or bases are added to a solution. It does this by replacing a strong acid or base (one that ionizes completely) with a weak acid or base (one that does not completely ionize).

$$NaHCO_3 + HCl \longrightarrow H_2CO_3 + NaCl$$

Sodium bicarbonate buffer + Hydrochloric acid → Carbonic acid + Sodium chloride

Notice that the strong hydrochloric acid was replaced by the weaker carbonic acid, thus minimizing the change in pH that would have been induced had HCl been added to the solution in the absence of buffer. The carbonic acid/bicarbonate *buffer system* also minimizes the effect of added base on the pH of the solution.

$$H_2CO_3 + NaOH \longrightarrow NaHCO_3 + H_2O$$

Carbonic acid + Sodium hydroxide → Sodium bicarbonate + Water

# Organic Chemistry

The chemistry of organic compounds is based on the ability of *carbon atoms* to form chains and rings with other carbon atoms. Carbon, which has six electrons (two in the first shell and four in the second shell), requires four more electrons to complete its outer shell; hence, it is said to have four *bonding sites*.

**Hydrogen** (one bonding site): H : H, or H—H, or $H_2$

**Oxygen** (two bonding sites): H : Ö : H, or H—O—H, or $H_2O$

**Nitrogen** (three bonding sites): H : N : H (with H above N), or H—N—H (with H above N), or $NH_3$

**Carbon** (four bonding sites): H : C : H (with H above and below C), or H—C—H (with H above and below C), or $CH_4$

Carbon atoms can be covalently bonded to each other by sharing one pair of electrons (single bond) or by sharing two pairs of electrons (double bonds). Carbon-carbon double bonds are called sites of *unsaturation*, since they do not have the maximum number of hydrogen atoms.

H—C—C—H (each C bearing two H) or $CH_3CH_3$ H₂C=CH₂ or $CH_2CH_2$

Carbon atoms can be covalently bonded together to form long chains or rings.

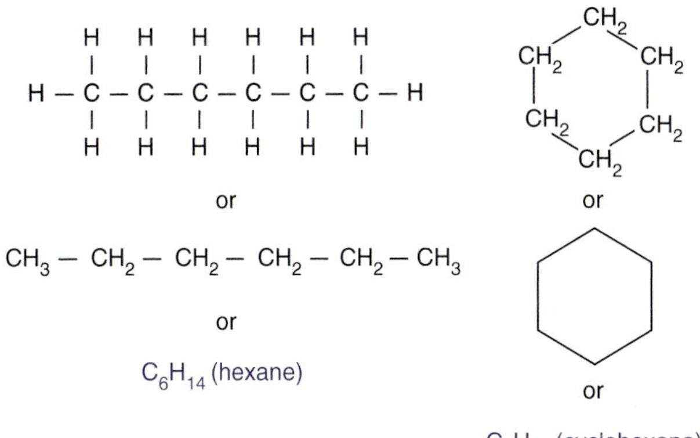


Notice that in the shorthand structural formulas for cyclic carbon compounds, the carbon atoms are represented by the corners of the figure and hydrogen atoms are not indicated.

Cyclic carbon compounds based on the structure of benzene are known as *aromatic* compounds. The common feature of their structural formula is the presence of three alternating double bonds in a six-carbon ring. This structural formula is in a sense misleading, since all the carbons in the aromatic ring are equivalent; hence, double bonds can be indicated between any two carbons in the ring.

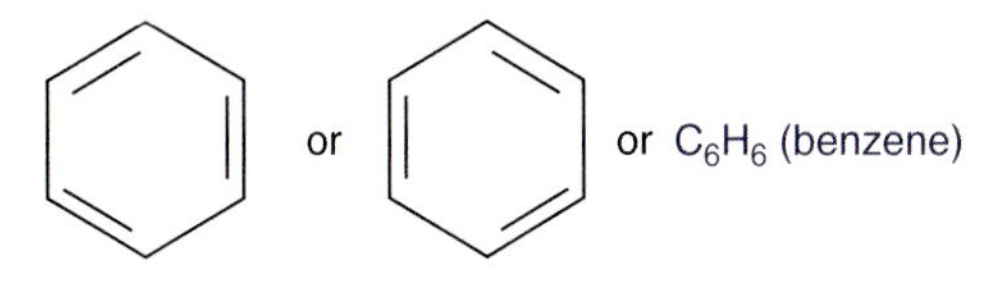


## Functional Groups

When carbon atoms are bonded together to form chains or rings, the remaining free bonding sites are available to combine with hydrogen atoms or with other compounds known as *functional groups*. These functional groups are generally more chemically reactive than the hydrocarbon backbone.

Some classes of organic compounds are named according to their functional groups. *Ketones*, for example, have a carbonyl group within the carbon chain, whereas *aldehydes* have a carbonyl group at one end of the chain. *Alcohols* have a hydroxyl group at one end of the chain, whereas *acids* have a carboxyl group at one end of the carbon chain (see p. 410).

Molecules that are identical in terms of the type and arrangement of their atoms but which differ with respect to the spatial orientation of key functional groups are called **optical isomers.** This name derives from the fact that these isomers can rotate plane-polarized light to the right or to the left, depending on the orientation of the functional group. The two optical isomers of the simple sugars and the amino acids are named *D* (right-handed) or

# Fisher General Scientific (SEA)

USA: (800)766–7000
FAX: (800)926–1166
www.fisher1.com
Canada: (800)2FISHER, or
(800)234–7437
www.fishersci.com

### Atlanta

3970 Johns Creek Court
Suite 500
Suwanee, GA 30024
(770)871–4500
FAX: (770)871–4600

### Chicago

4500 Turnberry Drive
Hanover Park, IL 60103
(630)259–1200
FAX: (630)259–4444

### Houston

P.O. Box 1546
9999 Veterans Memorial Drive
Houston, TX 77251–1546
(281)405–4000 or (800)766–7000
FAX: (281)878–2407 or (800)926–1166

### Los Angeles

2761 Walnut Avenue
Tustin, CA 92780
(714)669–4600
FAX: (714)669–1613

# Hardy Diagnostics

1430 McCoy Lane
Santa Maria, CA 93455
(800)266–2222
www.hardydiagnostics.com

# Kon's Scientific (for frogs)

P.O. Box 3
Germantown, WI 53022–0003
(414)242–3636

# Medical Analysis Systems, Inc.

Lincoln Technology Park
542 Flynn Road
Camarillo, CA 93012
(800)582–3095
(805)987–7891
www.mas.inc.com

# Niles Biological (for frogs)

9298 Elder Creek Road
Sacramento, CA 95829
(916)386–2665
www.nilesbio.com

# Sargent–Welch

P.O. Box 5229
Buffalo Grove, IL 60089–5229
(800)727–4368
www.sargentwelch.com

# Sigma Chemical Company

P.O. Box 14508
St. Louis, MO 63178
(800)325–3010
E-mail: custserv@sial.com
www.sigma-aldrich.com

# Stanbio Laboratory, Inc.

1261 North Main Street
Boene, TX 78006
(830)249–0772
(800)531–5535
www.stanbio.com

# VWR Scientific Products, Sargent Welch

P.O. Box 5229
Buffalo Grove, IL 60089–5229
(800)932–5000
FAX: (800)477–4897
E-mail: sarwel@sargentwelch.com
www.sargentwelch.com
www.vwrsp.com

### Corporate Headquarters

1310 Goshen Parkway
West Chester, PA 19380
Orders: (800)932–5000
(610)431–1700
FAX: (610)431–9174
www.vwrsp.com

Sales and Inventory Locations

### Philadelphia Regional Distribution Center

200 Center Square Road
Bridgeport, NJ 08014
(800)932–5000
(856)467–2600
FAX: (856)467–5499
www.vwrsp.com

### Atlanta Regional Distribution Center

1050 Satellite Boulevard
Suwanee, GA 30024
Orders: (800)932–5000
(770)495–1000
FAX: (770)232–9881
www.vwrsp.com

### Chicago Regional Distribution Center

800 East Fabyan Parkway
Batavia, IL 60510
Orders: (800)932–5000
(630)879–0600
FAX: (630)879–6718
www.vwrsp.com

### San Francisco Regional Distribution Center

3745 Bayshore Boulevard
Brisbane, CA 94005
Orders: (800)932–5000
(415)468–7150
FAX: (415)468–1105
www.vwrsp.com

# Ward's Biology

P.O. Box 92912
Rochester, NY 14692–9012
(800)962–2660
FAX: (800)635–8439
www.wardsci.com

# Warren E. Collins, Inc.

220 Wood Road
Braintree, MA 02184
(800)225–5157

# Appendix 3

## Multimedia Correlations to the Laboratory Exercises

The laboratory experience may be enriched with the use of computers that can receive data from the ongoing exercise, store and collate this data, and help students to analyze it. Computer-assisted data acquisition and analysis can be performed, for example, using equipment made available from Biopac Systems, Inc., Intellitool from Phipps & Bird/Intelitool, and iWorx. Where appropriate, instructions for the use of this equipment is included with the individual exercises in this laboratory guide. However, for the convenience of planning the laboratory curriculum, the use of this equipment for all of the laboratory exercises is summarized here.

The laboratory may also be a good time to incorporate supplementary computer-assisted instruction into the physiology curriculum. For example, there are computer programs that include instruction and animations that supplement more theoretical information. Two such programs are *A.D.A.M. InterActive Physiology*, from A.D.A.M. and Benjamin/Cummings, Publishers, and *MediaPhys*, from McGraw-Hill Publishers. The exercises that correlate with these programs are listed here.

## Section 1

**Exercise 1.3: Homeostasis and Negative Feedback**
- *MediaPhys 2.0*: Topics 1.3–1.6

## Section 2

**Exercise 2.6: Diffusion, Osmosis, and Tonicity**
- *MediaPhys 2.0*: Topics 3.9–3.24

## Section 3

**Exercise 3.1: Recording the Nerve Action Potential**
- A.D.A.M. *InterActive Physiology* (Nervous System I): The Action Potential (orientation, anatomy review)
- *MediaPhys 2.0*: Topics 3.27–3.34; Topics 4.4–4.21

**Exercise 3.2: Electroencephalogram (EEG)**
- Biopac Student Lab Lessons 3 and 4
- A.D.A.M. *InterActive Physiology* (Nervous System I): Ion Channels; The Membrane Potential

**Exercise 3.3: Reflex Arc**
- Intelitool: Flexicomp

## Section 4

**Exercise 4.1: Histology of the Endocrine Glands**
- *MediaPhys 2.0*: Topics 12.17–12.51

## Section 5

**Exercise 5.1: Neural Control of Muscle Contraction**
- A.D.A.M. *InterActive Physiology* (Muscular System): The Neuromuscular Junction
- *MediaPhys 2.0*: Topic 5.10

**Exercise 5.2: Summation, Tetanus, and Fatigue**
- Intelitool: Physiogrip
- A.D.A.M. *InterActive Physiology* (Muscular System): Contraction of Motor Units; Contraction of Whole Muscle
- *MediaPhys 2.0*: Topics 5.16–5.18.

**Exercise 5.3: Electromyogram (EMG)**
- Biopac: Student Lab Lessons 1 and 2
- Intelitool: Flexicomp
- A.D.A.M. *InterActive Physiology* (Muscular System): The Neuromuscular Junction; Contraction of Motor Units

## Section 6

**Exercise 6.1: Red Blood Cell Count, Hemoglobin, and Oxygen Transport**
- *MediaPhys 2.0*: Topics 10.37–10.44

## Section 7

**Exercise 7.1: Effects of Drugs on the Frog Heart**
- A.D.A.M. *InterActive Physiology* (Cardiovascular System): Cardiac Cycle

**Exercise 7.2: Electrocardiogram (ECG)**
- Biopac: Student Lab Lessons 5 and 6
- Intelitool: Cardiocomp
- A.D.A.M *InterActive Physiology* (Cardiovascular System): Cardiac Action Potential
- *MediaPhys 2.0*: Topics 8.17 and 8.18

**Exercise 7.3: Effects of Exercise on the Electrocardiogram**
- Biopac: Student Lab Lesson 7
- Intelitool: Cardiocomp
- A.D.A.M. *InterActive Physiology* (Cardiovascular System): Cardiac Output

**Exercise 7.4: Mean Electrical Axis of the Ventricles**
- Biopac: Student Lab Lesson 6
- Intelitool: Cardiocomp

**Exercise 7.5: Heart Sounds**
- Biopac: Student Lab Lesson 17
- A.D.A.M. *InterActive Physiology* (Cardiovascular System): Cardiac Cycle
- *MediaPhys 2.0*: Topics 8.19–8.23

Ex

# S

Ex

Ex

Ex

Ex

# Index

*Note:* Page references followed by the letters *f* and *t* indicate figures and tables, respectively.

# B

# C

# D



# E

# F

# G



# H

# I

# J



# K



# L



# M

# N



# O

# P

# Q



# R

# S

# T

# U



# V



# W



# Y



# Z